Global Biogeography

FURTHER TITLES IN THIS SERIES

1. A.J. Boucot
   EVOLUTION AND EXTINCTION RATE CONTROLS

2. W.A. Berggren and J.A. van Couvering
   THE LATE NEOGENE - BIOSTRATIGRAPHY, GEOCHRONOLOGY AND PALEOCLIMATOLOGY OF THE LAST 15 MILLION YEARS IN MARINE AND CONTINENTAL SEQUENCES

3. L.J. Salop
   PRECAMBRIAN OF THE NORTHERN HEMISPHERE

4. J.L. Wray
   CALCAREOUS ALGAE

5. A. Hallam (Editor)
   PATTERNS OF EVOLUTION, AS ILLUSTRATED BY THE FOSSIL RECORD

6. F.M. Swain (Editor)
   STRATIGRAPHIC MICROPALEONTOLOGY OF ATLANTIC BASIN AND BORDERLANDS

7. W.C. Mahaney (Editor)
   QUATERNARY DATING METHODS

8. D. Janóssy
   PLEISTOCENE VERTEBRATE FAUNAS OF HUNGARY

9. Ch. Pomerol and I. Premoli-Silva (Editors)
   TERMINAL EOCENE EVENTS

10. J.C. Briggs
    BIOGEOGRAPHY AND PLATE TECTONICS

11. T. Hanai, N. Ikeya and K. Ishizaki (Editors)
    EVOLUTIONARY BIOLOGY OF OSTRACODA. ITS FUNDAMENTALS AND APPLICATIONS

12. V.A. Zubakov and I.I. Borzenkova
    GLOBAL PALAEOCLIMATE OF THE LATE CENOZOIC

13. F.P. Agterberg
    AUTOMATED STRATIGRAPHIC CORRELATION

Developments in Palaeontology and Stratigraphy, 14

Global Biogeography

John C. Briggs

Arnoldsville, Georgia, USA

1995
ELSEVIER
Amsterdam – Lausanne – New York – Oxford – Shannon – Tokyo

ELSEVIER SCIENCE B.V.
Sara Burgerhartstraat 25
P.O. Box 211, 1000 AE Amsterdam, The Netherlands

Library of Congress Cataloging-in-Publication Data

Briggs, John C.
Global biogeography / by John C. Briggs.
p. cm.
Includes bibliographical references (p.) and index.
ISBN 0-444-88297-9
1. Biogeography. 2. Paleobiogeography. 3. Biological diversity.
I. Title.
QH84.B745 1995
574.9--dc20 95-30597
CIP

ISBN: 0-444-88297-9

This book is printed on acid-free paper.

Printed in The Netherlands

To my wife Eila for her patience

Preface

> In publishing these observations, together with the conclusions that I have drawn from them, my purpose is to invite enlightened men who love the study of nature to follow them out, verify them, and draw from them on their side whatever conclusions they think justified.
>
> J.B. Lamark, *Zoological Philosophy*, 1809

Five hundred million years (Ma) ago, a diversity of higher (metazoan) life forms was present in the sea. One hundred Ma later, an invasion of the land was well underway. Since then, there has been a generally increasing trend in the species diversity (richness) of both habitats. The trend has been interrupted, most notably in the marine environment, by a series of extinction episodes during which species diversity was temporarily reduced. None of these episodes occurred with catastrophic suddenness, consequently most biotic communities were able to accommodate in an ecological and evolutionary sense. The time spans occupied by the extinctions and the subsequent recovery of diversity were on the order of millions to tens of millions of years.

For the past 65 Ma, until the 20th Century, the multiplication of species continued until Cenozoic diversity became about double that of the preceding Mesozoic. Most of the increase may be attributed to the evolutionary success of vascular plants and the animals associated with them. Human beings became a distinct species about 3 Ma ago, thus arriving on a planet with a rich biota of some 12 million or more species. One important facet of biogeography is the study of the geographic distribution of diversity. The superior species diversity of the tropics has been known since the days of Charles Darwin and Alfred Russel Wallace. But, it is only recently that the true global pattern has begun to emerge.

As Wilson (1991) has noted, there appears to be a consensus among many tropical biologists that the rain forests probably hold 10 million or more species. Briggs (1991a, 1994) made rough global estimates of about 12 million metazoan species on land, as opposed to less than 200 000 in the sea. Rain forests occupy only about 6% of the land and, of the surface of the entire globe, about 2%. This leads to the astonishing conclusion that more than 80% of all living species may be concentrated in the rain forests that occupy only 2% of the earth's surface.

We live in a critical time in the history of the earth's inhabitants. Before human impact, rain forests covered a total area of about six million square miles. They have now been reduced to an area of about three million square miles (Wilson, 1991). By 1990, they were being destroyed at a rate of about 55 000 square miles per year (an area larger than the state of Florida). The rate of extinction may have already reached 50 000 species per year. This is a conservation problem of such magnitude that most of our other environmental concerns appear insignificant.

Once a species becomes extinct, it is lost forever. It cannot be reconstituted and will not evolve again. The catastrophic decline in species numbers, now taking place in the rain forests, can be slowed and, at some point, perhaps even stopped if there is sufficient human interest in doing so. But, we cannot reverse the process. No matter what we do, our children, our children's children, and future generations as long as humans exist on earth, will not see a recovery from the loss of species that has already taken place. The evolutionary process that produces new species is simply too slow. By the time a significant recovery in global species diversity takes place, the human species will have been long extinct.

In the ocean, the ecological equivalent of the rain forest is the coral reef habitat. The coral reefs of the East Indies (a triangular area including the Philippines, the Malay Peninsula, and the archipelago extending out beyond New Guinea) and associated benthic habitats, support the greatest species diversity in the marine world. The high human population of the East Indies has had a devastating impact on the reefs. They are being damaged by soil runoff from deforestation, sewage pollution, oil pollution, coral mining, and uncontrolled fisheries for curios and the aquarium trade. In the Philippines, less than one-third of the reefs remain in excellent or good condition (Gomez, 1988).

I place a great deal of emphasis on the species level in the taxonomic hierarchy. The higher levels are artificial constructs, useful in explaining evolutionary relationships, but they do not exist as natural populations. Each living species constitutes a unique population differing from all others in its structure, behavior, physiology, and genetic makeup. Each represents the end result of an evolutionary process that took at least thousands or, more often, millions of years to complete. Their annual extinction by the tens of thousands means that the world is steadily becoming more impoverished.

Much of the paleontological literature is devoted to lurid descriptions of so-called historic "catastrophes", sometimes referred to as "mass extinctions" or "mass killings". For example, some authorities claim that the Cretaceous-Tertiary (K/T) extinction episode of about 65 Ma ago was caused by a giant asteroid or comet that wiped out 50% to 80% of the earth's species. As has been noted elsewhere (Briggs, 1991b), such claims are highly speculative. That extinction took place over the course of hundreds of thousands to about 7 Ma, depending on the species group concerned.

The loss of global species diversity during the K/T extinction was small, since few terrestrial organisms were affected. It could have been less than 1%. The other extinction episodes also occurred very gradually, and the recovery to normal diversity levels generally took even longer. The truth is that the only rapid, catastrophic extinction the world has ever seen is the one going on right now. The tempo of the current losses is such that ecological and evolutionary adjustments are impossible.

A biogeography with a historic emphasis should, perhaps, not be so concerned with current happenings. But, the present extinction is a unique event that will have formidable biogeographic and evolutionary consequences. Certainly, there have been impacts by extraterrestrial bodies throughout Phanerozoic time. However, we have no evidence that these disturbances, by themselves, were important factors in the decline of global diversity. Massive volcanic eruptions can also occur over short periods of time and, in conjunction with other environmental changes, could have had a detrimental effect on diversity. But, it is difficult to make a connection between these sudden, high-energy events and the paleontological evidence of very gradual changes in species diversity.

Biogeography begins with the process of speciation which occurs as the result of a restriction in gene flow between segments of an interbreeding population. Such a restriction may be caused by a habitat separation in the same geographic area or by the presence of a barrier between the segments. The first type is called sympatric speciation and the second allopatric. Speciation by allopatry may occur in two ways: a barrier may be transgressed by individuals who succeed in colonizing the opposite side or a barrier may arise to separate a previously continuous population. These two are recognized as speciation by dispersal and speciation by vicariance. During the past 20 years, there has been much fruitless argument about which of the latter two has been the most important.

Although the longevity of species is quite variable, it is probably safe to say that most living species are not over 5 Ma old and that the great majority of modern genera are Tertiary in origin, making them less than 65 Ma old. Most of the families in such relatively well known groups as the birds, mammals, and flowering plants are not older than Cretaceous (65–130 Ma) in age. This means that for widespread species, genera, and some families, we should look for Quaternary or Tertiary means of dispersal or vicariance, rather than invoking continental separations that took place in the Mesozoic.

A complete biogeography should present: (1) a brief history of the science, (2) an account of the principal physical events in earth history, (3) a chronology of the distributional and evolutionary responses of the biota, (4) an outline of Recent biogeographic regions in the principal habitats (marine, freshwater, terrestrial), and, (5) a prognosis for the future. This book will strive towards these objectives. At the same time, I hope that it will give the reader an enhanced appreciation for the history of our planet and of the 12 million other species that share it with us.

John C. Briggs

Acknowledgements

The research for this book began at the Department of Marine Science, University of South Florida, St. Petersburg, Florida. Upon my retirement, the work was continued at the Museum of Natural History, University of Georgia, Athens, Georgia. The latter institution provided a faculty appointment which included office space and access to their fine Science Library. I am particularly indebted to Joshua Laerm, Director of the Museum.

I wish to thank Joan Hesler for the fine typing job and her patience with the continuous stream of additions and corrections. I am especially indebted to S. Gregory Tolley who proofread the entire manuscript. Ernst B. Peebles read the manuscript as it was being produced and provided helpful comments and corrections. J. Chad Edmisten did a fine job in preparing many of the figures included herein.

Contents

CHAPTER 1

History of the science

Read not to contradict and confute, nor to believe and take for granted, nor to find talk and discourse, but to weigh and consider.

Francis Bacon, *Of Studies*, 1605?

For the past 30 years, the time during which the geophysical concept of continental drift has become fully accepted, there has developed a need for biogeographers to take a fresh look at their discipline in the light of past changes in the relationships of the land masses and oceanic basins of the world. This new plate-tectonic framework has now become adopted and biogeography is undergoing a change from an emphasis on modern distributional patterns to a greater appreciation for the historical development of such patterns.

In order to realize the importance of the new plate-tectonic approach, one should take the time to place it in the context of significant changes that have occurred in the past. As is true in many cases, unless one is familiar with a discipline's historical progression, one cannot appreciate its present position in the stream of events, nor predict its future course.

IN THE BEGINNING

In the 17th century, the task of biogeographers was a relatively simple one. The book of Genesis told how all men were descended from Noah and that they had made their way from Armenia to their present countries. Since there had been a single geographical and temporal origin for man, the consensus was that this was also true for all animals and that they had a common origin from which they too had dispersed (Browne, 1983). So scholars like Athanasius Kircher (1602–1680) and his contemporaries set themselves the task of working out the details of the structure of the Ark so that it could accommodate a pair of each species of animal. It is interesting to see that this exercise of deducing the structure, and eventual grounding place, of the Ark has been repeated dozens of times in the past 300 years. In the year of 1985, there were news reports of five different expeditions busily combing the slopes of Mt. Ararat for the remains of the Ark.

Since well before Kircher's time, travelers and explorers had been bringing back to Europe thousands of specimens representing unknown species of animals. As these were described, religious scholars were obliged to find room for them aboard the Ark. No one seemed to have worried about the thousands of species of plants that could not have survived the Deluge. Matthew Hale (1609–1676) erected theoretical land bridges to explain how animals must have reached the New World from the Old. The bridges disappeared when no longer needed and the New World species supposedly became transformed after living in their new surroundings. By the time the 18th century arrived, the idea of the Ark had to be abandoned by people who were informed on the subject of natural history. However, the concept of the Deluge was still strongly entrenched so that a reasonable substitute for the Ark had to be found.

The person who came to the rescue was a young man in Sweden named Carl Linnaeus (1707–1778). He was a deeply religious person who felt that God spoke most clearly to man through the natural world. In fact, it has been said that Linnaeus considered the universe a gigantic museum collection given to him by God to describe and catalogue into a methodical framework (Browne, 1983). Linnaeus proceeded to solve the Ark problem by telescoping the story of the Creation into that of the Deluge. He proposed that all living things had their origin on a high mountain at about the time the primeval waters were beginning to recede. Furthermore, he proposed that this Paradisiacal mountain contained a variety of ecological conditions arranged in climatic zones so that each pair of animals was created in a particular habitat along with other species suited for that place.

As the flood waters receded, Linnaeus envisioned the various animals and plants migrating to their eventual homes where they remained for the rest of time. For him, species were fixed entities that stayed just as they were created. Linnaeus also emphasized that each species had been given the structure that was the most appropriate for the habitat in which it lived. This insistence on a close connection between each species and its habitat, exposed Linnaeus to criticism by other scholars. How could the reindeer, which was designed for the cold, have made its way across inhospitable deserts to get from Mt. Ararat to Lapland?

Some 18th century naturalists were not as ready to abandon the concept of the Ark. Thomas Pennant (1726–1798), in his *Arctic Zoology* in 1784–1785, observed that the animals of the Ark must have migrated across Asia and then to the Americas via the Bering Strait region. Thus, Pennant was the first to call attention to a migration route that proved to have a great historical importance.

The Comte de Buffon (1707–1788), who published his great encyclopedia, *Histoire Naturelle* in 1749–1804, was influential in persuading educated people to give up the Garden of Eden concept and also the idea that species did not change through time. He apparently believed that life originated generally in the far north during a warmer period and had gradually moved south as the climate got colder. Because the New and Old Worlds were almost joined in the north, the species in each area were the same. But, as the southward progression took place the original populations were separated. In the New World, some kind of a structural degeneration took place which caused those species to depart from the primary type. In regard to mammals, Buffon observed that those of the New and Old World tropics were exclusively confined to their own areas. This has been subsequently referred to as "Buffon's Law" and interpreted to mean that such animals had evolved in situ and had not migrated from Armenia (Nelson, 1978). In regard to South America and Africa, Buffon thought that they were at first connected by a land bridge which later became inundated.

As the result of the influence of Buffon and others, the idea of a single biblical center for all species was replaced by the idea of many centers of creation, each species in the area where it now lived (Browne, 1983). This, and the Linnaean concept of the importance of species as identifiable populations that existed in concert with other species, encouraged naturalists to think in terms of groups of species characteristic of a given geographic area. Linnaeus and his students and others began to emphasize the contrasts among different parts of the world by publishing various "floras" and "faunas." Johannes F. Gronovius published his *Flora Virginica* in 1743; Carl Linnaeus his *Flora Suecica* in 1745, *Fauna Suecica* in 1746, and *Flora Zeylandica* in 1747; Johann G. Gmelin his *Flora Sibirica* in 1747–1769; and Otto Fabricus his *Fauna Groenlandica* in 1780.

From the viewpoint of the mid-18th century, it may be seen that biogeography underwent a fundamental change during the preceding 100 years. Naturalists were at first occupied with the problems of accommodation aboard the Ark and the means by which animals were able to disperse to the various parts of the world following the Deluge. The Ark concept gave way to the Paradisiacal mountain which in turn yielded to the idea of creation in many different places. At the same time, the Linnaean axiom of the fixity of species through time was replaced by one of change under environmental influence. Finally, naturalists began to study the associations of plants and animals in various parts of the world and, in so doing, began to appreciate the contrasts among different countries.

Johann Reinhold Forster (1729–1798) was a German naturalist who emigrated to England in 1766. From 1770 to 1772 he published several small works including a volume entitled *A Catalogue of the Animals of North America*. In 1772, he, together with his son Georg, was given the opportunity to accompany Captain Cook on his second expedition to the South Seas. This was a 3-year circumnavigation of the globe. Upon their return, Forster published his *Observations Made During a Voyage Round the World* in 1778. In this work, he presented a worldwide view of the various natural regions and their biota. He described how the different floras replaced one another as the physical characteristics of the environment changed. He also called attention to the way in which the type of vegetation determined the kinds of animals found in each region.

Forster compared islands to the mainland and noted that the number of species in a given area was proportionate to the available physical resources. He remarked on the uniform decrease in floral diversity from the equator to the poles and attributed this phenomenon to the latitudinal change in the surface heating of the earth. He found the tropics to be beautiful, rich, and enchanting – the area in which nature reached its highest and most diversified expression (Browne, 1983). Forster, more than any of his predecessors, understood that biotas were living communities characteristic of certain geographical areas. Thus the concept of natural biotic regions was born.

As knowledge of the organic world increased and greater numbers of species became known, naturalists tended to specialize in the study of either plants or animals. For some reason, it was the early botanists who took the greatest interest in biogeography. Karl Willdenow (1765–1812) was a plant systematist and head of the Berlin Botanical Garden. In his 1792 book *Grundriss der Krauterkunde*, he outlined the elements of plant geography. He recognized five principal floras in Europe and, like Forster, was interested in the effect of temperature on floral diversity. To account for the presence of the various botanical provinces, Willdenow envisioned an early stage of many mountains surrounded by a global sea. Different plants were created on the various peaks and then spread downward, as the water receded, to form our present botanical provinces.

19TH CENTURY

Willdenow's most famous student was Alexander von Humboldt (1769–1859). Von Humboldt has often been called the father of phytogeography (Brown and Gibson, 1983). In his youth he was impressed and influenced by his friendship with Georg Forster. Von Humboldt felt that the study of geographical distribution was scientific inquiry of the highest order and that it could lead to the disclosure of fundamental natural laws (Browne, 1983). He became one of the most famous explorer-naturalists and devoted much of his attention to the tropics of the New World. As a part of his great 24 volume work *Voyage aux Regions Equinoxiales du Noveau Continent* (1805–1837, with A.J.A. Bonpland), von Humboldt included his *Essai sur la Geographie des Plantes* (1805). The latter work, his best contribution to biogeography, was inspired as a result of his climbing Mt. Chimborazo, an 18 000-foot peak in the Andes. There he observed a series of altitudinal floral belts equivalent to the tropic, temperate, boreal, and arctic regions of the world.

The next significant step in the progress of biogeography was made by a Swiss botanist named Augustin de Candolle (1778–1841). In 1820, he published his important *Essai Elementaire de Geographie Botanique*. In that work he made a distinction between "stations" (habitats) and "habitations" (the major botanical provinces). De Candolle's work had a significant influence on such important figures as Charles Darwin, Joseph Hooker, and his own son Alphonse. The elder de Candolle was a close friend of von Humboldt and was surely influenced by him. Three years later, in 1823, Joachim F. Schouw published his *Grunzuge einer allgemeiner Pflanzengeographie*. This was a classification of the floristic regions of the world with descriptions of the various plant communities.

The study of extinct floras got underway with the work of Adolphe Brongniart who published his *Histoire des Vegetaux Fossiles* in 1828. He was followed by Alphonse de Candolle. Both men believed that life first appeared as a single primitive population evenly distributed over the entire surface of the globe. This uniform population was supposed to have gradually fragmented into many diverse groups of species (Browne, 1983). In the meantime, Georges Cuvier had begun his work on fossil vertebrates and many others soon followed. Cuvier was not only a paleontologist but was undoubtedly the worlds most influential zoologist in the first half of the 19th century. He was a determined advocate of the Linnaean doctrine of the fixity of species. In order to account for the discovery of fossils representing unknown species, Cuvier conceived of a series of great catastrophes, the most recent being the biblical Deluge. The new species that subsequently appeared supposedly came from parts of the world previously unknown. Cuvier's catastrophes consisted of regional extinctions. It was William Buckland, in his 1823 work *Reliquiae Diluvianae*, who advocated worldwide catastrophes as the causes of breaks in the fossil record. Each catastrophe required a new creation (Hoffman, 1989a).

James Cowles Prichard began to publish his *Researches into the Physical History of Mankind* in 1813. But his most extensive biogeographical discussion was in the second edition of 1826. Prichard was also a catastrophist but he had interesting explanations for two difficult biogeographic problems: the distribution of island species and the occurrence of disjunct species (Kinch, 1980). Remote islands were a problem because they often supported endemic species which appeared to be the result of special creations. Prichard said that such endemic species were in reality only relict populations of species that once had a more extensive distribution. He also felt that disjunct distributions did not have to be explained by acts of special creation, at least as far as plants were concerned. He pointed out that seeds could be transported by winds, rivers, ocean currents, and by animals. And he noted that the seeds of some plants could be transported in the digestive tracts of birds.

From a distributional standpoint, the first effective connection between fossil and contemporary patterns was made by Charles Lyell (1797–1875). In his *Principals of Geology* (1830–1832 and subsequent editions), Lyell undertook extensive discussions on botanical geography, including the provinces of marine algae, and on the geographical distribution of animals. In addition, he analyzed the effects of climatic and geological changes on the distribution of species and the evidence for the extinction and creation of species.

As Browne (1983) has pointed out, Lyell's suggestion that the elevation and submersion of large land masses resulted in the conversion of equable climates into extreme ones, and vice versa, according to the quantity of land left above sea level, was most important. This view meant that floras and faunas had to be dynamic entities capable of expanding or contracting their boundaries as geological agents altered topography and climates. So Lyell, the champion of gradual change to the earth's surface, brought to biogeography a sense of history and the realization that floral and faunal provinces had almost certainly been altered through time.

As Kinch (1980) has noted, many early 19th century naturalists expressed in their work the belief that the universe displayed evidence of God's design. They did not, however, always agree as to the kind of design that was displayed in nature. One kind has been called purposeful or utilitarian. The utilitarian group made use of supernatural interventions to explain the distribution of life. Thomas Pennant felt that the migration of animals from Asia to America was divinely directed in order to prevent the Old World from becoming overpopulated. Richard Brinsley Hinds, a surgeon-naturalist aboard the H.M.S. *Sulphur* in 1836 to 1846 (Thomas, 1969), thought that the animals spread from one center, the resting place of the Ark. Plants, however, were supposed to have simultaneously covered the land surface of the globe. The latter assumption was based on the biblical statement that the earth brought forth vegetation.

Other naturalists felt that God had expressed an idealist design which could be discerned as orderly patterns in nature. Two well known botanists who searched for such patterns were Alexander von Humboldt and Robert Brown. They thought that certain plant groups occurred in a given ratio to others and that this ratio would vary in a predictable manner from one geographical area to the next. Another idealist design was called the quinary system. This was introduced by William Sharp MacLeary in 1819–1921 (Kinch, 1980). William Swainson decided to use the quinary system in his biogeographic work. He divided the earth into five major biogeographic regions which coincided with his view of the five regions of mankind. The entomologist, Edward Newman, advocated a septenary system in 1832 but this and the quinary system did not last beyond the advent of Darwinism.

Edward Forbes (1815–1854), despite his short life, made important contributions to both terrestrial and marine biogeography. He accounted for the evident relationship between the floras of the European mountain tops and Scandinavia by supposing very cold conditions and land subsidence in the recent past. His map of the distribution of marine life together with a descriptive text which appeared in Alexander K. Johnston's *The Physical Atlas of Natural Phenomena* (1856) was the first comprehensive work on marine biogeography. In it, the world was divided into 25 provinces located within a series of nine horizontal "homoizoic belts." A series of five depth zones was also recognized. In the same year, Samuel P. Woodward, the famous malacologist, published part three of his *Manual of the Mollusca* which dealt with the worldwide distribution of that group.

In 1859, Forbes posthumous work *The Natural History of European Seas* was published by Robert Godwin-Austen. In this work Forbes observed that (1) each zoogeographic province is an area where there was a special manifestation of creative power and that the animals originally formed there were apt to become mixed with emigrants from other provinces; (2) each species was created only once and that individuals tended to migrate outward from their center of origin; and (3) provinces, to be understood, must be traced back like species to their origins in past time. Another important contribution was made by James D. Dana who participated in the United States Exploring Expedition, 1838–1842. Through observations made on the distribution of corals and crustaceans, he was able to divide the surface waters of the world into several different zones based on temperature and used isocrymes (lines of mean minimum temperature) to separate them. His plan was published as a brief paper in the *American Journal of Science* in 1853.

The first attempt to include all animal life, marine and terrestrial, in a single zoogeographic scheme was by Ludwig K. Schmarda in his volume entitled *Die Geographische Verbreitung der Tierre* (1885). He divided the world into 21 land and 10 marine realms. However, it was P.L. Sclater who divided the terrestrial world into the biogeographic regions that, essentially, are still in use today. This was done in 1858 in a small paper entitled *On the General Geographical Distribution of the Members of the Class Aves*. Despite the fact his scheme was based only on the distributional patterns of birds, Sclater's work proved to be useful for almost all groups of terrestrial animals. This has served to emphasize that biogeographic boundaries, found to be important for one group, are also apt to be significant for many others.

When the young Charles Darwin visited the Galapagos Islands in 1835, he was struck by the distinctiveness, yet basic similarity, of the fauna to that of mainland South America. When Alfred Russel Wallace traveled through the Indo-Australian Archipelago, some 20 years later, he was puzzled by the contrasting character of the island faunas, some with Australian relationships and others with southeast Asian affinities. After considerable thought about such matters (many years on Darwin's part), each man arrived at a theoretical mechanism (natural selection) to account for evolutionary change. The key for both Darwin and Wallace was the realization that distributional patterns had evolutionary significance.

The announcement of their joint theory by Darwin and Wallace in 1858 in the *Journal of the Linnean Society of London* and, especially, the publication of Darwin's *Origin of Species* in 1859, changed the thinking of the civilized world. Darwin included two important chapters on geographical distribution in his book. In discussing biogeography from the viewpoint of evolutionary change, Darwin made three important points: (1) he emphasized that barriers to migration allowed time for the slow process of modification through natural selection; (2) he considered the concept of single centers of creation to be critical, that is, each species was first produced in one area only and from that center it would proceed to migrate as far as its ability would permit; and (3) he noted that dispersal was a phenomenon of overall importance.

In regard to the third point, Darwin observed that oceanic islands were generally volcanic in origin and must have accumulated their biota by dispersal from some mainland source. He felt that the presence of alpine species on the summits of widely separated mountains could be explained by dispersal having taken place during the glacial period when such forms would have been widespread. More important, he suggested that the relationships that biologists were then finding between the temperate biotas of the northern and southern hemispheres were attributable to migrations made through the tropics during the glacial period when world temperatures were cooler. Finally, he noted that the preponderant interhemispheric migratory movement had been from north to south and suggested that this was due to the northern forms having advanced through natural selection and competition to a higher stage of dominating power.

When Darwin was going through the long process of formulating his theory, his closest confidants were Charles Lyell and Joseph D. Hooker. Hooker, a great plant collector and systematist, having accompanied Sir James Ross on his Antarctic Expedition (1839–1843), was particularly interested in southern hemisphere botany. Hooker felt that Darwin was perhaps too dependent on dispersal in accounting for disjunct relationships. In describing the flora of New Zealand in 1853, Hooker speculated on the possibility that the plants of the Southern Ocean were the remains of a flora that had once been spread over a larger and more continuous tract of land than now exists in that part of the world. In modern terms, he was suggesting a vicariant rather than a dispersal history for the subantarctic floras.

While Darwin went on to investigate many other aspects of evolutionary change, Alfred Russel Wallace applied himself primarily to biogeography. Finally, in 1876, Wallace published his monumental two volume work *The Geographical Distribution of Animals*. In that work, he reached a number of conclusions about biogeography that are still worth reviewing. For example, he pointed out that (1) paleoclimatic studies are very important for analyzing extant distribution patterns; (2) competition, predation, and other biotic factors play important roles in the distribution, dispersal, and extinction of animals and plants; (3) discontinuous ranges may come about by extinction in intermediate areas or patchiness of habitats; (4) disjunctions of genera show greater antiquity than those of a single species, and so forth for higher categories; (5) the common presence of organisms not adapted for long distance dispersal is good evidence of past land connections; (6) when two large land masses long separated are reunited, extinction may occur because many organisms will encounter new competitors; (7) islands may be classified into three major categories, continental islands recently set off from the mainland, continental islands long separated from the mainland, and oceanic islands of volcanic and coralline origin; and (8) studies of island biotas are important because the relationships among distribution, speciation, and adaptation are easier to see and comprehend.

Wallace did considerable traveling in the Indo-Australian region and was particu larly concerned about the location of the dividing line between the Oriental and Australian faunas. As George (1981) has noted, Wallace, by 1863, had decided that the line should run from east of the Philippines south between Borneo and Celebes and then between Bali and Lompok. It was illustrated in his 1876 work and later in his book *Island Life* in 1880. Although Wallace, in his 1910 book *The World of Life* changed his mind about the affiliation of Celebes, his original line is the one generally called "Wallace's Line." It is represented in his regional scheme which is close to that proposed earlier by Sclater.

Also in 1876, Ernst Haeckel published his curious work *The History of Creation, or the Development of the Earth and its Inhabitants by the Action of Natural Causes*. He placed the location of Paradise, where life was created, in Lemuria, a continent (now sunk beneath the sea) in the center of the Indian Ocean. He provided a map showing how humans and the rest of the biota reached other parts of the world. In a later edition of his book in 1907, he relocated Paradise to the north and east of India. The Lemuria concept was resurrected by Hermann von Ihring in 1927.

Following the publication of Wallace's works, many biogeographers repeated his distribution plan without any major new interpretations. In 1890, E.L. Trouessart published his *La Geographie Zoologique* which examined both terrestrial and marine patterns. In 1895, Frank E. Beddard came out with *A Textbook of Zoogeography*. In 1907, Angelo Heilprin published a volume entitled *The Geographical and Geological Distribution of Animals*. The latter introduced some minor changes to Wallace's map and also reviewed the information then available about the distribution of fossil forms. Also a number of works, dealing with the establishment of hypothetical land bridges and the rise and fall of mid-ocean continents, were published. But as our knowledge of sea-floor history increased, these theories were discarded.

Plant geography also made significant progress, building on the foundation provided by Wallace and Darwin. The German school was particularly active as indicated by the 1872 work of A. Grisebach, *Die Vegetation der Erde nach ihrer Klimatischen Anordnung*, the 1879–1882 work of A. Engler, *Versuch einer Enwicklungsgeschichte der Pflanzenwelt*, and the 1890 book by O. Drude, *Handbuch der Pflanzengeographie*. These were followed shortly by the 1896 treatise of J.E.B. Warming, *Lehrbuch der Okologischen Pflanzengeographie*, and the 1898 publication of A.F.W. Schimper, *Pflanzen-Geographie auf physiologischer Grundlage*.

20TH CENTURY

A comprehensive historical biogeography, *Entwicklung der Kontinente und Iher Lebwelt*, was published by Theodor Arldt in 1906. The biological relationships among the continents were illustrated by many phylogenetic trees. In order to account for such relationships, Arldt assumed a history of stable continents that were connected by land bridges. Later, these land bridges were converted into imposing mid-oceanic continents by Hermann von Ihring in his 1927 work *Die Geschichte des Atlantischen Ozeans*. He delineated an Archatlantis in the North Atlantic, an Archhelenis in the South Atlantic, and a

Lemuria (including Madagascar and India) in the western Indian Ocean. Supposedly, these continents existed from the Cretaceous to the Eocene.

A significant advance in biogeography took place in 1915 when William Diller Matthew (1871–1930), a geologist and paleontologist, published his article on *Climate and Evolution*. Matthew was an expert on fossil mammals and his 1915 work was devoted primarily to emphasizing the importance of the northern hemisphere (the Holarctic Region) in the evolution and dispersal of that group. However, the most important aspect of that work has turned out to be Matthew's statement of his theory about centers of dispersal. He said, "At any given period, the most advanced and progressive species of the race will be those inhabiting that region; the most primitive and unprogressive species will be those remote from this center. The remoteness is, of course, not a matter of geographic distance but of inaccessibility to invasion, conditioned by the habitat and facilities for migration and dispersal."

Also in 1915, the botanist J.C. Willis began to develop his "age and area" hypothesis. Eventually a book *Age and Area: a Study in Geographical Distribution and Origin of Species* was published in 1922. In essence, the hypotheses stated that the areas occupied by a given group of allied species would depend on the ages of the species in that group; that is, the older the species, the greater the occupied area. While this theory was, at first, given considerable attention, it was soon realized that many species occupying limited areas were geographical relicts. The result being that one could find two kinds of species confined to small areas, newly evolved ones that had not yet become widespread and older ones that were formerly dispersed but had become extinct in most parts of their range.

Progress in our knowledge about distribution patterns in the marine environment was made by Arnold Ortman when he published his *Grundzuge der Marinen Tiergeographie* (1896). The following year, in 1897, Philip L. Sclater published a paper on the distribution of marine mammals. In 1935, Sven Ekman completed the huge task of analyzing all of the pertinent literature on marine animal distribution and published his results in a book entitled *Tiergeographie des Meeres*. In 1953, a second edition was printed in English. Modern books on marine zoogeography have been published by John C. Briggs, *Marine Zoogeography* (1974), Geerat J. Vermeij, *Biogeography and Adaptation* (1978), S. van der Spoel and A.C. Pierrot-Bults (eds.), *Zoogeography and Diversity in Plankton* (1979), Oleg G. Kussakin (ed.), *Marine Biogeography* (1982, in Russian), and A.C. Pierrot-Bults, et al. (eds.), *Pelagic Biogeography* (1986).

In the 1920s and 1930s a new development took place which combined the rapidly evolving field of ecology with biogeography. The beginning was marked by the appearance of Friedrich Dahl's *Grundlagen einer okologischen Tiergeographie* in 1921, and Richard Hesse's *Tiergeographie auf okologischer Grundlage* in 1924. These efforts were apparently in response to a need to examine the geographical distribution of plant and animal communities on a local and worldwide scale. A revised English edition of Hesse's book was prepared by W.C. Allee and Karl P. Schmidt and published in 1937. This was followed by a second edition in 1951. Other works that have carried on this approach are Marion I. Newbigin's *Plant and Animal Geography* published in 1936, V.G. Gepner's *General Zoogeography* (1936, in Russian), Frederic E. Clements and Victor E. Shelford's *Bio-ecology* in 1938 (which introduced the biome concept), and the work by L.R.

Dice *The Biotic Provinces of North America* in 1943. Among such works, that of Robert H. MacArthur and Edward O. Wilson, *Island Biogeography* (1967), deserves special mention. Its explanation of the relationship between colonization and extinction and its analysis of the species-area concept, had a stimulating impact on both biogeography and ecology. Other modern examples of the combined approach are the books by P.M. Dansereau, *Biogeography: An Ecological Perspective* (1957), Brian Seddon, *Introduction to Biogeography* (1971), C. Barry Cox, Ian N. Healey, and Peter D. Moore, *Biogeography* (1973), and James H. Brown and Arthur C. Gibson, *Biogeography* (1983).

In 1944, a significant work on phytogeography, *Foundations of Plant Geography*, was published by Stanley A. Cain. His analysis of fossil distributions and his discussion of the center of origin concept have been most useful to later workers. A work of similar importance for those interested in the distribution of animals was published by Philip J. Darlington, Jr. in 1957. Although *Zoogeography: the Geographical Distribution of Animals* was based only on patterns demonstrated by the terrestrial and freshwater vertebrates, it represented an important milestone because it was the first time in the 20th century that all of the information about those animal groups had been gathered together. Since the data on fossil vertebrates are, in general, better than those for the invertebrate groups, Darlington's book had great significance for historical biogeography.

Darlington (1957) emphasized that the major worldwide patterns of vertebrate animals indicated a series of geographical radiations from the Old World tropics. Such radiations were considered to take place because competitively dominant animals were continually moving out from their tropical centers of origin. In a later article, Darlington (1959) observed, "The history of dispersal of animals seems to be primarily the history of successions of dominant groups, which in turn evolve, spread over the world, compete with and destroy and replace older groups, and then differentiate in different places until overrun and replaced by succeeding groups."

THE ADVENT OF CONTINENTAL DRIFT

It was not until the late 1950s that the idea of historic continental movement began to be taken seriously by large numbers of earth scientists. Much earlier, between 1910 and 1912, Frederick B. Taylor, H.D. Baker, and Alfred L. Wegener had all advanced views about continental drift similar to those that are held today. However, at that time, the earth's crust was almost universally considered to have a solid structure without movement.

Between 1915 and 1929, Wegener published four editions of his book *Die entstehung der Kontinente und Ozeane* including an English edition (*The Origin of Continents and Oceans*). These works created considerable controversy but most geologists and geophysicists were still not convinced. Research into paleomagnetism then began to offer some supporting evidence for drift. In 1960, Harry H. Hess made the suggestion that the sea floors crack open along the crest of the mid-ocean ridges, and that new sea floor forms there and spreads apart on either side of the crest. Robert S. Dietz named this process sea-floor spreading and coupled with it the suggestion that old sea floor is absorbed beneath zones of deep ocean trenches and young mountains.

J. Tuzo Wilson (1963, 1973) noted that oceanic islands tended to increase in age away from the mid-ocean ridges and that certain "hot spots" existed where strings of volcanic islands had been formed. These and other discoveries led to the modern view of plate tectonics which holds that the earth's crust is divided into a mosaic of shifting plates in which the continents are embedded. We now have available many reconstructions of continental relationships covering the last 700 million years. The history of the development of the plate-tectonic theory has been authoritatively written by H.W. Menard (1986).

The plate-tectonic revolution in earth science had a gradual but decisive effect on biogeography. Previously, it had been necessary to discuss the historical relationships of the biogeographical regions and their biotas within the framework of stable continents. Now that biologists were released from this constraint, there were varied reactions. In 1965, Darlington published his book *Biogeography of the Southern End of the World.* He was able to contrast the life of southern South America, southern Africa, India, Australia, New Zealand, and Antarctica, and decided that these lands had once been situated much closer together. Darlington also noted that successive new groups of plants and animals had been invading the southern ends of the world over a long period of time. But counter invasions from south to north were exceedingly rare.

The subject of life on oceanic islands, in both the terrestrial and shallow marine environments, has been of interest to biogeographers since before Darwin's time. Although the literature on this subject is widely scattered, there is one comprehensive book that is still of great value. That is *Island Biology* by Sherwin Carlquist (1974). It is a sequel to an early work *Island Life* published in 1965. Islands have often been considered areas where one may find the results of natural experiments in dispersal, colonization, and evolution, that have been carried on for various times depending on the ages of the islands. Because the geographic areas are generally limited and the populations relatively small, islands are more easily studied than most mainland areas.

A number of important phytogeographic works were published beginning in the 1960s. Among them were H.A. Gleason and A. Cronquist, *The Natural Geography of Plants* (1964), A. Takhtajan, *Flowering Plants: Origin and Dispersal* (1969), R. Good, *The Geography of the Flowering Plants* (1974), P. Stott, *Historical Plant Geography* (1981), and J.D. Sauer, *Plant Migration: the Dynamics of Geographic Patterning of Seed Plant Species* (1988). The latter work presents a good modern analysis of historic changes in plant distribution patterns.

In 1969, Miklos D.F. Udvardy published his *Dynamic Zoogeography* which emphasized the importance of dispersal under different climatic conditions but did not attempt to assess continental drift. This work makes it clear that much of the non-English, European literature on biogeography has not been appreciated by authors working in the New World. Udvardy's book is a good entree into that literature as far as animal distribution is concerned. The 1978 book, *Biogeographie* by P. Bănărescu and N. Boscaiu is also helpful in this regard. Paleontologists interested in paleogeography were the first to take full advantage of the new plate-tectonics framework. This resulted in the publication of four important works in the 1970s. These are F.A. Middlemiss and P.F. Rawson (eds.), *Faunal Provinces in Space and Time* (1971), N.F. Hughes (ed.), *Organisms and Continents Through Time* (1973), A. Hallam (ed.), *Atlas of Paleobiogeography* (1973), and J. Gray and A.J. Boucot (eds.), *Historical Biogeography, Plate Tectonics, and the Changing Environment* (1979). The text, *Biogeography* (1979) by E.C. Pielou devoted considerable attention to plate tectonics.

THE RISE OF VICARIANISM

The most important and controversial development of the decade of the 1970s was the enthusiastic promotion of the theory of "vicarianism" Vicariance refers to the biogeographic patterns produced by a particular kind of allopatric speciation in which a geographic barrier develops so that it separates a formerly continuous population. This distinguishes vicarianism from the kind of allopatric speciation which takes place as the result of migration or dispersal of individuals across an existing barrier to colonize the other side. Although these two kinds of allopatric speciation had been recognized for many years, the advocates of vicarianism. came to feel that their viewpoint had been neglected and the vicarianism was *the* important process in producing evolutionary change.

A connection between vicarianism and plate tectonics was established by envisioning, before the separation of the continents, a "hologenesis," a kind of primitive cosmopolitanism based on a theory espoused by Rosa (1923). Hologenesis, where species were supposed to have been created with cosmopolitan ranges, may be contrasted with the center of origin concept (Darwin, 1859) where species originated in a limited area and then spread as far as their capabilities would permit.

As their enthusiasm for a supposedly new concept (which actually may be traced back to the works of Adolph Brongniart and Alphonse de Candolle) grew, the proponents of vicarianism emphasized that it was really vicariance that produced geographical differentiation and multiplication of species while dispersal produced only sympatry. In the best explanation of the mechanics of vicarianism, Croizat et al. (1974) stated, "The existence of races or subspecies that are separated by barriers (vicariance) means that a population has subdivided, or is subdividing, not that dispersal has occurred, or is occurring across the barriers." Belief in vicariance led its disciples to maintain that centers of origin do not exist since, to recognize such centers, they would have to concede that species are capable of dispersing from their places of origin to establish themselves elsewhere, the usual result being, after a period of time, allopatric speciation by migration rather than by geologic change. Consequently, Croizat et al. said, "We reject the Darwinian concept of the center of origin and its corollary, dispersal of species, as a conceptual model of general applicability in historical biogeography."

In the late 1970s and 1980s many journal articles were published about the pros and cons of vicarianism. In 1981, three books appeared, one edited by Gareth Nelson and Donn E. Rosen, *Vicariance Biogeography: Critique*, one written by Gareth Nelson and Norman Platnick, *Systematics and Biogeography*, and the third written by E.O. Wiley, *Phylogenetics. The Theory and Practice of Phylogenetic Systematics*. It has been implied that one must use the vicarianist approach if one is to examine distributions in the light of continental drift and that the biogeographical regions of Wallace and Sclater are no longer useful (Nelson and Platnick, 1980). We were told that the endemism apparent at various oceanic islands of the Pacific can be explained by vicarianism rather than by dispersal (Springer, 1982).

In the meantime, before vicarianism had gotten underway, a book by Willi Hennig, *Phylogenetic Systematics* (1966), was published. This was the second edition of a book originally published in German in 1950. By the 1970s, this work began to have a significant impact on the methodology employed by people who did systematic work. Hennig provided a set of rules for the practice of systematics which have collectively been called "cladism." These rules have generally been helpful but the one that applies to biogeography has turned out to be suspect. It states that species possessing the most primitive characters are found within the earliest occupied part of the area, i.e., the center of origin for that group. Although this rule was at first enthusiastically adopted by some, very little biogeographical evidence has been found to support it.

In the 1980s, a change in the attitudes of the leading vicarianists became apparent. McCoy and Heck (1983) said that vicarianists now admit that allopatric speciation via dispersal can take place and only maintain that it is less important than vicariance. Cracraft (1983) indicated that some cladists were able to forego their center of origin concept in order to join forces with the vicarianists. But Humphries and Parenti (1986) considered dispersal biogeography to be an unscientific, ad hoc discipline that "... can never let us discover the history of the earth." In contrast, they considered vicariance hypotheses to be scientific because they were testable. It was stated that two tests may be applied to a vicariance hypothesis: add more tracks (reinforcement by other taxa that show the same pattern) and compare the hypothesis to a geological one. It must here be emphasized that any biogeographic hypothesis based only on the distribution and relationships of a single group of organisms is on shaky ground. The strongest hypotheses are those based on common patterns demonstrated by many different biotic groups and are, at the same time, consistent with a well substantiated geological history. It makes no difference whether the hypothesis involves vicariance or dispersal or both. Such "tests" (if they really can be considered as such) are certainly not the exclusive property of the vicariance method.

As things now stand, phylogenetic biogeography in the sense of Hennig (1966), including his centers of origin approach, is still being defended (Brundin, 1988). There is another concept called "panbiogeography" which emphasizes distributional patterns alone without any particular connection to phylogenetic patterns (Craw, 1988). Advocates of panbiogeography believe that ocean basins represent the natural biogeographic regions of the globe, and that land masses are biological and geological composites located at the boundaries of the basins (Grehan, 1991). They would substitute an ocean basin classification for the traditional terrestrial regions of Sclater and Wallace.

Parenti (1991) maintained that the biogeography of freshwater fish supported the panbiogeographic concept. However, she used secondary freshwater fish as examples. These are groups that have a high salinity tolerance and are able to migrate across saltwater barriers. Particular emphasis was placed on the global distribution pattern of the sicydiine gobies (family Gobiidae). But, the larval stages of this group are obligate marine dwellers and have probably been able to achieve their present patterns, which are ocean basin related, by dispersal through the sea. Distributions of primary freshwater fish, and almost all other groups of purely freshwater and terrestrial organisms, are not confined to ocean basins.

Vicariance biogeography has mostly become what is often called cladistic biogeography. The latter takes phylogenies, constructed according to the rules of Hennig, and combines them with the vicariance model. The vicariance model or paradigm embodies two general propositions (Wiley, 1988): (1) the best first-order explanation for an observed disjunct distribution between sister groups is that it represents a fragmentation of a widespread ancestral species, rather than a dispersal phenomenon from a more restricted "center of origin;" (2) the more cases of such observed patterns, the more likely the first proposition becomes.

The difficulty with the modern vicariance model is that it asks the investigator to perform an act of faith before he begins his analysis; he must believe in a given first-order explanation. The objectivity of such an approach is questionable. Suppose a researcher has analyzed the relationships within a given taxon so that he has been able to construct a phylogenetic tree or a cladogram. If the work is to have a biogeographic value, the geographic patterns of the species need to be related to the evolutionary ones. How should this be done? An objective approach would be to forego a priori assumptions and let the group being worked on, with comparisons to similar groups, suggest what the history might have been. Most likely, if the group is an old one, and if proper attention has been paid to impinging factors (paleontological, geological, climatological), the resulting hypothesis will involve both vicariance and dispersal. This uncommitted approach needs to have a name to distinguish it from the other kinds. The designation "eclectic biogeography" seems appropriate. It was Simpson (1980) who observed, "A reasonable biogeographer is neither a vicarianist nor a dispersalist but an eclecticist."

The vicariance model, as it now stands, places two important restrictions on the development of biogeography as a field of research. First, it recognizes only a certain kind of allopatric speciation as being of historical importance. Second, it does not recognize that centers of origin exist since, to do so, would admit that dispersal has taken place. Almost everyone interested in biogeography knows that allopatric speciation via dispersal can and does take place. And the speciation problem is not that simple. There is a growing body of evidence showing that both parapatric and sympatric speciation can take place in nature (Otte and Endler, 1989). Neither of the latter two processes require geographic barriers. The case for the existence of centers of origin has been stated by Briggs (1984, 1994).

THE PRESENT WORK

A precursor to the present work, *Biogeography and Plate Tectonics* (Briggs, 1987) was the first attempt to evaluate, on a broad scale, the biological effects of the Mesozoic continental dispersal. Paleogeographic maps were utilized in order to illustrate historic distributions. This approach emphasized the importance of fossil distributions in the formation of modern biogeographic patterns. Due to the lack of a satisfactory geographic base, it did not attempt to evaluate the effects of continental movement during the Paleozoic. Also, it did not discuss, in detail, distribution patterns in the marine environment. The present work examines the Paleozoic, provides new information on the Mesozoic, and discusses the distribution of many groups of marine animals. It also evaluates the effects of historic extinction episodes on biogeography and evolution.

This book, as well as its predecessor, has been written from the point of view that dispersal and vicarianism have each played an important role in historical biogeography. Most higher organisms have, as an integral part of their life history, a dispersal phase which allows them to spread out and occupy new territories. This enables each species to eventually move as far as its migratory ability and ecological versatility will permit. As a young species enlarges its territory, it will encounter barriers that will prohibit or delay its further expansion. Such barriers may comprise one or more of a variety of physical, chemical, and biological features of the environment.

It is important to realize that barriers to migration are not often static and that, over time, most of them have been or will be changed. Sometimes the creation of a barrier will result in the interruption of the range of a species or a species complex. For example, when the Isthmus of Panama was finally completed in the Pliocene, it separated the tropical marine environment of the New World into two parts, one inhabiting the Eastern Pacific and the other the Western Atlantic. This is considered to be a vicariant event, in that it prevented gene flow between the two parts and caused the separated populations to embark on their own evolutionary courses. But, at the same time, the isthmian connection provided a dispersal corridor between North and South America for terrestrial and freshwater organisms, also with profound evolutionary (and ecological) consequences.

In a similar manner, when the land connection across the Bering Strait was first made in the late Cretaceous, it separated the marine populations of the Bering Sea–Arctic Ocean but connected terrestrial North America to Asia. From a marine standpoint, the creation of Beringia was a vicariant event but from a terrestrial viewpoint it provided a dispersal opportunity. The tectonic uplift of a mountain range can constitute an important barrier for lowland forms but simultaneously may present a migratory corridor for species of the high-altitude biota.

Dispersal is an everyday occurrence undertaken by succeeding generations of almost all species while vicarianism is an event of much greater rarity since it must involve the creation of a barrier to separate existing populations. A most important point is that when vicariance does take place it appears to offer, at the same time, unusual dispersal opportunities for some groups of species. So dispersion may be looked upon as a continuing, inexorable process while vicariance, when it occurs in one habitat usually stimulates dispersal in another. This is particularly true in regard to continental movement with its making and breaking of land and sea barriers.

For the past 20 years, a significant portion of the theoretical literature on biogeography has been devoted to argument about the efficacy of vicarianism compared to dispersalism. It is important that biogeographers attempt to appreciate the biosphere as a whole instead of concentrating too heavily on a single habitat. It is time to discard preconceptions about what might have happened in the history of a given taxon. Let us use the clues that can be found within the relationships of the group itself and in the history of its territory. This is eclectic biogeography and it means freedom from preconceived ideas.

PART A – Historical biogeography

CHAPTER 2

Precambrian and Early Paleozoic

Have the changes which lead us from one geological state to another been, on a long average, uniform in their intensity, or have they consisted of epochs of paroxysmal and catastrophic action, interposed between periods of comparative tranquility? These two opinions will probably for some time divide the geological world into two sects, which may perhaps be designated as the Uniformitarians and the Catastrophists.

William Whewell in his *Review of Lyell's Principles*, 1832

Paleozoic map reconstructions often differ widely from one another depending on the interest of the author. The three disciplines concerned are paleomagnetism, paleoclimatology, and biogeography. All three may contribute significantly to our knowledge of latitudinal position. Assuming a co-axial, geomagnetic dipole field in the geological past, paleomagnetic inclinations should yield direct information about paleolatitudes. Paleoclimates are often indicated by deposits such as coal swamps, glacial tillites, carbonate sediments, and evaporites. Certain groups of animals and plants, particularly in the marine environment, are typically restricted to either warm or cool habitats; and many of these restrictions have apparently held through time. Consequently, biogeographic patterns can help to separate the low latitude tropics from the high latitude temperate regions.

Theoretically, biogeography should be critical for the determination of longitudinal separations, since neither paleomagnetism nor paleoclimatology have, in this case, much predictive value. However, in the marine environment, one must be careful to judge the longitudinal separation of land masses from the relationships of the shelf biota, rather than organisms from the deep sea or pelagic habitats. Furthermore, certain groups of shelf organisms, due to their long-lived, pelagic larvae, can maintain a circumglobal continuity even when continents are widely separated. The making and breaking of continental connections is most easily determined by fossil evidence from terrestrial organisms, but unfortunately that fossil record does not begin to become useful until late in the Paleozoic; even then, the marine record is far better.

Comparison of Paleozoic base maps, produced by a variety of authors with different viewpoints, can be an exercise in frustration. For example, the maps produced by Bambach et al. (1980), which have been widely used for paleobiogeography, are much different than those incorporating the latest paleomagnetic results (Van der Voo, 1988; Kearey and Vine, 1990). Boucot (1988), in his thorough review of Devonian biogeography, arbitrarily settled on a pangaeic paleogeography because none of the available reconstructions agreed with his fossil distribution patterns.

More recently, considerable advances in paleogeography have been made. Among these, the most useful are: (1) P.A. Zeigler's (1989) account of the evolution of Laurussia, (2) the Paleozoic base maps published by Scotese and McKerrow (1990), and (3) the continental drift maps produced by the Paleomap Project of the International Lithosphere Program (Scotese, 1992). While these maps, constructed from paleomagnetic and geological data, are important for the purpose of determining the relative positions of the continental blocks, their value to historical biogeography is limited. They do not indicate shore lines and generally use a Mollweide projection, which produces an elliptical shape with severe distortion at higher latitudes. The Lambert equal-area projections are the most satisfactory for biogeographic purposes. They allow a minimal distortion of continental shapes and have the additional advantage of showing both poles on the same map (see the Appendix for the complete Phanerozoic series, Maps 1–16).

Highly metamorphosed sedimentary rocks about 3.8 billion years (Ga) in age from western Greenland have yielded some indications of photoautotrophic biological activity in the form of distinctive carbon isotope ratios (Goodwin, 1991). The oldest known microfossils are 3.5 Ga stromatolites from western Australia. Thus life evolved in the early part of the Archean Eon which extended from 4.0 Ga to 2.5 Ga. A rapid increase in atmospheric oxygen began about 2.0 Ga. This development theoretically made possible the organization of the eucaryotic cell about 1.4 Ga ago. The theoretical timing of subsequent evolutionary events in relation to the levels of atmospheric oxygen was described by Cloud (1983).

During the early development of life forms, fragments of the Precambrian continental crust were supposedly being gathered into a Proterozoic supercontinent. Piper (1987), on the basis of paleomagnetics and bedrock geology, recognized the existence of a supercontinent from about 950 to 700 Ma ago. After that time, the various components supposedly began to separate from one another. However, some authors, utilizing the same type of data, have argued that the Proterozoic provinces developed largely in their present relative positions. Goodwin (1991) has warned that any Precambrian paleoreconstruction is to be used with extreme caution.

Despite equivocal information about the existence of a supercontinent, the concept has received wide acceptance. Worsley et al. (1986) proposed a "supercontinent megacycle" composed of four phases: fragmentation, maximum dispersal, assembly, and stasis. These events were stated to recur over a period of 500 Ma. Gurnis (1988) and Nance et al. (1988) followed with an explanation of the mechanism. Supposedly, the supercontinent interferes with the escape of heat through the earth's mantle. This causes overheating below, which eventually fragments the supercontinent. The fragments (continents) drift and then congregate over colder, downwelling mantle. Then the overheating begins anew. The Proterozoic supercontinent has been given the name "Rodinia" and the corresponding superocean called "Mirovia" (McMenamin and McMenamin, 1990).

The work of Hartnady (1991) and others refer to a continuous operation of plate tectonics over the greater part of Earth history. This is supposed to have led to the episodic disruption and re-assembly of continental fragments into a number of varied mosaical configurations at different times. These recurring episodes have been called the "Wilson Cycle." Just how far back in history such an idea can be carried is problematical, since

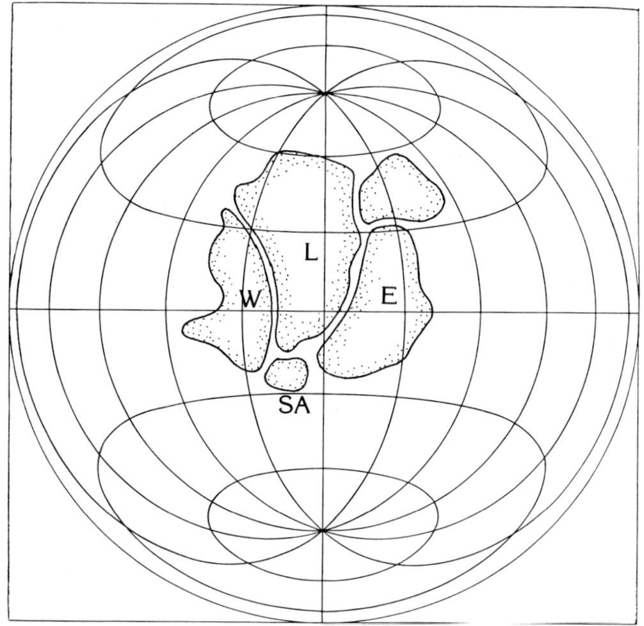

Fig. 1. Rhodinia before 750 Ma ago. East (E) and West (W) Proto-Gondwana are separated by Laurentia (L). Southern Africa (SA) is situated below. After Hartnady (1991).

the Archean heat flow was much greater and the continental crust much thinner than in the Proterozoic (Goodwin, 1991).

A pre-750 Ma configuration of Rodinia (Fig. 1) has been published by Hartnady (1991). He preferred to use the name "Ur-Gondwanaland" but this designation is undesirable because Gondwana has been traditionally used to designate a southern conglomeration, not all continents together. His arrangement shows an East and West Proto-Gondwana separated by Laurentia. As continental dispersal took place, West Proto-Gondwana swung around with a southern African block in the hinge position. This motion, shown in Fig. 2, was apparently completed by about 650–600 Ma ago. This is quite different from previous models which showed the Proterozoic continent as a single entity.

Another mechanism, if true, could have had an important effect on continental dispersal in the Paleozoic and Mesozoic. That is the theory of an expanding earth, apparently first proposed by Lindemann (1927) and soon espoused by others. Carey (1956, 1976) amassed an imposing amount of data to support the theory. His banner was picked up by Owen (1983) who has forcefully continued the campaign. But, McElhinny et al. (1978), using paleomagnetic techniques to determine the paleoradius of the earth, had concluded that 400 Ma ago the radius was 1.02 ± 2.8% of the present radius. Therefore, only a very slight expansion or contraction could be tolerated by this analysis. Kearey and Vine (1990) stated that the expanding-earth hypothesis clearly does not stand up to direct testing. Cox (1990) remarked that the expanding earth belongs in the same category as the flat earth.

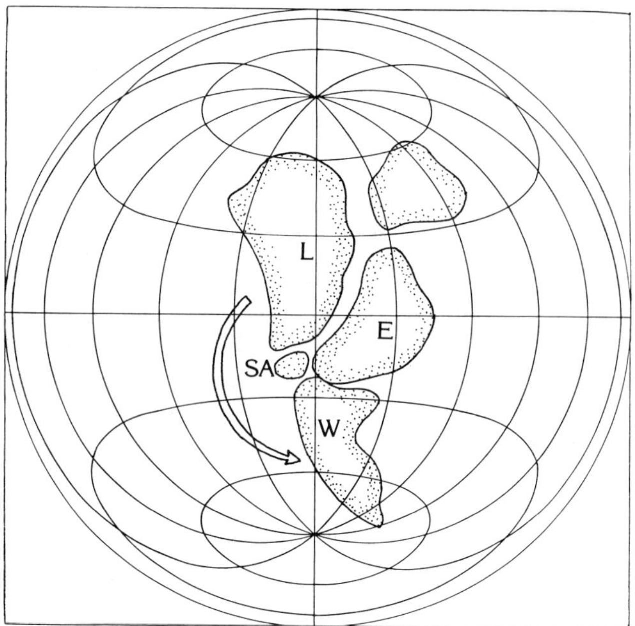

Fig. 2. Rhodinia about 650–600 Ma ago. West Proto-Gondwana (W) has swung to the south. After Hartnady (1991).

PRECAMBRIAN

In the late Precambrian the global climate was cold and extensive ice sheets were present. There may have been as many as four major glaciations, lasting for some 400 Ma (Harland, 1983). The final one possibly took place about 650–620 Ma ago. The first unequivocal metazoan fossils appear stratigraphically above the tillites of the latter (Fedonkin, 1990). These animals comprised the "Ediacaran Fauna" which developed in the interval between the glaciation and the beginning of the Cambrian about 570 Ma ago. The fauna had a generally broad geographical distribution with many identical forms occurring at distant localities. Fedonkin concluded that such patterns indicated cosmopolitanism and evidently low rates of evolution. However, McMenamin (1982) had noted that some of the North American fossils were different than those of Baltica/Gondwana. He argued that the Ediacaran fauna first evolved on Gondwana and that Baltica must have been close by. The breakup of Rodinia may have begun as much as 700 Ma ago (Hartnady, 1991). The map of Scotese and McKerrow (1990) for about 600 Ma ago indicates that considerable continental separation had already taken place. On the other hand, Kearey and Vine's (1990) map for 600 Ma ago still maintains the pangaean configuration.

The period from about 610–570 Ma ago is sometimes termed the Vendian. At that time, there existed a group of microfossils with organic walls called acritarchs. They appear to be similar to modern dinoflagellate cysts and are thought to represent the resting stages of planktic, eucaryotic marine algae (McMenamin, 1990). The diversity of these microfossils underwent a severe decline during the Middle to Late Vendian, which some paleontologists think may represent a major extinction. Acritarch diversity did not recover to early Vendian levels until the Lower Cambrian. This extinction event has been linked to the final (Varangian) glaciation of the Precambrian by Vidal and Knoll (1982).

Numerous soft-bodied, apparently metazoan, fossils are known from the Vendian. Many of them were frond or disc-shaped. *Pteridinium* (Fig. 3) had a body consisting of elongate tubes joined together like the partitions in an air mattress. Others were thin, pancake-shaped organisms reaching one m or more in diameter. Some had a simple, sac shape. McMenamin (1989a) pointed out that their large surface area to volume ratio suggested they were well suited for chemo- or photoautotrophic feeding. Heterotrophy may have become increasingly important near the end of the Precambrian. Stromatolites, built by filamentous algae, reached their peak in abundance and diversity about 900 Ma ago. After this time, they underwent a marked decline, which may reflect the appearance of algae-eating metazoans.

The largest known members of the Ediacaran fauna were about twice as large as the largest animals of the Cambrian. Was Vendian life an idyllic existence when animals were at peace with one another and ate only plants? McMenamin and McMenamin (1990) wrote about the "Garden of Ediacara," suggesting that its downfall was caused by the evolution of predatory animals. The transition, near the Vendian-Cambrian boundary, from large soft animals to small forms protected by shells and jointed exoskeletons amounted to a basic restructuring of community relationships.

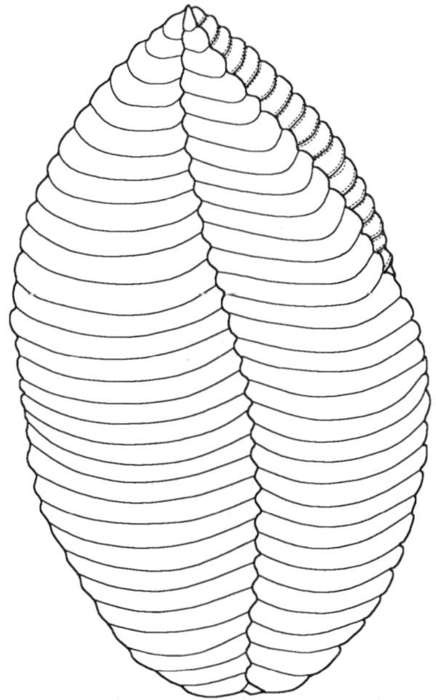

Fig. 3. *Pteridinium*. After McMenamin (1989).

CAMBRIAN PERIOD

The advent of the Cambrian Period meant the beginning of the rapid, sometimes called explosive, evolution of an astounding variety of animal groups, possibly as many as 100 different phyla. This was also the beginning of important, global physical changes which undoubtedly had a forcing effect on the marine biota. In 1984, Fischer called attention to the historic relationship among biotic crises, climate, sea level, and volcanism. He recognized two supercycles in Phanerozoic history, each of them lasting about 300 Ma. He noted that a global "icehouse" condition existed in the late Precambrian. This gave way to a "greenhouse" in the Ordovician, Silurian, and Devonian (Fig. 4). A succeeding icehouse was present from the Carboniferous through the mid-Permian. The next greenhouse held sway through the Jurassic, Cretaceous, and the first part of the Cenozoic. Our present icehouse climate began in the middle Eocene.

As the ice sheets melted in the late Precambrian and early Cambrian, the sea level rose. The tectonic plates bearing the continental crust began to move apart. These movements were accompanied by an increase in volcanism. As the plates moved apart, oceanic spaces were created between continental fragments. As movement continued, mantle

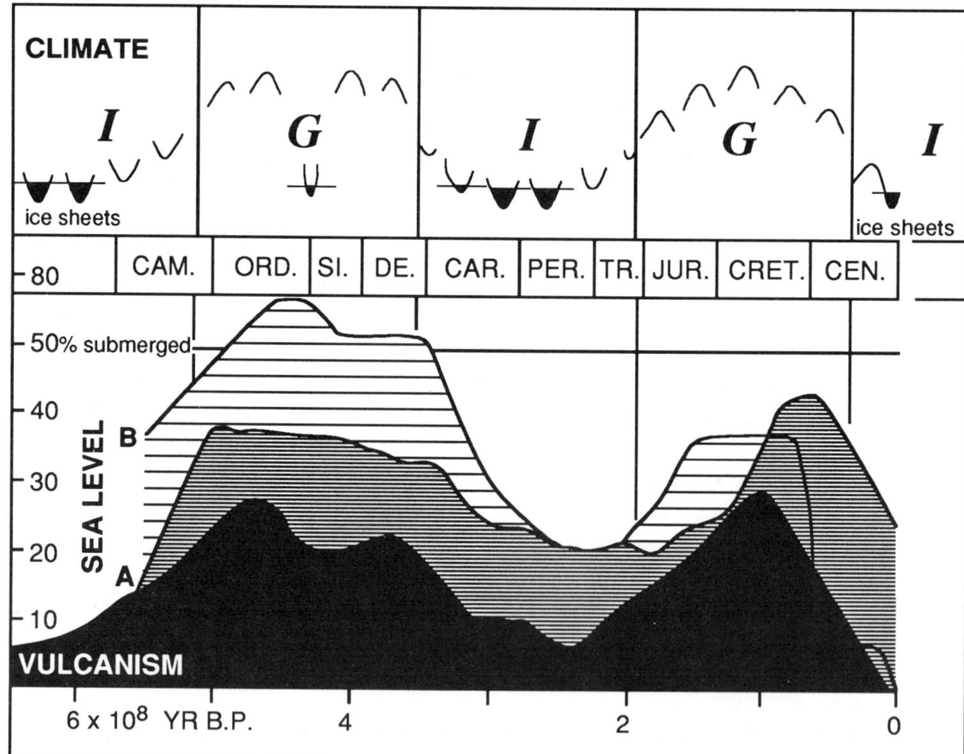

Fig. 4. The two supercycles of Phanerozoic history. I, icehouse state; G, greenhouse state. Sea levels: A, from Vail et al. (1978); B, from Hallam (1992). Modified from Fischer (1984).

upwelling took place on the ocean bed between the continents. The hot mantle material was extruded in the form of huge submarine ridges along the boundaries between plates. The ridges left less room in the oceanic basin. Continental flooding was caused by a combination of ice melt and plate separation.

Increased volcanism, which accompanied enhanced plate movement, injected CO_2 into the atmosphere. The rise in atmospheric CO_2 caused retention of much of the heat lost to space by radiation, so the surface of the earth became warmer. As the sea level rose, the relative size of the continents became smaller. This interfered with the absorption of CO_2 into the lithosphere by the weathering process. Consequently, the atmosphere-hydrosphere system retained a high level of CO_2. Another gas of biological importance is oxygen. There seems little doubt that free oxygen was present in the atmosphere by about 2 Ga ago (Holland, 1984). As photosynthetic organisms became more numerous, the O_2 level rose until it reached the threshold required for respiration by multicellular organisms.

The rise in sea level created large epicontinental seas. This permitted the rivers to discharge their nutrients into shallow waters where they could be recycled by wind-driven currents. The continental rifting itself, which provided a greatly increased shore line, plus the flooded areas, produced an enormous increase in available habitat to the continental-shelf biota. The warmer water, increased O_2, expanded shelf habitat, and improved nutrient supply were all positive factors in an evolutionary sense. But, these influences by themselves do not explain the large soft-body to small hard-body transition. Nor do they explain the rapid changes in diversity which produced so many animal phya within a relatively brief period.

Animals would not have undertaken the difficult genetic and metabolic task of covering themselves with shells or jointed armor unless they were under selection pressure by something that wanted to eat them. Another way of coping with predation is escape to a different habitat, i.e., benthic to pelagic or benthic to burrowing beneath the surface. No one knows when sexual reproduction, with its enhancement of individual variation by means of genetic recombination, took place. It could have been the most important factor in the stimulus of Cambrian evolutionary diversity. All of these foregoing events, a combination of internal and external factors, probably had some bearing on the outcome.

In the earliest Cambrian (the Tommotian Stage), moderate-diversity (five to about fifteen species) of small shelly fossils evolved. These are found in most parts of the world where suitable fossilization has occurred. At the same time, the strange vase-shaped archaeocyathans (Fig. 5) appeared. They somewhat resemble sponges but probably belong in a phylum of their own. These animals, plus calcareous algae, formed wave-resistant reefs called bioherms. A little later, in the next stage of the Cambrian, high-diversity shelly faunas, are found. For many years, these remains were thought to represent very small animals. Now, thanks to discoveries in China and Greenland, two larger animals are known that carried the small shells and plates on their bodies (Gould, 1990).

The Burgess Shale represents a fortunate accumulation of fine grain sediment in an anoxic environment where scavengers and decay bacteria were absent. This mid-Cambrian deposit has yielded fossils of the earliest-known carnivores. The largest was *Anomalocaris* (Fig. 6) which grew up to 1 m in length. Conodonts, important pelagic

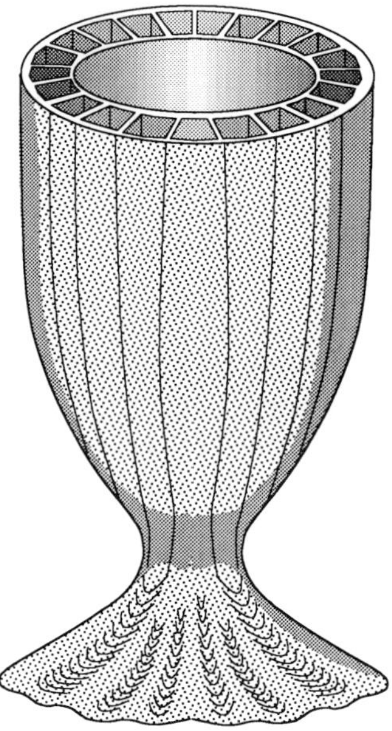

Fig. 5. An archaeocyathan. After McMenamin (1989).

predators, were also found. In addition, there was a remarkable variety of echinoderm classes, brachiopods, molluscs, sponges, and annelid worms.

The first trilobites appeared in the Atdabanian Stage about 550 Ma ago. A rapid rise in archaeocyathid diversity approximately coincided with the rise in the early, olenellid trilobites. For a period of about 10–15 Ma, the reefs were built and dominated by the archaeocyathids while the level bottoms were dominated by the olenellids (Fagerstrom, 1987). During the Elankian Stage, the archaeocyathid reefs underwent an abrupt decline, although the animals themselves persisted on level bottom communities until the late Cambrian. Very few reef structures have been found in Middle Cambrian to early Ordovician deposits.

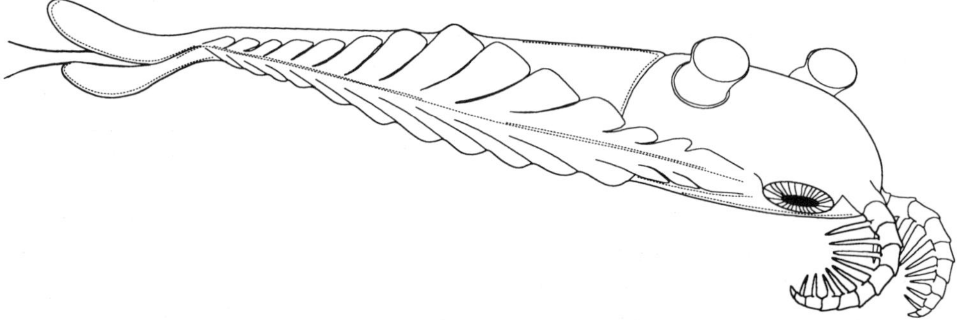

Fig. 6. *Anomalocaris*. After McMenamin (1989) and other sources.

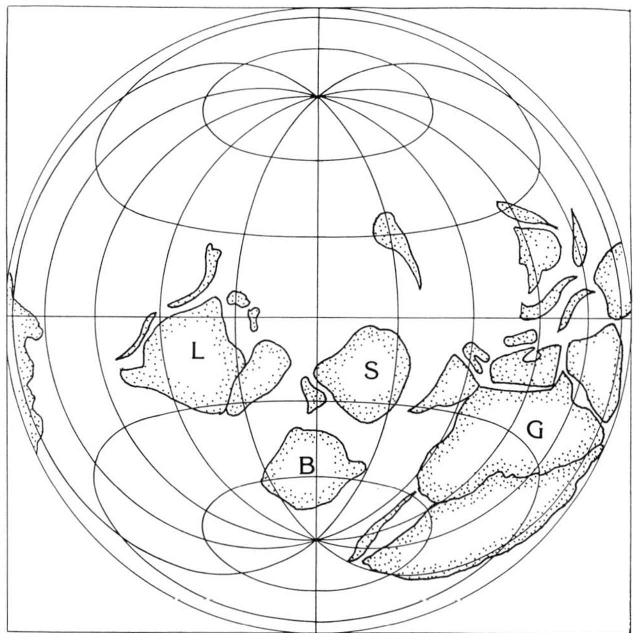

Fig. 7. The Olenellid Province included Laurentia (L), Siberia (S) and Baltica (B). The Redlichiid Province was confined to Gondwana (G). After Fortey and Owens (1990).

Trilobites, flattened arthropods with a segmented exoskeleton, are the best known animal group of the Cambrian. It is interesting to see that, from their beginning in the earliest Cambrian, there is clear evidence of biogeographic differentiation. An Olenellid Province was recognized for much of Europe, Siberia, and North America; a Redlichiid Province included eastern Asia, China, and Australia (Palmer, 1973; Fortey and Owens, 1990). The provinces are illustrated on the Cambrian base map (Fig. 7). One may now distinguish the separate continents of Laurentia, Baltica, and Siberia; these three constituting the Olenellid Province. The Redlichiid Province was confined to Gondwana, although parts of China may have existed as nearby terranes (Burrett et al., 1990).

An illustrated cladogram showing early trilobite phylogeny and biogeography has been published (Fig. 8) by Fortey and Owens (1990). Trilobite ancestors probably had soft bodies and an extensive Precambrian history. It is likely that the two early Cambrian provinces became formed in the Precambrian when the northern continents and Gondwana first split apart. For the middle and late Cambrian, Palmer (1973) advocated a more complicated scheme involving four provinces for the continental seas and three others for exposed shorelines.

Certainly, by the late Cambrian the northern continents had moved farther apart and this may have promoted additional provincialism. Conodont distribution in the late Cambrian indicates the presence of two faunal regions (Bergström, 1990), a tropical Midcontinent Region and a cool Atlantic Region (Fig. 9). The distributions of trilobites and conodonts should, however, not be closely compared because the first were primarily

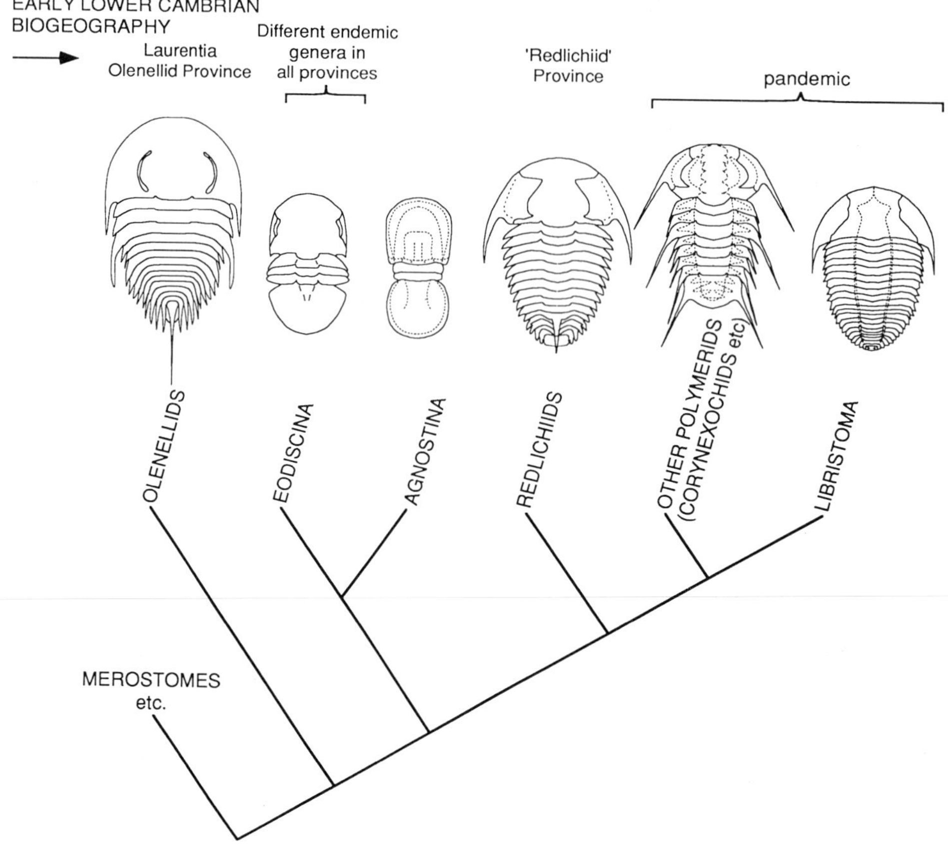

Fig. 8. Early trilobite phylogeny and biogeography. After Fortey and Owens (1990).

benthic and the second primarily pelagic. Conodont fossils consist of jaw-like structures that resemble the teeth of modern hagfish (Fig. 10). They may have been the earliest vertebrates.

The Cambrian fauna has been recognized as the first of the three great evolutionary faunas of the Phanerozoic (Sepkoski and Miller, 1985). It consisted largely of trilobites, inarticulate brachiopods, monoplacophorans, hyolithids, and eocrinoids. The Cambrian-Ordovician boundary is sometimes recognized as the time of a major extinction event. In their review of the fossil record of the arthropods, Briggs et al. (1988) concluded that the boundary did not represent a major extinction event for trilobites at the family level. There was apparently a high generic turnover, but taxonomic pseudoextinctions were considered to be a problem. The recognition of "biomere" boundaries within the Cambrian, as major extinctions, were viewed to have similar difficulties. Palmer (1984), in discussing trilobite communities of the middle and late Cambrian, referred to three disasters and recoveries. During each extinction, it was the shelf trilobite communities that suffered the most. The subsequent radiations originated from deep water or high-latitude

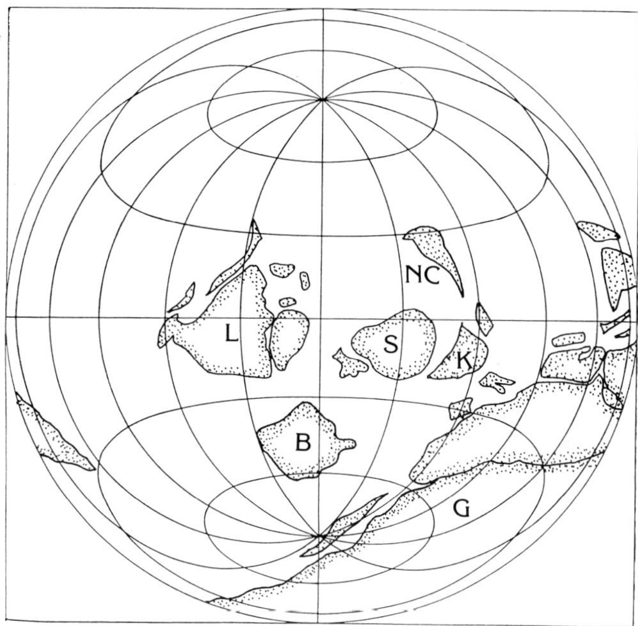

Fig. 9. The Midcontinental Faunal Region included Laurentia (L), Siberia (S), Kazakhstania (K) and North China (NC). The Atlantic Faunal Region included Baltica (B) and probably parts of Gondwana (G). After Bergström (1990).

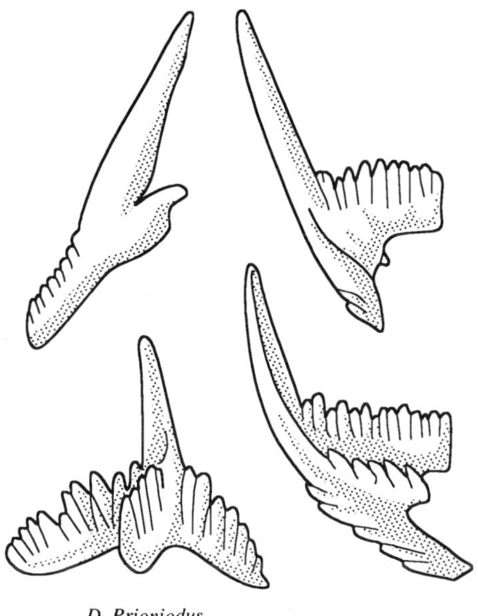

D. Prioniodus

Fig. 10. Conodont genus *Prionoidus*. After Stearn and Carroll (1989).

groups that had survived the extinctions (Stanley, 1989). Considering the fact that most of our information about diversity changes in the late Cambrian is based on trilobite material, and that this group does not demonstrate a clearly defined boundary effect, the close of the Cambrian is not recognized as a major extinction.

ORDOVICIAN PERIOD

With the advent of the Ordovician, the dominant trilobite-rich community began to be displaced by a new brachiopod-rich assemblage (Sepkoski and Miller, 1985). This resulted in a restriction of the trilobite community to deeper benthic waters and the pelagic habitat. The development of the new community, plus the continued expansion of its predecessor, caused a rapid increase in diversity at all taxonomic levels. The evolutionary innovations of the Cambrian and Ordovician are particularly notable among the skeletonized invertebrates. These two periods introduced all 11 of the phyla, 54 of 56 classes, and about 152 of 235 recognized orders (Erwin et al., 1987). In contrast, the post-Paleozoic diversification, following the Permian-Triassic extinction, which was of a similar duration, produced no new phyla or classes and few new orders.

The biogeographic literature for the Ordovician is voluminous compared to that of the Cambrian. Fortunately, two important reviews, which included summaries of data based on several animal groups, have been published. Although Tuckey (1990a) concentrated on the bryozoans, he also utilized information on the brachiopods, trilobites, graptolites, conodonts, and corals. In their review of the Ordovician and Silurian faunas, Cocks and Fortey (1990) incorporated selected trilobites, brachiopods, and graptolites.

By the early Ordovician, trilobite evolution had proceeded to the extent that distinctive shelf, deep-water, and pelagic taxa could be identified. In their work, Cocks and Fortey (1990) drew an elegant comparison among the three groups. They showed that the shelf or platform assemblages were geographically separated into four distinct regions (Fig. 11) indicating the importance of both latitudinal and longitudinal barriers. In contrast, the pelagic groups demonstrated only a latitudinal separation between tropical and cooler waters. The deep-water trilobites apparently had cosmopolitan distributions.

Tuckey (1990a) recognized, for the early Ordovician, only two faunal provinces. But, this determination was made principally on bryozoan distribution and this animal group was not very diverse at that time. By the middle Ordovician, four provinces were evident. In the Caradoc Epoch of the late Ordovician, there were also four provinces (Fig. 12). These results are surprising because of the indicated relationship between the faunas of Laurentia and Australia and between Baltica and the southeast Asian terranes.

The conodonts show a very high provincialism through the Ordovician with two regions, each being divided into provinces (Bergström, 1990). Considering that conodonts occupied the pelagic as well as neritic habitats, the high amount of provincialism might not have been expected. Considerable work has been done on Ordovician graptolites. They were colonial hemichordates related to the living genus *Rhabdopleura*. In the Cambrian, they lived attached to the sea floor. In the Ordovician, they invaded the pelagic habitat and underwent a burst of evolutionary radiation.

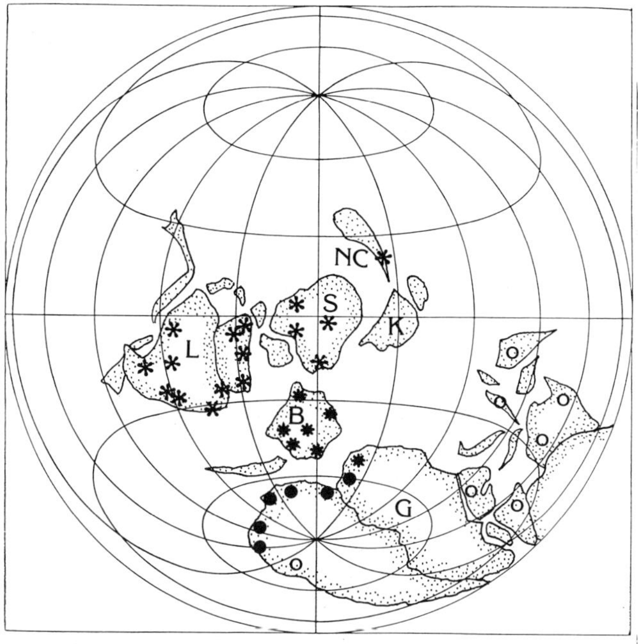

Fig. 11. Four early Ordovician continental-shelf assemblages, each indicated by a different symbol. Their positions indicate the presence of biogeographic barriers. After Cocks and Fortey (1990).

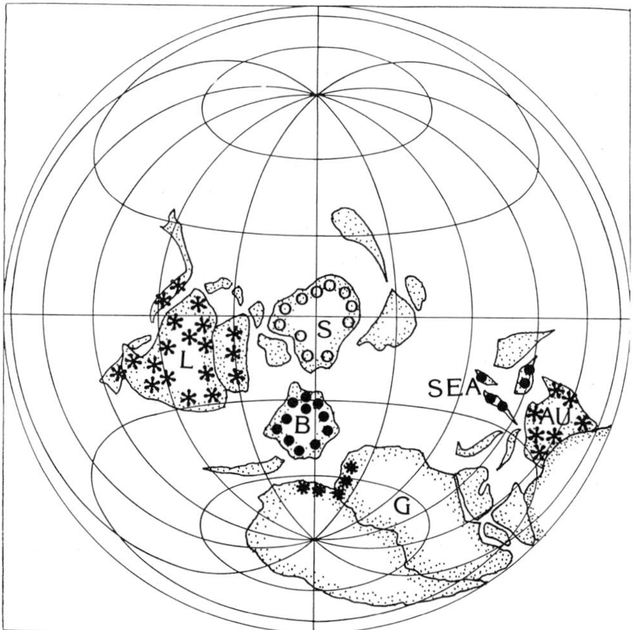

Fig. 12. Four provinces illustrated by bryozoan distributions: North American (✻), Siberian (O), Baltic (●) and Mediterranean (✳). Note the puzzling relationships between Baltica (B) and the southeast Asian terranes and between Laurentia (L) and Australia (AU). After Tuckey (1990a).

Unlike the groups so far examined, graptolite biogeography shows that significant changes took place between the early and late Ordovician (Finney and Xu, 1990; Berry and Wilde, 1990). In the early part of that Period, it was possible to distinguish two principal regions which contained some 12 provinces. The two regions were a warm water "Pacific" and a cool water "Atlantic." Provincialism decreased somewhat in the Middle Ordovician. In the latter part of that Period, a glaciation, accompanied by high-latitude temperature declines, apparently resulted in the disappearance of the Atlantic fauna. Evolutionary flowering among graptolites has been linked to eustatic sea-level rise and large extinctions to sea-level lowering (Berry and Wilde, 1990).

Recent works have also been devoted to the biogeography of the nautiloid cephalopods (Crick, 1990) and the brachiopods (Sheehan and Coorough, 1990). The latter group lost about 67% of its genera during the Ashgill extinction and provinciality declined. The loss of provinciality was apparently an effect caused by the superior survival of widely distributed genera.

END-ORDOVICIAN EXTINCTION

Examination of the maps which illustrate Ordovician distributions will show that considerable continental movement took place over the 70 Ma of that Period. Laurentia moved eastward, closing the distance to Siberia and Baltica. Baltica, which had been isolated to the south, moved to the north approaching Siberia. Siberia itself moved slightly to the north. Kazakhstan also shifted northward and closer to the eastern Asian parts of Gondwana. The vast continent of Gondwana, which was situated north of the polar region, swung southward to place its western part directly over the South Pole (Kearey and Vine, 1990). The glaciation at the end of the Period took place near the South Pole.

Over a period of about one to two Ma, near the end of the Ordovician, many taxa belonging to complex benthic and pelagic communities became extinct (Brenchley, 1989, 1990). There may have been two peaks of extinction. There were either two sea-level regressions or possibly one regression with the second extinction being caused by a transgression, which might have introduced anoxic conditions to the continental shelves. It has been suggested that the regression(s) were accompanied by climatic cooling which restricted the tropics and had a detrimental effect on diversity (Sheehan, 1982, Stanley, 1989).

Opinions differ in regard to the severity of the extinction. McLaren (1983) considered it to be neither sudden nor catastrophic, Stanley (1989) described it as one of the most severe ever to strike life in the oceans, while Brenchley (1989) termed it less severe than the end-Permian and end-Cretaceous events. Stanley stated that the tropical reef community was devastated but Fagerstrom (1987) had indicated that there was a far greater impact on the structure of the level bottom and pelagic communities than on the reef communities. The following groups apparently lost most of their genera: trilobites, conodonts, graptolites, primitive echinoderms, and corals (Brenchley, 1990). Eight echinoderm families disappeared (Paul, 1988).

The patterns of extinction and survivorship among the Ordovician trilobites were studied by Chatterton and Speyer (1989). They found that all taxa with pelagic adults had become extinct by the end of the Ordovician. Similarly, most of those with entirely pelagic larvae had also disappeared. They concluded that the preferential loss of certain life-history categories provided some information as to the causes of the Ashgillian extinction. In their opinion, this event was most likely the result of several factors including environmental perturbations, ecosystem breakdown, and biogeographic restriction. These changes are consistent with an extinction model involving global cooling and sea-level regression caused by glaciation. It apparently took 12–15 Ma for marine diversity to rebound from the extinction (Sepkoski, 1992).

SUMMARY

1. The hypothesis of a Proterozoic supercontinent, which may have existed from about 950 to 700 Ma ago, has gained wide acceptance despite very little evidence. This idea has prompted some to propose a supercontinent megacycle or "Wilson Cycle." The supposition is that, over a period of about 500 Ma, the continents will assemble together, fragment, disperse widely, and then reassemble.
2. The Proterozoic supercontinent (Rodinia), if indeed one existed, supposedly began to fragment about 750 Ma ago. The subsequent dispersal of the continental blocks may, according to some authorities, have been aided by an expanding earth.
3. The first undoubted metazoan fossils appear just above the tillites of the final Precambrian glaciation which took place about 650 to 620 Ma ago. These animals comprised the Ediacaran Fauna, which was present until the beginning of the Cambrian about 570 Ma ago.
4. The advent of the Cambrian Period meant the beginning of a rapid evolution resulting in an astounding variety of animal groups, possibly as many as 100 different phyla.
5. Two marine biogeographic provinces have been recognized for the early Cambrian; one for Laurentia, Baltica, Siberia, and Kazakhstan; and the other for Gondwana. By the late Cambrian, provincialism may have increased, for as many as seven trilobite provinces have been named. The distribution of the pelagic conodonts suggested the presence of two regions. During the Cambrian, the dominant trilobite communities suffered three extinction episodes. In each, the low-latitude shelf community suffered the most. The subsequent radiations originated from deep-water or high-latitude groups that had survived the extinctions.
6. In the Ordovician, the trilobite-rich communities began to be displaced by a new brachiopod-rich assemblage. This resulted in a restriction of the trilobite groups to deeper benthic waters and to the pelagic habitat.
7. The voluminous literature on the biogeography of Ordovician bryozoans, trilobites, graptolites, conodonts, and corals gives the impression of considerable provinciality. The shelf faunas were separated both longitudinally and latitudinally. The pelagic groups showed separations into tropical and cool-water faunas. The deep-water groups (trilobites) apparently had cosmopolitan distributions.

8. By the end of the Ordovician, the vast continent of Gondwana had moved southward over the south pole. An extensive southern hemisphere glaciation took place, the sea level dropped, and there was a severe extinction episode. There may have been two peaks of extinction lasting over a period of 1–2 Ma. Global cooling, sea-level regression, and possibly a transgression which introduced anoxic conditions, were the probable causes.

CHAPTER 3

Later Paleozoic

One of the most pervasive effects during times of biotic crisis is the massive disruption of tropical and low-latitude ecosystems, and the relative non-disturbance of high-latitude and polar ecosystems.

George R. McGhee, Jr., *Catastrophes in the History of Life*, 1989

SILURIAN

With the melting of the southern hemisphere glaciers, and possibly a resumption of plate movement, the sea level rose again but did not reach the historic high of the late Ordovician (Fig. 13). During the late Silurian, there was another decline but this did not result in a major extinction. The regression appeared to be relatively gradual and was not accompanied by a major glaciation (Hallam, 1984). The Silurian and Devonian periods were similar in that sea levels, despite some fluctuations, remained high and sea-surface temperatures were warm.

Although the end-Ordovician extinction was severe, most of the animal groups that had previously flourished, rediversified in the Silurian and Devonian. The trilobites were an exception for they did not manage to regain their former diversity. Huge organic reefs, much larger than those of the Ordovician, were constructed by tabulate and rugose corals and by stromatoporoids. Gastropod and bivalve molluscs diversified, with the latter entering Devonian freshwater habitats. The graptolites underwent a spectacular expansion in the early Silurian.

Recent works on the biogeography of several important Silurian groups are available. Boucot (1990a) provided a general overview of the Period with an emphasis on brachiopods. He delineated two major realms, a North Silurian Realm and a Malvinokaffric Realm. The former was subdivided into a North Atlantic Region with two provinces and a Uralian-Cordilleran Region. His principal map, which depicts the Upper Silurian, has the continents in a pangaeic arrangement, because he felt that such an amalgamation gave the best explanation for the biogeographic relationships. He also plotted the distributions on the Scotese and McKerrow (1990) base map. Inconsistencies appear troublesome in both versions.

A detailed analysis of bryozoan genera was published by Tuckey (1990b), who constructed distribution maps for each of the four Silurian epochs. For the Llandovery, he distinguished Baltic, North American-Siberian, and Mongolian provinces. In the Wen

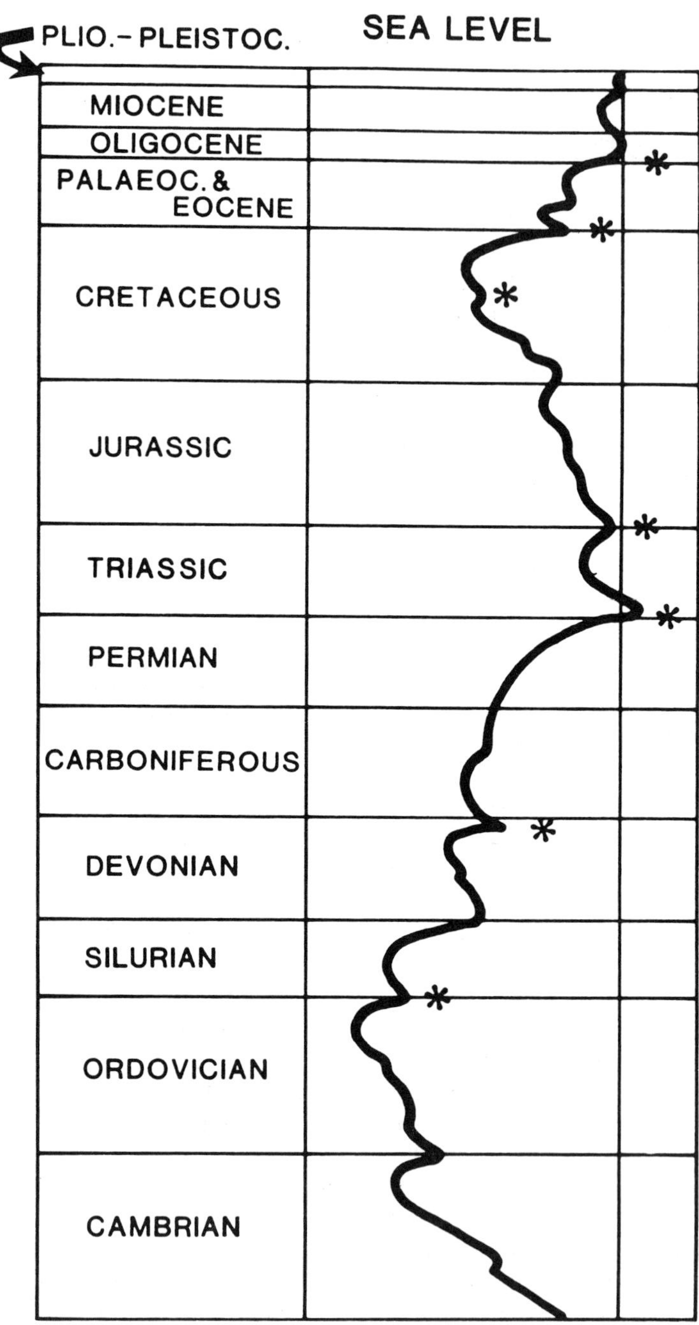

Fig. 13. Phanerozoic sea-level graph. After Hallam (1992). Asterisks indicate times of significant marine extinctions.

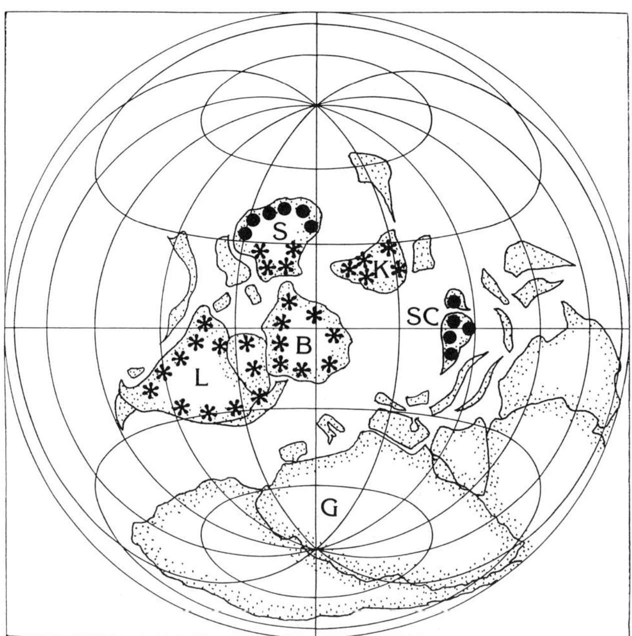

Fig. 14. With the contact between Laurentia (L) and Baltica (B), a combined North American-Siberian-Baltic Province (✻) became established. The distribution of the Mongolian Province (●) in northern Siberia (S) and South China (SC) appears to be incongruous. After Tuckey (1990b).

lock, the first two provinces merged (Fig. 14), and by the Ludlow Epoch, cosmopolitanism was the rule. The latter pattern continued into the Pridoli Epoch. The relationship between Siberia and South China, indicated on the map, seems incongruous but that distribution is also shown by the brachiopods (Boucot, 1990a). In fact, the brachiopods show one continuous province extending from South China to North China to Kazakhstan to Siberia. Such patterns may indicate that these continents and terranes were geographically closer than depicted on the maps.

The biogeography of Silurian stromatoporoids is of interest because they were stenothermic, tropical organisms found in shallow waters. In the late Ordovician, they showed a clear biogeographic differentiation. With the marine regression and colder temperatures of the Ashgillian glaciation, the provinces disappeared. There was some tendency towards provincialism in the later Silurian but it was not marked (Nestor, 1990). Nestor noted that stromatoporoid fossils had been found in the northern part of Siberia, so that the base map may show that continent lying too far to the north.

The ostracods recovered slowly from the end-Ordovician extinction (Berdan, 1990). By the Llandovery, three provinces could be distinguished, one in eastern Laurentia, one in western Laurentia, and a third in southern Baltica. Although these provinces became somewhat expanded, they stayed in the same general areas throughout the Silurian. The graptolite fauna was divisible into only two widespread bands, a Pacific warm-water region and an Atlantic cool-water region (Berry and Wilde, 1990). Among the conodonts, the marked provincialism of the Ordovician was replaced by a regional uniformity at the low- and mid-latitudes (Bergström, 1990). This agrees with the earlier finding of Charpentier (1984). Silurian chitinozoans were predominately cosmopolitan (Laufeld, 1979).

Our knowledge of the colonization of the land has been reviewed by Selden and Edwards (1989). Although terrestrial forms of algae and other micro-organisms may have occurred in the Proterozoic, there are no fossils to prove it. The earliest spores, thought to have been produced by vascular plants, came from sediments near the end of the Ordovician. Fragments of cuticle, which might have come from vascular plants have also been found in Ordovician deposits. However, it was not until the late Silurian that fossils of plants with vascular tissues were found. Such megafossils have turned up from widely scattered localities but their fragmentary nature has prevented a biogeographic analysis (Selden and Edwards, 1989).

The principal tectonic movement of the Silurian was the closing of the distance between Baltica and Laurentia. A consolidation between the two finally took place in the latest Silurian about 410 Ma ago. With this meeting, the new continent of Laurussia was formed. A major orogeny, the Arctic-North Atlantic Calidonides, took place along the zone of contact (Ziegler, 1989). At the same time, some continental terranes which had broken off from Gondwana moved towards Laurussia from the south. Gondwana itself also moved slowly northward.

DEVONIAN

For the Devonian, there is an enormous literature of biogeographical importance. Fortunately, the pre-1988 work is competently reviewed by Boucot (1988). The distributional patterns he recognized have been adopted by many others. One difficulty is that Boucot's realms and regions were drawn on maps that showed the continents in a pangaean accumulation. They have here been transferred to the Lambert equal area maps. Boucot emphasized that the Devonian was a unique period owing to the fact that, at the beginning, provincialism was highly developed, whereas at the end of the period, cosmopolitan conditions reigned.

For the later, early Devonian, Boucot (1988) delineated an Eastern Americas Realm, a Malvinokaffric Realm, and an Old World Realm (Fig. 15). The latter was divided into six regions. There is some indication that the three realms may represent different sea-surface temperatures. The Eastern Americas were probably warm-temperate, the Malvinokaffric cold-temperate, and Old World tropical. Boucot noted that his boundaries were drawn mainly from brachiopod data.

The Eastern Americas Realm, sometimes called the Appalachian Province (Stanley, 1989), and the Malvinokaffric Realm developed primarily within the confines of great epicontinental seas. The former lacks some of the common tropical groups such as calcareous algae, but it is the latter that is the most peculiar. The Malvinokaffric biota contained no stromatoporoids, conodonts, ammonoids, or graptolites and almost no corals or bryozoans. This realm can be recognized until the mid-Devonian when more cosmopolitan conditions began to develop (Boucot, 1988).

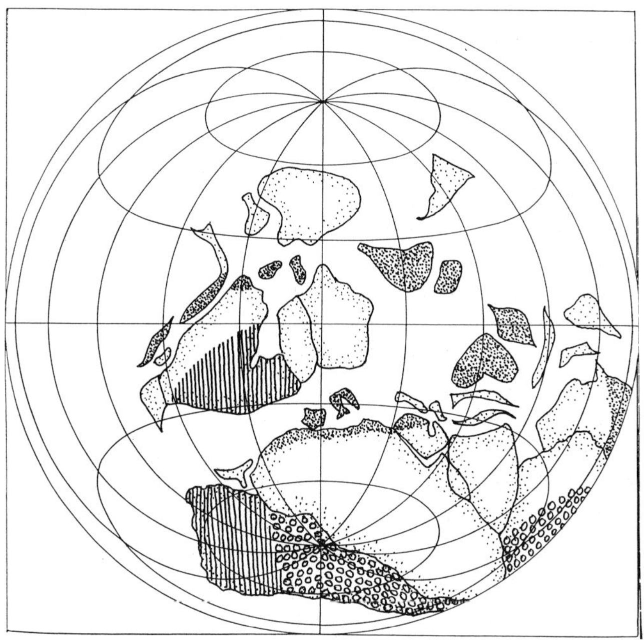

Fig. 15. The three later, early Devonian realms: Eastern Americas (|||||), Malvinokaffric (shaded with small open circles) and Old World (shaded with dots). After Boucot (1988).

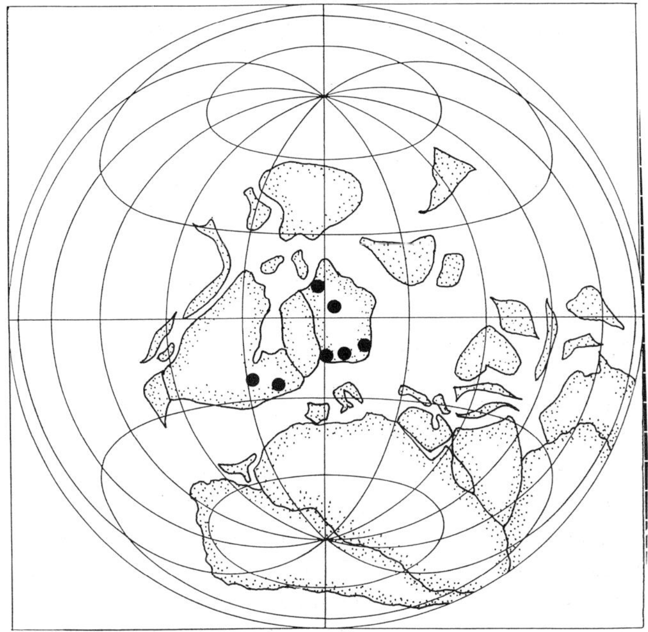

Fig. 16. The tropical distribution of the late Devonian malacostracan crustaceans (●). After Schram (1977).

The acme of trilobite success was in the Ordovician. Although there was a modest radiation in the Devonian, the pelagic forms were missing (Fortey and Owens, 1990). Their disappearance may have been caused by the Ashgill extinction but, possibly, the radiation of Silurian-Devonian predators such as the pelagic cephalopods and the jawed fish was important. The Malvinokaffric Realm proved to be a beneficial habitat which produced a spectacular range of trilobite morphologies. Elsewhere, trilobite genera appeared to be broadly distributed.

The higher malacostracan crustaceans (hoplocarids, eumalacostracids) apparently originated during the early Devonian in the tropical waters of Laurentia and Baltica (Schram, 1977). The fossil record suggests that these groups continued to be restricted to tropical marine waters during the late Devonian and Carboniferous. Their late-Devonian distribution has been depicted by Schram (Fig. 16). By the Permian, the malacostracans had spread into temperate marine and freshwater habitats.

Stromatoporoids were especially common from the middle to the late Devonian (Stock, 1990). In the early Devonian, provincialism at the generic level was apparent in the Old World and Eastern America realms. For the remainder of the period, most genera were cosmopolitan. The shallow, tropical calcareous algae are found on carbonate platforms in a "Palaeotethyan Realm" (Poncet, 1990). Their distribution was circumtropical. The Scotese and McKerrow (1990) reconstruction was criticized by Poncet because it would have the algae occurring as far as 55°N, where the temperature would have been too cold. The early and Middle Devonian gastropods are divisible into Old World, Eastern Americas, and Malvinokaffric Realms (Blodgett et al., 1990). Endemism in these areas increased from a moderate level in the early part of the period to a very-high level in the mid-Devonian.

By the beginning of the Devonian, there were in existence four major groups of agnathans or jawless vertebrates, and four major groups of gnathostomes or jawed fish. The placoderm fish were the most diverse with about 250 genera (Gardiner, 1990); they evolved in the mid-Silurian, reached their greatest diversity in the late Devonian, and disappeared in the late-Devonian extinction event. Young (1990) discussed all Devonian vertebrate occurrences and published a map of fossil localities. In addition, he made two distribution maps, one for the early Devonian and one for the late part of the Period. For the early Devonian, Young identified as provinces: Euramerica (Laurussia), Siberia, Kazakhstan, North China, South China, and Gondwana. He also suggested the possibility of a Tuva Province on the northeastern edge of Siberia.

For the late Devonian, Young (1990) mapped approximately the same provincial arrangement except that Turkey, Arabia, and Iran are identified as possible provinces along the northern edge of Gondwana. Young called attention to his previous work where he found evidence, from freshwater fish, of a late-Devonian connection between Gondwana and Laurussia. Young's evidence reinforces Ziegler's (1989) interpretation which shows a collision occurring about 365 Ma ago. Burrett et al. (1990) also call attention to the fact that the freshwater antiarch *Remigolepis* occurred in the latest Devonian in Laurussia, but earlier in South China, North China, and Australia (Fig. 17).

A seminal paper on patterns in vascular land-plant diversification was published by Niklas et al. (1985). A multitude of important events in vascular-plant evolution took place in the Devonian. The primitive rhyniophytes, which had risen in the Silurian, con-

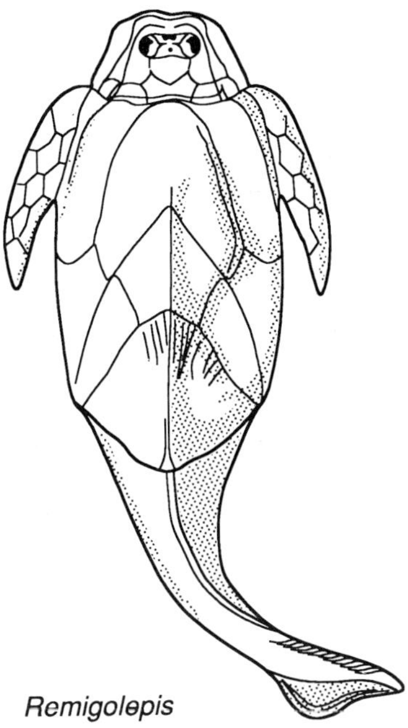

Fig. 17. The freshwater antiarch *Remigolepis*. After Burrett et al. (1990).

tinued until the late Devonian. Two other primitive groups, the zosterophyllophytes and trimerophytes, made their appearance in the early Devonian and were extinct by the end of the Period. Also in the early Devonian, the lycopods arose, possibly from a zosterophyll ancestor. It was the lyocopods that continued to diversify rapidly and become a major component of the subsequent Carboniferous swamp floras. By the late Devonian, the first pteridophytes, sphenophytes, and progymnospermophytes appeared. From the latter came the first seed plants, the pteridosperms. This was an important biogeographic development, for it meant that terrestrial plants were no longer reproductively dependent on swamps and other high-humidity habitats.

The early Devonian is about the earliest that one could attempt a phytogeography. Even then, the low diversity, the state of the systematics, and the fragmentary nature of the fossils, pose a difficult task. Raymond (1987) used a combination of generic analysis and morphological traits to construct distribution patterns. She found that the plant assemblages could be separated into three major units: a Equatorial-middle latitude unit, an Australian unit, and a Kazakhstan-north Gondwanan unit (Fig. 18). The Equatorial-middle latitude unit was broken into three subunits: a South Laurussian subunit, a Chinese subunit, and a Siberian-North Laurussian subunit.

In her earlier work on Devonian floras, Edwards (1973) found a distinction, in the middle Devonian, between northern and southern communities, but this disappeared by the late Devonian. Streel et al. (1990) did a microspore analysis of the mid-Devonian relationship between Laurussia and western Gondwana. They found a rather uniform vegetation prevailing at all latitudes. They concluded that no wide ocean could have

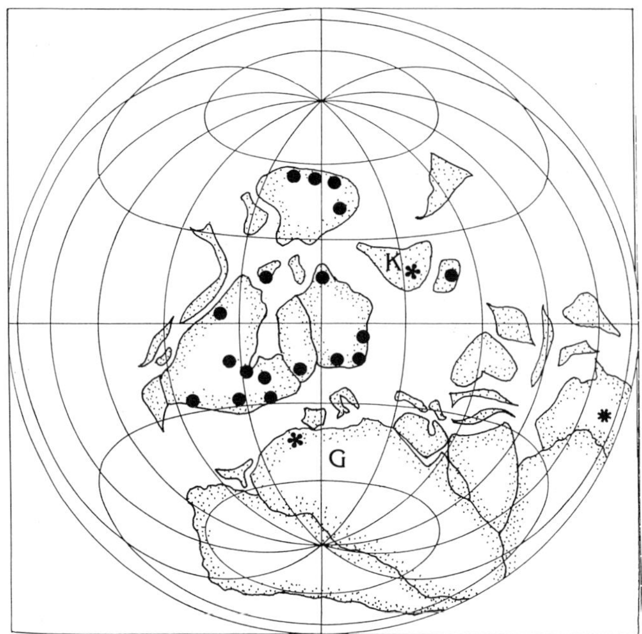

Fig. 18. Devonian terrestrial plant assemblages: Equatorial-middle latitude unit (●), Kazakhstan-north Gondwanan unit (*), Australian unit (✳). After Raymond (1987). Relationship between Kazakhstan (K) and Gondwana (G) appears to be incongruous.

separated the continents at this time. This conclusion appears to be consistent with the appearance of *Protosalvania* in Gondwana and Laurussia (Boucot, 1988).

The oldest, undoubted terrestrial-animal body fossils are found from the 400 Ma old deposits of Aberdeen, Scotland, 390 Ma old material from Alken-an-der-Mosel, Germany, and from 380 Ma old rocks of Gilboa, New York (Selden and Edwards, 1990). The three localities all yielded comparable material. These are the earliest records of many arachnids, including trigonotarbids, a possible spider, mites, a pseudoscorpion, and arthropleurids; also, an order of centipedes as well as myriapods and collembolans. The fully terrestrialized features of these animals suggest an origin in the Silurian or earlier.

The first tetrapod remains appear in the Frasnian Stage of the late Devonian (Milner, 1990). *Ichthyostega*, the most primitive tetrapod known, retains many fish-like characteristics. It has left abundant remains in the Famennian red beds of East Greenland. Unlike the invertebrates, where terrestrialization was achieved by many unrelated groups, the land vertebrates were probably derived from one particular group of bony fish, the Osteolepiformes. Devonian tetrapod footprints have been found in Australia and Brazil. Milner referred to a partial skeleton, more advanced than *Ichthyostega*, from European Russia. So far, it appears that late-Devonian tetrapods existed in the paleoequatorial regions of Laurussia and Gondwana.

As may be noted by the positions of the continents on the base maps, there was during the Devonian, a general northward movement of all the continental masses. There was also a continued closure between Laurussia and Gondwana so that, by the late Devonian, a collision took place. This was the beginning of a broad connection that developed over the ensuing 55 Ma (Ziegler, 1989). The Arctic-North Atlantic Caledonides mountains formed a continental divide. The North China terrane moved close to, or may have contacted, Kazahkstan. The latter moved somewhat closer to Laurussia.

Although the Devonian close approximation, followed by a joining of Laurussia and Gondwana, appears acceptable on geological and biogeographic grounds, some of the recent paleomagnetic accounts differ widely. The scenario offered by Kent and Van der Voo (1990) consists of a collision between Laurussia and the northwest South American margin of Gondwana at about the time of the Silurio-Devonian boundary. This was supposedly followed by a great rebound which created, by the late Devonian, a huge ocean between the two continents. A similar interpretation was given by Kearey and Vine (1990).

Which version is correct? Was there a late-Devonian connection or a great rebound? Boucot's (1988) early, middle-Devonian map shows a close faunal relationship between northeastern South America (Gondwana) and eastern America (Laurussia). A freshwater antiarch appeared in the latest Devonian of Laurussia and earlier in Gondwana (Burrett et al., 1990). In terrestrial floras, Streel et al. (1990) concluded that there could not have been a wide-ocean separation in the mid-Devonian. The late-Devonian tetrapods existed in the equatorial regions of both continents (Milner, 1990). So far, the biogeographic data appear to favor the connection hypothesis.

FRASNIAN EXTINCTION

The second mid-Paleozoic extinction took place in the Frasnian Stage of the late Devonian. Fagerstrom (1987) described the extinction as extensive. It meant the end of virtually all the stromatoporoids, almost all the corals, many of the major brachiopod groups (Ager, 1988), and the tentaculitids. There was a substantial reduction in trilobite diversity and a high taxonomic turnover among the ammonoids. The higher taxa involved were generally important members of the level-bottom, pelagic, and reef communities. Frasnian reefs were large and widespread but those of the next stage (Famennian) were relatively rare. Both rugose and tabulate corals were decimated (Scrutton, 1988). Pedder (1982), in discussing the rugose coral record across the boundary, noted that only 4% of the shallow-(warm) water species survived compared to 40% of the deep-water species. Anstey (1978), who worked on the Paleozoic bryozoans, found that the morphologically complex taxa underwent a much greater rate of extinction during times of crisis. The tropical taxa were the most severely affected while, in contrast, the polar communities were largely unaffected (Stanley, 1989).

The first major foraminiferid radiation occurred in the Middle Devonian and culminated in the Frasnian Stage (Brasier, 1988). The extinction event eliminated forms of relatively advanced architecture while those of primitive or intermediate architecture survived. Supposedly, the foraminferan extinctions were caused by a rapid transgressive pulse which caused anoxia on the continental shelves and a drowning of reef-associated fauna. Among the cephalopods, major groups of nautiloids and ammonoids disappeared.

However, it should be noted that the cephalopods have always been very sensitive to environmental changes, with a total of eight extinctions and recoveries being recorded for the mid-Paleozoic (House, 1988). The trilobite evidence suggests a protracted decline in diversity throughout the Devonian with the final demise of only a few clades at the end of the Frasnian (Briggs et al., 1988).

The general marine-invertebrate data pertaining to the Frasnian extinction were reviewed by McGhee (1988, 1990a, 1990b). Instead of a single peak in the extinction rate, he found multiple periods of high extinction spanning an interval of 2–4 Ma. There was a sharp net loss in species diversity at the very end of the Frasnian but this appeared to be mainly attributable to a decline in species originations. Goodfellow et al. (1989) suggested that a sudden turnover of anoxic waters after a period of ocean stagnation was responsible for a "mass killing" at the boundary. They observed further that such a catastrophic mixing of bottom and surface waters could have been caused by the ocean impact of a large extraterrestrial body. This narrative appears to be speculative in view of the fact that there is, so far, no unequivocal evidence showing such a sudden extinction event. There is no doubt that the late-Devonian extinctions constituted a series of disastrous events for many animal groups. As noted by Hallam (1989b), anoxia appears to have been an important factor but it is unclear whether this condition was associated with sea-level transgression or regression. Global cooling has also been suggested (Stanley, 1989). The sea-level chart of Hallam (1992) shows a dramatic drop at about the end of the Frasnian Stage.

From an evolutionary standpoint, the data of differential survival, reviewed by McGhee (1989, 1990a), appear to be of unusual significance. There was a latitudinal effect in that the stromatoporoids, brachiopods, and foraminiferans that survived the extinction were mainly high-latitude forms adapted to relatively cool water. In the brachiopods, 91% of the primarily tropical families perished, but only 21% of the families containing cool-water species. In the Foraminifera, the high-latitude species showed a better survival and, as the tropical species receded, expanded their ranges into the low latitudes.

There was also a differential survival in a bathymetric sense (McGhee, 1989, 1990b). Among the rugose cnidarians, 96% of the shallow-water species were lost in the crisis compared to 60% of the deep-water species. At the Frasnian-Famennian boundary in eastern North America, many shallow-water sponges became extinct. Simultaneously, the hyalosponges (glass sponges) migrated from deep-water regions into the shallows and underwent a burst of diversification. Modern glass sponges are generally found at water depths in excess of 200 m.

CARBONIFEROUS-PERMIAN

The single, most helpful work on late-Paleozoic marine biogeography is that written by Bambach (1990). He established a set of biogeographic units for the early Carboniferous, late Carboniferous, early Permian, and late Permian on the basis of 30% generic endemism in each regional fauna. This high degree of endemism assured that the identified regions were very distinctive. Data were compiled on the distribution of six animal groups: tabulate corals, rugose corals, bivalve molluscs, strophomenid brachiopods,

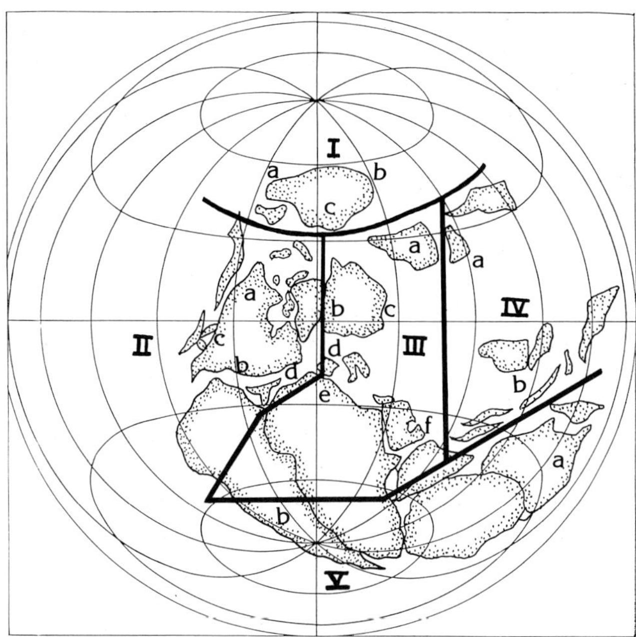

Fig. 19. Marine biogeography in the early Carboniferous. The Roman numerals refer to the five major realms and the lower case letters to the provinces within the realms (Table 1). After Bambach (1990).

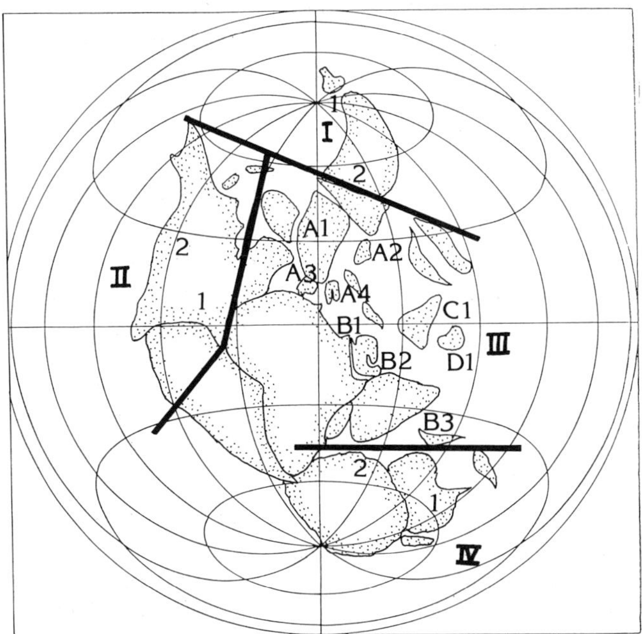

Fig. 20. Marine biogeography in the late Permian. The major realms have been reduced to four. The Tethys Realm (III) appears for the first time. It is subdivided into several regions indicated by capital letters. Provinces are identified by arabic numbers (Table 2). After Bambach (1990).

articulate brachiopods, and bryozoans. The distribution of each was delineated on four maps, one for each time period. The world was then divided into a number of realms, regions, and provinces which were displayed on four biogeographic maps.

Two of Bombach's 1990 maps are, with some modification, reproduced here. In the early Carboniferous, the world was divided into five realms, each with provincial subdivisions (Fig. 19). By the late Permian, the number of realms was reduced to four, and four distinct regions could be recognized within a large Tethyan Realm (Fig. 20). Comparison of the two maps reveals the westward movement of the east Asian terranes and the southward shift of Australia-Antarctica.

The major change in realms was the loss of the distinction between the tropical European and Chinese Realms of the Carboniferous and the emergence of the Tethyan Realm in the Permian. The pooled data (Tables 1 and 2) effectively demonstrate the latitudinal changes in generic diversity during the late Paleozoic. Very high diversities (over 100 genera) were found only on areas on continental blocks within 15° of the equator. In the early Carboniferous and early Permian, this high-diversity belt was bordered by bands of intermediate diversity (between 20 and 90 genera) extending far to the north and south.

TABLE 1

EARLY CARBONIFEROUS PROVINCIALITY (AFTER BAMBACH, 1990)

Area	Recorder genera/ endemic genera	Percent endemism
I. Siberian Realm		
A. Northeastern USSR	2/1	?
B. Mongolian Province	8/3	37.5
C. Siberian Platform Province	27/9	33
II. American Realm		
A. North Laurentia	2,4/1,6/1	?
B. South Laurentia Province	125/76	60.8
C. Cordillera	22/4	(18.2)
D. Maritime Canada	1/1	?
III. European Realm		
A. Kazakhstanian Province	37/16	43.2
B. West Baltican Province	9/4	44.4
C. East Baltican Province	73/33	45.2
D. Anglo-North German Province	77/55	71.4
E. North African Province	5/3	60
F. Iranian Province	4/2	50
G. Afghanistan-India	3/1	?
IV. Chinese Realm		
A. North China Province	26/12	46.2
B. South China Province	9/4	44.4
V. Austral Realm		
A. East Australian Province	29/16	55.2
B. Southern South America	1/1	?

TABLE 2

LATE PERMIAN PROVINCIALITY (AFTER BAMBACH, 1990)

Area	Recorded genera/ endemic genera	Percent endemism
I. Siberian Realm		
1. Northeast USSR	1/1	?
2. Siberian Platform	2/1	?
II. American Realm		
1. South Laurentian Province	10/10	100
2. Cordillera	5/1	(20)
III. Tethyan Realm (32 genera)		
A. European Region		
1. Arctic Canada-Greenland Province	6/4	66.7
2. East Baltican Province	23/14	60.8
3. Anglo-North German Province	9/4	66.7
4. France-Czechoslovakia	1/1	?
B. Southwest Tethyan Region (6 genera)		
1. Southern Europe	7	?
2. Iranian Province	5/3	60
3. Afghanistan-Indian Province	13/6	46.2
C. North China Region		
1. North China Province	20/9	45
D. Southeast Tethyan Region (9 genera)		
1. Southeast Asia-Timor Province	30/21	70
IV. Austral Realm		
1. Australo-New Zealand Province	7/3	42.9
2. West Australia	1/1	?

Low-diversity areas (less than 20 genera) were found at high latitudes. During the late Carboniferous, the high-diversity tropical belt persisted but those of intermediate diversity were eliminated. This was the time of the most extensive Paleozoic glaciation. It seems apparent that the change in the diversity gradient was a reflection of a change in the earth's climatic gradient.

In the early Carboniferous and late Permian an asymmetry appeared in the latitudinal range of intermediate diversities between the hemispheres. A similar asymmetry appeared in echinoderm distributions throughout the Paleozoic (Bambach, 1990). This pattern was probably related to a less severe, oceanic-climatic gradient in the northern hemisphere. The more continental southern hemisphere with its polar land areas would generate more severe climate. The paleoclimatology of Gondwana has been reviewed by Parrish (1990).

Some of the individual continental blocks showed diversity changes as they moved to lower latitudes. Diversity rose in East Baltica as it moved southward deeper into the tropics in the Permian, despite a cooling trend and a steepening of the world diversity gradient. The Southeast Asia-Timor block increased markedly in diversity as it moved northward in the Permian, even as the total world diversity began a sharp decline. Data showing a decline in world diversity during the late Paleozoic came from the six groups listed above plus the ammonoids and crinoids (Bambach, 1990). Family diversity decreased about 15% and generic diversity about 35%.

There was no marked change in either cosmopolitanism or endemism of the total fauna from the early Carboniferous to the early Permian (Bambach, 1990). This was an unexpected result, since previous authors had observed an increase in late-Paleozoic provinciality. There was, however, a detectable change in the ranges of non-endemic genera. That is, although the number of cosmopolitan genera remained fairly stable, there was a Permian drop in the number of biogeographic units inhabited per genus. This tended to make the provinces more distinct, even though their number did not change.

Ross and Ross (1990) conducted an extensive investigation of late-Paleozoic bryozoan biogeography. They compared their findings with a variety of other invertebrate groups including fusulinids, corals, cephalopods, and conodonts. They recognized the existence of cosmopolitan tropical faunas in the early Carboniferous. By the middle and late Carboniferous, the one tropical fauna became two due to the fusion of Gondwana with Laurussia. In the early Permian, three provincial tropical and subtropical faunas were found. In the early, late Permian the closure of Kazakhstan with Laurussia was seen to form two tropical realms, one on each side of Pangaea. Their chronology does not agree with the idea that the coalition of Gondwana with Laurussia began in the late Devonian (p. 43) nor with the Bambach (1990) finding that there was no increase in provincialism during the late Paleozoic. The latter subject was further clarified (or confused) by Hanger (1990) who found among brachiopods a decrease in provinces from ten in the early Permian to two in the late Permian.

In the terrestrial environment, the temnospondyls were the largest group of archaic amphibians (Milner, 1990). They first appear in the Visian Epoch of the early Carboniferous and survived until the mid-Cretaceous. They comprised about 40 families and some 160 genera, making them a larger group than all other earlier amphibians combined. Milner considered them to be a monophyletic group and included a cladogram of their relationships. It is probably this group that gave rise to the modern Lissamphibia.

Almost all temnospondyls from the early Carboniferous to the late Permian have been found between New Mexico and the Urals along a belt representing equatorial Pangaea (Milner, 1990). It appears that the Carboniferous phase of the radiation took place entirely within the Euramerican region of Pangaea, close to the equator. It matches part of the distribution of coals in the Carboniferous and may reflect the restriction of temnospondyls to hot humid environments. Most late-Permian temnospondyls from the Gondwana region have early-Permian sister families in the Euramerican region. This pattern appears to be consistent with a hypothesis of an early Permian range extension following the withdrawal of the Gondwanan glaciation. However, two southern families are anomalous in their appearance and have no close relatives among the equatorial faunas.

Many of the Paleozoic temnospondyl families occurred throughout the known Euramerican range and geographical endemism was exceptional if it occurred at all (Milner, 1990). Upper Permian temnospondyls are known from Europe, Africa, India, Australia, and South America and may be assumed to have been global in distribution. There is no evidence of any major late-Permian cladogenesis as the result of the range extensions but the Gondwana temnospondyls do show some evidence of morphological innovation. Some of the Gondwana-late Permian forms are highly derived and have been given distinct family status.

None of the monotypic temnospondyl families became extinct prior to the Permian, but most of them did not survive the end of the Period. The direct evidence is not sufficient to indicate whether the Carboniferous radiation gradually declined in diversity or whether it was partly truncated by the end-Permian extinction (Milner, 1990). All of the post-Permian temnospondyls and their descendants appear to have been of Gondwanan origin. This suggests a regional phenomenon, with extinction of the conservative equatorial forms.

The amniotes appear to be a monophyletic group that evolved from a single stock of primitive tetrapods during the early Carboniferous (Carroll, 1988). By the Upper Carboniferous, the amniotes had diverged into three major lineages: one that gave rise to mammals, a second to turtles, and a third that gave rise to the majority of the other reptilian groups and to the birds. The ancestors of mammals are commonly classified as reptiles, yet they are phylogenetically closer to modern mammals than to any of the modern reptilian orders.

Remains of the earliest known amniotes are found in two localities of early and middle-Pennsylvanian age in Nova Scotia, eastern Canada (Carroll, 1988). These fossils are not found in normal coal-swamp deposits, but within the upright stumps of the giant lycopod *Sigillaria*. The development of the amniote egg, with its extraembryonic membranes, liberated the vertebrates from reproductive dependence on standing water. This evolutionary accomplishment was the ecological equivalent of the development of the seed in vascular plants. In each case, the protection of the early embryo from excessive water loss permitted a great geographic expansion into dry, upland habitats.

The ancestors of mammals are usually classified in the subclass Synapsida within the Reptilia. However, those early genera can be recognized as being members of the same monophyletic assemblage as living mammals. The synapsids are included within two successive orders: the Pelycosauria, which are known from the base of the Pennsylvanian into the Upper Permian, and the Therapsida, which appear in the middle Permian and extend to the mid-Jurassic (Carroll, 1988). The pelycosaurs first appeared in the equatorial regions on the east side of Pangaea, United States to England, France, and Germany.

Of the succeeding therapsids, the dicynodonts were by far the most abundant terrestrial vertebrates in the late Permian and early Triassic (Carroll, 1988). King (1990) has written an informative book on this group. The earliest fossils were described from Russia, and date from about 260 Ma ago. In the very latest Permian, they became more widespread, reaching Africa, Asia, and Europe. They suffered a drastic decline at the end of the Permian, but recovered somewhat during the Triassic. It was another therapsid group, the cynodonts, that led to the true mammals. Cynodont fossils have been found in the latest Permian of Russia and southern Africa. The relationships of these groups of mammal-like reptiles have been diagrammed by Carroll (1988) (Fig. 21).

An assemblage of extinct orders and suborders can distinguish a clearly-marked Paleozoic insect fauna which had almost entirely disappeared by the end of the Permian (Wootton, 1990). From the Lower Permian onwards, alongside the Paleozoic fauna, the foundations of the modern fauna were being laid. Permian fossils of Orthoptera, Plecop-

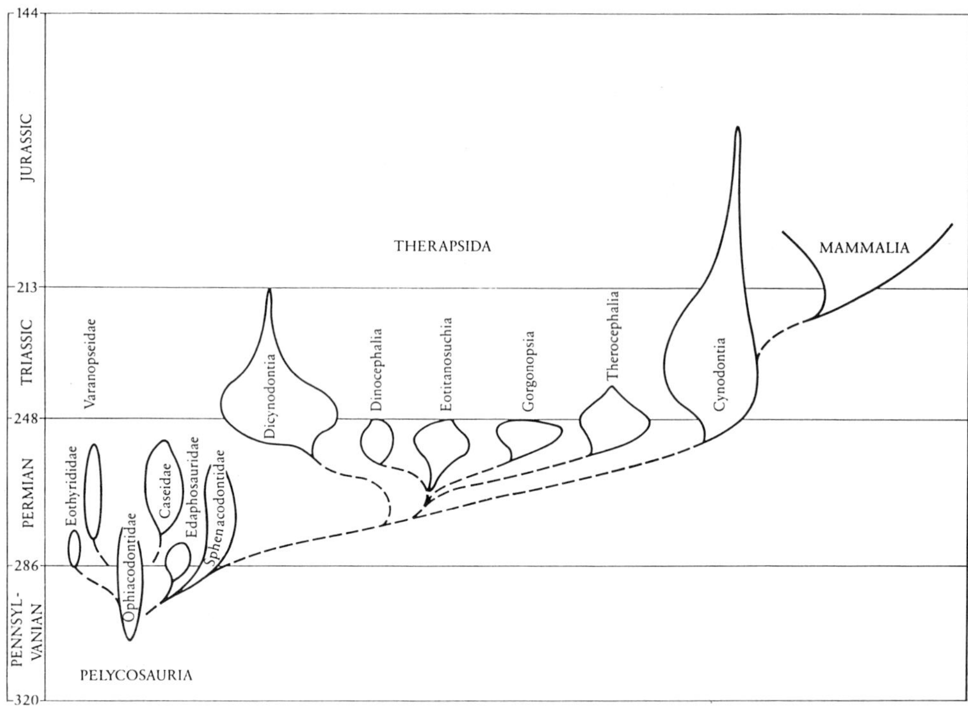

Fig. 21. Evolutionary relationships among the major groups of mammal-like reptiles. After Carroll (1988).

tera, Psocoptera, Thysanoptera, Hemiptera, Neuroptera, and Coleoptera have been found. No other terrestrial animal group has shown such rapid evolutionary radiation. By the end of the Paleozoic, modern types of insects inhabited a world occupied by very primitive amphibians, reptiles, and plants.

The Carboniferous is the time when most of the major plant groups represented in the Devonian developed specialized arborescent or rhizomatous lines (Niklas et al., 1985). Some of the lycophytes achieved heights in excess of 30 m. There was a rapid diversification of seed plants. Five orders of seed ferns developed, among them the well known *Glossopteris*, as well as the cordaites and the voltziacean conifers. The Permian marked the appearance of new seed- plant lineages including the cycads, cycadeoides, and ginkophytes (Fig. 22). An upland flora developed although it is not well represented in the fossil record. The rapid diversification of seed plants was accompanied by a massive radiation of insects. This radiation began in the early Carboniferous and reached a peak in the middle Permian (Labandeira and Sepkoski, 1993).

For the Lower Carboniferous, the research of Rowley et al. (1985) suggests the presence of five phytogeographic units: a Siberian; a Northern Mid-latitude unit found in Spitzbergen, Greenland, and the Urals; an Acadian found in eastern Canada, Wales, and Great Britain; a Southern Equatorial unit in South China and Morocco; and a Middle Gondwanan unit in Ghana and Kashmir. These units were succeeded by a Middle Carboniferous distribution which contained only three units: a Siberian; a large Low-latitude

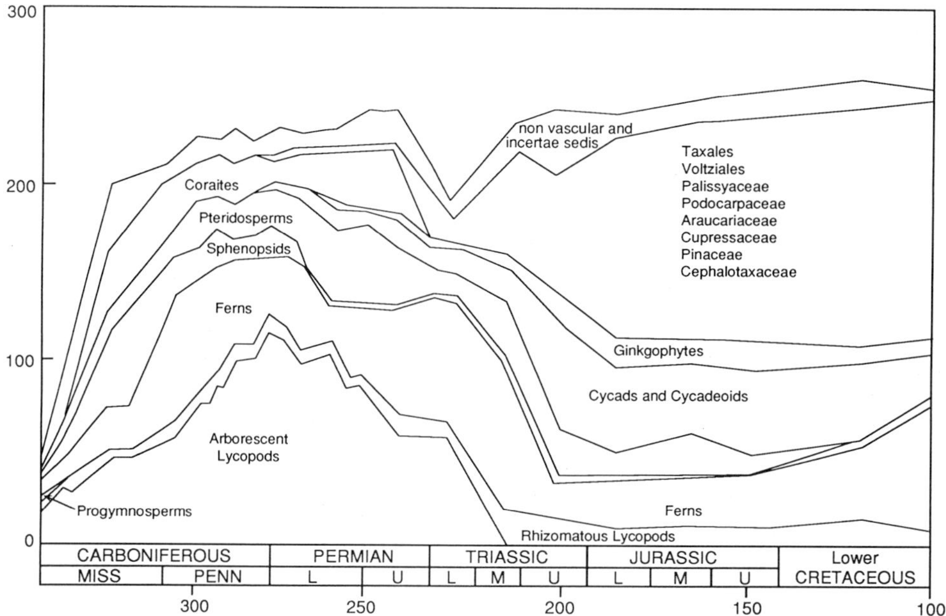

Fig. 22. The Permian appearance of the seed-plant lineages, cycads, cycadeoides and ginkophytes. After Niklas et al. (1985).

unit including Kazakhstan, Spitzbergen, the United States, and Europe; and a Gondwanan unit.

This decrease in biogeographic differentiation between the Lower and Middle Carboniferous probably resulted from the progressive collisions between Gondwana and Laurussia, Laurussia and Kazakhstania, and Siberia and Kazakhstania, leading to the formation of Pangaea (Rowley et al., 1985). The collision between Laurussia and Gondwana probably resulted in a northward deflection of equatorial water, resulting in a climatic amelioration. This, together with the erosion of mountain barriers surrounding the Acadian terrane, may have been causes for the decrease from three to one low-latitude phytogeographic unit.

The late Carboniferous was a time of climatic transition. A late Paleozoic episode of continental glaciation began in the early Carboniferous, peaked in the late Carboniferous, and declined in the Permian. Continental climates generally became drier throughout the late Paleozoic. A drying trend through the Carboniferous is indicated by an increase in the number and extent of evaporite deposits (Rowley et al., 1985). By the end of the Carboniferous, the interior of Pangaea was relatively dry, suggesting that monsoonal conditions had become established (Parrish, 1990).

Essentially, the three phytogeographic units, which had become apparent by the Middle Carboniferous, were continued into the Permian. For the latter Period, they have been recognized as the north-temperate Angaran Realm of Siberia and Kazakhstania; the tropical Cathaysian Realm of equatorial Gondwana, Laurussia, and the south Asian microcontinents; and the south-temperate Gondwanian Realm of central Gondwana

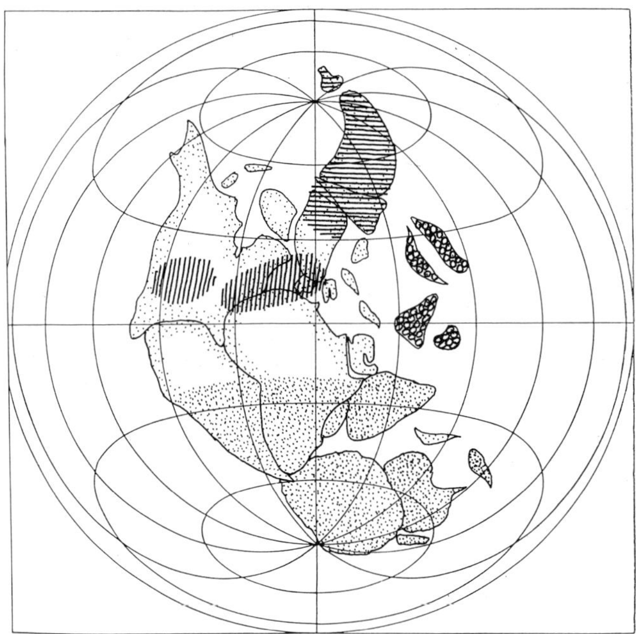

Fig. 23. Permian phytogeographic realms: north temperate Angaran Realm (≡), tropical Euramerican Realm (IIII), tropical Cathaysian Realm (shaded with small open circles) and south temperate Gondwanan Realm (shaded with dots). After Erwin (1993).

(Ziegler, 1990). Erwin (1993) recognized two tropical realms (Fig. 23). The major barriers to floral interchange were two great subtropical deserts. The southern desert may have extended across Gondwana and is represented by evaporites in Brazil and Arabia. The northern desert is known from evaporites of the western United States and northern Europe.

Ziegler (1990) has, in addition to recognizing the three major Realms, attempted to plot the distribution of Permian plants in nine `biomes' on the basis of temperature zones and rainfall amounts. Biome is an ecological term referring to communities characteristic of certain climatic conditions. For example, in the modern world one can distinguish among a tropical everwet biome, a tropical and subtropical summerwet biome, and a winterwet biome. Whether such differences can accurately be distinguished in the fossil flora of the Permian, is somewhat problematical. But, a separation into tropical, temperate, or upland and desert floras is certainly possible.

Many of the continental collisions which began in the Carboniferous were completed in the Permian. The western half of Pangaea was assembled and the new supercontinent moved steadily northward (Scotese, 1992). The configuration of the blocks that were to form the eastern half of Pangaea is still problematic. Some authors have described as many as eleven terranes in the assembly of Asia. Scotese and McKerrow (1990) describe the allochthonous terranes of Asia as comprised of two groups: Cathaysian and Cimmerian. The Cathaysian terranes (South China, North China, and Indochina) probably rifted from Gondwana during the mid-Paleozoic and occupied an equatorial position in the Permian. The second group of terranes (Turkey, Iran, Tibet, Shan Thai-Malaya) are considered to have formed the elongate continent of Cimmeria. The rifting of Cimmeria from Gondwana took place in the Permian.

END-PERMIAN EXTINCTION

Although the Permian/Triassic (P/T) boundary had been recognized for many years as a time of severe extinction, the probable magnitude of this event was emphasized when Raup (1979) estimated that as many as 96% of all marine species may have died out. His procedure was to determine the percent of extinction for the higher taxonomic groups by reference to the general fossil record. Then, using the rarefaction curve technique, he calculated the percent of species extinction that must have been responsible for the disappearance of the higher groups. It has since been found that the fossil record as a whole may not be dependable for such broad analyses (Smith and Patterson, 1988). However, this does not detract from the fact that the P/T boundary marks a major extinction episode. Erwin (1990) characterized it as the most severe of the Phanerozoic and noted that it eliminated twice as many families as the second largest, the end-Ordovician extinction. The duration of the P/T extinction was about 3–8 Ma.

Taxa which became extinct include tabulate and rugose corals, conularids, eurypterids, leperditiid ostracods, several gastropod groups, goniatitic ammonites, orthid and productid brachiopods, blastoids, several crinoid subclasses, and the few remaining trilobites. Many other groups suffered sharp drops in diversity, including the echinoids, ammonoids, bryozoans, foraminiferans, and fish. Jablonski (1986) analyzed the survival patterns of articulate brachiopods and noted that 75% of the tropical families became extinct, while only 56% of the higher-latitude families died out. Van Valen (1984) observed that the probability of extinction had gradually decreased throughout the Paleozoic but this trend was interrupted by the late-Permian extinction. That extinction was so drastic that it apparently reset the clock of community evolution so it resumed at a faster rate.

In the late Permian, reefs of imposing size existed (Fagerstrom, 1987). They were constructed primarily by skeletal and non-skeletal algae of various types, calcareous sponges, bryozoans, and some brachiopods. These reefs supported a high diversity of animals. The Capitan Limestone reefs of the southwestern United States were found to contain about 350 species of reef dwellers. The end-Permian extinction event affected all of the reef-building higher taxa and resulted in the complete collapse of the structure of all existing reef communities. The annihilation was so extensive that it took 8 Ma before reef communities began to be reestablished. There are almost no early Triassic records of any skeletal, colonial reef-builders such as algae, poriferans, corals, or bryozoans. About 76% of the existing bryozoan families became extinct in the late Permian (McKinney, 1985). In regard to the foraminiferans, Brasier (1988) noted evidence of a prolonged decline with preferential extinction of specialized forms and the survival of simpler ones.

It has been noted that, at the beginning of the Ordovician, the trilobite-rich, Cambrian

Fauna began to be replaced by a brachiopod-rich, Paleozoic Fauna. As the succession took place, the former was restricted to deeper benthic waters and the pelagic habitat (Sepkoski and Miller, 1985). The Paleozoic Fauna attained its maximum diversity from the late Ordovician to the Devonian and then began a long decline. During the Carboniferous and Permian, this decline was matched by a slow expansion of the Modern Fauna. The Modern Fauna is dominated by gastropod and bivalve molluscs, osteichthyian and chondrichthyian fish, gymnolaemate bryozoans, malacostracans, and echinoids (Sepkoski, 1990). At the end of the Permian, the Modern Fauna suffered a lesser extinction, relative to the Paleozoic Fauna, and went on to become the dominant group thereafter. Many of the surviving members of the Paleozoic Fauna became relegated to deeper and colder waters.

In the terrestrial environment, there was a gradual decline in vascular plant diversity, from the early Permian to the mid-Triassic, but no evidence of a major extinction at the end of the Permian (Knoll, 1984). In Gondwana, the dominant *Glossopteris* flora of the Permian was replaced by a *Dicroidium* flora in the Triassic. Where fossiliferous sequences are complete enough, it appears that these assemblages are separated by intermediate floras that contain elements of both groups. The family-level diversity of insects apparently dropped at the end of the Permian, although there is a dearth of Lower Triassic insect-bearing deposits (Labandeira and Sepkoski, 1993).

For the terrestrial vertebrates, Padian and Clemens (1985) and Colbert (1986) noted that the most conspicuous change took place in the therapsids or mammal-like reptiles. This group dominated late-Permian history but, with the transition from Paleozoic to Mesozoic times, suffered a drastic reduction. Benton (1988) observed that 15 of the 20 families that existed in the final stage of the Permian died out. Among the true reptiles, more than one-half the families and over two-thirds of the genera disappeared (Olson, 1989). On the other hand, Sigognau-Russell (1989) considered the P/T extinction of terrestrial vertebrates to be an artifact of taxonomic inflation. The beginning of the Triassic was marked by the rise of the archosaurs, the reptiles that were to dominate the Mesozoic. Colbert (1986) and Charig (1984) suggested that the origin of this new and vigorous group may have had something to do with the decline of the therapsids. Certainly, the archosaurs introduced a new and very successful suite of skeletal adaptations that greatly increased their agility.

Another important vertebrate group was the labyrinthodont amphibians. They were among the first of the tetrapods and for a long time in late-Paleozoic history they dominated the land (Colbert, 1986). During Permian times, their dominance was challenged by the early reptiles but the labyrinthodonts continued on through the Triassic and Jurassic as a diminishing group occupying ecological niches that had not been fully exploited by the reptiles. They disappeared from most of the world by the late Jurassic but a labyrinthodont apparently persisted in Australia until the early Cretaceous (Rich and Rich, 1989). A similar history, Paleozoic origin and survival in decreasing numbers through part of the Mesozoic, is shown by the primitive cotylosaurs and protorosaurians.

Padian and Clemens (1985) called attention to Pitrat's (1973) original work on the family diversity of marine and freshwater fish leading up to and through the P/T boundary. The marine families underwent a sharp decline at the boundary but the freshwater (and euryhaline) group did not. In fact, the latter had undergone a less severe decline some 15–20 Ma earlier and, at the boundary, were in the process of increasing their diversity, a trend that continued on into the mid-Triassic. Consequently, one might observe that the terrestrial and freshwater (vertebrate) biota underwent considerable evolutionary change but the changes were apparently gradual with Paleozoic groups slowly dying out and being replaced by new groups in the Triassic.

In the latter part of the Paleozoic Era, the tectonic plates bearing Gondwana and seven other continental blocks slowly converged to form a single Pangaea. When the formation of Pangaea was completed, plate movement ceased and the ocean basin became relatively quiet. The hot mantle material, which had been extruded in the form of spreading ridges, began to cool and contract. As a result, the ocean basin became deeper and the sea level fell. Most of the regression, in which the sea level may have fallen as much as 280 m (Hallam, 1989b), took place during the final stage of the Permian, a period lasting about 5 Ma (Holser et al., 1986). The formation of ice sheets also has a direct effect on sea level, but the P/T crisis apparently occurred after the Permian glaciations were over (Fischer, 1984). However, Stanley (1988) found some evidence of polar glaciation in the late Permian and has emphasized the possible role of temperature reduction in this as well as the Ashgillian and Frasnian events. Maxwell (1989) also emphasized the possible effect of temperature reduction by referring to "a period of global refrigeration." But evidence for a global-scale temperature drop at the end of the Permian appears to be lacking (Erwin, 1990).

Erwin (1993) published a book in which he designated the P/T event as "the mother of mass extinctions." In that work, he discussed in detail the various extinction hypotheses and presented them in tabular form (Table 3). Some are less important than others because they are supported by very little data. Erwin selected those that were, in his view, the most important and presented them in graphic form (Fig. 24). It may be noted that the

TABLE 3

CLASSIFICATION OF EXTINCTION HYPOTHESES ACCORDING TO WHETHER THEY PREDOMINANTLY EFFECT MARINE OR NON-MARINE SETTINGS, OR BOTH, AND WHETHER THEY SHOULD PRODUCE A RAPID OR GRADUAL EXTINCTION

Suggested causes	Marine	Non-marine	Rapid	Gradual
Nutrient reduction	X			X
Decline in provinciality	X			X
Trophic resource instability	X	?	?	X
Habitat diversity	X			X
Ecosystem collapse	X	?	?	X
Extra-terrestrial impact	X	X	X	
Global cooling	X	X		X
Salinity	X			X
Species-area effects	X			X
Oceanic anoxia	X		X	
Atmospheric anoxia	X	X		X
Pyroclastic volcanic eruptions	X	X	X	
Flood basalts	X	X	X	
Trace element poisoning	X		?	?

In several cases some effects are not known or not well understood. See text for details. (After Erwin 1993).

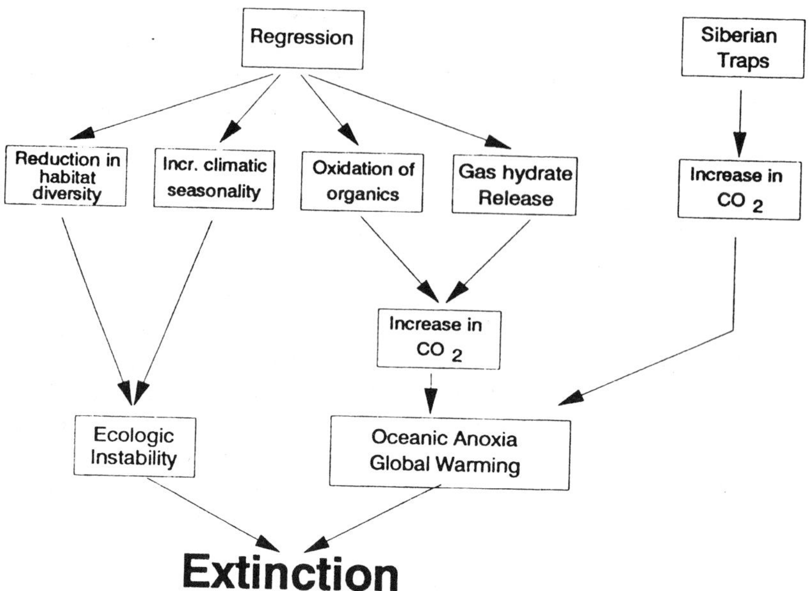

Fig. 24. Factors leading to the extinction episode at the close of the Permian. After Erwin (1993).

two primary causes are sea-level regression and the Siberian Traps. The latter is in reference to volcanic eruptions that released flood basalts over large sections of Siberia. Although this model is informative, it may be best to consider the terrestrial and marine environments separately. An important extinction cause in one may have little effect in the other.

As the continents of the early Permian closed with one another, they brought together terrestrial faunas which had developed in isolation for many millions of years. The resulting competition probably resulted in an extensive loss of species (Briggs, 1987). The accompanying climatic changes must have been severe. Pangaea was evidently dominated by a highly seasonal, monsoonal circulation (Parrish, 1987). This condition strongly affected the zonal circulation in both hemispheres. The result was a large dry region that included equatorial latitudes and the subtropics. For much of the world, this meant a sharp contrast to the previous moist, maritime climate. The changes in the flora and fauna must have been considerable. About 49 percent of the tetrapod families disappeared (Benton, 1988).

In the marine environment, the continental amalgamation resulted in a significant reduction of the world shoreline and, consequently, a loss of marine provinces (Schopf, 1980). This, combined with the lost of shelf area due to sea-level regression, probably had a drastic effect on shallow-water diversity. Beurlen (1956) proposed that the late-Permian crisis was caused by a marked reduction in oceanic salinity. This theory was further elaborated by Fischer (1964) and Flessa (1980). Supposedly, as the Permian seas withdrew, large shallow basins were left. Evaporation in the basins produced large amounts of brine which drained into the oceanic abyss and stagnated. This led to a great reduction of salinity in the circulating part of the ocean, resulting in an extinction of stenohaline organisms. Stevens (1977) estimated that the accumulation of salt deposits during the Permian was equal to at least 10% of the volume of salt now in the oceans. But Benson (1984) maintained that this salinity reduction was not enough to cause a general reduction of the normal marine faunas.

Although a dominant role for temperature has been advocated in mass extinctions (Stanley, 1984, 1987), its involvement at the P/T boundary seems problematical. The earth was apparently in a warming phase following the mid-Permian glaciation. Considering the magnitude of the drop in sea level, it is likely that the rivers began depositing their sediments, together with phosphates, nitrates, and other nutrients, directly into the deep sea. This would vastly decrease primary productivity in shallow waters. Gruszczyński et al. (1989) suggested that an additional cause of extinction was a decrease in atmospheric oxygen caused by the oceanic burial and oxidation of large amounts of organic material. The disappearance of the continental shelves, plus a reduction in shallow-water marine provinces from 14 to 8 (Schopf, 1980), could go a long way towards explaining the gradual but enormous loss in diversity towards the end of the Permian. A salinity reduction effect should not be discounted, for great salt deposits did take place towards the latter part of the Permian (Stevens, 1977).

In Erwin's (1993) model (Fig. 24), the box entitled "reduction in habitat diversity" refers to the loss of habitat in the epicontinental seas. It assumes that the marine regression took place too rapidly for the endemic organisms to migrate and adapt. Also, not mentioned so far is the box called "gas hydrate release." Gas hydrates contain large quantities of methane which may be released during sea-level regression. Methane is a greenhouse gas that could have exacerbated the climatic effects of CO_2. The CO_2 increase would trigger global warming and oceanic anoxia.

Erwin (1993) did not consider species-area effects to be particularly important and left them out of his model. However, works on the Recent marine fauna (Abele, 1982; Briggs, 1985) indicated that the species-area relationship is applicable to large areas of continental shelf. In the case of the P/T regression, where the sea level dropped about 280 m and probably exposed almost all of the continental shelf, this factor alone could have been responsible for the elimination of most of the neritic biota. It is also important to note that the tempo of the decline in organic diversity was largely coincident with that of the marine regression.

SUMMARY

1. In the Silurian, with the melting of the southern hemisphere glaciers, the sea level rose once again but did not reach the historic high of the late Ordovician. The Silurian and Devonian periods were similar in that sea levels, despite some fluctuations, remained high and the surface temperatures were relatively warm.
2. The bryozoan genera indicate a well-established provincialism in early Silurian which declines to a general cosmopolitanism by the end of the Period. The stromatoporoids, conodonts, and chitinozoans show broad, low-latitude distributions

throughout the Period. On the other hand, the ostracods and brachiopods appear to have maintained a moderately high level of provincialism at low latitudes.

3. Fossils of terrestrial vascular plants appear in the late Silurian, although spores possibly produced by such plants have been found in late Ordovician sediments. The Silurian fossils are too fragmentary to be useful for biogeographic purposes.
4. For the marine invertebrates, the early Devonian appeared to be a time of high provinciality. Both latitudinal and longitudinal provinces are recognized. In regard to the former, tropical, warm-temperate, and cold-temperate associations appear to have been present. Through the Devonian, provincialism decreased as the continents moved closer together.
5. An analysis of early-Devonian, terrestrial-plant distributions revealed three major units: an Equatorial-middle latitude unit, an Australian unit, and a Kazakhstan-north Gondwanan unit. In the Middle Devonian, there was still an apparent distinction between the vegetation of Gondwana and the northern continents, but this disappeared in the late Devonian.
6. The oldest terrestrial-animal body fossils have been found in mid-Devonian deposits. They included many arachnids, centipedes, myriapods, and collembolans. The fully terrestrialized features of these animals suggest an origin in the Silurian or earlier. The first tetrapod remains appear in the Frasnian Stage of the Devonian. It appears that late-Devonian tetrapods existed in the paleoequatorial regions of both Laurussia and Gondwana.
7. The biogeographic information lends support to the hypothesis of a late-Devonian collision between Laurussia and Gondwana. The marine invertebrates indicate a close relationship between northeastern South America and eastern Laurussia, a freshwater antiarch fish is common to both areas, the terrestrial floras are similar, and the early tetrapods appear in both continents.
8. The second great extinction episode of the Paleozoic took place, over a period of 2–4 Ma, in the Frasnian Stage of the late Devonian. Again, it was the tropical, shallow-water fauna that suffered the brunt of the extinctions. The high-latitude forms, adapted to relatively cool water, and the deep-sea inhabitants survived much better.
9. For the Carboniferous and Permian, the marine invertebrate faunas are quite well known and their distributions have been worked out in detail. The data show that there was a steady decrease in generic diversity through the late Paleozoic, with a sharp drop (33%) from the early to the late Permian. Significant latitudinal changes in diversity took place in accordance with global climatic changes. There was no marked change in either cosmopolitanism or endemism from the early Carboniferous to the early Permian.
10. In the terrestrial environment, the temnospondyls were the largest group of archaic amphibians. They first appeared in the early Carboniferous and survived until the mid-Cretaceous. Almost all temnospondyls, from the early Carboniferous to the late Permian, have been found along a belt representing equatorial Pangaea.
11. Remains of the earliest known amniotes have been found in two localities of early and middle-Pennsylvanian age in Nova Scotia. They are not located near the

normal coal deposits. The development of the amniote egg, with its extraembryonic membranes, liberated the vertebrates from reproductive dependence on standing water. This accomplishment was the ecological equivalent of the development of the seed in vascular plants. In each case, the protection of the early embryo from excessive water loss permitted a great geographic expansion into dry, upland habitats.

12. The ancestors of mammals are usually classified in the reptilian subclass Synapsida. But, they are members of the same monophyletic assemblage as living mammals. The synapsids are included within two successive orders, the Pelycosauria and the Therapsida. The pelycosaurs first appeared in the equatorial regions on the east side of Pangaea (United States to England, France, and Germany). Of the succeeding therapsids, it was the cynodonts that led to the true mammals. They have been found in the latest Permian of Russia and southern Africa.
13. A clearly-marked, Paleozoic insect fauna had almost entirely disappeared by the end of the Permian. The modern insect fauna began to develop in the Lower Permian and, by the end of that Period, many extant orders could be identified. These advanced groups of insects inhabited a world occupied by very primitive amphibians, reptiles, and plants.
14. Beginning with the Carboniferous, most of the groups of terrestrial plants developed specialized lines; some of the lycophytes achieved heights in excess of 30 m. There was a rapid diversification of seed plants, with a notable invasion of upland habitats. In the Lower Carboniferous, five major phytogeographic units have been identified. By the mid-Carboniferous, the five units were reduced to three, probably because of the progressive continental amalgamation of Pangaea. Three major plant realms have also been recognized for the Permian.
15. The effects of the P/T extinction episode were most drastic in the shallow waters of the marine tropics. In fact, very little shelf habitat remained, so this factor alone could have been responsible for most of the loss. During the Carboniferous and Permian, there had been a gradual expansion of the Modern Fauna which was accompanied by a decline of the Paleozoic Fauna. In the P/T extinction, the Modern Fauna was less affected so that its ascendancy may have been accelerated by the extinction.
16. On land, the P/T extinction was severe among some groups. The therapsids lost 15 of 20 families and the reptiles more than one half of their families. On the other hand, there is no evidence of a major extinction among the terrestrial plants and the freshwater fish. The primary causes of the extinction were probably the final accretion of the continents, cessation of plate movement, climatic extremes, and marine regression.

CHAPTER 4

Early Mesozoic

Lamarck ... concluded that a continuum existed between simple and complex organisms, and, furthermore, that the simpler organisms existed in the fossil record at dates earlier than those of the complex ones. This was diametrically opposed to the classical view which held that perfection originated first and that poor quality control was responsible for forms of life departing from the original plans.

Richard H. Benson, *Perfection, Continuity, and Common Sense in Historical Geology*, 1984

TRIASSIC

When Siberia and Kazakhstan were accreted to the eastern margin of Laurussia during the late Permian-early Triassic, the formation of Pangaea was almost complete. Shortly afterward, Gondwana derived terranes, forming parts of China and southeast Asia, converged to become parts of the supercontinent (Map 9). Smaller terranes became incorporated on the northwest margin of Pangaea. This series of collisions resulted in major orogenies which, in turn, had important effects on climate and biogeography.

In the marine environment, it took a long time for the biota to recover from the P/T extinction. There are no early Triassic reefs anywhere in the world. In fact, there are almost no early Triassic fossil records of colonial, skeletal organisms such as calcareous algae, sponges, corals, or bryozoans (Fagerstrom, 1987). Ammonites were the dominant group, bivalves were widespread but not very diverse, and brachiopods were the third most common invertebrate group. Foraminiferans and all the echinoderm taxa were extremely rare. The most striking feature of the early Triassic (Scythian) marine fauna is its apparent impoverishment. Kummel (1973) delineated the distribution of the bivalve genus *Claraia* and many ammonoid genera. He found a very widespread distribution of genera and species, and the greatest diversity in the Tethys region, probably indicating that the higher latitudes were somewhat cooler.

For Upper Triassic heteromorph ammonites, Wiedmann (1973) described an essentially tropical, Tethyan-circum-Pacific distribution. This pattern presents a notable contrast to that found for the late-Triassic bivalve genus *Monotis* (Westermann, 1973). The latter apparently originated in the north circumpolar region then spread to very high southern latitudes. Eventually, three different species groups were established, one along the east coast of Pangaea, another on the west coast, and a third in the Tethys Sea. The extensive latitudinal distributions were taken to imply equable ocean temperatures.

On land, the first positive dinosaur remains are found in the Carnian Stage of the Triassic (Padian, 1986). At its beginning, the Triassic was populated mainly by vertebrate groups that had evolved in the Paleozoic and had survived the extinction at the end of the Permian. The therapsids were the largest and most diverse tetrapod group at the beginning of the period. Their prominence decreased during the Triassic until only a couple of lineages survived at the end of that time. Other Paleozoic holdovers included various amphibian groups, the Procolophonia, and some small diapsids. According to present knowledge, some 19 groups of tetrapods, recognized as distinct families or higher taxa, are endemic to Triassic strata. It seems as if these groups ecologically replaced the early Triassic therapsid faunas just as they were, in turn, replaced by the dinosaurian faunas towards the end of the Triassic.

The main lines of dinosaurian evolution were already apparent during the Carnian Stage, with representatives of all three major lineages (theropods, sauropodomorphs, ornithischians) being present (Benton, 1993). Although Carnian dinosaurs were geographically widespread, they were relatively small – less than 6 m long – and comprised less than 6% of the tetrapod fauna. By the Norian Stage, a few Ma later, dinosaurs had become larger and more abundant. Depending on locality, they comprised 25–60% of the land vertebrates. Also, they became very widely distributed. More than 75 late-Triassic dinosaur localities are known in North America, Europe, Asia, South America, Africa, and Australia (Weishampel, 1990).

In general, early-Triassic tetrapod assemblages were dominated by taxa with extensive geographic ranges (Shubin and Sues, 1991). Although there appears to be minor differences between the Laurasian and Gondwanan temnospondyls, it is the Australian fauna that seems to be the most peculiar (Thulborn, 1986). Its early-Triassic fauna has a more primitive aspect, since it consists largely of labrynthodont amphibians and very few reptiles. This suggests that Australia may not have been completely fused with the rest of Gondwana.

By the middle Triassic, almost all tetrapod families appeared to be cosmopolitan, with no apparent latitudinal variation (Shubin and Sues, 1991). However, by the late Triassic some differentiation could be seen. Three general associations were observed: the Germanic Basin, North America and Western Europe, and South America. The Maleri formation of India apparently shared some genera with the South American assemblage and others with the Laurasian groups. The same mixture was found in Madagascar. Shubin and Sues concluded that both India and Madagascar possessed faunas with predominately northern relationships.

A different late-Triassic arrangement was proposed by Rage (1988). He provided evidence for a distinctive province that occupied northern Gondwana and southern Laurasia. It was characterized by the frequent association of metoposaur amphibians and phytosaur reptiles that were not found in South America and southern Africa. This Peri-Tethyan Province extended from western Europe to India and Madagascar (Fig. 25). It probably also extended to southern China and Thailand, since phytosaurs (but not metoposaurs) have been found there. A Southern Gondwanan Province was recognized for South America and southern Africa. By the early Jurassic, these provinces had disappeared.

A late-Triassic tetrapod biogeography has also been published by Kalandadze and Rautian (1991). Their analysis of intercontinental relationships was based on the distributions of thecodonts, dinosaurs, therapsids, and mammals. India was found to be most

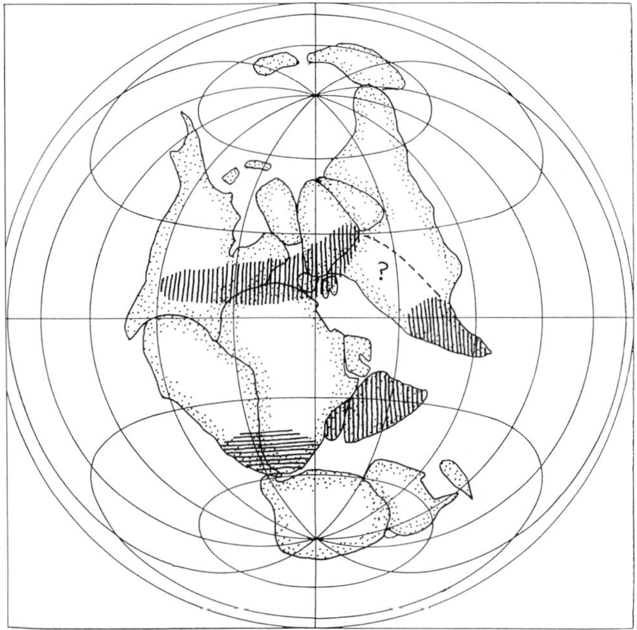

Fig. 25. Late Triassic provinces: Peri-tethyan Province (IIII), Southern Gondwanan Province (≡). After Rage (1988).

closely tied to northern Europe and Thailand. Earlier, Chatterjee and Hotton (1986) had emphasized the northern affinities of the Indian Maleri fauna. They also noted the presence of this fauna in China and concluded that, in the Upper Triassic, India was a crossroad of faunal migration between Africa and Asia, just as it is today.

In the terrestrial flora, the Triassic marks the extinction of the arborescent lycopods and cordaites, and a precipitous decline in some fern families (Niklas et al., 1985). The main thrust of adaptive radiation occurs within the coniferophytes, cycads, and cycadeoids. A decrease in pteridophyte species and a concomitant rise in the various seed-plant groups, which was initiated in the Lower Permian, was continued through the Triassic (Fig. 22). By the end of that Period, seed plants accounted for about 60% of the world's flora.

In his general review of Mesozoic phytogeography, Vakhrameev (1991) noted that three paleofloras emerged in the early Triassic: an Angara Region in the far north contained mostly pteridosperms, ferns, conifers, and a characteristic lycopsid; a Euroamerican Region contained a Voltzian flora (after the conifer genus *Voltzia*); and a Gondwanan Region with a *Dicroidium* flora. Essentially, these appear to be a continuation of the three phytogeographic units that were recognized for the Permian (Ziegler, 1990). Although Shubin and Sues (1991) referred to the presence of a southern-hemisphere plant assemblage in India, a comparison of the Triassic floras of India and Antarctica reflected considerable differences (Bose et al., 1990). The floras of the two areas were apparently not closely related.

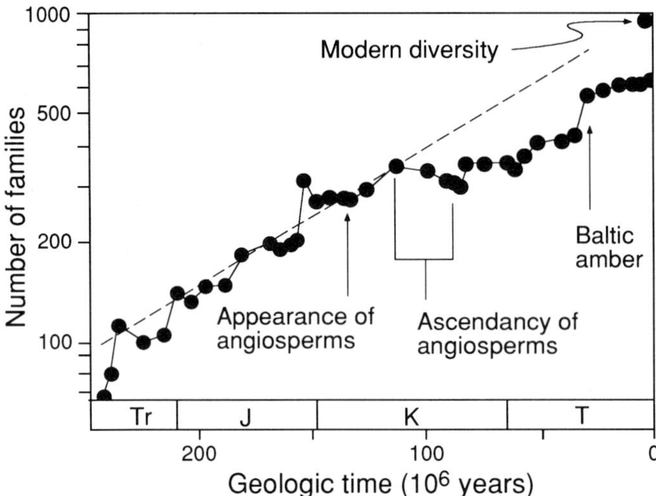

Fig. 26. Apparent exponential increase in insect families from the Triassic into the early Cretaceous. After Labandeira and Sepkoski (1993).

Within the Triassic, there occurred an enormous radiation of insect families. This was the beginning of an apparent exponential diversification which continued into the early Cretaceous (Labandeira and Sepkoski, 1993) (Fig. 26). The evolutionary impetus may have been the availability of a variety of seed plants, together with the return of favorable climatic conditions following the end-Permian extinction.

There is some information available on the distribution of the freshwater conchostracan branchiopods (Tasch, 1987). While South America, Africa, and India share two genera, there is no such relationship between India and Antarctica nor between India and Australia. These data, together with those on the flora and tetrapod fauna, raise a question about the traditional southern placement of the Indian continent. Information on the later Mesozoic relationships will help solve this problem.

The southeast Asian area, including the present Malay Peninsula and most of the large islands of Borneo, Celebes, Sumatra, and Java, evidently represents a mosaic resulting from an amalgamation of many small terranes. During the early Paleozoic, these fragments were parts of, or were adjacent to, eastern Gondwana. Later in the Paleozoic, at least two groups of terranes departed Gondwana and moved northward. It was at first supposed that these groups did not become parts of southeast Asia until the Cretaceous (Audley-Charles, 1987; Scotese, 1991). However, these migrations have been reevaluated with the aid of new paleontological data which indicate that the movements of the suspect terranes took place much earlier.

Metcalfe (1991) determined that the continental core of southeast Asia was comprised of four principal blocks: South China, Indochina, East Malaya, and Sibumasu. The first three were rifted from Gondwana in the Silurian or Devonian and amalgamated to each other during the early Carboniferous. Sibumasu remained attached to Australian Gondwana until the late, early Permian when it rifted away as part of the elongate Cimmerian continent. Collision between Sibumasu and the earlier blocks, forming the continental core of southeast Asia, may have occurred in the early Triassic.

The core of southeast Asia, forming a large island, was supposed to have collided with North China in the latest Triassic to early Jurassic (Metcalfe, 1991). The final suturing of southeast Asia to Eurasia was suggested to have occurred in the early to middle Jurassic. However, this final amalgamation must have taken place by the late Triassic. Both Rage (1988) and Kalandadze and Rautian (1991) reported the presence, in Thailand, of late-Triassic tetrapods related to those found in India and northern Europe.

Late-Triassic extinction

Major extinction events took place towards the close of the Triassic and possibly at other times in that Period. In the marine reef habitat, losses among the algae were minor but the calcareous sponges suffered greatly (Fagerstrom, 1987). The sphinctozoans, which had been major reef-builders, ceased to become important in this activity. By the late Norian Stage, scleractinian corals had replaced the sphinctozoans and inozoans as the dominant reef-builders. The Norian extinction represented a setback for the scleractinians and reef construction did not get underway again until the Pliensbachian Stage of the early Jurassic, some 15 Ma later. For five of the nine bivalve orders, the generic extinction rate ranged between 40 and 62% (Hallam and Miller, 1988). The families and genera of ammonoid cephalopods showed a steep decline towards the end of the Triassic (House, 1988). The main invertebrate groups affected by the extinctions were the cephalopods, gastropods, brachiopods, bivalves, and sponges (Benton, 1990a). In the fish, the older paleopterygians decreased in diversity due to decreasing origination rates but the overall diversity in the early Jurassic was about the same as it had been throughout most of the Triassic (McCune and Schaeffer, 1986).

Although the Norian is often referred to as the Stage in which the greatest Triassic extinctions took place, Johnson and Sims (1989) found evidence of heavy marine-invertebrate extinctions in the Carnian. Most recently, Benton (1990b) concluded that there were three separate extinctions, none clearly larger than the other two. The first occurred during the Scythian Epoch over a period of 5–6 Ma, the second from the Scythian to the end of the Carnian Stage (15–19 Ma), and the third during the Norian (12–17 Ma). Thus there may have been three protracted extinctions which, together, occupied most of the Triassic.

In the terrestrial environment, an analysis of the fossil plants was published by Ash (1986). He concluded that neither megafossils nor palynomorphs showed a significant change across the Triassic-Jurassic boundary. In general, the fossils indicated that plants were slowly evolving with one assemblage gradually giving way to another and that the floras were gradually becoming more modern. The palynological data presented by Boulter et al. (1988) show a considerable drop in diversity beginning in the Rhaetian Stage, but the decrease may be mainly due to a cessation of species formation instead of species extinction.

Towards the end of the Triassic, several new groups of vertebrates began to appear. These included crocodylomorphs, pterosaurs, lepidosaurs, two orders of dinosaurs, and several groups of therapsid synapsids, including the morganucodontids, generally considered the first true mammals. Also, the first salamander-like lissamphibians showed up. It has been observed that evolution of these new groups marked the origin of the modern vertebrate fauna (Padian, 1986). Charig (1984) emphasized that, at the beginning of the Triassic, the large land vertebrates were nearly all therapsids but, by the end of that period, they were nearly all archosaurs (dinosaurs and their kin). He suggested that the competitive success of the latter was due to locomotor improvements and a great increase in size. Colbert (1986) noted that the tetrapod extinctions that took place at or near the end of the Triassic could be placed in three categories: (1) the disappearance of long-established groups, (2) displacement of ancestors by their descendants, and (3) extinctions of small "holdovers" by newly arisen competitors. Altogether, about 22% of the tetrapod families disappeared (Benton, 1988). Shoemaker and Wolfe (1986) estimated that the extinction was distributed somewhere within an 18 Ma interval. Olsen et al. (1987), suggested that a rapid or catastrophic change took place at the boundary but Padian (1988) felt that this conclusion was not justified.

JURASSIC

As southeast Asia was being formed from Gondwana terranes, large portions of Tethys Sea floor were subducted beneath the Asian continent. This subduction created a tension on Africa (Scotese, 1991). As a result, Africa began to pull away from North America and, at the same time, began a counterclockwise rotation towards the Tethys trench (Map 10). The latter movement initiated a separation between the southern parts of Africa and South America. As these separations took place, oceanic basins began to form in the North and South Atlantic.

Biogeographic evidence pertaining to the establishment of a Central Atlantic Seaway between Europe and America has been published by Hallam (1983). He noted that, although marine deposits first appeared on the margin of the Gulf of Mexico in the early Jurassic, there was no evidence of a seaway extending as far as Europe. But, by the Toarcian Stage about 180 Ma ago, there were indications of molluscan traffic along the seaway. At first, the traffic was eastward then, by the Bajocian Stage, the movements were mixed. Hallam emphasized that these mid-Jurassic continental separations were occupied by epicontinental seas which were sufficiently shallow to disappear at times of major regressions. Apparently, this intermittent marine separation between North America and Africa persisted through the late Jurassic, since there was still some migration of terrestrial animals between North America and Africa.

Early-Jurassic ammonoid patterns in the northern hemisphere were studied by Smith and Tipper (1986). They were able to identify distinct boreal and tethyan faunas and their zones of mixture. They were also able to constrain the latitudinal positions of certain suspect terranes on the west coast of North America. For the terranes of Wrangellia (Queen Charlotte Islands) and Stikinia (northern British Columbia), it was determined that they came from the northern rather than the southern hemisphere, and from the eastern rather than the western Pacific.

In regard to the belemnites, Doyle (1987) was able to see signs of a developing provinciality in the late part of the Lower Jurassic, when distinct Arctic taxa became apparent. By the middle Jurassic, Boreal and Tethyan realms had become established, and the Boreal Realm was divisible into informal Boreal-Atlantic and Arctic provinces. The development of cooler sea-surface temperatures in the polar regions by the mid-Jurassic, may be contrasted with the apparent cosmopolitan warm temperatures of the Triassic and early Jurassic.

The middle-Jurassic madreporan coral genera apparently had a cosmopolitan distribution, except for the high-latitude boreal areas where water temperatures may have been too cool (Beauvais, 1979). Certain basins were identified where the species endemism was 25 to 50%. These were the Anglo-Paris Basin, the Alsace-Swabian-Jura Basin, the Morocco Basin, the Bolivian Basin, and the Indian-Malagasy Basin. By the late Jurassic, the madreporans had disappeared from North America and northern Europe. These changes were related to a southern displacement of the Boreal Realm and the opening of the North Atlantic Ocean.

In the terrestrial environment, the distribution of dinosaurs continued to be very broad in the early and middle Jurassic, but the fossil records are fewer (Weishampel, 1990). By the late Jurassic, the dinosaur lineage had flowered into many branches and the fossils were relatively plentiful. The early archosaurs gave rise to two groups in which active flight evolved. The most primitive pterosaurs began to appear in the late Triassic and became common throughout the Jurassic. The first fossils of *Archaeopteryx*, a genus that unites the reptiles and birds, were found in the Upper Jurassic deposits of southern Germany. If it were not for the preservation of its feathers, *Archaeopteryx* would have been described as a dinosaur. None of its skeletal features are uniquely avian (Carroll, 1988).

Lizards and sphenodontids are closely related and may share a common ancestry (Carroll, 1988). The earliest fossil lizards are known from the Upper Permian, so the sphenodontids are probably of the same age. The latter are known from the Upper Triassic of Europe and the middle and late Triassic of Africa, India, and South America. A single species of the genus *Sphenodon* lives on small islands off the coast of New Zealand. Elsewhere, all sphenodontids had become extinct by the Cretaceous. Its survival in New Zealand is as remarkable as that of the primitive frog family Leiopelmatidae (p. 71). Their presence suggests an early dispersal avenue, from South America via Antarctica, that did not include Australia.

A recent work on the development of paleoseaways around Antarctica (Lawver et al., 1992) indicates that New Zealand was most closely tied to Antarctica from the late Jurassic to the early Cretaceous. There may have even been a narrow isthmus (Fig. 27). This connection could have permitted *Sphenodon*, a leiopelmatid frog, and possibly a dinosaur to enter New Zealand. The latter is known from caudal vertebrae taken from a late Cretaceous deposit (Molnar, 1981). At the same time, epicontinental seas may have isolated Australia.

The Upper Permian and Triassic lizards, called the Eolacertilia (Estes, 1983), emerged in South Africa and later appeared in England and North America. A possibly related form was found in the Upper Triassic of southern Asia. The fossil record of the true lizards began in the late Jurassic and is thus separated from that of the Triassic eolacertilians by more than 50 Ma. The most significant aspect of Jurassic lizard distribution is that all the major groups (infraorders) of true lizards are represented in the late Jurassic

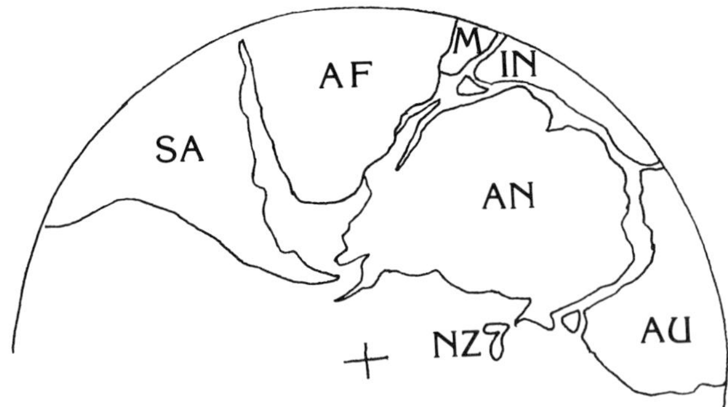

Fig. 27. South polar view of southern lands in the late Jurassic. After Lawver et al. (1992). Note the difference between this version and that given on the global view (Map 9).

of western Europe. Estes suggested that the formation of these four groups may have been aided by the presence of barriers in the form of epicontinental seas.

Despite some indications of continental separation in the Upper Jurassic, Cox (1974) and Rage (1988) noted that there was still a great similarity between the dinosaur faunas of Laurasia and Gondwana. While the relationship between North and South America was relatively weak (as it was in the Triassic), the strongest ties were between North America and Africa, probably via Europe. The northern relationships of India, which were noted in the Triassic discussion, seem to have become reinforced by the discovery of the sauropod dinosaur, *Barapasaurus* (Chatterjee and Hotton, 1986). It is closely related to sauropods from Tibet and China.

The Kota Formation of India, where *Barapasaurus* was found, also produced a symmetrodont mammal, very similar to those known from Great Britain and North America. These early mammals have not been recorded from the southern hemisphere. Also, a pterosaur from that formation belongs to the same genus as a European form (Chatterjee and Hotton, 1986). The fossil fish found in the Kota Formation appear to have strong European affinities (Patterson and Owen, 1991). One may conclude that the Jurassic vertebrate fauna of India is definitely Laurasian.

It now seems possible on biogeographic grounds, to make a determination of the Pangaean (Triassic-Jurassic) position of India. Briggs (1989) adopted the Pangaean reconstruction of Weijermars (1989) which depicted India lying to the south as a wedge between southern Africa on one side and Antarctica-Australia on the other. It is now apparent that such a position is no longer tenable. Previously, Chatterjee and Hotton (1986) had concluded, on biogeographic and geophysical evidence, that India must have been situated adjacent to the Somali Peninsula of northeast Africa and separated from Asia by a narrow Tethys Sea.

In the most recent reconstruction, Chatterjee (1992) reiterated the northern position of India in the Jurassic. This orientation is depicted on appendix Map 10. At this time, India had not started its northward progression and counterclockwise rotation. This location of India may be contrasted with the southern position illustrated by Scotese (1992) and most other geophysical reconstructions (Fig. 28). It seems apparent that Madagascar separated

Fig. 28. Late Jurassic location of India as depicted by Scotese (1992).

from Africa in the Middle Jurassic about 160 Ma ago (Rabinowitz et al., 1983). Madagascar was originally in a northerly position against the coast of Tanzania and, after separation, moved southward.

The ostariophysan fish comprise a group of related orders and families that are characterized by the possession of a Weberian apparatus for the transmission of sound impulses from the swimbladder to the inner ear. This evolutionary innovation has enabled the ostariophysans to become very successful, in fact, to dominate the freshwater habitats to which they have been able to gain access. This group contains about 28% of the known fish species in the world and about 72% of the freshwater species (Nelson, 1984). They are considered to be "primary" freshwater fish (Myers, 1938) that have evidently been confined to freshwaters throughout their history.

Of all terrestrial animal groups, the ostariophysan fish are probably the most useful for the determination of historic continental relationships. They must remain in freshwater and cannot be transported by winds, floating debris, birds or other agencies. Although the earliest ostariophysan fossils are from the Lower Cretaceous, they are considered to be a relatively primitive bony fish group and probably were extant by the Upper Jurassic. There are three principal groups of ostariophysans, the characoids (characins), cyprinoids (carps and minnows), and siluroids (catfish).

The characoids, generally considered to be the most primitive ostariophysans, are presently found in the Neotropics, where they are represented by 15 families, and in Africa, where there are four families. Early Tertiary fossils have been reported from western Europe (Patterson, 1981). There are 13 families of siluroids in the Neotropics, six families in Africa, three in the Oriental Region, and one in North America. The cyprinoids are most diverse in the Oriental Region where they are represented by six families. They are also present in lesser numbers in Europe, North America and Africa but not South America. The ostariophysans are obviously a monophyletic group that must have achieved their present, almost worldwide, distribution by means of terrestrial continental connections.

It has been suggested that the early characoids, followed closely by the first siluroids, originated in the Oriental Region in the Upper Jurassic (Briggs, 1979). From tropical Asia, these groups probably spread to Europe via an occasional trans-Turgai connection.

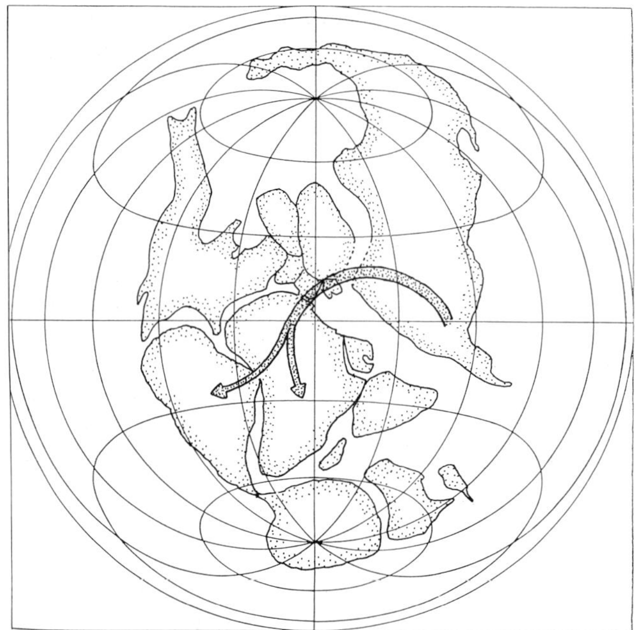

Fig. 29. Suggested late Jurassic dispersal route of primitive ostariophysan freshwater fishes (characoids and siluroids) from a possible center of origin in southeast Asia.

Once in Europe, they could have migrated around the west end of the still incomplete Tethys Sea (Fig. 29) and entered Africa. Their entry into South America was probably from Africa by means of a peninsular connection that was in existence in the Upper Jurassic to the early Cretaceous. The cyprinoid fish, the most advanced ostariophysan group, evolved later and evidently did not enter Africa until the Miocene, too late to continue on to South America.

An alternative view of ostariophysan history (Novacek and Marshall, 1976) assumed an origin in South America followed by dispersals to Africa and the northern continents. Two factors seem to argue against this hypothesis: (1) in other groups of primary freshwater fish, Africa has a far more diverse fauna than South America, so was more likely a donor instead of a receptor, and (2) the parasitic nematode family Camallanidae (infecting mainly freshwater fish) is represented in South America by a depauperate and primitive fauna (Stromberg and Crites, 1974); the South American fauna most likely came from Africa along with the invasion of ostariophysan fish. It has also been suggested that the early ostariophysans lived in eastern Gondwana and were carried to Asia by East Asian terranes and India (Bănărescu, 1990). However, the lack of Mesozoic fossils in either Australia or India make this an unlikely history. Ostariophysans apparently did not invade North America until the Tertiary.

The fossil history of the salamanders (Caudata) extends back to the middle Jurassic, so their biogeography is correlated with the fractionation of Pangaea. It has been suggested that the earliest protosalamanders were distributed throughout the northern humid zone at the beginning of the Jurassic (Duellman and Trueb, 1986). They were probably excluded from low latitudes by arid climates, and were therefore unable to reach the southern hemisphere. Duellman and Trueb noted that the opening of the North Atlantic Ocean in the early Jurassic would have separated the protosalamanders into two stocks, one in Asia and one in Euramerica. But this separation would have been effected by the advent of the Turgai Sea, not the North Atlantic (Map 10). Supposedly, the stock in Asia gave rise to the Karauroidea (an extinct suborder) and the population in Euramerica formed all other salamanders.

The three modern salamander suborders were supposed to have been formed by the breakup of the northern continents during the early and middle Jurassic (Duellman and Trueb, 1986). However, further continental separation in the north did not occur until the Cretaceous (Midcontinental Seaway) and the early Tertiary (opening of the North Atlantic). If the three surviving suborders were formed by the middle Jurassic, it was not because their populations were divided by rifting continents. The earliest fossil salamander dates from the middle Jurassic of Europe and appears to be a prosirenid (Sirenoidea), so the hypothesis of suborder differentiation by that time appears to be reasonable. Formation of modern salamander families probably took place in the Cretaceous, and may be related to the presence of three northern continents in the mid-Cretaceous (see following chapter).

The frogs (Anura) clearly have a Jurassic history and may even extend back to the Triassic. A Triassic fossil from Madagascar (*Triadobatrachus*) provides a plausible link between Lower Permian dissorophids and primitive frogs (Carroll, 1988). The earliest known true frog is *Vieraella* from the Lower Jurassic of Argentina (Fig. 30). It belongs to the family Leiopelmatidae which also has two living genera, one in North America and

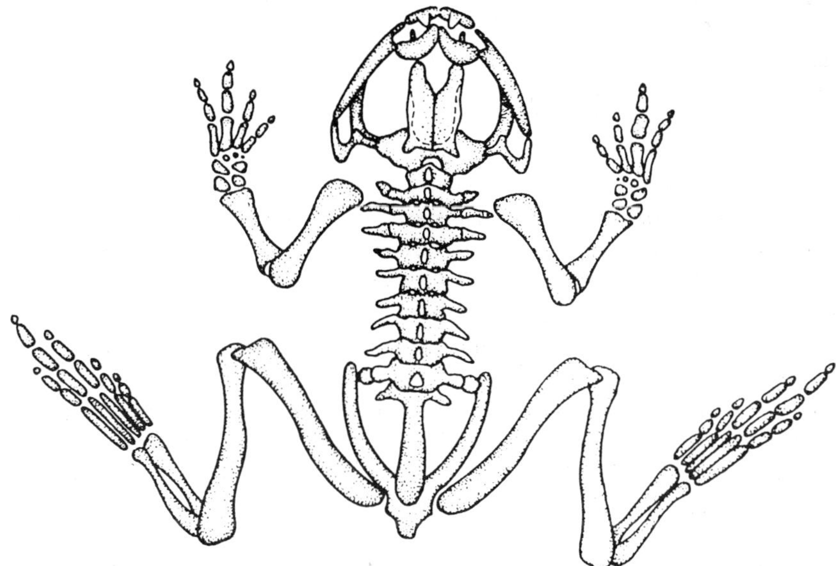

Fig. 30. The Lower Jurassic frog *Vieraella*. It closely resembles modern frogs except for the retention of vestigial ribs. After Stearn and Carroll (1989).

one in New Zealand. Another group, the pipoids, comprising three related families, also existed in both Laurasia and Gondwana. The extinct paleobatrachids are known from the uppermost Jurassic to the Pliocene of Europe and the late Cretaceous of North America (Duellman and Trueb, 1986). Rhinophrynids are known from the late Paleocene through the Oligocene of North America; a single living species exists in Mexico and Central America. Pipids are known from the Cretaceous of South America and Africa and live there now.

The pipoid stock leading to the formation of the northern families and the southern Pipidae was probably separated in the late Jurassic-early Cretaceous by the final split of Euramerica from Africa (Duellman and Trueb, 1986). Another northern family, the Discoglossidae is known from the late Jurassic of Europe and the late Cretaceous of North America. It seems to have diversified mainly in the European area where four genera survive. The other Euroamerican families that probably arose in the Jurassic-Cretaceous are the Pelobatidae and the Pelodytidae.

It is the radiation of frogs in the southern hemisphere that has produced, by far, the greatest diversity. In addition to the leiopelmatids and pipids, another ancestral anuran stock was apparently in Gondwana by the middle Jurassic (Duellman and Trueb, 1986). Prior to the separation of South America and Africa, a common anuran fauna consisted of pipids, bufonids, leptodactylids, and microhylids. Also, apparently restricted to Africa, were ranids, hyperoliids, and rhacophorids. After the separation, evolution continued on both continents and other families emerged. Four families apparently reached Madagascar and India from Africa. These early dispersals were followed by others in the Cretaceous and Tertiary.

Three groups of insects, whose larvae are restricted to freshwaters, provide significant information on Mesozoic continental relationships. These are the mayflies (Ephemeroptera), stoneflies (Plecoptera), and caddisflies (Trichoptera). Although the adults are capable of flight, they are short-lived and weak flyers. All three are primitive insect orders, undoubtedly present in Pangaean times. Each appears to demonstrate a basic north-south dichotomy at the family level, possibly reflecting the Laurasian-Gondwana split (Edmunds, 1982; Bănărescu, 1990). In the southern hemisphere, the strongest present day relationships are among southern South America, Australia, and New Zealand. This strong amphinotic pattern exists along with a marked Afro-Indian pattern. In comparison, the ties between South America and Africa are relatively weak. These relationships are reiterated by many other terrestrial groups (Briggs, 1987). This points to the maintenance of an amphinotic migration route which must have persisted long after the northward departure of Africa-India (and its separation from southern South America).

In regard to the terrestrial flora, Vakhrameev (1991) found that he could divide the earth into four, latitudinal belts, a moderate-warm Siberian-Canadian Region, a subtropical Euro-Simian Region, a tropical Region, and a subtropical Austral or Notal Region. These belts were noted to have shifted during the Jurassic in response to the changing climate. Thus the relatively cool climate of the early and middle Jurassic (Fig. 31) may be contrasted with the warmer late Jurassic (Fig. 32).

Vakhrameev (1991) noted that the genus *Pachypteris* was a good indicator of the subtropical zone, but was absent in the tropics. This genus therefore had an antitropical distribution. The Austral Region, however, differed from its northern counterpart due to

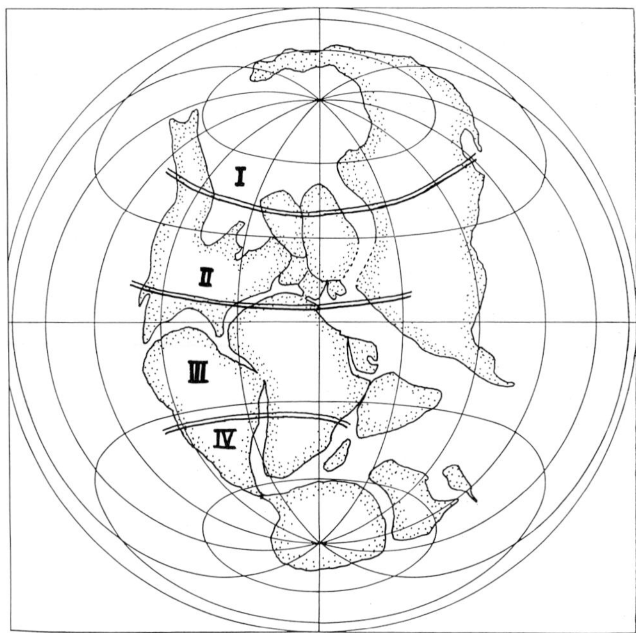

Fig. 31. Climatic belts indicated by vegetation in the early and middle Jurassic. They are Moderate-warm (I), Northern Subtropical (II), Tropical (III) and Southern Subtropical (IV). After Vakhrameev (1991).

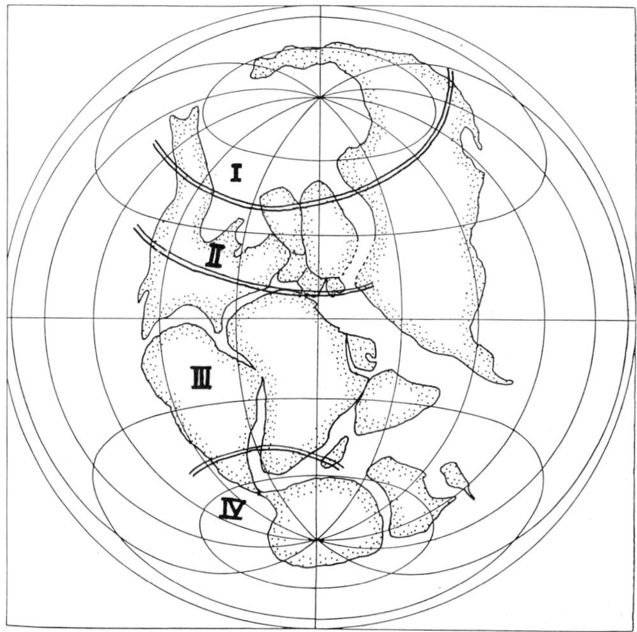

Fig. 32. Climatic belts for the late Jurassic. Note that the tropical belt (III) has become expanded while the cooler belts have become restricted. After Vakhrameev (1991).

the absence of ancient Pinaceae and the development of the Podocarpaceae and Araucariaceae. Towards the end of the Jurassic, a southern Pentoxylales flora evolved in India, New Zealand, and southern Australia. Vakhrameev, as well as Krassilov (1972), felt that this flora provided good evidence for a southern location for the Indian continent.

On the other hand, Smiley (1979) concluded that the Jurassic and early-Cretaceous floras of India were generically more closely related to the floras of Asia than to those of Gondwana. He found that 29 of 38 Jurassic genera (36%) occurred on other Gondwana continents, in contrast to 48 genera (60%) that were present in the Jurassic floras of Eurasia. Only five of the genera occurring in Gondwana areas were restricted to that part of the world. He considered the composition of the flora to indicate that India was close to its present position (Smiley, 1979). The inconsistency of the interpretations of the Indian flora may be contrasted with general agreement about the northern (Laurasian) relationship of its terrestrial fauna.

Jurassic extinctions

Two extinction events are often recognized for Jurassic times. The first of these took place in the Pleinsbachian Stage about 187 Ma ago. Although the Pleinsbachian has been recognized as a minor, global marine extinction (Sepkoski, 1986a), Hallam (1986) pointed out that this event was focused in western Europe. Here the seas retreated and black muds accumulated under conditions of reduced oxygen concentrations. About 80% of the bivalve species in that area were eliminated. In regard to reef communities, Fagerstrom (1987) noted that the Pleinsbachian and Toracian were recovery periods for the reef-building corals that had been damaged in the Norian extinction. In the tetrapods, there appears to have been about a 30% reduction in family diversity at the end of the Lower Jurassic (Colbert, 1986). This may or may not have been concurrent with a possible extinction in the marine environment. None of the orders to which the eliminated tetrapod families belonged became extinct at the time.

Another extinction is sometimes recognized as occurring in the Tithonian Stage at the end of the Jurassic about 146 Ma ago. As with the Pleinsbachian event, Hallam (1986) determined that this was a regional rather than a global extinction. Five bivalve families became extinct but all were small, containing only a few genera (Hallam and Miller, 1988). Most of the extinctions took place among the ammonoids and to a lesser extent, bivalves and corals. Fagerstrom (1987) called attention to a remarkable change in the composition and structure of the reef-building guilds. The cyanophytes became considerably less important. Numerous sponge taxa became extinct so that, after the Jurassic, neither the hexactillinids nor the demosponges remained important reef-builders. The Inozoa and the Stromatoporoidea also decreased and thereafter did little reef-building.

According to Colbert (1986), there was a significant extinction of tetrapod families at the end of the Jurassic amounting to about 36% of the fauna. But the compilation of Benton (1988), based on more recent data, does not indicate a significant decrease. One may conclude that the end-Tithonian event was not a great disaster in terms of the loss of major evolutionary lines. Bakker (1978) detected a change in dinosaur feeding behavior at the beginning of the Cretaceous. About this time, a radiation of big, low-browsing ornithischian dinosaurs took place. This was also the time that the radiation of angiosperm plants got underway. It is possible that the new availability of low, fast-growing plants had an important effect on dinosaur evolution.

SUMMARY

1. A very slow recovery from the P/T extinction meant a highly impoverished marine fauna in the early Triassic. Colonial skeletal organisms such as calcareous algae, sponges, corals, and bryozoans were completely absent. Latitudinal geographical distributions were extremely broad, probably indicating warm sea-surface temperatures from pole to pole.
2. On land, the early-Triassic was populated mainly by vertebrate groups that had evolved in the Paleozoic and had survived the extinction at the end of that time. At first, the therapsids were the most diverse tetrapod group. They were replaced by a series of groups endemic to the Triassic. The latter were, in turn, replaced by the dinosaurs towards the end of the Period.
3. The early Triassic tetrapod faunas had extensive geographic ranges. The only continent that exhibited some difference was Australia, where the fauna had a more primitive aspect. This suggests that Australia may not have been completely attached to the rest of Pangaea. By the late Triassic, some local differentiation could be detected. The tetrapods of India showed mostly northern relationships.
4. Three phytogeographic units were apparent by the early Triassic: an Angara Region, a Euroamerican Region, and a Gondwanan Region. These appear to be a continuation of the three regions that were recognized for the Permian.
5. Although the end of the Triassic has often been considered to be a time of great extinction, it now seems apparent that there were three protracted marine extinctions that, together, occupied most of the Triassic.
6. In contrast, the terrestrial plants showed little evidence of extinction, but gradually evolved towards more modern types. Towards the end of the Triassic about 22% of the tetrapod families disappeared, but the extinctions were accompanied by the evolution of new, possibly more competitive groups. There was apparently no rapid or catastrophic change.
7. The late Triassic and Jurassic were times of continental assembly and disassembly. An Indochina block and other terranes were added to the eastern part of Pangaea, and other terranes became attached to the northwest coast. At the same time, Madagascar separated from Africa and Africa began to pull away from North and South America.
8. Distribution patterns of marine ammonoids and belemnites appear to indicate the development of cooler sea temperatures towards the polar regions by the mid-Jurassic. In contrast, cosmopolitan warm temperatures prevailed in the Triassic and early Jurassic. Madreporan coral distributions indicate that the cooling trend continued into the late Jurassic.
9. The earliest fossil lizards (Eolacertilia) are known from the Upper Permian. The sphenodontids are probably the same age; they survive in New Zealand. Their

presence, plus that of the frog family Leiopelmatidae, indicates an early dispersal avenue from South America to New Zealand. The true lizards appeared in the late Jurassic when they are represented by the four major infraorders. All have been found within a relatively small area of western Europe.

10. In the Jurassic, dinosaur distributions continued to be very broad. North American forms continued to be related to those of Africa, probably via Europe. The northern relationships of India were reinforced by discoveries of a fossil dinosaur, a symmetrodont mammal, and freshwater fish.
11. Ostariophysan fish, possibly the best animal group for the determination of past continental connections, probably began their evolution in the Upper Jurassic. They were apparently able to spread from East Asia to Europe, to Africa, and then to South America, before these continents were completely separated. Their dispersal to North America took place in the Tertiary. Alternative views of ostariophysan history assume origins in South America or in eastern Gondwana.
12. The salamanders (Caudata) apparently began their development in the early Jurassic. The early protosalamanders were probably distributed across the humid zone of Laurasia. This pattern was disrupted by the formation of the Turgai Sea in the middle Jurassic. The Asiatic form eventually became extinct and all modern salamanders evolved from the Euramerican stock.
13. The frogs (Anura) extend back to at least the early Jurassic and possibly the Triassic. The most primitive family, the Leiopelmatidae, is known from a Lower Jurassic fossil in Argentina and two living genera, one in North America and the other in New Zealand. Another early group, the pipoids, also existed in both hemispheres. These two groups were apparently separated, in the late Jurassic-early Cretaceous, by the split between Euramerica and Africa. In the meantime, another stock had evolved in Gondwana which eventually led to an enormous southern hemisphere diversity.
14. The Jurassic terrestrial flora may be divided into four longitudinal belts. The shifting of these belts during the Period, appears to be a reflection of an increasingly warm global climate. This progression is at odds with the marine patterns which seem to indicate an increasingly cooler climate.
15. Two authors suggested that the late Jurassic flora of India showed that continent should be located in far south Gondwana. A third author concluded that its floral relations were primarily Asiatic.
16. Two minor extinctions took place, one in the Pleinsbachian Stage of the Jurassic about 187 Ma ago, and the other in the Tithonian Stage about 146 Ma ago. Both may be considered regional, rather than global in scope. Their effects were primarily confined to shallow, marine waters.

CHAPTER 5

Late Mesozoic

One great thought prevails in natural historical studies, the study of laws regulating the geographical distribution of natural families of animals and plants upon the whole surface of the globe.

Louis Agassiz, *On the Succession and Development of Organic Beings*, 1842.

CRETACEOUS

The Cretaceous was a Period during which greenhouse conditions existed. As was noted for the early Paleozoic, a rise in sea level was probably caused by the divergence of tectonic plates in the ocean. This caused the mid-ocean ridges to become enlarged vertically and horizontally, which displaced water from the oceanic basins to form epicontinental seas. The reduction in global land surface meant that less CO_2 was removed from the atmosphere by the weathering process. In addition, the increased plate movement was accompanied by increased volcanism which contributed more CO_2 to the atmosphere. As a result, the level of CO_2 in the atmosphere-hydrosphere system increased (Fischer, 1984; Marshall et al., 1988). The rise in atmospheric CO_2 caused retention of much of the heat now lost to space by radiation, so the average surface temperature of the globe continued to rise.

The extensive inland seas, the moist tropical climate, and the high CO_2 level produced a luxuriant growth of plant life. This great primary production on land and in the epicontinental seas provided fodder for a large mass of herbivores which, in turn, permitted the evolution of a variety of carnivores. The greenhouse condition of the Cretaceous lasted more than 60 Ma, allowing time for the evolution of ecosystems dependent on a high level of primary production.

In the sea, the epicontinental flooding allowed the rivers to deposit their loads of sediments and nutrients in the shallow waters of the continental shelves. This permitted the nutrients to be recycled by vertical, wind-driven currents and caused the production of phytoplankton at a continuously high rate. The evolution of coccolithophorids in the Triassic and diatoms in the early Cretaceous may have increased the, per unit of area, primary production of shelf waters about eight times (Walsh, 1988). Considering that, during the middle and late Cretaceous, extensive continental seas covered the interior of North America, southern Europe, and large portions of Australia, Africa, and South America, primary production in the shallow sea may have been several times that which exists today. By the mid-Cretaceous, the phytoplankton supported a huge mass of zooplankton, which was, in turn, fed upon by many large carnivores.

On land, the high primary production resulted in a diverse fauna of large herbivorous dinosaurs that were preyed upon by large carnivorous dinosaurs. Coexisting were other food chains supporting much smaller animals (mammals, reptiles, amphibians, insects and other invertebrates). The predominant food chain in the sea was longer, due to the fact that the primary producers were mostly small, one-celled organisms. They were eaten by zooplankters which were then consumed by fish and larger invertebrates. Both primary and secondary carnivores were eaten by very large fish and marine reptiles. Although other, more complicated food webs existed in both environments, the main flow of energy through the systems was by means of short food chains that supported many large-sized animals.

Because they have undergone less metamorphism and erosion than older geologic systems, Cretaceous deposits are represented on modern continents by an extensive record of shallow marine and nonmarine sediments and fossils. In addition, Cretaceous sediments and fossils are widespread in the deep sea, in contrast to the sparse deep-sea records for the Triassic and Jurassic periods; most of the deep-sea sediments older than the Cretaceous have been swallowed up along subduction zones.

During the Cretaceous, a series of continental movements took place which, to a large extent, resulted in the establishment of biogeographic patterns that are still evident in our modern faunas and floras. These movements resulted in the formation of the completed Tethys Sea which stretched all the way from east Asia to western America, the separation of Africa and South America, the isolation of Australia and New Zealand, the formation of Beringia, and the departure of Africa from Europe. These tectonic shifts were accompanied by rising sea levels which flooded major portions of the continents.

Marine patterns

In the marine environment, the classic treatise by Kauffman (1973) on Cretaceous bivalves still sets the standard for meaningful paleobiogeographies. He was able to delineate three principal realms, a tropical Tethyan Realm, a North Temperate Realm, and a South Temperate Realm. Within these major divisions, he recognized a series of regions, provinces, subprovinces, and endemic centers (Fig. 33). These units can be compared to modern biogeographic zones, since they were defined in terms of generic endemism: endemic centers 5–10%, subprovinces 10–25%, provinces 25–50%, regions 50–75%, and realms >75%. Kauffman's map was drawn with the continents in their Recent positions. The configuration of the Tethys Realm is here depicted on a late Cretaceous map (Fig. 34).

In his analysis, Kauffman (1973) traced the evolutionary histories of the biogeographic units within each of the realms. These have made it possible to gauge the effects of continental movement and sea-level rise on the bivalve fauna of the shelf. Within the Tethyan Realm, the history and distinctiveness (percent of endemism) of the provinces and subprovinces reveal important biogeographic events (Fig. 35). Among these, it may be noted that a North Indian Ocean Subprovince developed very early in the

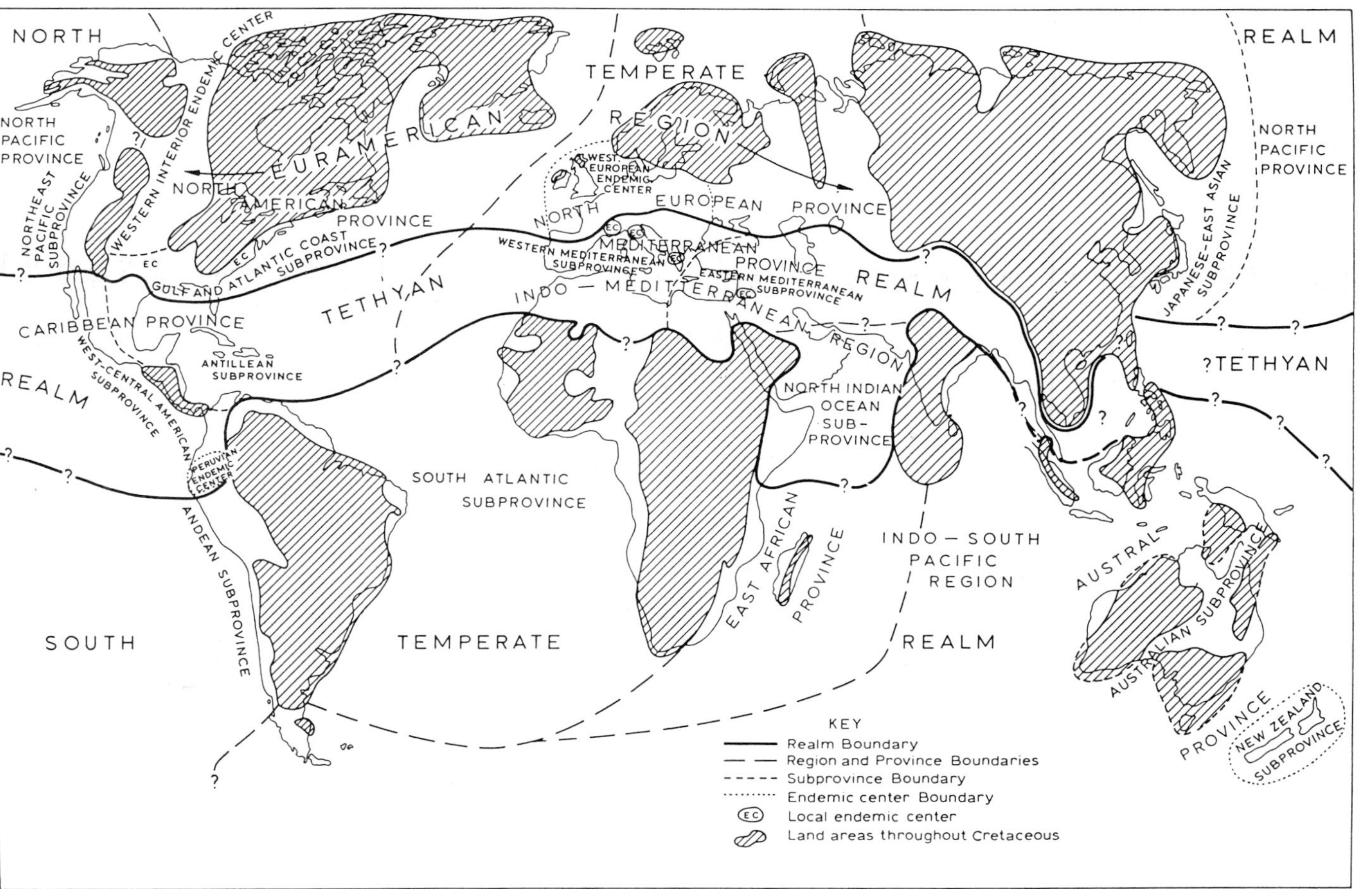

Fig. 33. Average distribution of Cretaceous biogeographic units based on the Bivalvia. Base map indicates Recent configuration of continents. After Kauffman (1973).

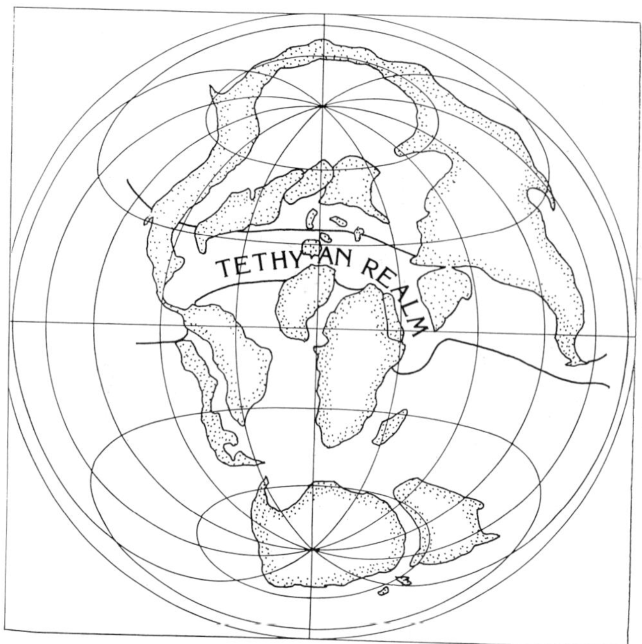

Fig. 34. Tethys Realm depicted on a late-Cretaceous globe. Bivalve relationships, together with evidence of terrestrial animal migrations, indicate that a complete isthmus probably connected North and South America.

Cretaceous. The newly determined position of India against the northeast African coast, as opposed to being located far out in the ocean, permitted India to act as a partial barrier behind which the Subprovince could develop. Without such a barrier, the existence of that unit would be difficult to explain.

Another important event was the establishment, about 132 Ma ago, of separate Eastern and Western Mediterranean Subprovinces. This subdivision strongly suggests the development of an isolating mechanism near the eastern end of the Mediterranean. By Aptian time, about 124 Ma ago, a distinct Caribbean Province became apparent. This was the result of the Atlantic Basin spreading to the extent that the Caribbean shallows were geographically isolated and gene flow from the east was interrupted. Towards the late Cretaceous, the West Central American and Antillean subprovinces developed as separate entities. This was probably in response to a Central American archipelago, which gradually built up during the Cretaceous to briefly become a complete isthmus towards the end of the Period.

In a similar manner, the history of the North Temperate Realm (Kauffman, 1973) can be followed (Fig. 36). At the beginning of the Cretaceous, one may distinguish a Euramerican Region divisible into a North American Province and a European Province. The former province gradually developed increased endemism and, with the increased continental flooding of the Albian-Cenomanian, became divided into a Western Interior Endemic Center and a Gulf-Atlantic Coast Subprovince. The European Province was at

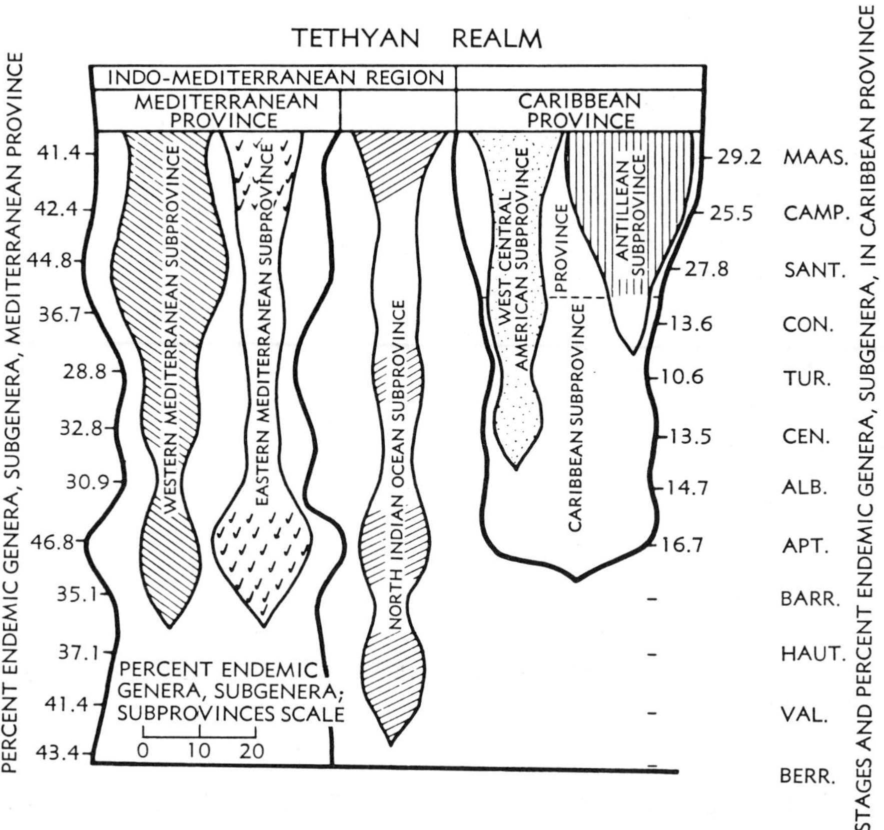

Fig. 35. History of the Tethyan Realm. After Kauffman (1973).

first very distinctive but then declined due to an increase in the number of widespread Euramerican taxa.

At the beginning of the Cretaceous, the North Pacific Province is divisible into well defined Japanese-East Asian and Northeast Pacific subprovinces. But, as the North Pacific Ocean became more constricted, these separate faunas came into greater contact and the endemism of each declined (Kauffman, 1973). The greatest decline was during the Albian-Cenomanian transgression.

The South Temperate Realm was divided into a South Atlantic subprovince and an Indo-Pacific Region (Fig. 37). The former began its development in the late Albian Stage and continued through the remainder of the Cretaceous. Its independent evolution was made possible by the opening of the South Atlantic Ocean. The Indo-Pacific Region consisted of an Austral Province, an Andean Subprovince, and an East African Province. Within the Austral Province, an Australian Subprovince was well defined in the early Cretaceous, but later gave way to a New Zealand Subprovince. There was some suggestion that the Austral Province was a series of small, isolated endemic centers (New Zealand, Australia, New Caledonia, etc.) with little in common. The East African Province

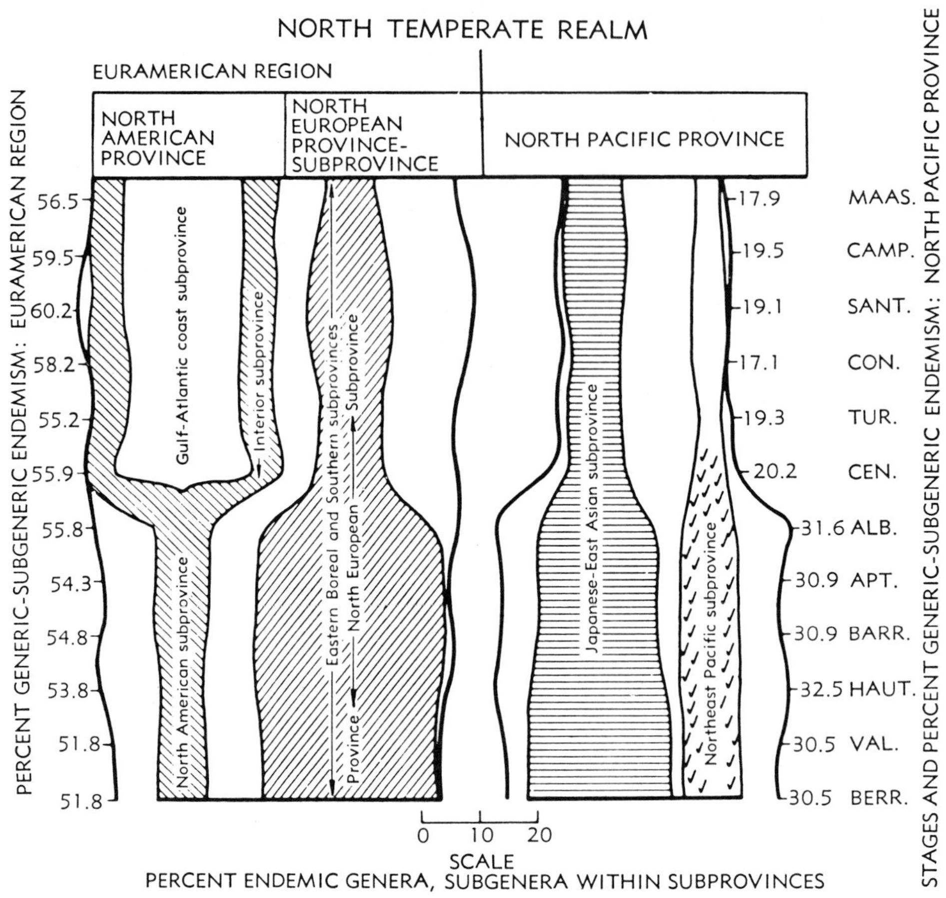

Fig. 36. History of the North Temperate Realm. After Kauffman (1973).

was strongly developed in the early Cretaceous but declined to an endemic center by the late Cretaceous. The Andean Subprovince came into being during the Cenomanian Stage. Kauffman (1973) warned that, while the Cretaceous bivalves of New Zealand and Australia are well known, Africa and South America are less known and stratigraphic control and regional correlation is poor.

Subsequent, but less detailed, works on Cretaceous marine invertebrates have tended to corroborate Kauffman's (1973) conclusions. In regard to Tethys Sea history: ostracod distribution indicated a separate New World (Gulf of Mexico) fauna by Albian time (Babinot and Colin, 1988); foraminiferan faunas separated after Albian time (Dilley, 1973); coral-rudist populations in Aptian-Albian (Coates, 1973); and the brachiopods by the late Cretaceous (Sandy, 1991). All of these separations, including that of the bivalves by Aptian time, indicate responses by various groups to the Cretaceous expansion of the North Atlantic Ocean.

The paleobiogeography of early Cretaceous corals has been analyzed by Beauvais (1992). Coral fossils were not found in the first (Berriasian) stage of the Cretaceous, but appeared in limited numbers during the Valanginian. Although the mid-Atlantic was still not very wide at that time, it was possible to recognize separate, American and Tethyan Realms. In the pre-Barremian, a significant radiation took place among the scleractinians and they became latitudinally widespread. Two provinces were identified in the Tethyan Realm and three in the American Realm. This general pattern was maintained through the early Cretaceous, except for the Albian appearance of a Malagasy Subprovince. Beauvais observed that the Tethys appears to have been a dispersal center for corals. The first early Cretaceous species appeared in Europe and persisted into the Cenomanian in the remote areas such as East Africa, Lebanon, and India.

Recent reviews of inoceramid and belemnite biogeography have been published. In the early Cretaceous, the inoceramids occurred mainly in temperate seas (Dhondt, 1992), but tended to become cosmopolitan by the Albian-Cenomanian. By the Turonian, they

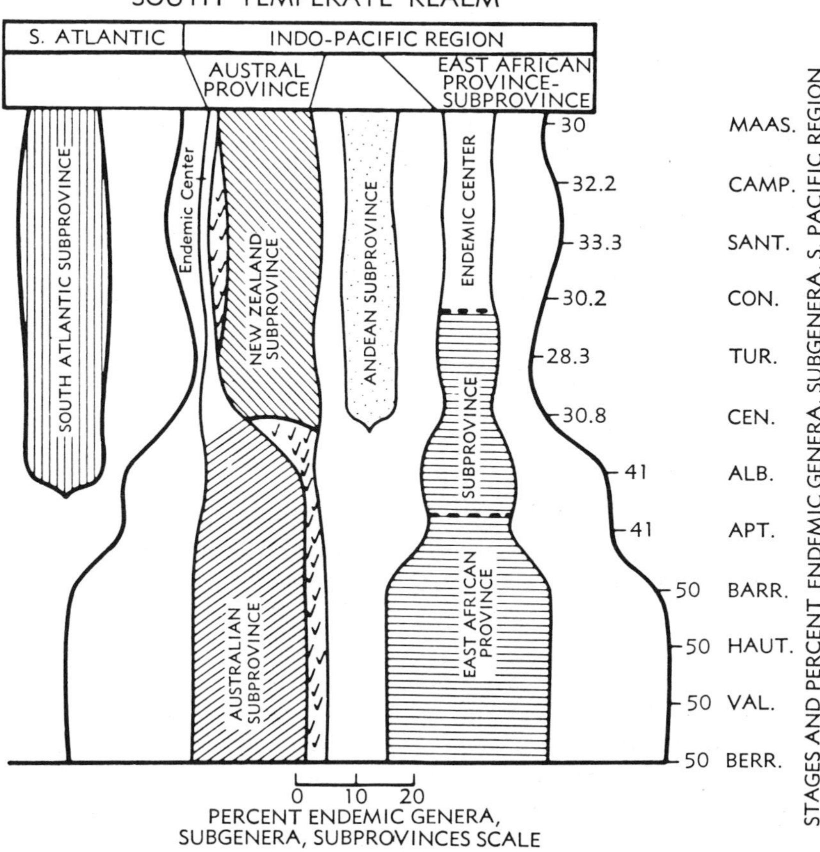

Fig. 37. History of the South Temperate Realm. After Kauffman (1973).

had invaded the south Atlantic. The final, surviving species of the Upper Cretaceous were found in the tropics. In the belemnites, the Boreal and Tethyan faunal realms of the Jurassic were continued into the Cretaceous (Doyle, 1992). An Austral Realm became apparent during the late Barremian to early Aptian. After the Cenomanian, the Tethyan species died out, leaving only the two cool-water realms to extend through the remainder of the Cretaceous.

The final separation between South America and Africa began in the early Cretaceous, when extensive evaporite deposits took place between the two continents (Stanley, 1989). These were caused by occasional incursions of seawater followed by periods of desiccation. Some fossil distributions provide indications as to when a permanent sea passage became established. Halocyprid ostracods, along with certain planktonic foraminiferans and nannofossils, strongly favor a marine north-south connection as early as the Lower Albian (Colin and Andreu, 1990); Masse (1992) indicated that a variety of Tethyan forms migrated southward during the Albian; Rawson (1981) found migrations by late Albian ammonites; Smith (1984) noted that a typical North African echinoid fauna had reached Angola in the Albian; and Sandy (1991) described a brachiopod migration during the late Albian, Santonian, and Campanian from northwestern Europe to the South Atlantic.

Ostracod distributions for the late Jurassic to very early Cretaceous (Berriasian) appear to show a continuous sea lane from the southern tip of South America northward up the east coast of Africa, and eastward along the north coasts of Antarctica and Australia (Dingle, 1988). Also on ostracod evidence, India was placed in a southern hemisphere province in the Valanginian-Aptian but, by the late Cretaceous, there were strong affinities to the north Tethyan fauna (Babinot and Colin, 1988). The ostracods of the late Cretaceous have been divided into Tethyan and South Temperate realms (Babinot and Colin, 1992). Within the Tethyan Realm, three provinces were identified: South European, North-Central American, and Afro-Arabian.

The history of the Western Interior Seaway (Mid-Continental Sea) has been reviewed in conjunction with work on the Cretaceous crabs (Bishop, 1986). In North America, epicontinental seas gradually transgressed northward from the Gulf of Mexico and southward from the Arctic Ocean, finally joining in early, late Albian time. A regression then occurred and the connection was broken in the late, late Albian. A second transgression reestablished the seaway by Cenomanian time. It remained open until a final withdrawal at the end of the Cretaceous (Map 12). Provincial differences in the crab faunas appear to have existed for the North Atlantic Coastal Plain, the Mississippi Embayment, the Interior Seaway, and the Pacific Slope. However, Bishop warned that these differences could be artifacts of a poor and widely scattered record.

Another north-south epicontinental seaway, the Turgai Sea, apparently formed in the late, middle Jurassic (Hallam, 1981) and continued in various configurations throughout the Cretaceous (Vakhrameev, 1987). This area was recognized as part of the North European Province on the basis of bivalve zoogeography (Kauffman, 1973). Its existence is also documented by distributions of fossil belemnites (Stevens, 1973) and brachiopods (Ager, 1973).

Until recently, evidence for the existence of a third north-south seaway, extending from the Arctic Ocean southward between Europe and North America, was unsubstantial. Although some geophysical works had illustrated such a passage, there was little fossil evidence to reinforce the hypothesis. Consequently, Euramerica was considered to have been a single, undivided continent throughout the Cretaceous (Briggs, 1987). A reconstruction, based on tectonic and sedimentary data (Doré, 1991), now suggests that a complex passage with major land barriers may have been established by rifting in the late Jurassic and early Cretaceous. Later on, an open seaway could have existed during the marine transgressions of the Aptian, Cenomanian, and possibly at other intervals of the late Cretaceous.

As Kauffman (1979a) has noted, the Tethys Sea was, prior to the Cretaceous, restricted to Mediterranean Europe and the Indo-Pacific. When the connection between Africa and the Iberian Peninsula was broken in the late Jurassic-early Cretaceous, the Tethys spread westward across the developing North Atlantic to the New World tropics, forming a circumglobal marine belt. Its shelf fauna developed a high level of generic and species diversity. For the later Cretaceous, Smith (1984) delineated two main centers of endemism for echinoids, one in the Caribbean-Gulf of Mexico area and the other in the Mediterranean-North European area. The first had a dominant influence on the composition of the New World faunas and the second provided elements to adjoining parts of the Old World.

It has been noted that ostracod fossils appear to indicate an early Cretaceous sea lane from southern South America northward up the east coast of Africa. A similar pattern seems to be indicated by the ammonoid and bivalve faunas (Riccardi, 1991). They show a passage (which at first may have been shallow and intermittent) extending all the way from south-western South America, across the South Atlantic, and up the African east coast. This connection was apparent in the Tithonian-Berriasian and became better established in the Valanginian-Hauterivian. This is consistent with a recent work on the development of paleoseaways around Antarctica (Lawver et al., 1992) which indicated an open passage occurring by 100 Ma ago.

Terrestrial patterns

Flora

On land, it may be seen that fundamental changes in the world's flora took place in the Cretaceous. As Krassilov (1981) has noted, a characteristic Mesozoic vegetation, dominated by conifers, cycadophytes, and ginkophytes, had emerged by the mid-Triassic. Although replacements of dominant species occurred, the general structure of plant communities remained constant until expansion of the angiosperms in Albian time. A cladistic analysis (Crane and Lidgard, 1990) suggested that angiosperms may have evolved as early as the Triassic, but did not radiate into prominence until the Aptian-Albian. At first, they were found at low latitudes only but, by the Cenomanian, had achieved worldwide ranges. By the Upper Cretaceous, they had become the dominant group of higher plants.

For the Jurassic and early Cretaceous, Vakhrameev (1991) recognized a series of four latitudinal belts defined by characteristic floras (Fig. 32). The northern hemisphere above the tropics was divided into "moderate-warm" and "subtropical" regions, while the extra-

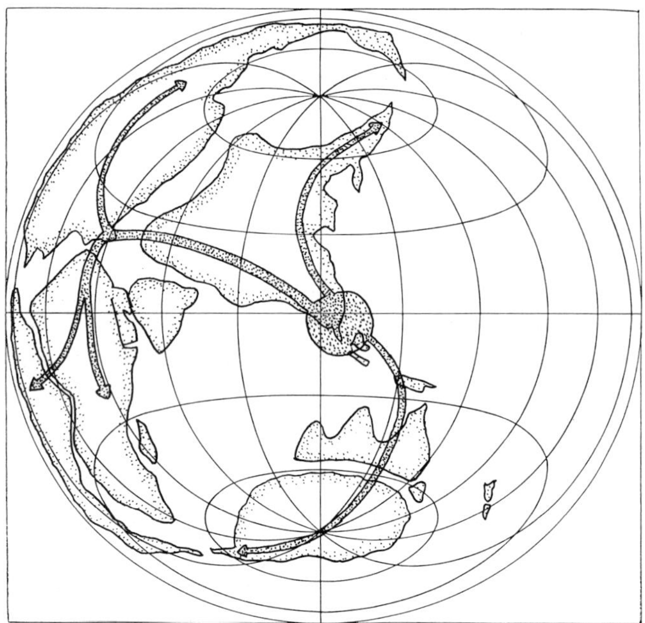

Fig. 38. Hypothetical dispersal of primitive angiosperms and other plant life from southeast Asia.

tropical part of the southern hemisphere contained only one subtropical region. By the Campanian, an additional southern hemisphere moderate-warm region was added. The latter was defined by the appearance of southern beech (*Nothofagus*) pollen.

Since the angiosperms are probably a monophyletic group, the question of their place of origin has received considerable speculation. Unfortunately, the fossil evidence is so sparse that it is of little help. The pertinent literature was reviewed by Briggs (1984) and a conclusion was presented: considering that (1) a great modern diversity of angiosperm plants occurs in southeast Asia, (2) the most primitive families are found either on the fringes of that area or in the adjacent mountains, and (3) many phylogenetic lines leading from primitive to advanced stages are found in that area, it was concluded that the center of origin for the flowering plants was located in that part of the world (Fig. 38).

The concept of an angiosperm origin in southeast Asia has been substantiated by the observations of Latham and Ricklefs (1993). They pointed out that there has been in eastern Asia, since before the Tertiary, a continuous corridor of mesic forest connecting tropical and temperate latitudes. They proposed that the colonization of temperate floras has occurred most frequently in that area, with subsequent production of new taxa which spread within the temperate biomes. They suggested that the relatively low diversity of angiosperm trees in temperate areas arose from the difficulty of colonization from the tropics. This means that the latitudinal diversity gradient may be due to historical and evolutionary factors rather than contemporary ecological interactions.

In her summary of the phylogeographic relationships of the Australian Cretaceous flora, Dettmann (1981) concluded that significant alliances existed between Australia and other regions in southern Gondwana. The early-Cretaceous microfloras from South America, India, New Zealand, and Australia were found to have many common features. She noted that angiosperms did not reach Australia until the mid-Cretaceous (Albian), about 100–108 Ma ago, and that dispersal routes into Australia would probably have involved Antarctica.

In his work on the dispersal of parts of Gondwana and the evolution of angiosperms, Audley-Charles (1987) portrayed parts of southeast Asia as occupying a position in the Tethys Sea intermediate between Australia and Asia. Since this was thought to be the situation in the early Cretaceous, he argued that these terranes could have acted as a Noah's Ark in carrying an evolving angiosperm flora from east Gondwana to Asia. However, as noted in the Triassic account, more recent fossil data appear to indicate that the rifting from Gondwana began in the Silurian or Devonian (Metcalfe, 1991). The final suturing of southeast Asia to Asia probably took place by the late Triassic, well prior to the spread of the angiosperm flora.

Furthermore, Truswell et al. (1987), who investigated the paleobotany of the Australian-southeast Asian region, pointed out that the Noah's Ark idea is not supported by fossil evidence. The first angiosperms in northern Australia arrived there later than they did in other parts of the world. Even the Myrtaceae, a dominant Australian family, evidently had a pan-tropical distribution in the late Cretaceous and did not reach Australia until the Paleocene. So, the original theory of Takhtajan (1969), which identified an angiosperm center of origin in southeast Asia, appears to be viable.

Another indication of the importance of the southeast Asian area may be found in the history of the southern beech family Nothofagaceae which is closely related to the northern hemisphere family Betulaceae (Nixon, 1989). The genus *Nothofagus* holds a premier position in the study of southern hemisphere plant evolution and biogeography. Many have attempted to reconstruct its history. A review of our current knowledge by Hill (1992) suggests the existence of an early Cretaceous fagalean complex in south Asia. From there, ancestral lineages of Fagaceae and Betulaceae may have spread to the north while the ancestor of *Nothofagus* moved south to Australia. By the late Cretaceous, the latter lineage had reached New Zealand, Antarctica, and southern South America. So, *Nothofagus* probably evolved from its fagalean ancestor in that amphinotic area. It was never able to reach Africa or India.

Other ancient angiosperm families including the Sapotaceae, Arecaceae, and Ulmaceae, first appeared in the Cretaceous of southeast Asia before spreading to Australia and other parts of the southern hemisphere (Truswell et al., 1987). Although plant migration into Australia from the North has predominated, there is some evidence of northward migration of late Cretaceous-Tertiary floras, prior to the mid-Miocene collision with the southeast Asian archipelago. Examples are the Casuarinaceae, *Dacrydium*, and certain Proteaceae.

The relationships between the paleofloras of Siberia and Alaska have been investigated by Smiley (1979) who made comparisons of numerous fossils. He found that the Lower Lena River material from Siberia showed a similarity, in the Neocomian-Albian, of 80 to 90% with fossils from northern Alaska. He felt that this close relationship implied a land connection between the two areas. He also suggested that this land bridge (Beringia) had existed, at least from later Paleozoic time. It has generally been assumed, but not clearly shown, that a continental convergence took place in the Bering Strait region, starting in mid- to late-Cretaceous time (Hallam, 1981).

After the Cretaceous land flora became established, it remained more or less homogeneous through the Cenomanian (Tschudy, 1984). Subsequently, broad floral provinces began to emerge. By the end of the Cretaceous, several provinces could be recognized: a Malesian-Central Atlantic or Pantropical Province occupied the land areas adjacent to the ancestral Tethys, and an Australian-Antarctic Province occupied the south circumpolar area. Palynological data from North America and Eurasia indicate two, almost mutually exclusive, provinces: one in Westamerica-Asia called the Aquilapollenites Province and the other in Euramerica called the Normapolles Province. The presence of these provinces reflects the late-Cretaceous condition of two northern hemisphere continents, separated by the Mid-Continental Sea in the west and the Turgai Sea in the east (Map 12).

There are some floral data that provide information on the Cretaceous position and relationships of India. These have been reviewed by Chatterjee (1992) and suggest that India was not an island continent. *Aquilapollenites* pollen has been recovered from the Lameta Formation, a strong similarity to a middle Cretaceous assemblage from the Peace River in northwestern Canada has been noted, and a Mongolian charophyte genus has been identified from the Takli Formation.

It has been commonly assumed that the extraordinary diversity of living insects is due to the diversity of angiosperm plants (Berenbaum and Seigler, 1992). This means that the fossil record should reflect a simultaneous diversity increase in both groups or that the plant radiation came first. However, Labandeira and Sepkoski (1993) have shown that a simultaneous rise did not take place and, furthermore, that insect diversity may have actually decreased in the mid-Cretaceous when flowering plants were undergoing their greatest increase. The rapid, perhaps exponential, increase in insects took place from the Triassic to the early-Cretaceous, while the ascendancy of the angiosperms did not begin until the early Cretaceous. So insect diversity did not benefit from the radiation of flowering plants. But one may hypothesize just the opposite, that angiosperm evolution was stimulated by the presence of a great variety of insects.

Freshwater fauna

Among the freshwater fish, the archaic family Polyodontidae (paddlefish) is of considerable interest, for there are but two living species (each in a distinct genus), one in the Mississippi River system and the other in the Yangtze system in China. Two extinct genera are known, one from the late Cretaceous of Montana and one from the Eocene of Wyoming (Patterson, 1981). So, an Asian-American relationship, probably via Beringia, is indicated.

The bowfins (Amiidae) are also an archaic, preteleostean family. Included are a number of late-Jurassic and Cretaceous genera from Europe, South America, Africa, and Eurasia. One living species exists in eastern North America, but two fossil species were common to North America and Europe in the early Tertiary. The gars (Lepisosteidae) are another old, preteleostean group. The earliest fossils are fragments from the Lower Cretaceous of West Africa (Wiley, 1976). Other fossil materials indicate a widespread distribution in North America, South America, Europe, Africa, and India. There are two living genera occurring in the southeastern United States, Cuba, and Mexico to Central America. Eocene remains of both gars and bowfins have been found on Ellesmere Island, northwest of Greenland (Patterson, 1981). These patterns emphasize the importance of the North Atlantic connection (Briggs, 1987).

The order Cyprinodontiformes comprises a large group of about 900 species of teleostean fish commonly known as killifish, topminnows, or toothcarps. Traditionally, the order has included five families, the oviparous Cyprinodontidae which is widespread in the New and Old Worlds and the New World viviparous families Anablepidae, Goodeidae, Jenynsiidae, and Poeciliidae. As a whole, the group is considered to belong to the "secondary" freshwater fish category (Myers, 1938), since many of the species live in brackish water and some can tolerate very high salinities.

Parenti (1981) considered the cyprinodontiform fish to have been a widespread Pangaean group with an origin in at least the late Triassic. However, this hypothesis would place them in a preteleostean fauna. A later origin, possibly in the mid-Cretaceous, would still make it possible for them to reach Africa from South America over a narrow saltwater gap. These fish are primarily tropical and warm-temperate so probably did not disperse over a northern route. They have succeeded in reaching such places as Madagascar and the West Indies, so can negotiate modest saltwater passages. Generic diversity is highest in the New World (28), next highest in Africa (11), lower in the Near East (4), and least in southeast Asia (1). Considering also that all of the viviparous groups with their many reproductive specializations are endemic to the New World, it has been sug-

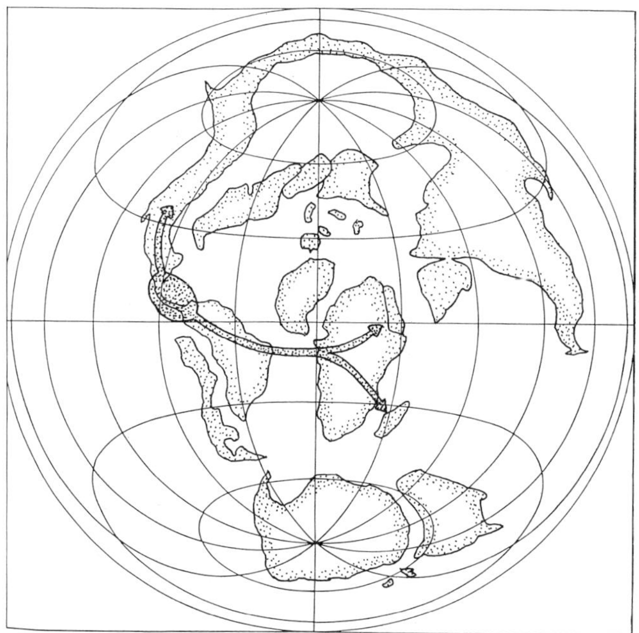

Fig. 39. Possible late-Cretaceous dispersal of cyprinodontiform fish following a mid-Cretaceous origin in the Neotropical Region.

gested (Briggs, 1984; Lundberg, 1993) that the cyprinodontiform fish originated in the New World, probably in the area from Mexico to South America (Fig. 39).

The family Cichlidae is another important group of secondary freshwater fish. The approximately 160 genera and 1,000 species are found primarily in Africa and the neotropics (South and Central America). Like the cyprinodontids, they have been able to invade the islands of Madagascar and the West Indies. The great majority of the genera and species are found in Africa, especially in the three great lakes (Victoria, Malawi, Tanganyika). Although Bănărescu (1990) has stated that ancestral cichlids were definitely present in Africa-South America before its breakup, this seems doubtful.

The Cichlidae is an advanced teleost (perciform) family. The earliest perciform fossil known is from the latest Cretaceous of North America, and the earliest cichlid is from the Oligocene of Africa (Van Couvering, 1982). As the marine distributions have shown, there is conclusive evidence of a continuous sea passage between Africa and South America in the Albian Stage of the mid-Cretaceous. Prior to that time, in the Aptian, intermittent marine incursions formed extensive evaporite deposits between the two continents. It is difficult to conceive of cichlids existing in the early Cretaceous, for no teleostean families have reliable early Cretaceous records (Patterson, 1993). The phylogenetic position of the Cichlidae suggests an origin in the late Cretaceous when the intercontinental separation was well developed. It appears that most of the evolution within the family took place in Africa and that it must have originated on that continent. Dispersal to South America probably took place over a modest saltwater gap during the late Cretaceous (Briggs, 1984; Lundberg, 1993).

Some of the fossil freshwater vertebrate fauna appears to indicate an early Cretaceous relationship between Africa and South America (Rage, 1988; Maisey, 1993). Among the fish, there were related genera of gars (Lepisosteidae), amiids, semionotids, coelacanths, and gonorynchiforms. Cretaceous fossils of the family Polypteridae, a living African group, have been found in Bolivia (Gayet and Meunier, 1991). Fossils of the lungfish *Neoceratodus africanus* have been reported from both continents. The Recent lungfish genera, *Lepidosiren* in South America and *Protopterus* in Africa, are related and were probably separated in the early Cretaceous. The pipid frog genus *Xenopus* occurred in the Cretaceous of both continents and still lives in Africa (Duellman, 1993). Two crocodilian genera were held in common. As noted by Maisey, the absence of any Lower Cretaceous records for the characoid and siluroid fish stands in contrast to their supposed Gondwanian origins.

Another old freshwater group with a probable Mesozoic dispersal history is the crayfish of the decapod infraorder Astacida. Two northern families have disjunct relationships between the Old and New Worlds. The Astacidae exists in Europe and the Near East as well as in northwestern North America (Hobbs, 1974). The Cambaridae is found in East Asia and in eastern North America plus Cuba. The southern family Parastacidae inhabits southern South America, Australia, New Zealand, and Madagascar. Supposedly, the northern intercontinental migrations took place via Beringia (Bănărescu, 1990). Except for the inclusion of Madagascar, the southern track of the Parastacidae is similar to those noted for the aquatic insects.

There are many other old groups of freshwater invertebrates potentially useful for biogeographic purposes, including additional groups of insects, many more crustaceans, mussels, gastropods, sponges, turbellarians, polychaete and oligochaete worms, leeches, and a variety of parasitic forms. Bănărescu (1990) has done a thorough job in summarizing the distributional data on these groups. It may be remiss not to give them more attention but, in most cases, their taxonomy is not as well understood as that of the groups covered in more detail.

All modern families of salamanders (Caudata) had probably emerged by the Cretaceous. Three families are known from the Upper Cretaceous and the remainder from the early Tertiary (Milner, 1983). Most of them are entirely or predominately restricted to one continent: four occur primarily in North and Central America, one in Asia, and one in Eurasia. The most primitive family, according to Milner's cladistic analysis, is the Sirenidae. It is restricted to southern and eastern North America. Next are two related families, the Cryptobranchidae and the Hynobiidae. The former is found as living species in eastern North America and East Asia and as Tertiary fossils in Europe and North America. The latter is confined to Asia, primarily East Asia. Milner considered these to be sister families with an origin in East Asia. The Cryptobranchidae probably spread eastward to North America via Beringia, and westward to Europe following the desiccation of the Turgai Sea in the Oligocene.

The salamander family Proteidae and its sister group, the extinct Batrachosauroididae, are probably Euramerican with an origin prior to the Upper Cretaceous (Milner, 1983). The former has living genera in Europe and in eastern North America. The remainder of the salamander families comprise the advanced "Neourodeles". The Amphiumidae, Dicampodontidae, and Ambystomatidae originated in and have remained endemic to North America. The Salamandridae probably originated in Europe but has two living genera in North America. The Plethodontidae evidently arose in North America but has one genus in Europe. The salamandrid dispersal probably took place via the North Atlantic connection and plethodontid migration via Beringia. The Cretaceous Mid-Continental Sea apparently had some effects on the distribution of various genera (as noted by Milner).

The caecilians (*Gymnophiona*) are wormlike creatures with no limbs or limb girdles. There are five families, three of which occur in South America with two being endemic. It has been surmised that most of the families evolved prior to the late Jurassic (Duellman and Trueb, 1986), but the single known fossil comes from the Paleocene of Brazil. The family distribution is primarily southern hemisphere, but three families occur in India. These were supposedly transported from Africa to Asia via the Indian plate. One group remained on the Seychelles when those islands became detached from India. One of the Indian families subsequently became more widespread in Asia. Also, a South American family dispersed to Central America, probably in the late Cretaceous. An African-South American origin for the caecilians, prior to the continental division, seems apparent.

As noted earlier, the evolutionary radiation of modern frog (Anura) families took place primarily in the southern hemisphere. The frog fauna of South America is incredibly rich with 18 families, 110 genera, and about 1334 species (Duellman, 1993). Seven of the families are endemic to the neotropics. In their attempt to trace the history of family distribution, Duellman and Trueb (1986) discussed the relationship of each southern land mass. They assumed that the frog fauna of Antarctica-Australia was definitely comprised of leiopelmatids and myobatrachids. Although the former surely reached New Zealand from South America via Antarctica, there are no indications that it ever existed in Australia. The myobatrachids, a relatively advanced family derived from the South American leptodactylids, probably dispersed to Australia at a much later date. The family Hylidae possibly followed the same route.

Before their separation, Africa and South America apparently shared a fauna of pipids, bufonids, leptodactylids, and microhylids (Duellman and Trueb, 1986). In addition, Africa apparently contained three more families (ranids, hyperoliids, and rhacophorids). From Africa, Madagascar-India received ranids, hyperoliids, rhacophorids, and microhylids. The four families had probably not yet evolved when Madagascar-India rifted from Africa in the mid-Jurassic. Their presence on India appears to indicate a later recontact with Africa. As India moved northward in the Cretaceous, it left behind the Seychelles islands populated with a hyperoliid frog genus and a stock which subsequently evolved into the family Sooglossidae. The presence of these animals, plus a caecilian, is strong evidence that the Seychelles were once part of India.

In the late Cretaceous, leptodactylids, bufonids, hylids, and microhylids dispersed from South to Central America (Duellman and Trueb, 1986). Additional intercontinental migrations took place in the Tertiary. In regard to the area of origin, it can be said that the anurans are certainly a monophyletic group. In the case of such groups that clearly arose in early Mesozoic times, there has been a tendency to assume that they were pangaean in origin and distribution. However, the fossil record and current biogeographic patterns do not support that argument. The older Laurasian frog groups are geographical relicts and are likely to have come from the south. The modern northern families have invaded from the south. There is also no reason to assume that all of Gondwana was involved. Virtually all the major steps in frog evolution took place in Africa-South America. That land mass evidently contributed frog taxa to the rest of Gondwana and to Laurasia. Instead of extremely broad areas, it appears that the centers of radiation for the lissamphibians may have included only Euramerica (salamanders) and Africa-South America (frogs and caecilians).

Higher vertebrates

Dinosaurs provide some indications that there may have been a late- Cretaceous, transtethyan route (Rage, 1988). The ornithopod dinosaur *Valdosaurus*, previously reported from the early Cretaceous of Europe, was discovered in the Aptian of Niger. A pachycephalosaurid dinosaur, known to be widespread in Laurasia, has been reported from the late Cretaceous of Madagascar. A late Cretaceous dispersal of tyrannosaurids apparently took place across the Beringian land bridge (Kelly, 1990). The early Cretaceous dinosaur fauna of Victoria, Australia, was dominated by small, endemic ornithopods (Rich and Rich, 1989). This archaic community indicates the presence of a barrier to other parts of Gondwana. For the latest Cretaceous, almost all dinosaur fossils are found in northwestern North America.

The definitive work on the fossil record and early distribution of lizards was accomplished by Estes (1983). The true lizards, or lacertilians, apparently underwent a dichotomy in the Jurassic when Laurasia and Gondwana began to separate. This vicariance produced a southern iguanian group and a northern group ancestral to all other lizards. In the Cretaceous, the remaining modern families appeared. The most primitive true lizards are placed in the family Iguanidae. This group initially occupied Africa-South America and, sometime in its early history, extended to Madagascar. The North American Iguanids appear to have resulted from dispersals from South America that began in the late Cretaceous.

In Africa, an iguanid group became isolated in the northeast region probably due to a barrier caused by an epicontinental sea. This "acrodont" group then evolved into the chamaeleonids in the west (Africa and Madagascar) and agamids in the east (India and southeast Asia). The latter were also able to reach Australia early in their history. The gekkonid group of lizards apparently originated in Asia. From there, one group (diplodactylines) became established in Australia and another (gekkonines) spread westward to India, Africa, and Madagascar. The latter eventually reached South America, probably by rafting (Estes, 1983).

While the iguanoids were radiating in Gondwana, the scincomorphs were diversifying in Laurasia. This development produced the lacertids in Europe and the teiids in North America. In the late Cretaceous, some of the teiid stock probably reached South America giving rise to the gymnophthalmids (Estes, 1983). The teiids also reached east Asia from North America. At about the same time, some of the lacertid stock reached India, possibly via north Africa. As Estes noted, the scincoids (Scincidae) are so diverse and cosmopolitan that it is difficult to determine their biogeographic history. They may have developed from an ancestral population in southeast Asia.

The origin and distribution of the anguimorph lizards is also not simple. By the late Cretaceous, all the known families were in North America but two groups (necrosaurs and varanids) were also in central Asia. This seems to indicate a long history in Laurasia. The necrosaurs may have originated in Asia, from varanoid ancestors, and dispersed across the Bering connection, together with the true varanids, during the late Cretaceous (Estes, 1983). Among the anguimorphs, only the varanids achieved a wide southern-continent distribution, so far as is known. They may have originally evolved as predators on small mammals prior to the evolution of snakes. The marine, varanoid mosasaurs were the most spectacular of all lizards. They are known only from the Upper Cretaceous, but nearly 20 genera are recognized (Carroll, 1988). The largest specimens exceeded 10 m in length. They thrived in the shallow waters of the epicontinental seas.

Snakes probably arose from an ancestral stock within the lizard groups. The oldest known fossil comes from the early Cretaceous. Nearly all Cretaceous specimens have come from the southern hemisphere, so snakes are usually considered to have originated somewhere in Gondwana (Rage, 1987). A dichotomy which produced two primitive groups, the boids and the aniliids, apparently took place in Gondwana. Subsequently, in the latest Cretaceous, the brief connection through Central America permitted both groups to enter North America. The succeeding events in snake evolution and dispersal took place in the Cenozoic.

The multituberculates were the most successful non-therian group of mammals. They are first known from the Upper Jurassic. By the Cretaceous, they had split into two groups, the Ptilodontoidea were primarily North American while the Taeniolabidoidea were predominately Asian (Carroll, 1988). Despite competition from the more advanced marsupials and placentals, the multituberculates continued to evolve well into the Tertiary. They were the longest-lived mammalian order, extending over 100 Ma to the Oligocene.

Very little is known about the distribution of the early therian mammals. Jacobs (1988), in describing fossil teeth from the early Cretaceous of Africa, suggested that such primitive mammals were on all continents before the breakup of Pangaea. Marsupials and placentals diverged from a common ancestor in the early or mid-Cretaceous. By the late Cretaceous, the distribution of the two groups showed significant differences. They were nearly equally abundant in North America and both occurred, in lesser numbers, in South America. But, at this time, marsupials were unknown in Asia while placentals were common. Neither group is known from the Cretaceous of Africa or Australia (Carroll, 1988). This pattern may be related to the existence of a widespread, Laurasian ancestor giving rise to the marsupials in Westamerica and placentals in Asia. With the advent of the Bering connection in the late Cretaceous, placentals may have been introduced to Westamerica to the eventual detriment of the marsupials. Mammal data do not support the existence of faunal links between South America and Africa during Cretaceous time (Bonaparte, 1990).

A summary of the terrestrial vertebrate exchanges between North and South America has been published by Gayet et al. (1992). They noted the following: in the Upper Cretaceous, hadrosaurid and ornithischian dinosaurs apparently migrated southward. A lepticitid mammal dispersed in the same direction in the Cretaceous or the Paleocene. More Cretaceous vertebrates traveled northward. These included a lepisosteid fish, a teiid lizard, boid and aniliid snakes, a titanosauroid dinosaur, and a caroloameghinid mammal. When one adds to this list, the probability that the northbound traffic included other old freshwater fish families, a caecilian, four families of frogs, and an iguanid lizard, the evidence for a late-Cretaceous isthmus becomes substantial. The hypothesis is reinforced by the recognition of distinct late-Cretaceous marine subprovinces of each side of a land barrier in the present location of Central America (Kauffman, 1973).

Australia

Despite the considerable amount of existing knowledge about the Cretaceous, the biogeographic relationships of two continents are inconsistent with almost all paleomaps based on geophysical determinations. These are Australia and India. The early Cretaceous relationships of Australia have been reviewed by Molnar (1992). Aside from the early Triassic, which was mentioned earlier, this is the only period for which more than two species of Australian terrestrial tetrapods are known. Eleven families of Cretaceous tetrapods are known from Australia, eight from South America, and none from India. Two of the Australian families are shared with Africa and at least one of them (probably both if fossil footprints are diagnostic) with South America. In contrast, Africa and South America share four of the families.

The fact that 33% of the Australian early Cretaceous families are endemic indicates, by itself, the presence of a barrier to the rest of the world. However, it is the relationship of the individual taxa that are the most revealing. One is a temnospondyl amphibian, a group that in the rest of the world did not live beyond the early Jurassic. Another is a monotreme, a group that may have become isolated in Australia by the early Jurassic (Carroll, 1988). The presence of *Allosaurus* means that it lived on in Australia at least five Ma after its disappearance elsewhere. The other Australian taxa also exhibit notable relict characters. Molnar (1992) suggested that some of the relicts entered Australia as early as the Permo-Carboniferous and others in the middle or late Jurassic. He concluded that none of them entered after the Jurassic.

Only a single tetrapod genus, the sauropod *Austrosaurus*, is known from the late Cretaceous of Australia and it belongs to a family not represented elsewhere at this time (Molnar, 1992). Freshwater lungfish of the genus *Ptychoceratodus* have been reported from the early and late Cretaceous of Australia as well as the late Cretaceous of South America and Madagascar (Rage, 1988). Another lungfish *Neoceratodus* occurred in the African and South American Cretaceous and still exists in Australia. The presumably, freshwater fish family Archaemaenidae was in both Australia and Antarctica.

Most of the foregoing information suggests that, during the Cretaceous, Australia did not form a terrestrial connection with other parts of Gondwana. Its isolation can probably be explained by the high stand of sea-level in the Cretaceous compared to the earlier Mesozoic. The continent would not necessarily have rifted from Antarctica, but became separated by a significant stretch of epicontinental sea. Large parts of Australia itself were submerged so the presence of a saltwater gap is quite probable. The presence of the freshwater fish may argue against this hypothesis, but many of the Mesozoic lungfish were marine so the genera discussed, as well as the archaemaenids, may have been euryhaline to the extent that a minor marine barrier would not have been significant.

India

One of the most challenging problems in historical biogeography is to provide a satisfactory account of the Mesozoic and early Cenozoic relationships of India. Ever since the classic depiction of continental drift by Dietz and Holden (1970), maps based on geophysical data have consistently shown India, as it moved northward from the Antarctic, to be an isolated continent located far out in the Tethys Sea. The departure from Antarctica and Africa supposedly took place in the mid-Jurassic about 165 Ma ago and the docking against Asia about 50 Ma ago. This would mean more than 100 Ma of isolation for India. If this had indeed happened, India would have developed a highly endemic fauna and flora.

As the result of work by Indian paleontologists (Sahni, 1984), it became apparent that the Cretaceous/Paleocene fauna of India did not reflect an extended isolation. This made it necessary to take another look at the paleoposition of India to see if it could be changed, yet still make sense in view of the geological features of the Indian Ocean floor. Chatterjee and Hotton (1986) then wrote a detailed article on the paleoposition of India. They suggested a predrift location adjacent to the Somali coast of northeast Africa. Briggs (1987) depicted a pangaean India lying far to the south between southeastern Africa and Antarctica. Subsequently, India was considered to have moved northward close to the African coast and to have been joined to northeast Africa (Somalia) in the late Cretaceous.

More recently, in a review of Indian historical biogeography (Briggs, 1989), Weijermars' (1989) version of global tectonics was adopted. This interpretation still considered predrift India to lie far to the south. However, for biogeographic reasons it was considered necessary for India to retain, during its northward movement, a close proximity to Africa. It was also noted that, in order to delineate a realistic India on a Mesozoic map, one must include a greatly expanded northern portion. During the collision with Asia, a continental shortening of 2600 ± 900 km occurred (Patriat and Achache, 1984). It was suggested that the Owen Fracture Zone, a north-south ridge that runs close to the tip of the Somali Peninsula, probably marked the track of India's western margin (Briggs, 1989). This would make possible the close relationship to Africa called for by the biological evidence.

A kinematic model for the evolution of the Indian plate has now been published (Chatterjee, 1992). Here, predrift India is shown close to Somalia and the Arabian Peninsula (Map 11). By the late Cretaceous, India had apparently moved counterclockwise and somewhat to the north. As it did so, the Seychelles Platform may have detached and Madagascar moved southward. While this movement by India caused a departure from Africa, it probably allowed greater India (the pre-Himalayan part) to make terrestrial contract with Asia (Map 12). Chatterjee proposed that, prior to India's main northward thrust, an extraterrestrial impact occurred on the west coast causing the detachment of the Seychelles and the basalt flooding of the Deccan Traps.

The proposed impact theoretically took place at the Cretaceous/Tertiary boundary about 65 Ma ago. Considerable evidence in favor of an impact at this time has been accumulated, but did it occur here or somewhere else? The case for this particular spot would be helped if a well-defined crater existed. Aside from the impact question, the movements of India as outlined by Chatterjee (1992) do appear to best fit the latest geological and biological data. The connection of Greater India to Asia in the latest Cretaceous could have interrupted the eastern end of the Tethys Sea, but if movement of India away from Africa had opened a new passage to the Indian Ocean, there could still have been an east-west continuity of the marine biota.

For the Cretaceous, some of the most compelling evidence for biological relationships has been published by Sahni et al. (1987) who investigated late Cretaceous-early Paleocene faunas from peninsular India. They found two frog families (Discoglossidae and Pelobatidae) which are essentially Laurasian. The discoglossid frog was determined to be closely related to a species from the Upper Cretaceous of Montana. Anguid lizards were found with close affinities to those of the Fort Union formation in Wyoming. Two dinosaurs belonged to genera that occurred in Madagascar. Prasad and Sahni (1988) described the first Cretaceous mammal from India. It represented a new genus with relationships to other genera in Morocco and North America.

Finally, as noted previously, India served as a Noah's Ark to transport a variety of animals from Africa to Asia. Apparently included were four frog families and three caecilian families (Duellman and Trueb, 1986). A fifth frog family may have been included, for an Eocene fossil from India belonged to the Myobatrachidae, a family that once inhabited Africa (Spinar and Hodrova, 1986). Possibly, a lacertid lizard was also a passenger (Estes, 1983). Davis (1979) suggested that the freshwater snails of the family Potamiopsidae may have been transported to Asia by the Indian plate.

Conclusions

The Cretaceous was a momentous Period in the history of the earth. With the exception of the mammals, this was the time when many of our modern families of animals and plants arose and when their basic distributions were determined. The making and breaking of continental connections were of great importance, although we still have a great deal to learn about their effects. The late Cretaceous closing of the Bering Strait provided a migration pathway for early mammals (Lillegraven et al., 1979); necrosaur, varanid, and teiid lizards (Estes, 1983); tyrannosaurid dinosaurs (Kelly, 1990); cryptobranchid and plethodontid salamanders (Milner, 1983); astacid and cambarid crayfish (Bănărescu, 1990); and polyodontid and lycopterid-hiodontid fish (Patterson, 1981). This was only the beginning of a migratory flood that continued through most of the Cenozoic.

The brief late-Cretaceous connection of North and South America via Central America had notable effects. It probably permitted hadrosauroid dinosaurs as well as condylarth and marsupial mammals to enter South America (Rage, 1988). Boid and aniliid snakes came northward (Rage, 1987); as did iguanid lizards (Estes, 1983); titanosaurid dinosaurs (Rage, 1988); and leptodactylid, bufonid, hylid, and microhylid frogs (Duellman and Trueb, 1986). This pathway was interrupted, probably in the Paleocene, but an island archipelago remained along which a number of Tertiary dispersals took place. Additional detail on these vertebrate exchanges has been provided by Gayet et al. (1992). Probable migratory traffic across Beringia and Central America is summarized in Table 4.

During the early Cretaceous, two important connections were broken. Africa and South America were separated, although, at first, only by intermittent epicontinental seas. This began an isolation for South America that produced a highly endemic fauna and flora. Africa, on the other hand, had more intercontinental contact (with Eurasia). The other severed connection was between Australia and Antarctica. This separation evidently lasted throughout the Cretaceous and helped to account for the survival in Australia of many phylogenetic relicts. India's initial connection to Asia may have taken place in the late Cretaceous, possibly followed by a hiatus until a final connection in the Eocene. Continuous fossil evidence is lacking.

In the marine environment, the two continental connections separated marine biotas that had been continuous for millions of years. In the North Pacific, this began the development of a rich, temperate biota that soon became different from and more diverse than that of the Arctic-Atlantic. In the New World tropics, the early isthmus divided the biotas into West-Central American and Antillian subprovinces within the Caribbean Province. The opening of the South Atlantic had profound effects, for it permitted an invasion of many species from the north. In contrast, the Cretaceous break between Australia and Antarctica had minor marine effects because the intercontinental separation was caused by a shallow sea that did not isolate the inshore biotas.

It has been proposed that the northward movement of India was responsible for the separation of the marine shelf faunas on the east and west sides of the Indian Ocean (Winterbottom, 1985; Hocutt, 1987; Springer, 1988). These two regions are currently placed in separate biogeographic provinces due to differences at the species level

(Briggs, 1974a). The basic flaw in the separation proposal is the idea that Cretaceous events could have been effective in the division of living species that are closely related. Species, especially in the marine tropics, simply do not live that long. The Cretaceous was the time when modern families, and perhaps a few long-lived genera, arose; almost no species are that old. Also, it now seems clear that India did not make an extensive voyage across the middle of the Tethys Sea, and thus could not have separated the biota in two parts.

TABLE 4

PROBABLE LATE CRETACEOUS MIGRATIONS

I. Beringia	Eastward	Westward
1. Freshwater fish		
Polyodontidae	X	
Hiodontidae	X	
2. Crayfish		
Astacidae	X	
Cambaridae		?
3. Salamanders		
Cryptobranchidae	X	
Plethodontidae		X
4. Dinosaurs		
Tyrannosaurus		X
5. Lizards		
Teiids...	X	
Necrosaurs	X	
Varanids	X	
6. Mammals		
Placentals?	X	
II. Central America	**Northward**	**Southward**
1. Freshwater fish		
Eight families (Bussing, 1985)	X	
2. Caecilians		
Caeciliids	X	
3. Frogs		
Four families (Duellman and Trueb, 1986)	X	
4. Lizards		
Teiids	X	
Iguanids	X	
5. Snakes		
Boids.	X	
Aniliids	X	
6. Dinosaurs		
Hadrosaurids		X
Ornithischians		X
Titanosaurids	X	
7. Mammals		
Lepticitids		X
Caroloameghinids	X	

CRETACEOUS EXTINCTIONS

Cenomanian-Turonian

At the close of the Cenomanian Stage, about 90 Ma ago, there occurred an unusual extinction event that correlates with a major sea-level rise, rather than a regression (Hallam, 1989b). The marine extinctions apparently took place in a series of steps, represented by narrow stratigraphic zones separated by intervals with little or no taxonomic change (Elder, 1989). The losses occurred along with a rise in the oxygen minimum zone and its progressive penetration into the epicontinental seas. Apparently some 35% of the molluscan genera, 70% of molluscan species, and 70% of the planktonic foraminiferans became extinct (Fagerstrom, 1987). Other taxa severely affected were ammonoids, bivalves, echinoids, dinoflagellates, malacostricans, and ostracods. On land, the tetrapod families lost very little diversity (Benton, 1988), indicating that there was not a notable terrestrial decline. The marine extinctions apparently took place over about a 2.5 Ma interval (Shoemaker and Wolfe, 1986).

Cretaceous/Tertiary boundary

For the past decade, the scientific and popular press have carried frequent articles about a catastrophic mass extinction that supposedly destroyed the majority of the earth's species, including the dinosaurs, approximately 65 Ma ago. Since 1980, more than 2000 papers and books have dealt with some aspect of a mass extinction at the Cretaceous-Tertiary (K/T) boundary. One authoritative estimate of the severity of the extinctions was that 60–80% of all the living species became extinct at this boundary (Raup, 1988). There appears to be a general acceptance of the fact that such a great catastrophe did occur. Most of the argument among scientists now is devoted to the determination of the cause.

Alvarez and Asaro (1990) provided evidence that a giant asteroid or comet committed this "mass murder," which they also referred to as a "sensational crime" killing half of all the life on earth. The other currently popular explanation is volcanism. A series of eruptions, over a period of tens of thousands to hundreds of thousands of years, has been said to have resulted in the demise of 60–75% of all species (Courtillot, 1990). Glen (1990) discussed these two hypotheses in an article entitled, "What killed the dinosaurs?"

The time-span problem

Let us first examine the evidence that extinctions were compressed into a small time span. Kauffman (1979b, 1984) has warned that, despite the general acceptance of a catastrophic terminal-Cretaceous extinction, this event has been poorly documented. He pointed out that, in the marine environment, more than 90% of the exposed boundary sequences of sedimentary deposits have major interruptions or intercalations of nonmarine sediments. The bigger the stratigraphic gap across the boundary, the greater the observed discontinuity between Cretaceous and Paleocene biotas, and the more sudden the apparent boundary catastrophe.

In a comprehensive review, Kauffman (1986) observed that most of the marine extinctions near the K/T boundary occurred in a series of steps, with earlier and more extensive extinctions among tropical taxa than among those in temperate areas. These extinctions apparently spanned a 2.50–2.75 Ma interval beginning before the boun dary and extending into the lowest Paleocene. Wiedmann (1986) also found a gradual decline, in most of the marine macroinvertebrate groups as time approached the boundary.

Other taxa, such as the marine reptiles, inoceramid bivalves, and some smaller groups of molluscs had lost most of their diversity several millions of years before the boundary. It has been noted that the ammonites, an important Mesozoic group, had only about 8–10 species extending into the late Cretaceous, and that the inoceramids had disappeared some 2 Ma prior to the boundary (Ward et al., 1991).

At one time it was thought that the marine planktonic groups had undergone an abrupt, catastrophic extinction (Zachos et al., 1989; Smit, 1990). However, Keller (1988; 1989) and coworkers (Keller and Barrera, 1990; Canudo et al., 1991) studied the planktonic Foraminifera from several continental slope to continental shelf sequences. These sections contained the most continuous sedimentation record known to date. Their studies have shown no trace of a mass extinction in this group. Up to one-third of the species disappeared below the K/T boundary and as many as one-third also survived into the early Tertiary. Keller and coauthors concluded that the K/T transition was not likely caused by a single instantaneous event, but rather by a set of complex and interrelated factors including changes in climate, sea level, and associated variations in ocean geochemistry. More recently, Olsson and Liu (1993) have argued that K/T extinction among the planktonic forams was more severe and geologically instantaneous.

Along the North Atlantic coastal plain in the United States, a recent study (Gallagher, 1993) demonstrated that the primary determinant of selective survival across the K/T boundary appeared to be a non-planktotrophic reproductive strategy. This seemed to be the case among molluscs, brachiopods, bryozoans, sponges, and corals. This observation may be consistent with the idea of a relatively sudden disappearance of planktonic taxa near the boundary. Another hypothesis states that a continuous volcanic event, beginning some 350 000 years prior to the boundary, may have produced so much CO_2 that the ocean surface became acidic to the extent that the calcareous plankton was eliminated (Hansen, 1990).

In the terrestrial environment, the emphasis has been on the supposedly sudden disappearance of the dinosaurs at the K/T boundary. But the results of research by several paleontologists have indicated that, instead of going out with a bang, the dinosaurs underwent a gradual loss of diversity over a long period of time. In their investigation of the Hell Creek Formation (Montana, North Dakota, South Dakota, and Wyoming), Sloan et al. (1986) found dinosaur extinction to be a gradual process that began 7 Ma before the end of the Cretaceous and accelerated rapidly in the final 0.3 Ma.

On the other hand, Sheehan et al. (1991) decided that the dinosaur extinction was sudden because their family diversity did not appear to drop during the final two to 3 Ma of the Cretaceous. But Dodson (1991), who had followed the generic diversity, noted that

73 genera were documented for the entire Maastrichtian Stage, but only 20 could be found in the late Maastrichtian. He concluded that there was an evident trend towards reduction in generic diversity. Some dinosaurs quite possibly remained alive as much as 3 Ma into the Paleocene (Van Valen, 1988).

The flying reptiles (pterosaurs) did not survive the end of the Cretaceous, but their extinction also did not occur in a catastrophic manner. They had been gradually diminishing in diversity for approximately 70 Ma throughout the Cretaceous (Carroll, 1988). By the end of that period, only a few members remained.

In regard to terrestrial vertebrates in general, the most plentiful fossils near the K/T boundary have been found in eastern Montana. The most recent analysis was made by Archibald (1992) who found an extinction rate at the species level of 35% to 42%. An earlier report by Archibald and Bryant (1990) had shown extinction rates that were slightly higher. Sheehan and Fastovsky (1992) had separated the freshwater and land-dwelling assemblages and also removed some rare taxa (due to possible sampling errors). The latter analysis indicated an 88% extinction in the land-dwelling groups, mostly dinosaurs and marsupials, and only 10% in the freshwater species.

As Sheehan and Fastovsky (1992) pointed out, it was the larger animals most directly dependent on primary production that suffered the most. Although these authors felt that this extinction pattern was compatible with the hypothesis of an asteroid impact, it, in fact, probably has a better relationship to the hypothesis of a gradual reduction in primary productivity due to climatic changes brought about by sea-level regression. There is no convincing evidence that the large land-dwellers underwent a sudden, catastrophic extinction. Data from this one locality may not be indicative of K/T events on a global basis.

It is interesting to compare the time span required for the extinctions with the time necessary for the recovery to normal diversity levels (when a recovery takes place). The restoration of the pelagic biota apparently took approximately 3 Ma (Lipps, 1986). Hansen (1988) studied the long-term effects of the K/T extinction on marine molluscs. Along the Gulf Coast of the United States, the boundary events reduced the diversity from approximately 500 species to just over 100. Afterward, the diversity slowly returned in a series of steps to approximately 400 species in the mid-Eocene. So it took almost 25 Ma for the diversity to build back to approximate its late-Cretaceous level.

The diversity problem

What proportion of Earth's species became extinct due to K/T boundary events? The data have come primarily from the marine communities of the shallow tropics, where almost all the extinctions occurred. But the answer to this question should not depend on data derived only from a few relatively nonspeciose groups sampled from the marine environment

The total number of marine metazoan species today is less than 200 000 (Briggs, 1994). In contrast, there are approximately 12 million species of terrestrial, multicellular animals. In addition to the animals, there are approximately 300 000 species of terrestrial vascular plants (Burger, 1981) compared to less than 100 marine species. Therefore, it seems that terrestrial organisms comprise at least 98% of the world's multicellular species.

There is little evidence that terrestrial species, except dinosaurs and marsupials (Sheehan and Fastovsky, 1992), suffered much extinction at the K/T boundary. Clemens (1986) noted that, among the other terrestrial vertebrates, all the families and genera of Cretaceous fish passed through the boundary and have been found in Paleocene or later deposits. All but one of the amphibian families and most of the genera did the same; so did all ten lizard families and most of their genera, both snake genera, and three of the four crocodilian genera. In the mammals, there was no reduction in generic diversity among the multituberculates and placentals (Savage, 1988).

The mammals began to radiate strongly in the late Cretaceous, well before the extinction episode (Benton, 1990b). Archibald (1991) observed that the reduction in marsupial diversity was correlated with the introduction of new placental mammals, and that the idea of massive extinction among the vertebrates was questionable.

For insects, little evolutionary change across the K/T boundary has been observed (Whalley, 1987). Briggs et al. (1988), in a review of arthropod diversity through time, found no evidence of a major insect extinction at the end of the Cretaceous. In their study of family diversity of insects through time, Labandeira and Sepkoski (1993) found no evidence of an end-Cretaceous decline. These findings are particularly important when one considers that more than 80% of all metazoan species are insects. If the insects did not suffer much from the K/T extinction, the global diversity could not have suffered very much.

In the angiosperm flora, there was a decrease in certain pollen species across the boundary and a corresponding increase in fern spores (Tschudy et al., 1984). Some megafossil evidence seemed to point towards a massive kill and recolonization in North America (Upchurch and Wolfe, 1987). However, Upchurch (1989) noted that some clades had become extinct in the Tertiary after declining in abundance at the boundary. Others had shown notable declines in the Maastrichtian, indicating some kind of gradual environmental change. In Australia, White (1990) could find no evidence of any K/T mass extinction in plants.

In Antarctica, there were no abrupt extinction events among the plant species, only a gradual reduction in diversity into the Paleocene which was related to climatic cooling (Truswell, 1990). From palynological research in North America, Sweet et al. (1990) found evidence of five sequential changes that occurred before and after the K/T boundary, and observed that such a pattern could not be explained by a single, extraterrestrial impact. But Nichols (1991), also on pollen evidence, found signs of an abrupt extinction that produced a major change. Although the angiosperm flora in North America may have suffered a reduction in biomass, there are no indications of a major decline in species diversity. The diversity (species richness) curve by Niklas (1986) shows no K/T interruption in the continuing trend towards greater diversity.

The K/T extinctions were severe in the tropical marine habitat. The peculiar bivalves called rudists disappeared, with their major extinction having taken place well before the boundary (Kauffman, 1979b). Two of the families and approximately 60% of the genera of scleractinian corals, along with a multitude of reef-inhabiting organisms, became extinct (Fagerstrom, 1987). The ostracods lost much of their diversity (Benson et al., 1985), as did the calcareous nannoplankton and foraminiferans. In addition, the ammonites, which had been in a gradual decline, became extinct. The great marine reptiles of the epicontinental seas, the plesiosaurs and the mosasaurs, were lost. The losses were not entirely confined to the tropics. Among the bivalve molluscs, with the exception of the rudists, extinctions also occurred at higher latitudes (Raup and Jablonski, 1993).

It appears that the marine metazoans approximately doubled their species diversity in the Cenozoic (Signor, 1990). The terrestrial, vascular plants and the terrestrial arthropods also may have approximately doubled their diversity (Niklas, 1986). Assuming that the metazoan species diversity of the late Cretaceous, on the land and in the sea, was approximately half that which exists today, marine species would have comprised less than 2% of the total species. Even if the K/T extinction destroyed half of all the marine species, the decline in the global species diversity would have been less than 1%.

But much of the marine fauna did not become extinct. In some places as many as one-third of the marine planktonic foraminiferan species survived the boundary (Canudo et al., 1991). Widmark and Malmgren (1988) found that benthic foraminiferans in the deep sea were scarcely affected by the boundary crisis. Of the species from the middle and lower slopes of the Central Pacific and South Atlantic, 75–90% survived the boundary. In the Antarctic, there was no significant drop in diversity across the K/T boundary (Thomas, 1990).

Extinction causes

In recent years, the almost exclusive focus on high-energy events, such as comet impacts and volcanic eruptions, has obscured the cause of the K/T extinctions. An extinction series sustained up to several Ma would require a continuous rain of impacts. There is only one iridium spike at the K/T boundary and, aside from the argument about whether it was produced by a volcano or a comet, a single high-energy event will not account for the evident extinction pattern. No single impact, by itself, could have been responsible for the extinction of the dinosaurs or the other organisms.

Towards the end of the Cretaceous, there was a significant global fall of sea level (Hallam, 1992). This regression may have been caused by continent uplift, a reduction of plate movement and shrinkage of the mid-ocean ridges, or both processes. Formation of ice sheets, a factor in other regressions, could not have been involved because the climate was too warm.

At approximately the same time, the major volcanism episode of the Deccan Traps took place in India. Judging from the volume of basaltic material produced, the Deccan eruption may have been among the most violent in Phanerozoic history. The resulting aerosol and dust veils must have caused temperature drops that may have each lasted several years. These short-term cooling events were followed by a longer-term warming due to the volcanically produced CO_2, which could only be removed by the slow weathering process. Hsü (1986) estimated that the warming phase lasted 30 000–50 000 years. The sulfate aerosols ejected by the large-scale eruptions probably produced acid rains that affected the surface of the land and sea.

The sea-level regression and volcanically induced events produced important changes in both marine and terrestrial environments. In the oceans, as the sea level receded towards the edges of the continental shelves, the nutrient materials brought in by the rivers began to be transported to the deep sea, the nutrient recycling process became less efficient, and epicontinental primary production dropped. The decrease in ocean surface temperature had its most devastating effect on the tropical biota, which is more stenothermic than that of the higher latitudes. In addition to having a direct effect on the tropical biota, the temperature decrease stimulated the oceanic thermohaline circulation, which increased the nutrient supply to the surface.

As predicted by the Hallock (1987) model, the increased nutrient supply probably had a detrimental effect on organic diversity. Primary production in the offshore environments must have increased, but probably not on the continental shelves. The intricate reef ecosystems and other tropical communities suffered great damage. They had to rebuild over the next several Ma from surviving remnants and from eurythermic organisms that invaded from cooler waters to the north and south.

In the terrestrial environment, with regression of the sea, a much greater land area was exposed and the continents stood higher. Both of these factors affected the prevailing weather patterns. Large continental masses, with high elevations relative to sea level, tend to develop dry seasonal climates over much of their central areas. Sometimes, depending on latitudinal position and prevailing winds, extensive deserts will form. In general, one may predict a drier climate with greater seasonal temperature extremes. This gradual climatic change, when combined with the short-term temperature drops and acid rain produced by volcanic activity, must have curtailed plant growth to the extent that primary production fell to a substantial fraction of its former level.

In the flora, it appears that many of the major changes were gradual as angiosperms replaced gymnosperms over the tropical and temperate parts of the globe. The former are better adapted to drier habitats, so the end-Cretaceous climatic change probably aided the evolutionary transformation. But there is also evidence of a sudden, widespread destruction of the higher plant life in North America over a small time interval at the K/T boundary. This destruction may have been the direct result of acid rains produced by volcanic activity. Sufficient vegetation appears to have survived to support the small vertebrates and invertebrates but not the great dinosaurs.

Conclusions

Most current articles dealing with the K/T extinctions focus on the argument between the impact hypothesis and the volcanic hypothesis. Underlying both scenarios is the implicit assumption that the end of the Cretaceous was marked by a catastrophic mass extinction. Terms such as *mass murder* and *mass kill* give the impression of a devastated globe where only small organisms in protected habitats managed to survive. How is it that such sensational, tabloid journalism has invaded the normally conservative scientific press?

The fossils have an interesting story to tell about the time over which the K/T extinctions took place, even though it is less spectacular than most current accounts. In both the marine and terrestrial environments, extinctions in major groups of animals began several Ma before the K/T boundary. For example, the ammonites began to die out approximately 6 Ma before and the dinosaurs approximately 7 Ma before. In the shallow, tropical marine waters, where the majority of the extinctions took place, the macroinvertebrates died out over a 2.50–2.75 Ma interval. The pelagic plankton species that disappeared did so over a shorter time, but their demise still took place over a period of hundreds of thousands of years. These changes were certainly not sudden nor catastrophic.

The magnitude of the K/T extinctions has been consistently exaggerated. Recent estimates range from a kill of half the species on Earth on up to 80%. Such estimates have no scientific basis. Some 98% of living species belong to groups (terrestrial arthropods, vascular plants, and nematodes) for which we have virtually no extinction information. They may have passed through the K/T boundary relatively unscathed. Most of the high-latitude and deep-water marine groups lost relatively few species. The loss of species diversity on a global basis could have been less than 1%.

There is only one fundamental global event that occurred gradually at a rate consistent with the tempo of the K/T extinctions: a regression of sea level probably caused by a cooling and shrinking of the mid-ocean ridges. The draining of the huge continental seas greatly reduced primary productivity on the continental shelf. The same process allowed the continents to become larger and to stand higher, creating deserts and dry, seasonal climates over large areas and greatly reducing the terrestrial primary production. The ultimate cause of the extinction of the large terrestrial and marine animals was probably a drastic decrease in primary production, which formed the base of the food pyramid in both environments.

The diversity decrease in the marine plankton and the drop in the terrestrial higher-plant biomass apparently occurred over a short period of time (hundreds of thousands of years). This was a time of major volcanic eruptions in India, which possibly caused fluctuations in atmospheric temperature and the production of acid rains. The short-term effect of volcanic dust and aerosols is to reduce global temperature, whereas the long-term effect of volcanic CO_2 production is to raise the global temperature. The acid rains could have had destructive effects on the surface of both land and sea. It is quite possible that the earth was struck by an asteroid at about the same time (Alvarez and Asaro, 1990). If so, its effects would have been added to those of the volcanic eruptions.

One may conclude that the K/T extinction episode was not a global catastrophe, since it had little effect on terrestrial diversity. Its destructiveness was largely confined to the shallow waters of the tropical oceans.

SUMMARY

1. The Cretaceous was a time of extensive inland seas, a moist tropical climate, and a high CO_2 level. These conditions permitted a high level of primary production on land and in the inland seas. It provided fodder for a large mass of herbivores which, in turn, permitted the evolution of a variety of carnivores.
2. A series of continental movements took place which, to a large extent, established biogeographic patterns that are still evident in our modern floras and faunas. These movements resulted in the completion of the Tethys Sea, the separation of

Africa and South America, the formation of Beringia, the isolation of Australia, and the rift of Africa from Europe.

3. Distribution patterns in the shallow seas are best illustrated by Kauffman's (1973) work on bivalve molluscs. Subsequent works on other marine invertebrates have tended to corroborate those patterns. They provided important information on the timing and subsequent effects of the continental movements.
4. Marine fossils appear to indicate the establishment of an early Cretaceous sea lane extending all the way from southwestern South America, across the South Atlantic, and up the African east coast. These data suggest a separation of Antarctica-Australia from the rest of Gondwana at an earlier time than usually recognized.
5. Fundamental changes in the world's flora took place in the Cretaceous. The earlier Mesozoic complex of conifers, cycadophytes, and ginkophytes began to be replaced, in the Aptian-Albian, by angiosperms. By the Upper Cretaceous, angiosperms became the dominant group of higher plants. Their geographic origin has been the subject of considerable speculation. Modern distributions suggest the southeast Asian area.
6. Floral patterns for the early and middle Cretaceous consisted of several latitudinal belts that were probably temperature controlled. By the late Cretaceous, four provinces could be recognized: a broad Pantropical Province, a southern Australian-Antarctic Province, an Aquilapollenites Province for Westamerica-Asia and a Normapolles Province for Euramerica.
7. Some archaic freshwater fish groups demonstrate a late-Cretaceous relationship across the North Pacific while others may have extended across the North Atlantic. The secondary freshwater fish order Cyprinodontiformes probably originated in the New World tropics and reached Africa in the mid-Cretaceous over a relatively narrow saltwater gap. The family Cichlidae probably arose in Africa in the late Cretaceous, then reached South America by negotiating a larger seawater gap.
8. Salamanders are primarily a Laurasian group. Four families occur mainly in North and Central America, one in Asia, and one in Eurasia. Their general areas of origin and subsequent dispersals demonstrate both North Atlantic and North Pacific connections. The Cretaceous Mid-Continental Sea affected the distributions of some of the genera.
9. The caecilians are primarily a Gondwana order, with three out of the five families occurring in South America. Two of the South American families are endemic. Three families occur in India, but they supposedly got there from Africa when India was close by. One family is found on the Seychelles, and was probably stranded there when that island group broke away from India.
10. The frogs are also primarily a Gondwanan order. Almost all the older families existed in South America-Africa. North America and Australia-New Zealand apparently received their early frogs from South America. India received its four families from Africa. Two frog stocks were left on the Seychelles when those islands were rifted from India. The frogs of Madagascar are also related to those of Africa.

11. Although Cretaceous dinosaur distribution generally illustrates the Laurasian-Gondwana dichotomy, there is evidence of some interhemispheric traffic. In the late Cretaceous, hadrosauroid and ceratopsian dinosaurs reached South America from the north by means of an isthmian connection, while the titanosaurids went the opposite direction. Two European genera apparently reached Africa by a transtethyan route. The dinosaur fauna of Australia comprised an archaic community indicating that it may have been isolated from the rest of the world.
12. The early history of lizards seems to indicate the initial presence of a Gondwanan iguanian group and a basic Laurasian group which was ancestral to all other lizards. The most primitive true lizards belong to the family Iguanidae, a group that arose in Africa-South America. Subsequent dispersals involved Central American, North Pacific, North Atlantic, and African-Indian connections. Snakes probably evolved in the southern hemisphere from a lizard ancestral stock.
13. Marsupial and placental mammals diverged from a common ancestor in the early or mid-Cretaceous. By the late Cretaceous their distribution patterns showed considerable differences. They were nearly equally abundant in North America and both occurred, in lesser numbers, in South America. But, at this time, marsupials were unknown in Asia while placentals were common. Neither group is positively known from the Cretaceous of Africa or Australia. This pattern may be related to the existence of a Laurasian ancestor giving rise to the marsupials in Westamerica and the placentals in Asia. With the advent of the North Pacific connection, placentals may have been introduced to Westamerica, to the eventual detriment of the marsupials.
14. In Australia, the high incidence of endemism involving geographical and phylogenetic relicts indicates that, during the Cretaceous, it was cut off from the rest of the world. There were apparently no animal migrations into Australia during that Period. The isolation was probably caused by epicontinental seas rather than tectonic movement.
15. Maps based on geophysical data have consistently depicted Cretaceous India being situated far out in the Tethys Sea. Almost all the biogeographic evidence indicates that this could not have been the case. Instead, India was probably located adjacent to the Somali Peninsula in northeast Africa.
16. The Cenomanian-Turonian extinction occurred during a major sea-level rise. Significant losses took place in the marine environment, but there was not a notable decline among terrestrial animals and plants.
17. The Cretaceous-Tertiary extinction episode, which took place about 65 Ma ago, had severe effects on the tropical marine habitat. Some groups, such as the rudists, inoceramids, ammonites, and the great marine reptiles (plesiosaurs and mosasaurs) became extinct. Others sustained severe losses. In contrast, the deep-sea and high-latitude faunas suffered very little.
18. On land, there is very little evidence that terrestrial species in general suffered much extinction at the K/T boundary. There is no indication that diversity among the insects, which comprise more than 80% of the total metazoan species diversity, suffered any decline. Among the vertebrates, which comprised a small fraction of the total species diversity, the losses were much higher among the land-

dwellers compared to the freshwater forms. It was the larger land-dwellers, with individually high demands on primary productivity (dinosaurs, marsupials), that suffered the most.

19. Considering all the affected groups, marine and terrestrial, it is apparent that the extinction process did not occur with catastrophic suddenness. The losses generally took place over periods of about one to several Ma. The time spans required for the recovery of diversity were even longer, ranging from about 3 to almost 25 Ma.
20. There was only one fundamental global event that occurred gradually at a rate consistent with the general tempo of the K/T extinctions: a regression of sea level probably caused by a cooling and shrinking of the mid-ocean ridges. The draining of the epicontinental seas and the associated climatic changes on land drastically reduced primary production. This production formed the base of the food pyramid in both environments. The large animals could therefore no longer be supported. The smaller forms survived because their demands on primary production were individually less.
21. Associated factors, which may have placed additional pressure on the biota, could have been volcanic eruptions and an asteroid impact. Some of the fossil materials (marine plankton and vascular plants) appear to indicate relatively short-term extinctions that are possibly attributable to such events. But, in most cases, the losses were gradual and extended over periods of millions of years.

CHAPTER 6

Paleogene

Almost all present-day insect families and many genera were well established at the close of the Cretaceous Period while many mammals and birds, on the distribution of which many prevailing zoogeographic conclusions are based, were only in the very early stages of their evolution.

G.W. Miskimen, *Zoogeography of the Coleopterous Family Chauliognathidae*, 1961

PALEOCENE

The fall of eustatic sea level at the end of the Cretaceous resulted in the disappearance of the epicontinental seas. This factor, plus the continued movement of tectonic plates, produced significant changes in continental relationships. In the northern hemisphere, the desiccation of the Mid-Continental Sea produced a single, huge continent, a Eurasiamerica. It was traversed by a single marine passage from the Arctic Ocean to the Tethys Sea, the Turgai Sea (Map 13).

India pivoted away from the African coast and its northern margin (the pre-Himalayan Greater India) probably contacted Asia (Chatterjee, 1992). This created a block to the eastern end of the Tethys Sea but, at the same time, a passage to the Indian Ocean was probably opened by India's eastward movement.

The northern and southern parts of New Zealand became widely separated from Antarctica with the rifting of the Campbell Plateau in the late Cretaceous (Lawver et al., 1992). This separation may have taken place earlier. Kauffman (1979b) found that, by the Cenomanian of the mid-Cretaceous, the New Zealand marine bivalve fauna had become highly distinctive. By the early Tertiary, New Zealand was broadly separated from both Antarctica and Australia. Smith (1984) argued that, throughout the Tertiary, New Zealand and Australia were as widely separated as they are today.

In the previous chapter, we noted that Australia was probably separated from the remainder of Gondwana throughout the Cretaceous and that the cause was possibly the high, eustatic sea level. Magnetic anomaly data appear to indicate that a rift zone began to open between Australia and Antarctica in the late Jurassic and that deep sea conditions extended along most of the rift by 80 Ma ago (Cande and Mutter, 1982). So, it seems that, during the high sea stand of the Cretaceous, a tectonic separation was underway. By the time of the major regression at the close of the Cretaceous, the two continental platforms had separated to the extent that a subaerial connection was probably still not available. However, Tasmania and the South Tasman Ridge formed island links that may have permitted some intermigration in the Paleocene.

In regard to Antarctica, it is probable that it drifted to the southwest as the separation from Australia took place (Scotese, 1991). At the same time, Africa moved to the north, closer to Europe and Asia. These movements, together with the development of the Antarctic Peninsula, made possible a closer relationship between Antarctica and South America. The isthmian connection between South and North America (via Central America), which formed an important migratory corridor in the late Cretaceous, became reduced to an island archipelago by the late Paleocene.

Marine patterns

In the shallow, tropical seas, the Paleocene was a time of recovery from the extensive extinctions near the K/T boundary. Among the pelagic plankton, several cool-water, high-latitude groups migrated to lower latitudes to replace the tropical forms that had become extinct (Gerstel et al., 1986; Lipps, 1986). In the planktonic foraminiferans, the earliest Paleocene fauna is comprised mostly of Cretaceous species that survived the boundary event (Canudo et al., 1991). For about 50 000 years, the new Tertiary fauna remained very sparse. This period was followed by one characterized by high faunal turnover and short-lived species. The latter were small, unornamented, and tended to be cosmopolitan in distribution. In the later Paleocene, they were replaced by larger and longer-lived forms. The restoration of the pelagic biota as a whole took approximately 3 Ma (Lipps, 1986).

In contrast to the pelagic community, the recovery of species diversity in the benthic habitat was apparently much slower. Along the Gulf Coast of the United States it took nearly 25 Ma for the molluscan diversity to build back to near its late-Cretaceous level (Hansen, 1988).

Near the end of the Paleocene, about 57 Ma ago, a rapid global warming and associated oceanographic changes caused a major extinction in the benthic fauna of the deep sea (Kennett and Stott, 1991). The event affected communities below 100 m and resulted in a 35–50% reduction in the foraminiferan taxa. Although the data came primarily from core samples in Antarctic waters, similar extinctions were noted in other parts of the world. Apparently the cause was a rapid global warming which produced an essentially isothermic water column. The high surface temperatures in the polar regions interfered with the normal thermohaline circulation and, with it, the oxygen supply to the deep sea. At the same time, warm, highly saline water from the Tethys Sea may have raised the temperature of the deep sea. During the time of the extinction, surface temperatures in the Antarctic had increased to about 18°C.

Terrestrial patterns

In North America, where the terrestrial vegetation underwent considerable destruction during the K/T extinction, a period of recovery took place which resembled a modern secondary succession (Upchurch and Wolfe, 1987). Precipitation levels greatly increased at low and middle paleolatitudes, and initiated widespread development of multistratal rainforests. Vegetation at high-middle latitudes was deciduous during much of the Paleocene. This was apparently due to the extinction of broadleaved evergreen taxa. However, such changes were not observed in Australia (White, 1990) nor in Antarctica (Upchurch, 1989) where there were no indications of a K/T extinction among the higher plants.

In the previous chapter, it was noted that some dinosaurs quite possibly remained alive as much as 3 Ma into the Paleocene (Van Valen, 1988). This possibility was investigated in detail by Charig (1989). He emphasized first, that in a cladistic classification, it is obligatory to include the birds within the theropod dinosaurs. Therefore, to a cladist, birds *are* theropod dinosaurs and, since birds are alive today, the dinosaurs never did become extinct. Next, Charig provided information about various places in the world where remains of the large dinosaurs occurred together with Paleocene fossils. He concluded that there was a modest amount of reliable evidence suggesting that in certain regions (notably Montana, New Mexico, Texas, Provence, and India) some large dinosaurs did survive into the Paleocene.

Although the Paleocene was a relatively brief 10 Ma stage at the beginning of the Tertiary, it was a time of faunal dispersals that had significant biogeographic and evolutionary consequences. Two intercontinental connections, first made in the late Cretaceous, were utilized by a variety of terrestrial and freshwater organisms. These were the Bering Land Bridge (Beringia) and the isthmian passage from South to North America. Also, the apparently sporadic connections between Africa and Europe were probably continued into the Paleocene. The North Atlantic connection became narrower but was probably uninterrupted.

In regard to the isthmian connection, it is possible that some of the migratory traffic by snakes, lizards, frogs, and mammals, described for the late Cretaceous, may have actually taken place in the Paleocene. The Paleocene separation of some of the hylid frogs is supported by immunological data (Duellman and Trueb, 1986). And iguanid lizards may have entered North America from the south during the Paleocene (Estes and Báez, 1985). In almost all cases, the herpetofaunal interchanges between North and South America were not accomplished directly, but through the medium of Central American radiations (Vanzolini and Heyer, 1985).

Freshwater fish invaded Central America from the south in late Cretaceous-Paleocene times (Bussing, 1985). This Old Southern Element included representatives of the families Characidae, Gymnotidae, Pimelodidae, Cyprinodontidae, Anablepidae, Poeciliidae, Atherinidae, and Cichlidae. The killifish family Poeciliidae, a secondary freshwater group, underwent a significant evolutionary radiation in Central America. Bussing, after studying the fish distribution patterns, concluded that substantial portions of the Central American archipelago were in place by Upper Cretaceous/Paleocene times.

Freshwater fish also exploited other intercontinental connections. The oldest known specimens of the primary freshwater family Catostomidae (suckers) are from the Paleocene of Alberta (Cavender, 1986). While the catostomids have undergone a significant radiation in North America (10 genera and about 55 species), they are known elsewhere from only a single extant genus in China. According to a cladogram published by Bănărescu (1990), the Chinese genus is the most plesiomorphic of the family. Since the family itself is part of a cypriniform group that originated in East Asia, the early catostomids were probably widespread and entered North America at least as early as the

Paleocene. Their decline in east Asia may be due to competition from the cyprinids, which diversified later.

Another primary freshwater family, the Ictaluridae, also has an extensive North American record beginning in the Paleocene (Cavender, 1986). This catfish (Siluriform) family is endemic to North America but it appears to be fairly closely related to the Asian-African family Bagridae. The present range of the bagrids extends north to the Amur River in Siberia. A bagrid catfish, or a form ancestral to the Bagridae and Ictaluridae, probably crossed Beringia in or prior to the Paleocene.

The freshwater fish family Esocidae (pikes) has a relatively good fossil record. The earliest fossils are from the Cretaceous of North America (Wilson et al., 1992) and the Oligocene of Europe (Cavender, 1986). There are five living species, one with a Holarctic distribution, one endemic to the Amur River basin in eastern Asia, and three confined to eastern North America. The presence of the oldest fossils and four of the five living species in North America indicates a probable origin in that area with subsequent dispersals to Asia. The species *Esox reicherti*, confined to the Amur River, may represent an early migration across Beringia and *E. lucius*, with its Holarctic range, a later one (Briggs, 1987).

The archaic fish family Hiodontidae has two living species in eastern North America, but the family also includes an extinct genus with three or four species of Paleocene to Oligocene age from western North America. The closest relatives of the hiodontids are the lycopterids, an extinct family from the late Jurassic and early Cretaceous of China and Siberia (Patterson, 1981). It is likely that a relative, perhaps an ancestor to the North American hiodontids, traversed Beringia in the late Cretaceous or early Paleocene.

The order Percopsiformes is another archaic group that includes three families, the Percopsidae, the Aphredoderidae, and the Amblyopsidae. All three have living species in eastern North America. This order is evidently related to the fossil family Stenocephalidae which is known from the Upper Cretaceous of Europe. Although the family is known from marine deposits, it is thought to be ancestral to the North American freshwater families (Nelson, 1984). The percopsids have an extended fossil record beginning in the Paleocene of Alberta (Cavender, 1986).

Although migrations across Beringia by various aquatic insects must have been numerous, only one such history has been worked out. Among the Trichoptera (caddisflies), a primitive generic line (*Sortosa*) originated in Asia, migrated to North America in the late Cretaceous, then to South America where it produced a descendent line. The derived genus (*Chimarra*) then migrated to North America, went back across the land bridge to Asia in the Paleocene/Eocene, and then in the late Eocene returned again to North America (Ross, 1958).

A fascinating tale of dispersal and replacement in the turtles has been told by Rosenzweig and McCord (1991). It seems that the Mesozoic Amphichelydia comprised a primitive group that were unable to tuck their heads and neck inside their shells. In the northern hemisphere, they were replaced by the Cryptodira, a group that had devised a method of retracting their heads and necks by vertical flexure. The earliest known cryptodires are from the Upper Jurassic of England. They are known from the Upper Cretaceous of North America and Asia. So, they possibly originated in Europe then spread to North America and Asia.

In western North America, the replacement of the Amphichelydia proceeded gradually at a rate of 8% or 9% per 5 Ma until the K/T boundary. The extinction event was associated with a 24% replacement. Although the pace of the replacement was accelerated, turtle diversity across the K/T boundary showed no significant change.

In the meantime, in the southern hemisphere the Amphichelydia were being replaced by the Pleurodira, a group of turtles that had developed a method of hiding their heads and necks under the sides of their shells (as opposed to a vertical flexure). In both North and South America, the replacement process went slowly, for the Amphichelydia held on until the end of the Eocene. But in Australia, fossils of both the Pleurodira and the Amphichelydia appeared in the Miocene. One might guess that they arrived there from South America in the Eocene along with the marsupials, but this is only conjecture.

In Australia, the Amphichelydia survived in abundance until they were finally replaced by the pleurodires in the Miocene. But, the former lived on Walpole Island near New Caledonia until probably extirpated by humans in the late Pleistocene. The fossils tell us that the amphichelydians were replaced by their more advanced relatives at different times in various parts of the world. The replacement also occurred in Africa but the timing is unknown. It illustrates that the replacement of a widespread group by its successor is often a lengthy process. In the case of the turtles it took about 150 Ma.

Among the placental mammals, there appears to have been a worldwide similarity of faunas, suggesting a high degree of cosmopolitanism and continental interconnection for the early and middle Paleocene (Gingerich, 1985). But the late Paleocene is a different story. Considerable endemism developed in North America, Europe, Asia, and South America. The faunas of Asia and South America became especially distinctive. These are indications that migrations via the connections of the North Atlantic, Beringia, and Central America had become difficult for intercontinental traffic.

Although such migrations had become difficult, they were not impossible. Three groups of highly specialized mammals (belonging to the orders Edentata, Notoungulata, and Dinocerata), known from the Paleocene of South America, appeared in the late Paleocene of North America (Gingerich, 1985). This dispersal apparently took place during a worldwide warming trend which made it possible for these tropical animals to continue far to the north. After invading North America, they must have proceeded across Beringia for all three groups have been discovered in later Paleocene deposits of East Asia.

The Beringia connection was also exploited by other groups of mammals (Webb, 1985a). In the mid-Paleocene, three genera of the order Acreoidi, known earlier in Asia, appeared in North America. There also appeared as many as three genera of Pantodonta, known from an earlier and richer record in Asia. By the late Paleocene, the first North American unintatheres, remarkable horn and sabre-bearing ungulates, appeared following an earlier and more diverse origin in Asia. Also found at this time, was a plesiadapid primate genus previously known in Europe. These new additions, plus those from South America, served to augment the products of an early Paleocene radiation of North American endemic mammals.

Another event, important in mammalian evolution, took place in the latest Paleocene. This was the entry of true rodents (paramyids) into North America from Asia. Until that time, the small herbivore niche had been filled by the multituberculates. The latter had been abundant since the late Jurassic making them a very long-lived mammalian order.

As soon as the rodents arrived in North America, the multituberculates began to decline. Moreover, it has been shown (Krause, 1986) that there was a distinct correlation between the decline of the multituberculates and the rise of the rodents. Due to structural similarities, it appears that the two groups were competing for the same resources. Krause concluded that they provided an example of active displacement through competition.

A more ancient mammalian group is the therapsids or mammal-like reptiles. They arose in the early Permian and were thought to have become extinct in the mid-Jurassic. But now, a Paleocene therapsid has been discovered in North America (Fox et al., 1992). Was the rodent invasion responsible for the final extinction of the therapsids as well as the multituberculates?

The Paleocene record of North American mammals documents more fully than in any other continent the early adaptive radiation of the placentals (Webb, 1985a). In the latest Cretaceous, seven placental genera and five marsupial genera were known (Savage, 1988). In the earliest Paleocene, the marsupial genera decreased to one while the placental genera increased to 22. The marsupial decline has been attributed to the so-called mass extinction at the K/T boundary (Sheehan and Fastovsky, 1992). But, the cause of the marsupial extinctions could have been competition from the placentals. The other two major mammalian groups, the multituberculates and the placentals, were apparently unaffected by the boundary events. Why should the marsupials have been particularly vulnerable?

The rapidity of evolutionary events among the placentals is illustrated by Van Valen's (1978) study of the rate of condylarth generic originations. In the earliest-Paleocene interval, the rate was on the order of 50 new genera per ancestral genus per million years. This explosive earliest interval may have generated at least five new orders: Condylarthra, Cimolesta, Insectivora, Dermoptera, and Carnivora. These were added to the four that had already appeared in the latest Cretaceous: Primates, Leptictida, Arctocyonia, and Taeniodonta. The rapid origin of so many orders implies an exceptionally rapid evolution of higher taxa, but as Webb (1985a) has cautioned, the distinctions among many of these early orders are remarkably subtle. By the late Paleocene, the apparent rate of placental diversification had slowed to about half its previous pace.

Finally, there is some evidence of occasional connections between Africa and Europe beginning in the Paleocene (Rage, 1988). Palaeoryctid mammals have been reported from the Paleocene of Morocco, lipotyphlan and creodont mammals from the Eocene of Algeria, and embrythopod mammals from the early Oligocene of Egypt. In the Eocene, sister groups of theridomyid rodents and macroscelid condylarths apparently existed in Africa and Europe. Rage suggested that these distributions may be attributed to an emergent Apulian Plate between the two continents.

EOCENE

Beginning in the early Eocene, renewed movements of the earth's tectonic plates, accompanied by a rising sea level, had important biogeographic consequences. In the North Atlantic region, earlier geophysical studies (Sclater and Tapscott, 1979; Barron et al., 1981) had shown the terrestrial areas of Europe, Greenland, and North America to have been separated by sea passages from the mid-Jurassic to the present. However, the biological data (Briggs, 1987) clearly showed that land connections must have persisted most of the time from the Mesozoic into the Cenozoic as late as the early Eocene.

Although a subaerial contact between India and Asia was probably made in the late Cretaceous (Chatterjee, 1992), the fusion between the two was not complete until the early Eocene. This amalgamation resulted in the elimination of the northern section of the Tethys Sea and the orogeny of the Himalayan mountain range. Prior to the early Tertiary, northern India was not a mountain range but consisted of extensive lowlands. Virtually all of the material comprising the Himalayas was once part of the Indian continent (Molnar, 1984). If one could spread out the crumpled Himalayas, they would cover a much greater area than they do now. By the Eocene, the occurrence of many Asian terrestrial vertebrate fossils in India attests to the availability of lowland migratory routes.

Central America, which had formed a more or less complete connection between North and South America during the late Cretaceous/Paleocene, became a series of disconnected islands (Map 14). They provided an isolated haven on which several groups of animals, that had originated in South America, underwent extensive evolutionary radiations. Examples are poeciliid fish, frogs, lizards, and snakes. The historical relationship between this Central American archipelago and the Greater Antilles has been the subject of considerable controversy.

A summary of the various theories about the tectonic history of the Caribbean has been published by Perfit and Williams (1989). The authors noted that most of the geophysical hypotheses assume that the Antilles began as an island chain occupying the present position of Central America. A new island chain then began to develop in the Pacific Ocean several hundred kilometers to the west. Supposedly, during the Paleocene, the Antilles migrated eastward across the Caribbean to eventually occupy their present position. Concurrently, the new island arc was also supposed to migrate eastward until it became the modern Central America. This scenario also suggests that certain elevated sections of the Greater Antilles may never have been completely submerged, so could have carried their biota on their journey across the Caribbean. Is this actually what happened? The biological data do not agree (p. 131).

For Australia, the Paleocene-Eocene apparently ended a period of isolation that had extended back to the Jurassic. Although a rift zone began to open between Australia and Antarctica in the late Jurassic (Cande and Mutter, 1982), the two continents did not drift widely apart. The southern tip of Australia (Tasmania) apparently did not have its continental shelf separated from that of Antarctica until after 40 Ma ago (Lawver et al., 1992). This situation, plus a probable archipelago connection between South America and the Antarctic Peninsula, evidently permitted considerable migratory movement. This filter route may have been available until the late Eocene when Australia began to move northward.

By the Eocene, Africa had moved closer to Spain leaving only a narrow passage for the Tethys Sea to reach the Atlantic Ocean. The Turgai Sea still separated Europe from Asia. Sea floor spreading between Greenland and Norway, sometime in the early Eocene, finally created a marine connection between the Arctic Ocean and the North Atlantic. The resulting rift between America and Europe created two northern continents, Europe and Asiamerica (Map 14).

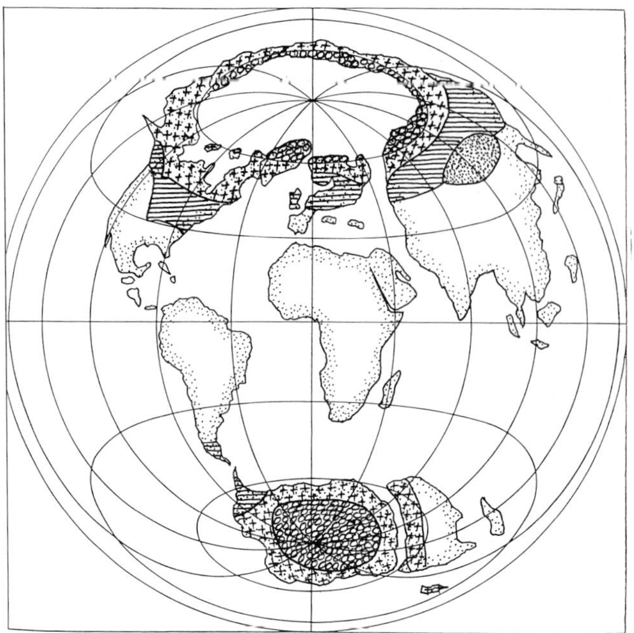

Fig. 40. Early middle-Eocene vegetation: Tropical forest (outlined with dots), Paratropical forest (shaded with horizontal lines), Broad-leafed evergreen (shaded with pluses), Broad-leafed deciduous (shaded with open circles), Woody savanna (shaded with dots). Modified after Janis (1993).

The climatic events of the Eocene/Oligocene are the key to the understanding of the biological events. This was a critical period in the earth's history, for it encompassed a fundamental change from the greenhouse condition of the Mesozoic and early Cenozoic to the icehouse of the later Cenozoic. The early Eocene was the warmest time interval of the Cenozoic. Warm marine conditions are indicated by oxygen isotope ratios of planktonic and benthic foraminiferans recovered from many locations (Sloan and Barron, 1992). On land, there were broad-leaved evergreen forests at high latitudes (Wolfe, 1987) and warm-temperate mammals and reptiles extended far north of the Arctic Circle. Janis (1993) has provided a global vegetation map for the early Eocene (Fig. 40). The subsequent global cooling and its effects extended through most of the Oligocene (details are given in the account of that Epoch).

Marine patterns

The high sea level and warm global climate of the early Eocene permitted a high primary production on the expanded continental shelves. Along the shorelines, multicellular algae spread over the rocky bottoms while the soft sediments of the bays were invaded by angiosperm seagrass and mangrove communities. For the first time since the Cretaceous, shallow marine waters were able to support large herbivorous and carnivorous animals.

The great marine reptiles were gone but their places were taken by the *Carcharodon* sharks, the early whales, and the first sirenians (dugongs, manatees, and sea cows).

The fossil record of mangrove vegetation dates back to the late Cretaceous (the mangrove palm *Nypa*) with two of the dicot genera appearing in the early Eocene. The group is far more diverse in the East Indies than in any other part of the tropics. That area was considered to be the original center of distribution by Chapman (1976). His conclusion has been somewhat modified by Ricklefs and Latham (1993). The latter observed that, although one could not pinpoint the origin of cosmopolitan genera, origins since the closure of the Tethys Sea appear to have been restricted to the Indo-West Pacific region and more specifically to southeast Asia.

The genus *Rhizophora* probably originated in the East Indies, spread west to east Africa and east to the New World, reaching the Atlantic through the Central American archipelago (Van Steenis, 1962). The genus probably reached the Eastern Atlantic from the western side of that ocean, since the three west African species are identical to those found in the Western Atlantic. Van Steenis' hypothesis is supported by the fact that certain small ground beetles of the mangrove swamps show the same relationship, indicating that they also probably reached west Africa by dispersal across the Atlantic (Bruneau de Mire, 1979).

The angiosperm seagrasses have a more widespread latitudinal distribution than do the mangroves. The four most primitive genera in the family Potamogetonaceae are primarily found in temperate waters and two of them have antitropical distributions (Hartog, 1970). Hartog as well as Phillips and Meňez (1988) suggested that the latter two had been displaced in the tropical waters by more stenothermic genera. Of the remaining five genera in the family, four are primarily tropical as are all three genera in the family Hydrocharitaceae. All seven of the genera occur in the Indo-West Pacific. It was supposed that the Western Atlantic and Eastern Pacific forms achieved their present distribution by means of a westward migration route through the Tethys Sea. A similar distribution history has been suggested for the seagrass-associated Foraminifera (Brasier, 1975).

Although seagrasses may have originated in the late Cretaceous, very little is known about their subsequent radiation. A well-preserved seagrass community has been discovered in the Middle Eocene Avon Park Formation of west Florida (Ivany et al., 1990). Occurring with the seagrass fossils were a variety of epibionts, molluscs, echinoderms, and dugongs. So, by the Eocene, we know that well-developed seagrass communities had been formed. There is little doubt that seagrasses contributed significantly to primary production. They succeeded in invading a habitat that had not been utilized by the large marine algae. The oldest recorded sirenians have been found in Lower Eocene rock from tropical marine habitats. It has been proposed that sirenians and seagrass provide an excellent example of co-evolution (Domning, 1981).

A similar evolutionary and distributional history for the cetaceans has been proposed by Gaskin (1982). The earliest archaeocetes (early and mid-Eocene) seem to have occurred only around the western Tethyan periphery of west Africa, India, and Egypt. By the Middle Eocene, somewhat less primitive archaeocetes had penetrated as far as present-day Texas and Louisiana. By the end of the Eocene, their remains could be found in northern Europe and western Canada. So, an origin in the western Tethys Sea seems to have been followed by a broad tropical dispersal and then a gradual adaptation to cooler waters.

In contrast to the foregoing groups, the sand-dollars (clypeasteroid echinoderms) have occurred as three different evolutionary types, each of them originating in a different part of the world (Smith, 1984). The oldest known fossil came from the Upper Paleocene of West Africa, but the group rapidly diversified in the Eocene and became worldwide by the Oligocene. The family Arachnoididae originated in the Australasian region, the Rotulidae originated and remained restricted to the west coast of Africa, and the Scutellidea originated in the circum-Mediterranean or Gulf and Caribbean region. While the first two achieved only limited distributions, the scutellids spread rapidly to become almost cosmopolitan.

The top predator of Eocene seas was the enormous white shark *Carcharodon* with a probable length of more than 10 m and jaws with a width of more than 2 m. The rapid evolution of the higher bony fish in the early Tertiary is illustrated by Eocene fossils from Monte Bolca, Italy. This area, and probably other parts of the Tethys Sea, supported a diversity of apomorphic families such as the Acanthuridae, Zanclidae, Siganidae (Blot and Tyler, 1990), and many others. It is likely that most of the higher fish families were extant by this time.

The earliest stomatopod crustacean, a member of the family Bathysquillidae, has been reported from the Eocene (Manning et al., 1990). The bathysquillids are the oldest stock of stomatopods. There are four contemporary species in three genera. They are found in tropical seas at depths of 200–1500 m.

Terrestrial patterns

Plants

A global warming initiated in the latest Paleocene culminated in an early Eocene thermal maximum (Wolfe, 1992). In northern North America (Wyoming) there were warm-temperate, broad-leaved evergreen forests and, in the lowlands of the Pacific northwest, there were tropical rain forests. During the middle Eocene, extensive volcanic eruptions and associated tectonic movements resulted in the creation of a major upland region in the northwestern interior. This development coincided with a general temperature decline following the early Eocene thermal maximum. The ensuing cool-temperate climate permitted the invasion of several conifer genera. These first appeared in the fossil record in large numbers about 47–45 Ma ago (Axelrod, 1990). In England, tropical and subtropical floral elements gave way to temperate forms at about the same time as in North America (Collinson et al., 1981).

When one compares the temperate vegetation of east Asia with that of eastern North America, some remarkable relationships become evident. This special affinity was first noted by Asa Gray (1859). Gray was encouraged to undertake his comparative study by Charles Darwin (Boufford and Spongberg, 1983). Gray not only called attention to the close systematic relationship of the two widely separated floras but suggested that an interchange between the two had taken place via Asia. Particularly noticeable are the large numbers of disjunct genera and families. Most of the disjunct genera have more species in eastern Asia, most of the disjunct families have more genera in eastern Asia, and in eastern Asia there are more than four times as many species in disjunct families than there are in eastern North America (White, 1983). Cheng (1983) undertook a comparative study of two areas within the disjunct regions that have equivalent climates, Hubei Province in China and the Carolinas in the United States. He found that 75% of the families were shared as well as a large number of genera.

Analyses of individual families and genera have provided important information about relative diversity and past migratory movements. In North America there are 23 genera and 106 species of orchids while eastern Asia has 80 genera and 350 species; more than two-thirds of the genera exhibit phylogenetic ties between the two regions (Sing-chi, 1983). The maple genus *Acer* has about 200 species worldwide but three-fourths of them occur in China (Ying, 1983). Wolfe (1981), through his examination of

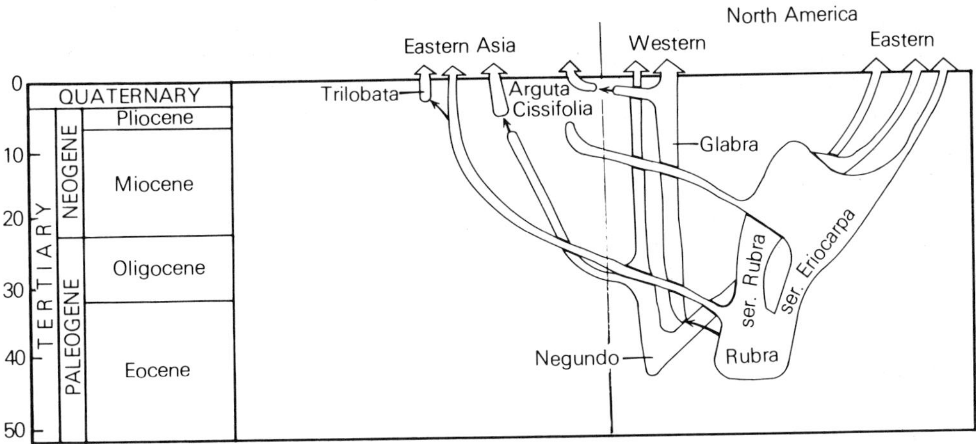

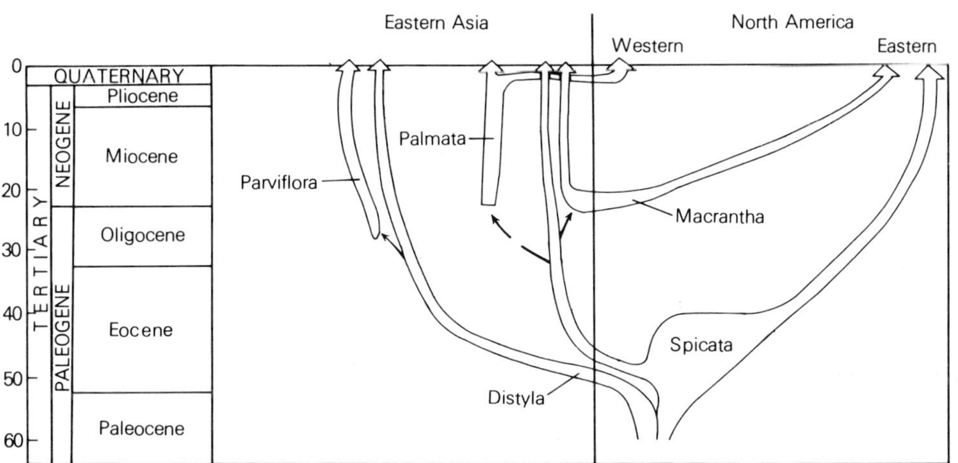

Fig. 41. Suggested distributional relationships among sections of the genus *Acer*. After Wolfe (1981).

fossil material, traced the development of *Acer* for the past 60 Ma. His analysis showed that the genus has a complicated history in which various species migrated back and forth across Beringia from the Eocene to the Quaternary (Fig. 41).

Similar histories to that of *Acer* are suggested by works on other plant groups. The genus *Magnolia* has eight species in the eastern United States, 18 others southward into tropical America, and about 50 in eastern Asia; seven of the 11 sections within the genus are entirely Asian and in two of the American sections there are Asian species (Little, 1983). In the hawthorn genus *Crataegus*, Phipps (1983) suggested that the ancestral crossing of Beringia took place in the Eocene from China to the New World. This was apparently followed by additional eastward migrations in the Miocene, and two possible subsequent migrations in the opposite direction. An investigation of the family Scrophulariaceae by De-Yuan (1983) indicated that six genera probably migrated from Eastern Asia to North America and that four genera must have taken the reverse route.

It is quite probable that the initial dispersals, which account for the basic relationship between eastern Asia and the eastern United States, took place in an easterly direction. It is also probable that most of them took place in the early Eocene when the warm climate at high latitudes permitted cool-temperate east Asian plants to spread into Siberia. For the angiosperms, it has been determined that their center of evolutionary radiation lies in the general southeast Asian area (Takhtajan, 1969; Briggs, 1984) (Fig. 38). From this region, a huge variety of angiosperms migrated northward then eastward across Beringia and across North America. In the earliest Eocene, before the North Atlantic connection was severed, many of them continued into Europe. In this manner, a Holarctic pattern was established.

In their work on angiosperm biogeography, Raven and Axelrod (1974) presented an analysis of North American-South American relationships. Those families emigrating from South America northward in the early Tertiary include the Cactaceae, Liliaceae-Allieae, Loasaceae, Martyniaceae, Nyctaginaceae, Tecophilaeaceae, and Zygophyllaceae. Early migrators in the opposite direction were the Boraginaceae, Clethraceae, Gentianaceae, Hydrophyllaceae, Scrophulariaceae, and others. The authors suggested that at least the great majority of temperate North American groups did not appear south of the Isthmus of Tehuantepec until the Upper Miocene, whereas South American plants became established in the tropical portions of North America from at least Eocene time onward. They concluded that the tropical biota of South America has progressively taken over areas of appropriate climate in North America (including Central America and the West Indies).

It seems apparent that, in the early Tertiary, Australia continued to pick up elements of its angiosperm flora from southeast Asia. The geographical relationship of the two areas in the late Eocene has been depicted by Audley-Charles (1987). There may have been a well-developed island chain extending from New Guinea to Malaya, Burma, and Indochina. As Truswell et al. (1987) have shown, a number of families first appeared in the Australian Eocene apparently having developed earlier in other parts of the world. Examples are the Arecaceae, Polygalaceae, Restionaceae, Bombacaceae, and Sapotaceae. The route to Australia may not have always been from the north.

There is a strong amphinotic, floral relationship among Australia, New Zealand, and South America involving some 50 to 60 genera (Moore, 1972). Although some of these genera have significant developments in both the New and Old Worlds, most of them appear to have migrated from east to west. The major radiation of such families as the Myrtaceae, Proteaceae, and Restionaceae evidently occurred in Australia. In the Proteaceae, all five of the subfamilies and most of the tribes are represented in Australia (Johnson and Briggs, 1975). Most of the ties to South America were probably interrupted when ice sheets began to form in Antarctica.

Mammals

The early ungulates are called condylarths. This was an assemblage of plesiomorphic ungulates which has been recognized primarily on the basis of the retention of primitive characters (Carroll, 1988). Condylarths ranged in time from the latest Cretaceous to the Eocene. By the latter epoch, they had given rise to several important groups of placental mammals such as artiodactyls, perissodactyls, proboscidians, cetaceans, and a variety of forms endemic to South America. The artiodactyls are the most diverse ungulate group with 79 living genera and a rich fossil record that goes back to the Eocene. They radiated extensively in all the northern continents during the early and middle Eocene. By the end of the Eocene, all of the major groups had emerged.

The artiodactyl suborder Suiformes includes the modern pigs, peccaries, and hippopotamus. The earliest suiform group that became abundant was the family Anthracotheriidae which apparently arose in the middle Eocene (Carroll, 1988). They were primarily an Old World group and may have originated in Asia. Later, they became widespread in North America from the early Oligocene to the early Miocene, and were found in Africa from the early Oligocene to the Pliocene. In Asia, they lasted until the Pleistocene. The piglike entelodonts were considered to have been Eocene immigrants from Asia to North America (Webb, 1985a).

The related selenodont artiodactyls included the Merycoidodontoidea, a diverse assemblage that arose in the late middle Eocene and remained entirely restricted to North America. The first modern family of selenodonts is the Camelidae, which is known from the upper Eocene. Camels arose in North America and were restricted there for most of the Tertiary. They reached Eurasia in the Pliocene to give rise to the living bactrian camel and the dromedary. Also, during the Pliocene, the related llamas (which also originated in North America) were able to invade South America. By the end of the Pleistocene, the North American camels had become extinct (Carroll, 1988). Most modern artiodactyls belong to a single group, the Pecora. Their major radiation apparently occurred somewhat later than that which gave rise to the pigs, hippos, and camels.

Perissodactyls are known from the earliest Eocene and their ancestry can be traced to the condylarth family Phenacodontidae. *Hyracotherium*, the best-known primitive perissodactyl, was common in the Lower Eocene. Most of its skeletal features indicate that it may be close to the ancestral pattern for all perissodactyls (Carroll, 1988). MacFadden (1992) considered this genus to lie at the base of the Equidae phylogenetic line. The Eocene was an important epoch for perissodactlyid evolution. No less than 14 families had differentiated by the end of that time. In contrast to the Artiodactyla, the subsequent history of this group shows a progressive reduction in diversity.

Among the perissodactyl lineages, four major groups (superfamilies) persisted from the Eocene to Pleistocene or modern times. The tapiroids were common to all the northern continents in the early Eocene. Subsequently, separate families differentiated in North America, Europe, and Asia (Carroll, 1988). *Helalates*, known from the middle Eocene of North America and the early late Eocene of Asia, is close to the origin of modern tapirs. Webb (1985a) considered this genus to be an immigrant from Asia to North America. In regard to the rhinoceratoids, ancestral rhinoceroses have been recognized from the early to the late Eocene in North America and Asia. A variety of lineages evolved in the late Eocene and early Oligocene. Two genera were common in both North America and Asia in the late Eocene, but subsequent evolution proceeded separately in each area. One family, the amynodontids, were considered to have migrated from Asia to North America (Webb, 1985a).

The chalicotheroids are represented in the early Eocene by fossils from both North America and Asia. The family Chalicotheriidae, recognized from the latest Eocene, continued into the Pleistocene in Asia and Africa. After the Eocene, most of the history of this group was restricted to the Old World (Carroll, 1988). In common with some of the tapiroids and rhinoceratoids, the chalicotherids are thought to have originally dispersed from Asia into North America (Webb, 1985a).

The equoids probably developed from *Hyracotherium* which is known from the base of the Eocene in both North America and Europe. European equoids became extinct early in the Oligocene, leaving only the family Equidae to represent this group for the remainder of the Cenozoic (Carroll, 1988). During the Eocene, the single lineage of equids is represented by *Orohippus* in the middle Eocene and *Epihippus* in the late Eocene (MacFadden, 1992). The major radiation of horses did not take place until the Miocene when extensive grass lands became available. This group provides one of the best examples of long-term evolutionary and biogeographical radiation.

The primitive condylarths, which gained access to Africa and South America, produced on those continents a variety of ungulates, some of which have good fossil records. In Africa, the elephants and their relatives within the order Proboscidea are the best known. There is a possible relationship to the hyraxes or conies and the sirenians. The oldest known proboscidians are from the early Eocene of southern Algeria. Their early history, until the end of the early Miocene, was confined to Africa (Carroll, 1988). Like the horses of the northern hemisphere, most of their evolutionary and geographical radiation did not take place until the Miocene.

On the isolated continent of South America, six unique ungulate orders evolved: Notoungulata, Astropotheria, Trigonstylopoidea, Xenungulata, Pyrotheria, and Liptoterna. Their ultimate ancestry is assumed to lie with the condylarths of the northern continents (Cifelli, 1983). Only the xenarthrans appear to have an early history outside South America. Included in this group are the pangolins or scaly anteaters. They survive today in Africa and southeast Asia, but a middle-Eocene pangolin is known from Europe. This distribution is difficult to explain if the xenarthrans underwent their early development in South America. It has been suggested that they may be the primitive sister group of all other placentals (McKenna, 1975). In that case, they could have originated outside South America. It has been observed that the ungulates of South America appear as a copy in miniature of the diversity seen in the rest of the world, with their own models of hippos, rhinos, horses, camels, elephants, and conies (Carroll, 1988).

There were only two major groups of carnivores in the early Cenozoic, the ancestors of the modern Carnivora and an archaic group, the Creodonta. There were two families of creodonts, both of which appeared in the late Paleocene and were common in the Eocene: the Oxyaenidae may have originated in North America while the Hyaenodontidae were more common in the Old World (Carroll, 1988). The Carnivora developed in the early and middle Paleocene. Two families are recognized by the Eocene: the Viverravidae and the Miacidae. Both were Holarctic in distribution, although the miacids may have first appeared in Asia before dispersing to North America and Europe (Gingerich, 1980).

The small herbivore niches are filled primarily by members of the single order Rodentia with the Lagomorpha occupying most of the remainder. The two orders probably shared a common ancestry among a group of early Cenozoic families from Asia that are united in the order Anagalida. The earliest lagomorphs have been found in the Paleocene of China. They are considered to have immigrated to North America in the Eocene (Webb, 1985a). Modern lagomorphs consist of rabbits, hares, and pikas.

Rodents may be considered the most successful of all mammals in terms of their broad distribution, diversity, and number of individuals. There are more than 1700 species in the recent fauna. It has been noted that, in the latest Paleocene, the rodents arrived in North America from Asia. Their invasion may have been responsible for the demise of the multituberculates and the last of the therapsids. The various rodent radiations and geographical movements have yet to be worked out in detail. They evolved rapidly and almost all the modern families are no older than the Miocene.

The Insectivora – including the modern shrews, moles, hedgehogs, tenrecs, and the African golden mole – are among the most primitive living placentals. The earliest insectivore, *Batodon* from the late Cretaceous of North America, may be related to *Solenodon* (a modern genus known from the West Indies). Eight fossil families are recognized for the Eocene (Carroll, 1988). In addition, the oldest members of two extant families, the Soricidae (shrews) and the Talpidae (moles) have been reported from the late Eocene. In contrast to their insectivore relatives, the bats (Chiroptera) evidently evolved rapidly in the early Cenozoic. The oldest complete skeleton is from the early Eocene. It shows almost all the advanced features of modern bats.

Primate evolution probably began with the rise of the order Plesiadapiformes which is represented by a single tooth from the late Cretaceous of Montana. Animals with similar teeth became common in the faunas of North America and Europe from the middle Paleocene to the middle Eocene (Carroll, 1988). This assemblage, which may include four or five families, is probably ancestral to all higher primates. Two major groups of more advanced primates appeared at the base of the Eocene, the Adapidae in Europe and North America and the Omomyidae in those areas plus Asia. The adapids resemble the living Madagascar lemurs in size and morphology.

The early Eocene is notable for the extensive evidence of immigration among the Holarctic continents (Webb, 1985a; Stucky, 1992). Many of the mammals that first appeared at this time in North America probably came from Asia. These included the first perissodactyls, artiodactyls, primates, and hyaenodontids. Even more striking is the extraordinary resemblance of the European mammal fauna to that of North America. At least half of the 61 land mammal genera of Europe had congeners in the North American fauna. This relationship was closer than any other time in history. In the early Eocene, North America and Asia retained about the same level of relationship as they had in the late Paleocene, but the relationship of the former to Europe became greatly enhanced (Webb, 1985a).

The close faunal relationship between North America and Europe was due to the development of a migratory corridor across the North Atlantic. Tectonic rifting between those continents had apparently made migrations difficult in the late Paleocene, for this was an important interval of faunal endemism (Gingerich, 1985). The connection, which probably occurred across Greenland and the Faroes, must have been heavily forested and supplied with freshwater streams. Within the early Eocene, the connection was broken so by the late Eocene, the resemblance between the two faunas had dropped to a remarkably low level. In the late Eocene, the Asiatic influence in North America was renewed. Immigrants from Asia included groups that were adapted to woodland savanna and habitats with scrubby vegetation (Webb, 1985a). Included were such animals as rabbits, lophodont rodents, entelodontids, and several families of artiodactyls and perissodactyls.

Although there are no South American Eocene fossils that would suggest an invasion of that continent from abroad, the caviomorph rodents of the early Oligocene were so diversified (at least seven families and 16 genera) that they must have been Eocene arrivals. The origin of the caviomorphs is unknown. Some researchers have favored Africa (by rafting) but most others suggested North America with perhaps some evolutionary radiation in Central America. The weight of the evidence seems to favor an invasion of South America by Middle American caviomorph ancestors in the mid-Eocene (Wood, 1985).

It has been noted that two families of advanced primates appeared at the base of the Eocene. The family Omomyidae was common in North America, Europe, and Asia. Twenty genera are known from North America alone. There is evidently a close link between the omomyids and the modern genus *Tarsius*, which is known from the Philippines and the East Indies. A fossil tarsiid has been described from Africa (Simons and Bown, 1985). Although the question of the immediate ancestor for the New World monkeys has still not been decided, there appears to be a majority opinion in favor of the Omomyoidea (Wood, 1985). It has been suggested that an omomyid ancestral stock underwent a waif dispersal from North to South America in the Eocene (Briggs, 1984). As with the caviomorph rodents, some opinion has been expressed in favor of a rafting dispersal from Africa. The earliest known South American monkey (*Branisella*) dates from the early Oligocene (Simpson, 1980).

In the Cretaceous account, it was observed that, with the advent of the Bering land connection, placental mammals may have been introduced into Westamerica. This might have worked to the eventual detriment of the marsupials. Between the late Cretaceous and the early Paleocene, five Westamerican marsupial genera were reduced to one (Savage, 1988). The survivors managed to reach Europe in the early Eocene, and, somewhat later, north Africa and Asia. This line became extinct by the Oligocene in the Old World and by the Miocene in North America.

Six late-Cretaceous marsupial groups have been reported from South America (Marshall and de Muizon, 1988). Three of the groups were shared with North America. This led to the recognition of a single marsupial faunal region encompassing North

America and west central South America. The oldest marsupial fossils are known from the Cenomanian to Campanian deposits in southern Utah. Marsupials may have originated in North America then spread southward (and then elsewhere), but this is by no means certain. In South America, they underwent a significant early Cenozoic radiation which culminated in the differentiation of 10 families. This development took place alongside the radiation of placental ungulates and edentates. One may ask, if placental competition was an important factor in the marsupial extinction in North America, why was it not effective in South America? The answer may be that only the two early placental groups managed to reach South America, and that they did not provide sufficient competition.

It was in South America that the marsupials enjoyed their longest tenure -from the late Cretaceous until modern times. And, according to the modern consensus, it was from South America that marsupials invaded Australia in the early Tertiary by way of Antarctica. This theory was reinforced by the discovery of a polydolopid marsupial from the late Eocene of the Antarctic Peninsula (Woodburne and Zinsmeister, 1984). In a serological study, Kirsch (1977) concluded that the Australian marsupial taxa were closely related to one another, indicating that they had all evolved from a few immigrant forms. In contrast, the greater differences among the South American forms suggested a longer evolution in that area. Kirsch felt that all of the events of marsupial evolution in Australia could have taken place during the last 50–60 Ma.

A single tooth from the Australian Eocene (about 54 Ma) has been identified as that of a condylarth-like placental mammal (Godthelp et al., 1992). Also, it now appears that monotremes existed in the Paleocene of South America (Pascual et al., 1992). While the marsupials and condylarths probably migrated from South America to Australia in the Paleocene-Eocene, the monotremes indicate an earlier relationship. An Australian monotreme has been described from the early Cretaceous (Archer et al., 1985). They may have become isolated in Australia by the early Jurassic (Carroll, 1988).

Birds

Despite the fact that birds have a history extending back to the Upper Jurassic, very little is known about their evolution and distribution prior to the early Cenozoic. Only two modern orders are recognized from the late Cretaceous. Remains from New Jersey and Wyoming appear to be members of the Procellariformes (albatrosses and petrels) and the Charadriiformes (gulls, auks, and other shore birds).

The ratites are a group of flightless birds that are considered to be interrelated, primarily due to their possession of a palaeognathous palate (Carroll, 1988). This structure is also found in the tinamous, which are flying birds living in South and Central America. Included in the ratite group are the kiwis, rheas, cassowaries, emus, and ostriches. Also included are the extinct dromorthinids of Australia, elephant birds of Madagascar, and moas of New Zealand. Although some of the ratites had been separated into distinct orders, Cracraft (1973) decided, on the basis of morphological and biochemical studies, that they all had a common ancestry and that the South American tinamous were closest to the ancestral type.

Cracraft (1973) proposed that the ratites evolved on Gondwana and had attained their present distribution as the result of continental drift. Sibley and Ahlquist (1981) confirmed, on the basis of their DNA-DNA hybridization work, that the ratites were a monophyletic group. They also adopted Cracraft's view that the ancestral ratite was distributed throughout Gondwana prior to that continent's fractionation. Sibley and Ahlquist devised a "molecular clock" in which genetic distances were largely calibrated against the geological dates for the Mesozoic breakup of Gondwana.

More recently, paleontological research (Olson, 1985; Houde, 1988) has revealed a series of grades in ratite evolution that occurred in North America and Europe from the late Paleocene to the middle Eocene. These grades, represented by eight species in three genera, appear to be ancestral to the modern ratites. They contradict Cracraft's assumption of a single flightless ancestor in Gondwana and they cast doubt on the validity of the avian molecular clock. The living and recently extinct ratites are evidently southern hemisphere relicts of a widespread group that probably originated in the early Tertiary.

The neognathous palate is a specialized structure of advanced birds relative to the palate of the Mesozoic birds and the ratites. It probably indicates that all orders in which it is present share a common ancestry (Carroll, 1988). The fossil record shows that most modern families were present by the end of the Eocene, which indicates a rapid radiation within the late Mesozoic and earliest Cenozoic.

Lizards

Among the lizards, the gecko family Eublepharidae is considered to have arisen in Asia in the late Jurassic (Grismer, 1988). A dichotomy between Old and New World groups probably occurred when North America was colonized in the early Cenozoic or late Cretaceous via Beringia. This is similar to the history of the family Agamidae (Estes, 1983). Fossils of the latter have been found in the late Cretaceous of Central Asia and in the Eocene of North America and Europe. The skinks (Scincidae) are also likely to have diversified first in southeast Asia and to have reached the New World via Beringia. All three families have undergone significant radiations in Australia but the times of their arrivals are unknown.

Four different lizard groups apparently dispersed across the North Atlantic connection from North America to Europe (Estes, 1983). The Anguidae probably originated in North America and moved across before the early Eocene connection was broken. An Eocene fossil is known from France and another from the Eocene of the Canadian Archipelago (Sullivan, 1979). The same appears to be true for the European necrosaurs of the Eocene which are related to the North American types. The helodermatids, which have an extended history in North America, have been found in the Paleogene of France. Finally, a single iguanid, known from the middle Eocene, reached Europe (Estes, 1983).

Freshwater groups

The teleost fish fauna of the early and middle-Eocene Green River Formation (Wyoming, Colorado, Utah) shows an interesting pattern of intercontinental relationships. Included are the genus *Eohiodon* of the Hiodontidae, *Diplomystus* (Ellimmichthyidae), *Knightia* (Clupeidae), *Amyzon* (Catostomidae), and two genera of the Percopsidae. The east Asian relationships of the hiodontids and catostomids, and the possible European affinity of the percopsids, have already been noted. Specimens of both *Diplomystus* and *Knightia* have been described from China (Grande, 1985). Thus, the Green River Formation exhibits a notable east Asian influence. This influence, which is indicative of migrations across Beringia, certainly began by the Paleocene and possibly as early as the late Cretaceous.

A study of the systematics of the family Percidae (Collette and Bănărescu, 1977) indicated the family probably originated in Europe and dispersed over the North Atlantic land bridge sometime between the end of the Cretaceous and the early Eocene. The Eocene interruption of contact between the two continents allowed their percid faunas to develop independently. Since that time, the endemic North American tribe Etheostomatini has undergone a remarkable radiation resulting in three genera and about 150 species. There is little doubt that the land bridge was supplied with freshwater streams. Ellesmere Island northwest of Greenland was apparently part of the bridge. Eocene fossils of other freshwater fish including gars (Lepisosteidae), bowfins (Amiidae), and pikes (Esocidae) have been found there (Cavender, 1986).

In the Paleocene account, it was noted that the primary freshwater fish (ostariophysan) family Ictaluridae (catfish) was already present in North America. Its progenitor probably came across Beringia. However, a distinctly more primitive catfish has been discovered in the Eocene of the Green River Formation (Grande, 1987). It has been described as representing a new family, the Hypsidoridae. It is considered to be a sister group to all other catfish except the archaic family Diplomystidae. This means that the two most primitive catfish families occur at the extreme ends of the New World range of the order, one (Diplomystidae) in southern South America and the other (Hypsidoridae) in northern North America. This kind of pattern with the most plesiomorphic taxa at the geographic periphery, especially at high latitudes, has been noted for many other groups.

Another ostariophysan fish group, the characoids, are generally considered to be more primitive than the catfish. Although the living members of the group are confined to the neotropics and Africa, fossils have been found in the Eocene and Oligocene of France (Bănărescu, 1992). It has been suggested that the characoids originated in Upper Jurassic of southeast Asia, spread westward to Europe, then to Africa and South America before those continents were completely separated (Briggs, 1979). Are the European fossils a remnant of an earlier Eurasian distribution or did early Tertiary characoids somehow make their way from Africa across the Tethys Sea? The latter course seems prohibitively difficult for a group of primary freshwater fish.

It has been noted that the salamanders (urodele amphibians) demonstrate some interesting transatlantic relationships. The North American genera of the family Salamandridae are considered to be a derived subgroup of a predominately Eurasian family. It has been suggested that they originated in Europe and dispersed to North America in the early Cenozoic (Milner, 1983). The family Proteidae has living genera in eastern North America and southeastern Europe. Fossil proteids are known from the Upper Paleocene of North America and from the Miocene of southwest Russia and Germany. An extinct related family, the Batrachosauroididae, is known from the Cretaceous to the Miocene of North America and the Paleocene and Eocene of Europe. Unlike the salamandrids, it appears that both the latter families may have originated in North America and then moved eastward.

Although most of the data pertaining to land-bridge connections comes from work on vertebrate animals, occasional works on freshwater invertebrates provide useful information. In his comprehensive work on the ecological biogeography of freshwater molluscs,

Taylor (1988) observed that the Atlantic and Pacific Oceans have had vastly different effects on distribution. The Atlantic cuts across genera and sets of genera, as if it came into being after the evolution of such groups. But the Pacific is not so crossed. Instead, it tends to be bordered by primitive and local families. In regard to early Tertiary North American relationships, Taylor listed six species (ranging from the late Cretaceous to the middle Eocene) with transatlantic affinities; and 10 taxa (Cretaceous to mid-Tertiary) with transpacific relationships. The mussels were notable in that the region west of the Rocky Mountains has a sparse fauna of Asiatic affinities. This may be compared to a rich eastern American fauna with European relatives. An equivalent pattern is shown by the gastropod family Viviparidae.

Among the aquatic insects, the ancient (Paleozoic) orders Plecoptera (stoneflies), Trichoptera (caddisflies), and the Ephemeroptera (mayflies) demonstrate marked amphipacific and amphiatlantic patterns at the family and generic levels. While these modern distributions certainly indicate past dispersals, there is very little fossil evidence to show when they may have occurred.

Australia

It is important to try to piece together the tectonic events which made possible the transantarctic migrations by vagile terrestrial organisms. In the early Cretaceous, about 110 Ma ago, East Antarctica was considerably northward of its present position and West Antarctica was a series of islands (Lawver et al., 1992) (Fig. 42). By the late Cretaceous, about 70 Ma ago, the two parts had fused together and the continent had moved to the southeast (Fig. 43). At this time, Australia was well separated from Antarctica but Tasmania and the South Tasman Ridge formed a connecting link. The tip of the Antarctic

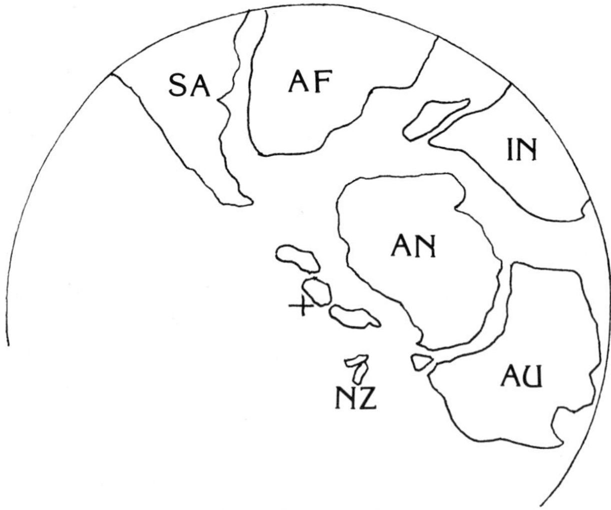

Fig. 42. South polar view, early Cretaceous. After Lawver et al. (1992).

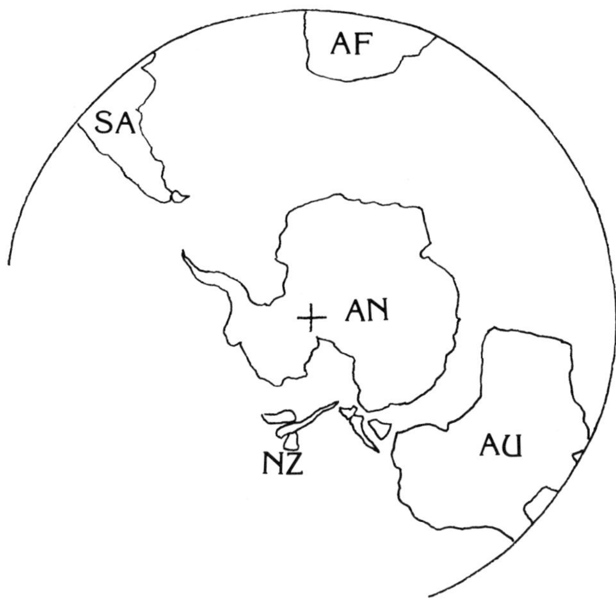

Fig. 43. South polar view, late Cretaceous. Modified after Lawver et al. (1992).

Peninsula was situated close to Tierra del Fuego, separated by a shallow strait. These positions remained relatively static until the late Eocene, about 40 Ma ago, after which Australia moved rapidly northward.

The notable sea-level drop at the end of the Cretaceous (Hallam, 1992) allowed a considerable exposure of the continental shelves which lessened the distance between shorelines and probably permitted the subaerial emergence of islands. This presented a migratory window of opportunity from about 60 Ma to 40 Ma ago. The advantage was probably taken by South American marsupials, a condylarth, and many other migrants. This resulted in the formation of what is often called an "amphinotic track". The strongest ties are between South America and Australia followed by New Zealand. More distant relations are shown by southern Africa, Madagascar, and New Caledonia (Briggs, 1987).

Australian-South American relationships are demonstrated by a large fraction of the flora and fauna. In addition to the mammals, the list includes other vertebrates such as the xiphodont crocodiles (Pleistocene of Australia and Eocene of Patagonia), the frog family Leptodactylidae, the turtle family Chelidae, ratite birds (cassowaries, emus, dromorthinids), and mound birds. The invertebrates include freshwater mussels, crayfish and aquatic insects such as the mayflies, stoneflies, caddisflies, and others. Also included are a host of terrestrial invertebrates such as land snails, oligochaete worms, and most major groups of insects and spiders. Similarly, some of the older gymnosperm and angiosperm families show the same amphinotic track. With the exception of many angiosperm species which are highly vagile, the systematics of the foregoing groups indicate ties at the generic, subfamily, or family levels. This means that most of the migratory traffic probably took place in the early Tertiary. For most of the groups, the northward movement of Australia and the presence of Antarctic ice sheets about 23 Ma ago, meant the end of the migratory pathway.

New Zealand

Like Australia, New Zealand probably received its earliest vertebrates (a leiopelmatid frog genus and the tuatara, *Sphenodon*) in the Triassic or Jurassic. But during the early Tertiary, it too received an infusion of biota from South America via Antarctica. But, unlike Australia, it received no vertebrate animals except birds and possibly some lizards. The ratite birds, moas and kiwis, probably arrived in the early Tertiary. Their migration was possibly facilitated because their ancestors were probably flying birds related to the South American tinamous. New Zealand's three endemic families of passerine birds also possibly represent early Tertiary colonizations (Bull and Whitaker, 1975). The same may be true of the lizard family Gekkonidae, which is represented by three endemic genera.

It should be noted that not all of the amphinotic migratory traffic moved from west to east. The freshwater crayfish family Parastacidae probably originated in Australia, with subsequent dispersals to New Zealand, South America, Madagascar, and the Aru Islands (Briggs, 1987). From southeast Asia to the general Australian-New Zealand region, there is a remarkable concentration of primitive conifer and angiosperm families. Most of these families are more diverse in the antipodes than in South America. It is likely that many of them spread from east to west.

Madagascar

It has been noted that Madagascar separated from Africa in the middle Jurassic about 160 Ma ago. As the result of its long isolation, the island has a depauperate or disharmonic fauna. Its present relationships to Africa indicate that the bulk of the fauna and much of the flora probably reached there in the early Tertiary. The existence on Madagascar of many endemic families and subfamilies provide a good indication of a long history on the island.

As with Australia and New Zealand, a ratite bird family probably reached Madagascar in the early Tertiary. This colonization produced the family Aepyornithidae (elephant birds) with 12 species. There are three endemic families of higher birds. Five endemic families of lemurs are present. Fossil "lemuroid" primates, probably ancestral to the Madagascar families, were common in the Eocene of North America and Europe and some have been reported from Africa (Martin, 1990). From Africa, one or more of the lemuroid groups managed to reach Madagascar. Other probable examples of early Tertiary immigrations are the present endemic families and subfamilies of frogs, lizards, rodents, aquatic insects, terrestrial insects, spiders, and angiosperm plants. Although most of these archaic groups have kinship ties to Africa, several of them do not. Amphinotic relationships may be found among the crayfish, some aquatic insects, terrestrial insects, and vascular plants.

India

We have seen that India probably made its first subaerial contact with Asia in the late Cretaceous. The earliest large mammal fossils from the Tertiary of the Indian subcontinent come from the middle to late Eocene. A variety of localities have yielded taeniodonts, creodonts, titanotheres, anthracotheres, artiodactyls, paramyid rodents, and many others. All of this array have European, North American or African affinities. There are no remains of endemic forms that might have evolved on an isolated Indian continent (Chatterjee and Hotton, 1986). The history of the mammalian invasion has been recapitulated by Kurup (1974). It appears that migrations to India came through two gateways, one in the Assam region to the east and one to the northwest. To the north, the rising Himalaya formed a barrier restricting entry from that direction. The migratory flow through the Assam gateway appeared to be the greatest.

The migratory path of freshwater fish was studied by Hora (1937) who concluded that this fauna originated mainly in South China and spread westward. Hora (1939) recorded the presence of several Asian freshwater fish families from Indian "beds that cannot possibly be older than Lower Eocene". Patterson and Owen (1991) felt that his identifications were doubtful. However, Gayet (1987) found cyprinid fossil teeth from the early or middle Eocene of Pakistan (probably part of the Indian plate). The family Cyprinidae is a primary freshwater group unknown from Africa before the early Miocene. So, the invasion of the Indian subcontinent in the Eocene by an Asian freshwater fish group appears to have taken place. Undoubted fish or mammalian remains from the Paleocene of India have not been found. So, there may or may not have been an immigration hiatus from the late Cretaceous to the Eocene.

Antillean relationships

Biologists and geologists have attempted to account for the presence of the Antillean terrestrial and freshwater biota by proposing three different methods of access: (1) a land bridge (or bridges) from the mainland to the islands; (2) movement of the islands themselves, with at least a partially intact biota, eastward from a position adjacent to the mainland; and (3) fortuitous overseas transport over a long period of time. The history of these hypotheses and, a summary of the pertinent geological data, have been given by Briggs (1987) and in more detail by Williams (1989).

At the present time there is no serious support for the land bridge hypothesis. For the past 15 years, the argument has been between the remaining two. The island-rafting hypothesis began with a paper by Rosen (1976) who presented a vicariance model of Caribbean biogeography. This model, which has been supported by several geophysical studies, envisions the existence in the late Cretaceous/early Tertiary of a proto-Antilles. As already noted (p. 115), this archipelago, lying in the position of the present Central America, was supposed to have migrated across the Caribbean to form the modern Antilles.

The alternative hypothesis (Briggs, 1987) says that Central America stayed put and that the Greater Antilles are oceanic islands that never have had mainland connections.

This hypothesis has often been considered simply a dispersal, as opposed to a vicariance, theory. But, in reality, it combines elements of both. Almost every biotic group, that is well represented in the West Indies, shows considerable evidence of local speciation. Patterns of several groups suggest an initial colonization followed by migrations to other parts of the archipelago. But also, gaps between islands were created or disappeared according to the historic fluctuations in sea level. Volcanic action and erosion affected the sizes and shapes of the islands. Such changes produced barriers to interrupt gene flow and induced allopatric speciation; removal of barriers allowed dispersal to occur. The geography of the speciation that has taken place among several West Indian groups – such as the anolid lizards, eleuthrodactylid frogs, and poeciliid fish – provides ample evidence that both dispersal and vicariance have played important roles.

How does one choose between the migrating island and oceanic island hypotheses? In recent years, the systematics of the island groups, particularly the vertebrate animals, has become reasonably well-known. Yet, the people responsible for the research on the individual groups disagree in their interpretation of the history of the islands. This emphasizes the fact that it is very difficult, if not impossible, to reconstruct the history of a geographic area from the distribution of a single group of animals or plants. One must attempt to make such judgments on the basis of the entire biota, even though portions of it may be poorly known. It is, of course, beneficial to have dependable geological or geophysical data; but, for the West Indies, there are so many conflicting physical theories that the history of the region is probably best determined from its biology.

The peculiar characteristics of oceanic islands were emphasized by Darwin (1859) who devoted 21 pages to this subject in his *Origin of Species*. In regard to the West Indies, it was Matthew (1915, 1918) who pointed out that the vertebrate fauna, fossil and recent, represents only a few selections from the continental faunas of either North or South America and that it is an unbalanced assemblage that had apparently arrived at diverse times. Myers (1938) called attention to the complete absence of primary freshwater fish. These, and many more recent works, have emphasized the depauperate, disharmonic nature of the Antillean fauna. Perfit and Williams (1989) have pointed out that, among the vertebrates it is the birds, bats, lizards, rodents, snakes, and frogs (in descending order) that have shown the greatest potentiality for cross-water colonization. Ungulates, carnivores, and salamanders show the least. The native West Indian fauna includes none of the latter three.

The West Indies evidently began to accumulate their biota as early as the Eocene. At least, this is probably true of Cuba and Hispaniola. Dominican amber, apparently from the Upper Eocene, has included specimens of three extant genera, a frog (*Eleuthrodactylus*), and two lizards (*Anolis* and *Sphaerodactylus*) as well as various invertebrates (Williams, 1989). In general, the amount of evolutionary change in the Antilles does not indicate a very old biota. The great majority of the endemic taxa are at the species level with only a few such genera. Molecular data suggest that the West Indian vertebrates originated via overseas dispersal (Hedger et al., 1992).

On the other hand, if we look at the extent of evolutionary radiation that has taken place in Central America, we see evidence of an older assemblage. There is a high degree of endemicity at the generic and suprageneric levels. As Savage (1982) has emphasized, the herpetofauna of Central America is very distinctive: of 169 genera, 32 are endemic and 21 others are essentially restricted to the region. A number of other genera, now widespread in North America, probably originated in Central America (Vanzolini and Heyer, 1985). Among the freshwater fish, there is an Old Southern Element that probably entered Central America from the south in late-Cretaceous or early Tertiary times (Bussing, 1985). Of a total of 39 genera, 18 are still endemic and five others have remained primarily in that area. The family Poeciliidae has clearly undergone the bulk of its evolutionary radiation in Central America (Rosen and Bailey, 1963). From there, occasional migrants apparently managed to reach the Antilles.

The island rafting hypothesis of Rosen (1976) proposed that the Antilles carried their biota along as they moved from their original position between North and South America. If that really happened, it would mean that the Antilles are much older than their Central American successor. It is the older island group that will demonstrate the greatest evolutionary change in its endemic biota. There *is* a distinct contrast, but it is not the Antilles that demonstrates the greater evolutionary change, it is Central America. Furthermore, there is evidence from studies on the lizards, frogs, and freshwater fish, that their ancestral groups first entered Central America in the Cretaceous/Paleocene. This invasion could not have taken place if Central America was only just developing as an island arc several hundred kilometers out in the Eastern Pacific. This information, plus the obvious disharmonic nature of the Antillean fauna, strongly indicates that Central America has remained about where it was formed and that the Antilles have always been oceanic islands.

OLIGOCENE

It is difficult for the biologist to separate the Oligocene from the Eocene because events, that had a detrimental effect on biotic diversity, began in the middle Eocene and extended on through most of the Oligocene. Plate tectonics, during this period, were apparently active. The Atlantic Ocean continued to widen, India continued to push its way into Asia, and – perhaps most significant – Australia departed from the vicinity of Antarctica. It seems likely that the development of a major seaway between the latter two continents caused a climatic change which resulted in the initial formation of Antarctic ice sheets (Bartek et al., 1992).

As the ice accumulated in the South Polar Region, the eustatic sea level fell. As the sea level fell, the continents began to stand relatively higher. At the same time, considerable volcanism and continental uplift was taking place. The Laramide orogeny was occurring in western North America, the Andean orogeny in South America, and the building of the Himalayas in southern Asia. It seems that the mountain building accentuated the effects of the lowering sea level, so that the global climate became much cooler and drier. This may have led to more ice accumulation in Antarctica. As more water was taken out of the hydrologic cycle, the sea level became still lower (and the climate still colder).

The marine regression had significant biogeographic effects. The Beringia connection between Asia and North America became broader, so that it presented an easier migratory route. The Turgai Sea, which had separated Europe and Asia since the mid-Jurassic, dried up permitting a mixture of the terrestrial biotas. From a global perspective, this meant the creation of one continuous northern continent and allowed many vagile organisms to achieve Holarctic distributions.

Sloan and Barron (1992) performed a modeling experiment which suggested that the two factors of changing sea-surface temperature and continuous tectonic uplift were responsible for the change to a cooler and drier continental climate. Alternatively, one might argue that the cooler and drier continental climate was primarily caused by the uplift and that the climatic change resulted in the accumulation of more polar ice and colder seawater. The idea of the tectonic forcing of late Cenozoic climate also has been advocated by Raymo and Ruddiman (1992). The sinking of the cold, dense seawater began a thermohaline circulation and created the psychrosphere, the cold bottom layer of the ocean.

Two other events may also have had important effects on ocean temperature and circulation. After the breaking of the North Atlantic land bridge in the early Eocene, spreading continued between Norway and Greenland until the formation of a deep-water passage permitted cold, saline water to flow into the North Atlantic (Williams, 1986). Concurrently, the gradual cooling and subsidence of the Walvis/Rio Grande Rise across the South Atlantic allowed a deep-water passage to the Southern Ocean. Consequently, the deep sea was supplied with cold, oxygenated water from both the Antarctic and the Arctic/North Atlantic.

At one time, research people thought that Antarctic ice sheets did not form until Miocene time. There was a general belief that the Miocene ice sheets had formed in response to the initiation of a deep circumpolar current following the opening of Drake Passage. But we now have evidence that ice sheets may have first formed as early as 42 Ma ago (Keller et al., 1992). At first, the Antarctic ice may have waxed and waned with minor climatic cycles, but by late Oligocene time it appears that a major ice sheet grounding event (an icepack) took place (Bartek et al., 1992). Drake Passage evidently did not open sufficiently to permit circulation of a deep current until the early Miocene (about 22–23 Ma ago).

The record of global cooling in the ocean can be followed by examining the oxygen isotope curves for both the shallow, planktonic foraminiferans and the deep benthic forms (Keller et al., 1992). A major cooling took place near the middle/late Eocene boundary, a further drop began near the Eocene/Oligocene (E/O) boundary, and temperature decline continued through most of the Oligocene reaching a minimum at about 29 Ma ago. Altogether, high-latitude surface, and intermediate water at both high and low latitudes, decreased by about 12–15°C. Low-latitude surface temperatures declined about 6–8°C. Comparison of a low-latitude faunal turnover in the Indian Ocean (latitude about 2°S) with the high- latitude temperature record, showed a close correlation between the onset of Antarctic glaciation, at about 42 Ma ago, and the beginning of the faunal change.

The severe climatic changes caused a series of extinctions which considerably reduced species diversity. The close of the Eocene or the E/O boundary had been traditionally recognized as the time of a great, catastrophic extinction. It is considered to be one of the seven major extinctions in the Phanerozoic history (Briggs, 1990). The impression of a single event is reinforced when family or generic extinction rates are portrayed as a single, sharp peak on a graph with geological time as the horizontal axis (Sepkoski, 1989). However, for the past 12 years, works on a variety of biotic groups have demonstrated that there was not a single massive extinction episode, but an extended series of minor events. In the marine environment, these faunal turnovers began in the middle/late Eocene and lasted until the early/late Oligocene, a period of some 14 Ma (Keller et al., 1992).

Marine patterns

The decline of seawater temperatures, the stronger thermohaline circulation, and the sea-level regression had generally detrimental effects on species diversity. As the biotic declines took place, new species evolved and others invaded from areas that had suffered less depletion. The extinctions took place in a series of pulses evidently related to relatively rapid drops in ocean temperature.

In their study of low-latitude planktonic foraminiferans, Keller et al. (1992) identified two major faunal turnovers, one near the middle/late Eocene boundary and the other spanning the early/late Oligocene boundary. In the first episode, over 80% of the middle-Eocene fauna became extinct. This decline was accompanied by the rise of a new fauna consisting primarily of intermediate-dwelling forms plus a few deep-dwelling species. Only a few surface-dwelling species survived to reach the E/O boundary at 34 Ma ago. This turnover took place over a span of about 4 Ma. Near the E/O boundary, a new surface group evolved. In the second episode, the extinction again consisted primarily of surface-dwelling forms. Their decline began at approximately 30.5 Ma and they disappeared altogether at 28–29 Ma. At the same time, a major global cooling took place which reached its maximum at 29 Ma.

During the past 10 years, there has been considerable speculation that extraterrestrial impacts caused a catastrophic extinction at the E/O boundary. While there are microtektite layers near this time and two iridium peaks, which probably represent impacts (McLaren and Goodfellow, 1990), there is no evidence of any direct association with the two gradual extinction episodes. Nor is there evidence that these impact events brought about the climatic change, since the global cooling preceded the time of the apparent impacts (Keller et al., 1992).

A study of the lower bathyal benthic foraminiferal faunas from Maud Rise (Weddell Sea, Antarctica) revealed three gradual but stepped extinctions (Thomas, 1992). These occurred at about 46.4–44.6 Ma, 40–37 Ma, and 34–31.5 Ma ago. The gradual nature of the faunal changes suggested that they resulted from the gradual cooling of surface waters which continued to sink to bathyal depths. In regard to the calcareous nannoplankton, Aubry (1983) noted a stepwise extinction that affected the tropical species and caused an equatorward migration of species from higher latitudes. More recently, Aubry (1992) found evidence of major turnovers near the middle/late Eocene boundary and in the early Oligocene. There was only a weak change in the latest Eocene. She found an excellent correlation between the timing of the changes and the timing of cooling events as inferred from isotopic studies. She also noted that eutrophication may have adversely affected diversity.

A comprehensive study of diatoms from the Southern Ocean, low-latitude Atlantic and Pacific, the Labrador Sea, and the Norwegian-Greenland Sea was carried out by Baldauf (1992). He detected a middle-Eocene turnover involving about 30% of the species, a latest Eocene/earliest Oligocene 40% turnover, and a late Oligocene/early Miocene 50% turnover. A final extinction and replacement took place in the late/early Miocene. In general, these events were responses to oceanographic reorganization including intensification of latitudinal and vertical gradients, and development of the psychrosphere.

In the Western European ostracod faunas, the Lower Oligocene community of 33 species contained only two that were known from older rocks (Keen, 1990). Of the remainder, approximately half could be phylogenetically related to local, late-Eocene species, but the other half had no known ancestors in the area. Keen concluded that the great faunal break could be explained only by the arrival of new immigrants which rapidly colonized the area, along with accelerated evolution of the indigenous species. While there were important differences between the Eocene and Oligocene faunas at the species level, there was no evidence of a mass extinction.

For the molluscan fauna of the southeastern United States, Dockery (1984) described pulses of extinction due to episodes of climatic cooling and increased water turbidity. In a more detailed study of molluscs in the same area, Hansen (1992) was able to follow sea level, surface temperature, and species diversity from the Paleocene to the Oligocene. He found little correlation with sea-level change but a pronounced temperature effect. The E/O extinction of molluscs was found to be stretched over the entire late-Eocene interval.

A global study of the echinoids was undertaken by McKinney et al. (1992). They found that echinoid diversity showed an Eocene peak bounded by much lower diversities in the Paleocene and Oligocene. The low Paleocene diversity was thought to represent a lag phase as speciation gradually replaced diversity lost at the end of the Mesozoic. The low Oligocene diversity was attributed to global climatic deterioration. A diversity peak occurred early in the middle Eocene followed by a small to moderate loss. A second, much larger diversity drop occurred during the terminal Eocene when reductions of over 50% occurred in all orders. After the Eocene, there was a general loss of diversity attributed to climatic cooling. An important regional exception to low Oligocene diversity was Australia. As that continent moved into warmer waters, it picked up Tethyan species that had been known from the Mediterranean and intermediate areas.

The origin and early distribution of the whales was described in the Eocene account (p. 117). Contrary to many other marine animal groups, there is no evidence of a major extinction near the end of the Eocene (Fordyce, 1992). The early Oligocene was a time of dramatic changes in cetacean history. By then, there had developed four families and the dichotomy between Mysticeti and the Odontoceti had taken place. By the late Oligocene, there were 11 families and perhaps 36–50 species. Cetacean evolution seems to have been driven by the availability of new feeding opportunities which were made possible by ocean cooling and increased thermohaline circulation. The sirenians had also appeared by the early Eocene and the extinct desmostylians are known from the Oligocene. The pinnipeds (seals, sealions, and relatives) appeared at about the Oligocene/Miocene boundary.

Terrestrial patterns

In the north, central United States, Retallack (1992) followed the changes in climate and vegetation across the E/O boundary. There, he found evidence that moist forests of 38 Ma ago gave way to dry forests by 34 Ma, to dry woodland by 33 Ma, to wooded grassland by 32 Ma, and to large areas of open grassland by 30 Ma ago. The fossil soils (palesols) provided indications of a stepwise climatic deterioration from about 40 Ma ago to 29.5 Ma ago. The deterioration was not only a matter of cooling but also of drying in the interiors of the large, mid-latitude continents.

A study of the low-biomass vegetation of the Oligocene was conducted by Leopold et al. (1992). They examined fossil floras from the Great Plains of the United States and also material from Russia and China. In the United States, they found evidence of cooling and, beginning in the late Eocene, establishment of a woody savanna vegetation. They found no direct evidence of grassland in the Oligocene. In the Oligocene of Russia, forest vegetation was widespread except in Kazakhstan where more open woodlands occurred. In north China, there was a large arid area to the west and northwest. A low-biomass vegetation consisted of shrubs with scattered trees or woodlands. Grasses and herbaceous plants were unimportant or lacking.

Floral changes in western North America were followed by studying the physiognomy of leaf assemblages (Wolfe, 1992). These indicated a marked decline in mean annual temperature which was accompanied by an increase in the mean annual range. This deterioration occurred about 1 Ma following the end of the Eocene. Extinctions in the tropical vegetation were the most severe. Less damage was done to the warm-temperate forms and still less to the cold-temperate species. Regional extinction at high latitudes was particularly marked. Similarly, Collinson (1992) examined macrofloral evidence from Czechoslovakia and Germany. She observed changes from a dominantly evergreen subtropical vegetation in the late Eocene to mixed-evergreen and deciduous in the early Oligocene. These changes were considered to be the result of a cooling climate which began in the early/late Eocene.

In the southern hemisphere, an Oligocene palynoflora was examined in southeastern Australia (Nott and Owen, 1992). In that area, the cool, humid climate of the Paleocene became warm to subtropical in the mid-Eocene. A marked change towards a cooler climate took place after the middle Eocene. The change was highlighted by dramatic increases in *Nothofagus* and gymnosperms and a decrease in the Proteaceae. In the lowland areas, drier conditions appear to have begun by the late Oligocene. But this was not the case in the southeastern highlands where a moist, closed rainforest vegetation still existed. The global vegetation of the early Oligocene has been illustrated in map form (Fig. 44) by Janis (1993). Comparison to the early Eocene map (Fig. 40) shows the dramatic contrast between the two epochs.

The availability of considerable fossil material, especially from the Badlands of South Dakota, has made it possible to put together a reasonably good account of mammalian events in North America. As Webb (1985a) has stated, "With the onset of the Oligocene came the most impressive faunal turnover in the whole age of mammals." This event becomes still more remarkable when one sees it repeated on every continent that offers a full record of its mammalian history. A great change in the North American rodent fauna

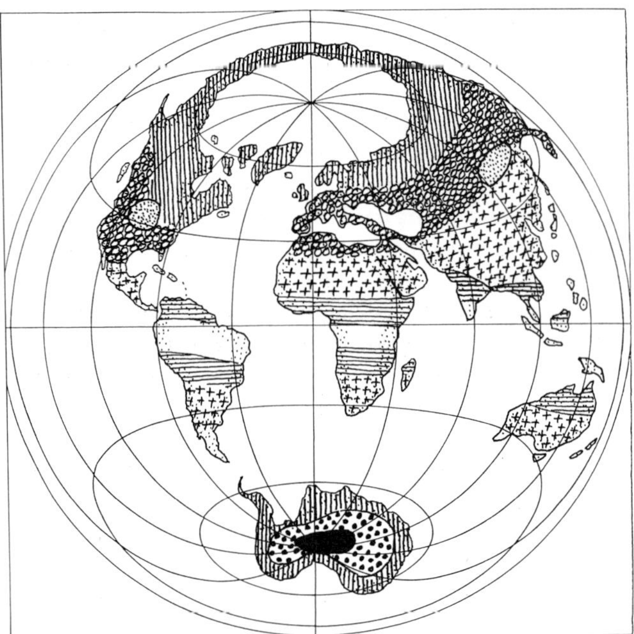

Fig. 44. Early Oligocene vegetation: Tropical forest (outlined with dots), Paratropical forest (shaded with horizontal lines), Broad-leafed evergreen (shaded with pluses), Broad-leafed deciduous (shaded with open circles), Mixed coniferous-deciduous (shaded with vertical lines), Woody savanna (shaded with dots), Tundra (shaded with filled circles) and Ice (solid black area). Modified after Janis (1993).

took place with appearance of such modern families as the heteromyids, geomyids, castorids, sciurids, cylindrodontids, and cricetids. Among the large mammals, were the first appearances of the Canidae, Felidae, Mustelidae, Tapiridae, Rhinocerotidae, Anthracotheriidae, and Tayassuidae. Many of the new groups arrived from Asia via the Beringia connection. Among the groups most probably traceable to Asiatic stock are the Castoridae, Anthracotheriidae, and Tapiridae. Some of the new North American rodent families are suspected to have originated in Central America. These include the Heteromyidae and the Geomyidae.

An environmental deterioration is reflected in the general shift from small mammal/arborial faunas in the mid-Eocene to large mammal/savanna faunas in the Eo-Oligocene (Webb, 1985a). Among the most characteristic mammals of the early Oligocene are the herding ungulates such as *Mesohippus* and *Leptomeryx*. They were accompanied by a variety of small herbivores such as *Palaeolagus* and *Ischyromys*. Their presence indicates the continued expansion of woodland savanna habitats and increasingly cool and arid climates.

The most recent review of the North American Eocene/Oligocene mammalian faunas is that of Stucky (1992). Many clades of mammals, which originated in the late Cretaceous or early Paleocene became extinct in North America by the end of the Oligocene. These include the multituberculates, leptictids, cimolestins, dermopterans, creodonts, euprimates, mesonychids, arctocyonids, condylarths, uintatheres, and palaeodonts. But the reductions took place gradually through the middle Eocene and Oligocene. During the early Eocene, these groups comprised nearly 50% of all genera but, by the early parts of the last Oligocene stage, they amounted to only 2%. Prothero (1986) felt that the extinctions took place in five steps spaced out over a period of 10 Ma.

Changes in the British mammalian communities across the E/O transition have been followed by Hooker (1992) and the evolution of mammalian faunas in Europe has been reviewed by Legendre and Hartenberger (1992). The latter authors recognized three main extinction/origination events. The first one of earliest Eocene age corresponds to the arrival in Europe of the Perissodactyla, Artiodactyla, and Rodentia. At the same time, the diversity of the Condylarthra and Multituberculata decreased dramatically. The second event occurred near the early/middle Eocene and consisted of a minor immigration wave for Primates, Chiroptera, Artiodactyla, and Rodentia. But, at present, this event is not well understood.

The third event described by Legendre and Hartenberger (1992) is situated at the E/O boundary and corresponds to the "Grand Coupure". It was first described in the Paris Basin by Stehlin (1909). It is now recognized on a continental scale in Europe. It resulted from the combination of an important immigration event with an extinction event. Regional studies at the species level show that the turnover was very high. In the Quercy area for example, up to 60% of the local species disappeared. The survivorship curves emphasize that all cohorts were involved with no evident selectivity, and some cohorts were affected before the main event so that the change was not geologically instantaneous.

The Grande Coupure was characterized by the disappearance of apatemyids and primates and by the beginning of the diversification of modern insectivores, rodents, and fissiped carnivores. The arrival of new perissodactyl families may also be noted, although this group did not show much radiation. The artiodactyl decline did not exactly coincide with this event but seems somewhat delayed. Finally, during the Oligocene, a minor extinction event occurred which involved only old Eocene cohorts.

The Asian Oligocene mammalian faunas, like the European and American, are much advanced over those from the Eocene (Wang, 1992). During the transition from the late Eocene to the early Oligocene, a large number of the archaic forms became extinct and many new taxa appeared. A terminal Eocene event occurred in Asia but the change was not as abrupt as in Europe and North America. The extinction of the archaic forms occurred in a stepwise manner. Three families and 45 genera became extinct in the late Eocene, but seven of the ancient families did not become extinct until the end of the early Oligocene. Some of the groups, such as the Condylarthra, survived in Asia longer than in Europe and North America. The Brontotheriidae not only lingered but also diversified in the early Oligocene.

The new advanced forms appeared gradually. Some new families including the Entelodontidae, Cricetidae, and Ochotonidae made their first appearance in the late Eocene, earlier than in Europe and North America. One living family (Castoridae) appeared in the early Oligocene. Among 31 new genera, 11 showed up in the early/early Oligocene, 18 in the middle/early Oligocene, and 2 in the latest/early Oligocene. These Oligocene mammals were mostly Asian endemics; few appear to have migrated from other continents. With the beginning of the Oligocene and the retreat of the Turgai Sea, more and more species migrated from Asia to Europe. This was one cause of the Grande Coupure.

Exchanges between Asia and North America apparently continued throughout the Oligocene (Wang, 1992). Several genera were common to both areas in the early Oligocene. The number of such genera increased in the middle Oligocene. *Desmatolagus* may have migrated from Asia into North America while *Palaeocastor*, *Agnotocastor*, and *Prosciurus* dispersed in the opposite direction. A number of genera were shared in the late Oligocene. *Haplomys* and *Meniscomys* of North America may have been derived from the Asian *Haplomys* and *Promeniscomys*. As noted (p. ??), Webb (1985a) also discussed the Asiatic influence on the composition of the North American mammalian assemblage.

Many African mammals belong to endemic groups unknown outside Africa in the early Tertiary (Rasmussen et al., 1992). These include the Tenrecoidea, Chrysochloridae, Macroscelidea, Proboscidea, Embrithopoda, Hyracoidea, phiomyid rodents, anthropoid primates, and others. Only a small handful of Eocene and Oligocene taxa are shared with other continents. Unlike the Holarctic continents, where the E/O boundary has been associated with significant shifts in faunal composition, the data from Africa indicate little evidence of faunal turnovers. Many genera seem to have continued across the boundary time. Africa maintained a relatively stable, tropical climate which allowed for the continuation of its diversity.

For the reptiles and amphibians of North America, the late Eocene through the early Oligocene represented a period of great change (Hutchison, 1992). The overriding factors appear to have been integration of the main streams into ocean-directed rivers, sedimentation filling of the early Tertiary lacustrine basins, decrease in the number of permanent streams, and a general increase in aridity. The changes in the freshwater systems were accompanied by a decrease in mean atmospheric temperature, but this was not so severe that it eliminated the tortoises and large-bodied lizards. The available evidence suggests that diversity declined over a period of approximately 16 Ma and cannot be attributed to a single catastrophic event. The tempo of the change appears to have quickened across the E/O boundary.

In Europe, more than three fourths of the late Eocene herpetofauna had died out by the time of the E/O boundary (Rage, 1986). In the early Oligocene, oriental migrants settled in western Europe and probably contributed to the extinction of the remaining Eocene forms. The end of the Turgai Sea in the Oligocene was an exceedingly important event for the primary freshwater fish (Bănărescu, 1992). None of the primary freshwater fish families now in Europe has a fossil record on the continent prior to the Oligocene. Thus the families Cyprinidae, Siluridae, and Cobitidae, which dominate the European fresh waters, are Asian migrants which probably dispersed westward in the Oligocene.

The fish, and much of the invertebrate freshwater fauna of Europe, apparently followed a northern migration route across Siberia rather than across central Asia (Bănărescu, 1992). All of the southern areas of the continent, including northwestern Africa, Anatolia, and the Middle East have Holarctic affinities.

SUMMARY

1. The fall of eustatic sea level at the end of the Cretaceous resulted in the desiccation of the Mid-Continental Sea and thus formed a single northern continent, an Eurasiamerica. A single marine passage, the Turgai Sea, extended from the Arctic Ocean to the Tethys Sea. India pivoted away from the African coast and its northern margin probably contacted Asia.
2. New Zealand was probably well separated from Australia and Antarctica. Australia remained close to but was separated from Antarctica. Africa moved closer to Europe and Asia. By the late Paleocene, the connection between North and South America, via Central America, became reduced to an island archipelago.
3. In the shallow, tropical seas, the Paleocene was a time of recovery from the major extinction that took place near the K/T boundary. Among the plankton, cool-water, high-latitude groups migrated to lower latitudes to replace the tropical forms that had become extinct. In this habitat, the recovery process took about 3 Ma. Recovery in the benthic habitat was evidently much slower. It took nearly 25 Ma for a molluscan community to approach its late-Cretaceous diversity level.
4. Near the end of the Paleocene, about 57 Ma ago, a rapid global warming, and associated oceanographic changes, caused a major extinction in the benthic fauna of the deep sea. Evidently, high surface temperatures in the polar regions interfered with the normal thermohaline circulation.
5. On land, where the North American terrestrial vegetation underwent considerable destruction during the K/T event, a period of recovery took place which resembled a modern secondary succession. No such changes were found in Australia nor Antarctica where there were no indications of a K/T extinction among the higher plants.
6. There are indications, among the patterns of frogs, lizards, freshwater fish, and mammals that a migration corridor between South and North America still existed, at least in the early Paleocene. Three groups of specialized mammals, that apparently originated in South America, are known from the late Paleocene of North America. These are the orders Edentata, Notoungulata, and Dinocerata.
7. It is clear that two primary freshwater fish families managed to migrate from Asia to North America at least as early as the Paleocene. The Catostomidae is represented by an archaic, relict genus in Asia. After a radiation in North America, one genus dispersed back across Beringia to Siberia. The Ictaluridae is endemic to North America but related to the Asian family Bagridae.
8. The freshwater fish families Esocidae and Hiodontidae also have Asian-American relationships. The esocids probably originated in North America, then made two westward crossings of Beringia. A generic line of caddisflies moved from Asia to North America, to South America, again to North America, and finally back to Asia.
9. An ancient group of turtles with a worldwide distribution was gradually replaced by two new groups, one of which arose in the northern hemisphere and the other in the southern hemisphere. Although the process was somewhat accelerated by the K/T extinction, it still took about 150 Ma.

10. Among the mammals, a high degree of cosmopolitanism was evident in the early and middle Paleocene but this gave way to considerable endemism in the late Paleocene. While intercontinental traffic in general appeared to slow down, the Beringia route was still comparatively active.
11. After invading North America from the south, the edentates, notoungulates, and dinocerates proceeded to migrate across Beringia, for fossils of all three groups have been discovered in the late Paleocene of Asia.
12. The Paleocene fauna of North America was apparently enriched by genera from Asia belonging to the orders Acreoidi and Pantodonta, and by unintathere ungulates. Plesiadapid primates may have reached North America from Europe.
13. An important event in mammalian biogeography took place when the first true rodents invaded North America from Asia. The smaller herbivore niche, occupied by multituberculates in North America, was probably taken over by the rodents. There seems to be a good correlation between the rise of the rodents and the demise of the multituberculates.
14. The marsupial decline, which became apparent in the North American Paleocene, has been attributed to the K/T boundary event. But the marsupial extinctions could have been due to competition from the placentals. Neither the multituberculates nor the placentals were adversely affected by the K/T extinctions.
15. There is some evidence of occasional Paleocene/Eocene connections between Africa and Europe. These involved palaeoryctid, lipotyphlan, creodont, and embrythopod mammals. In the Eocene, sister groups of theridomyid rodents and macroscelid condylarths apparently existed in Africa and Europe.
16. Beginning in the early Eocene, renewed movements of the earth's tectonic plates, accompanied by a rising sea level, had important biogeographic consequences. India became completely fused with Asia, the North Atlantic land bridge was severed, and an amphinotic filter route allowed exchanges of terrestrial organisms among South America, Antarctica, Australia, and New Zealand.
17. In the sea, for the first time since the Cretaceous, shallow marine waters became productive enough to support large herbivorous and carnivorous animals. These included the great white sharks, the first whales, and sirenians (dugongs, manatees, seacows). Mangrove swamps and angiosperm seagrasses became widespread.
18. On land, the early Eocene thermal maximum permitted tropical and temperate vegetation to spread to high latitudes. A highly diverse, temperate angiosperm vegetation, that had arisen in East Asia, was able to spread to Siberia, across Beringia, and across North America. Some of the species reached Europe to establish a Holarctic distribution. After radiation in the New World, many North American groups made reciprocal migrations back to Asia.
19. South American plants made their way up the Central American archipelago and established themselves in tropical North America by at least Eocene time. Australia continued to pick up elements of its angiosperm flora from southeast Asia. At the same time, a strong amphinotic floral relationship, involving 50–60 genera, developed among South America, Australia, and New Zealand.
20. The condylarths were early ungulate, placental mammals with an almost world-

wide distribution. By the Eocene, they had given rise to a variety of groups such as the artiodactyls, perissodactyls, proboscidians, cetaceans, and a variety of forms endemic to South America.

21. The early Eocene is notable for extensive evidence of migration among the Holarctic continents. Many of the mammals that appeared in North America at this time, probably came from Asia. Included may have been the first perissodactyls, artiodactyls, primates, and hyaenodontids.
22. Even more striking is the extraordinary resemblance of the European early Eocene mammal fauna to that of North America. At least half of the 61 land mammal genera of Europe had congeners in North America. This relationship was made possible by a land connection which probably occurred across Greenland and the Faroes. The connection was broken in the early Eocene.
23. In the late Eocene, the Asiatic influence in North America was renewed. The emigrants included mammals adapted to woodland and savanna habitats. Included were rabbits, lophodont rodents, entelodontids, and several families of artiodactyls and perissodactyls.
24. South America was probably invaded from Central America by ancestors of the caviomorph rodents. At about the same time, it is likely that there was also an invasion of omomyid primates from North America. The omomyids are possibly ancestral to the modern South American monkeys.
25. In South America, the marsupial mammals enjoyed their longest tenure, from the late Cretaceous to modern times. They probably reached Australia from South America via Antarctica in the Paleocene-Eocene. A late-Eocene fossil of a polydolopid marsupial has been found on the Antarctic Peninsula. A monotreme and a condylarth probably made the same migration.
26. The ratite birds (including the modern ostriches, cassowaries, kiwis, and rheas) were at one time thought to have been a Mesozoic group that had been dispersed by continental drift. Fossil materials now indicate that the ratites originated in the early Tertiary of North America and Europe. They exist as southern hemisphere relicts of a previously widely distributed group.
27. Three lizard families apparently originated in Asia and entered North America via Beringia. All three families (Eublepharidae, Agamidae, Scincidae) eventually became very widespread. Before the North Atlantic connection was broken, four lizard groups evidently migrated from North America to Europe.
28. The teleost fish fauna of the Eocene Green River Formation (Wyoming, Colorado, Utah) contains four genera that show Asiatic relationships. Also present was a primitive catfish family with closest relationships to a southern South American family. Another freshwater fish family, the Percidae, apparently reached North America from Europe in the early Eocene.
29. The salamander family Salamandridae probably entered North America from Europe in the early Eocene. Two other families of this group (Proteidae, Batrachosauroididae) seem to have crossed the North Atlantic in the opposite direction. Freshwater molluscs and aquatic insects also exhibit these historic amphipacific and amphiatlantic patterns.
30. It is suggested that the Andes orogeny, and associated tectonic changes in the

Paleogene, produced a filter bridge (an archipelago) from South America to Australia and New Zealand. This permitted, for the first time in almost 100 Ma, a significant biotic exchange.

31. Much of Madagascar's peculiar fauna and flora probably reached that island in the Paleogene. The various colonizations took place by means of overseas transport and included a wide variety of vertebrates, invertebrates, and plants.
32. Although India probably first made subaerial contact with Asia in the late Cretaceous, it is the Eocene fossils that reveal a large variety of mammals, freshwater fish, and plants with Eurasian, North American, and African relationships. There are no remains of a highly endemic biota that would be expected had India been isolated for 100 Ma or more.
33. The biology of the West Indian islands indicates that they did not first develop in the position of the present Central America and then migrate, with their biota, to their present positions in the Caribbean. The West Indian biota is younger and less differentiated than that of Central America. It also exhibits the disharmony that one expects to find on oceanic islands that have never been a part of the mainland.
34. Beginning in the middle/late Eocene and continuing to the late Oligocene, some important tectonic movements took place. These included the drift of Australia away from Antarctica and the continued push of India into Asia. At the same time, the major orogenies of the Laramide, the Andes, and Himalayas took place. The buildup of ice in Antarctica lowered the sea level and accentuated the climatic effects of the mountain building. The lowered sea level caused the desiccation of the Turgai Sea and allowed Europe and Asia to became one continent.
35. On land, the global climate became colder and drier. In the sea, the cold dense water near the poles descended to form the psychrosphere, the present cold bottom layer of the ocean. As the temperature declined, the biotic diversity in both environments also declined. The temperature drop continued through most of the Oligocene, reaching a minimum low about 29 Ma ago.
36. The severe climatic deterioration caused a broad range of extinctions. Because of these, the E/O boundary had been traditionally recognized as the time of a great, catastrophic extinction. However, during the past decade especially, a variety of works have demonstrated that there was not a single mass extinction, but an extended series of minor events that lasted some 14 Ma.
37. In the marine environment, pulses of extinction took place which appeared to be related to relatively rapid drops in ocean temperature. These have been found in both benthic and pelagic habitats. At the E/O boundary, the extinctions appear to have been relatively minor. There is no evidence of a direct association with postulated extraterrestrial impacts.
38. The whales suffered no evident extinction. The early Oligocene was a time of a dramatic evolutionary radiation. By the late Oligocene, there were 11 families and perhaps 36–50 species. Cetacean evolution seems to have been driven by new feeding opportunities made possible by ocean cooling and increased thermohaline circulation.
39. Terrestrial vegetation in the northern hemisphere has provided evidence of a

gradual cooling and drying which proceeded at a rate comparable to the oceanic cooling. Moist forests gave way to a low-biomass vegetation where savannas and open woodland were common.

40. With the advent of the Oligocene, a change in the mammalian fauna took place which has been called, "... the most impressive turnover in the whole age of mammals." In North America, a host of new mammal families appeared. Fossils of many modern families were found for the first time, with some having arrived from Asia.
41. An environmental deterioration is reflected in the general shift from small mammal/arborial faunas in the mid-Eocene to large mammal/savanna faunas in the Eo-Oligocene. This change is consistent with the evidence from plant studies which indicated increasingly dry habitats. By the end of the Oligocene, many of the older clades, that had originated in the late Cretaceous or early Paleogene became extinct in North America.
42. An important mammalian event in Europe took place near the E/O boundary. This has been called the "Grande Coupure" and represents an important immigration together with an extinction. The causes were probably climatic change together with an invasion from Asia, which was permitted by the retreat of the Turgai Sea.
43. The changeover of the mammalian fauna in Asia was not as abrupt as in Europe and North America. The extinction of the archaic forms occurred in a stepwise manner. The new advanced forms appeared gradually. Some of them made their first appearance earlier than in Europe and North America.
44. The highly endemic Eocene/Oligocene mammalian fauna of Africa showed little evidence of faunal turnover. Africa maintained a relatively stable, tropical climate which evidently allowed for a continuation of its diversity.
45. The reptiles and amphibians of North America underwent great changes attributable to the loss of freshwater streams and lakes and the general temperature decline. The loss in diversity occurred over a period of about 16 Ma and could not have been caused by any single event.
46. In Europe, more than three fourths of the late-Eocene herpetofauna had died out by the time of the E/O boundary. Oriental migrants invaded in the early Oligocene and probably contributed to the extinction of the remaining Eocene forms.
47. The end of the Turgai Sea in the Oligocene was an important event for the freshwater fish. None of the primary freshwater families now in Europe has a fossil record on the continent prior to the Oligocene. The three families which dominate the European fresh waters today are clearly Asian immigrants that probably arrived in the Oligocene.

CHAPTER 7

Neogene

Even in well-documented sequences, species and genera appear suddenly in the fossil record. This pattern may be attributed to sudden evolution within the area being sampled, but it can almost always be accounted for by migration from some other part of the world.

Robert L. Carroll, *Vertebrate Paleontology and Evolution*, 1988.

MIOCENE

Northward movement of the African plate against Eurasia had important physiographic and biogeographic consequences. The contact with Asia via the Arabian Peninsula in the early Miocene provided a route for the interchange of terrestrial and freshwater organisms. At the same time, the intercontinental fusion terminated the Tethys Sea and created the Mediterranean basin (Map 15). In the late Miocene, further closure between Africa and Eurasia caused the orogenesis of the Alps and other mountain ranges to the east. These elevations separated the Paratethys basin from that of the Mediterranean. Most of the European freshwater runoff was captured by the Paratethys. The tectonic movement possibly closed the Strait of Gibraltar and isolated the Mediterranean. Because the evaporation rate exceeded the annual precipitation, the Mediterranean underwent a Messinian salinity crisis. Following this stage, the Paratethys broke through to the Mediterranean basin, establishing a series of freshwater lakes. Finally, the Gibraltar portal opened wider, once again establishing normal marine conditions.

As the Australian plate moved northward during the Miocene, its northern portion, including New Guinea and associated islands, contacted the end of the Indo-Malayan chain. Part of Sulawesi (Celebes) was formed from Australian Plate materials. All of the islands to the east of Sulawesi are evidently on the Australian Plate (Audley-Charles, 1987). The formation of the Indo-Australian Archipelago permitted the invasion of Australia-New Guinea by Asiatic terrestrial organisms adept at island hopping. These included numerous species of frogs, lizards, and snakes. There was a lesser amount of counter invasions.

In the marine environment, the northward movement of the Australian Plate restricted the flow of warm currents between the Pacific and Indian oceans. The restriction of this "Indonesian Seaway" blocked the westward movement of Pacific water and caused an increase in the Pacific Equatorial Countercurrent and Undercurrent. At the same time, the shoaling of the Central American Seaway interfered with the flow of Atlantic water into the Pacific. This also probably helped to strengthen the Equatorial Countercurrent. Considering that a stronger countercurrent structure retained more warm water within the tropics, the ultimate effect must have been an increased high-latitude cooling.

Formation of the Red Sea, a part of the East African Rift System, probably began to take place as the result of tectonic movement that began in the mid-Eocene (Braithwaite, 1987). By the Miocene, a basin had formed which, at first, was connected to the Mediterranean. The basin was then isolated and extensive evaporite deposits took place. There may have been further intermittent connections to the Mediterranean. Towards the end of the Miocene, a terrane uplift caused a lasting separation between the two seas. At about the same time, there was a renewed separation of the African and Arabian plates creating a deep trough that opened into the Indian Ocean. This allowed an invasion of an Indo-West Pacific biota.

In the New World, the continued movement of Central America against the Cocos Plate contributed to the building of an apparently complete peninsula from North America all the way to Panama. The existence of such a peninsula is indicated by fossils of North American mammals that occur in the Miocene Cucaracha Formation of Panama. This improved north-south connection is also suggested by an increase in the migration of terrestrial vertebrates. Two families of ground sloths, a leptodactylid frog, and a teiid lizard probably migrated from South America northward, while a turtle, a raccoon, and colubrid snakes went the opposite direction.

Although a global change to a cooler and drier climate began in the Eocene, it took the terrestrial biota a long time to accommodate in an evolutionary sense. It took some 20 Ma for angiosperm plants to develop forms that could take full advantage of spaces that were left by the retreating forest. When a variety of grasses and herbaceous plants finally appeared in the Miocene, adaptive changes in animals such as insects and mammals took place very rapidly.

Marine patterns

In 1985 the Geological Society of America published an important volume devoted to the paleoceanography and paleobiology of the world ocean during the Miocene Epoch. This multi-authored work represents a significant advance in our knowledge of Miocene biogeography. The distribution of Pacific Ocean deep-sea benthic foraminiferans during the Miocene indicated that the early Miocene, deep ocean was relatively warm with a poorly developed, vertical thermal gradient (Woodruff, 1985). Near the middle Miocene, several important changes took place: (1) an increase in productivity along the ocean margins, (2) an increase in productivity along the equator, (3) a cooling and shallowing of the deep-ocean water mass, and (4) an intensification of the low-oxygen zone after 8–10 Ma. These changes appear to be related to continued expansion of the Antarctic ice sheet, an increase in the thermohaline circulation, and further development of the latitudinal thermal gradient.

The distribution of planktonic foraminiferans in the Pacific was investigated by Keller (1985). She found that the early and middle Miocene equatorial Pacific was dominated by a warm surface-water group, which showed a distinct east-west provincialism. During the late Miocene, the surface group declined and the east-west provincialism disappeared. It was concluded that this faunal change may have been associated with a major Antarctic glaciation and a strengthening of the general gyral circulation, and an increase of the Equatorial Countercurrent due to the closing of the Indonesian Seaway. The geographic distribution of inferred surface, intermediate, and deep-water dwellers in the late Miocene, was found to be very similar to modern temperature profiles.

Pacific planktonic foraminiferans were also studied by Kennett et al. (1985). They found the greatest contrast to be between the early and middle Miocene. They concluded that the increased Antarctic glaciation and high-latitude cooling caused the development of the Equatorial Undercurrent system when the Indonesian Seaway closed, and a general strengthening of the gyral circulation and the Equatorial Countercurrent. Significant changes in radiolarian populations appear to be correlated with an increase in Antarctic ice (13–15 Ma ago), closure of the Indonesian Seaway (11–12 Ma ago), and some Messinian (6–7 Ma ago) effect (Romine and Lombari, 1985).

Planktonic foraminiferans of the South Atlantic Ocean were examined by Hodell and Kennett (1985). They recorded a number of changes which seemed to indicate the initiation of the Benguela Current between 16 and 8 Ma ago. Also during this time, the ^{18}O data reflected an increase in global ice volume and/or a cooling of surface waters. Global isotopic data suggested that the surface temperature response to increased glaciation was a warming of the tropics, little or no change in the mid-latitudes, and a cooling at high latitudes. The net effect was a marked intensification of the planet's latitudinal temperature gradient.

The changing relationships of the tropical Pacific and Atlantic between the late Miocene (8 Ma ago) and the present have been followed using radiolarian distribution patterns (Romine, 1985). At 8 Ma ago, the flow of Atlantic water through the Central American Seaway disrupted the flow pattern of the Equatorial Countercurrent. This caused warm water, which is now returned to the Eastern Pacific, to be diverted into the northward-flowing western limb of the subtropical gyre. This resulted in increased transport of warm water to the north and warmer sea-surface temperatures in the North Pacific. As the rising Panama isthmus became a barrier to surface water flow, the interoceanic flow decreased and the North Pacific became cooler. This may have accelerated the global cooling trend.

Considerable changes also took place in the Atlantic. During the Messinian Stage towards the end of the Miocene, the Mediterranean Sea was possibly cut off from the Atlantic. This was due to the northward movement of the African plate towards Gibraltar and a lowering of sea level due to a renewed accumulation of ice in the Antarctic. Since the evaporation rate is much greater than the precipitation in that part of the world, the Mediterranean began to dry up and, in the process, thick layers of evaporites were deposited on the sea floor. The effects on the marine fauna were severe and over most of the area only a few species, able to withstand hypersaline conditions, survived (Landini and Sorbini, 1989). Supposedly, the period of evaporite deposition lasted about 500 000 years, but occasional fossils of fish typical of normal seawater (Sorbini, 1988), indicate that the desiccation process may not have been continuous. This suggests that the Gibraltar Strait might not have been constantly closed over that period. A model suggesting a continuous inflow and outflow during the Messinian has been developed (Sonnenfeld and Finetti, 1985).

Towards the end of the Messinian, the freshwater Paratethys drained into the Mediterranean basin so, for a brief time (about 0.1 Ma), both seas were reduced to a network of low salinity lakes. This has been called the "Lago Mare" phase of the Mediterranean. Presumably, this phase allowed Paratethyan freshwater fish to enter the Mediterranean and to ascend its tributaries (Bianco, 1990). The Messinian crisis ended when the Straits of Gibraltar opened and normal seawater conditions were reestablished. The remainder of the Paratethys (Black and Caspian basins) also became flooded with seawater causing a catastrophic extinction of the endemic freshwater biota.

Molluscan provinces for the low- and mid-latitude parts of the New World Miocene have been defined by Petuch (1988). For the Western Atlantic, he recognized a Transmarian Province from North Carolina to Labrador; a Caloosahatchian Province from North Carolina to Florida and the Gulf of Mexico; a Gatunian Province for the Caribbean, Antilles, and northern South America; and a Pernambucan Province below about Cape Sao Roque, Brazil. He considered the Gatunian Province to also include, as a separate component, the Eastern Pacific from California to Peru. The recognition of a single molluscan province for both sides of the developing Isthmus of Panama indicates a fairly homogeneous tropical fauna.

The maximum homogeneity between the molluscan faunas on each side of the developing isthmus has traditionally been considered to have occurred during the middle Miocene, but more recent estimates have indicated the Pliocene (Jones and Hasson, 1985). Many of the hermatypic coral genera extant in the Indo-West Pacific disappeared from the Eastern Pacific in the Paleocene and Eocene but survived in the Caribbean until the Miocene. The Miocene benthic microfaunas of the Caribbean and Pacific coasts of northern South America show that a single faunal province extended throughout Trinidad, northern Venezuela, Columbia, and Ecuador.

Terrestrial patterns

Plants

It has been suggested that the Neogene might be described as the Age of Herbs (Stanley, 1989). In general, herbs are small, non-woody plants that die back after releasing their seeds. The phenomenal success of herbs and grasses is due to the worldwide climate deterioration which began in the Eocene and continued through the Miocene. For example, the Compositae, an important family of herbs, appeared only about 20–25 Ma ago, yet now is represented by about 13 000 species. There are about 10 000 species of grasses. Both groups can tolerate rainfall levels that are too low to sustain forests or woodlands.

The expansion of grasslands and savannas during the Miocene greatly increased the diversity of certain animal groups. The proboscideans in Africa and the horses of the family Equidae in North America underwent notable radiations. A huge increase in bovid diversity took place, especially in Africa. The deer family Cervidae first appeared in the early Miocene of Europe, Asia, and Africa. The North American antilocaprids emerged at about the same time. The proliferation of grasses and herbs also gave an impetus to the radiation of small rodents and insects. The insects fueled diversity increases in frogs,

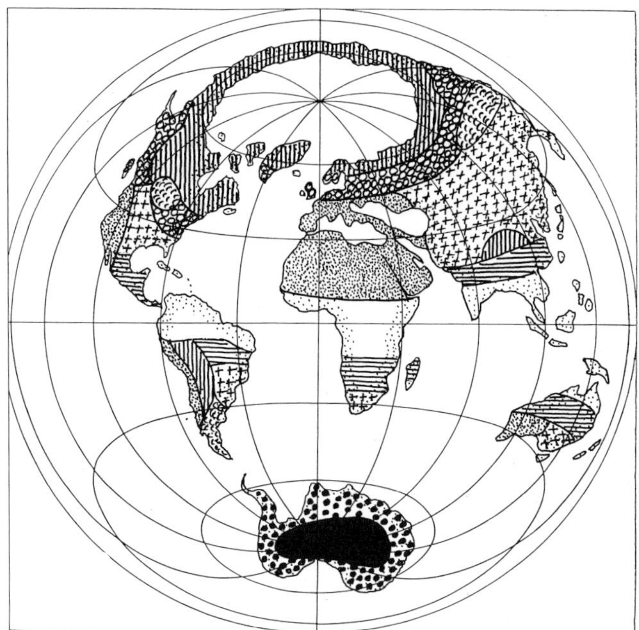

Fig. 45. Late-Miocene vegetation: Tropical forest (outlined by dots), Paratropical forest (shaded with horizontal lines), Broad-leafed evergreen (shaded with pluses), Broad-leafed deciduous (shaded with open circles), Mixed coniferous-deciduous (shaded with vertical lines), Woodland/thorn/scrub/chaparral (shaded with dots), Tundra (shaded with filled circles), Grassland (shaded with small curves) and Ice (solid black area). Modified after Janis (1993).

lizards, and passerine birds. Similarly, snakes benefitted from the more numerous rodents and frogs. And predatory cats, dogs, and birds benefitted at the highest trophic level.

Indeed, it may be seen that the global climatic change to colder and drier conditions at high latitudes, caused an extensive series of changes throughout the food web. These changes, which got well underway in the Miocene, resulted in the evolution of our modern terrestrial flora and fauna. Forests were replaced by tall-grass prairie, short-grass prairie, or desert depending on the amount of annual precipitation. At lower latitudes, tropical rain forests were often replaced by savanna or thorn scrub communities. A map (Fig. 45) of the global vegetation in the late Miocene by Janis (1993) illustrates the change to drier conditions at higher latitudes. Drought conditions in Africa were especially severe and many taxa suffered extinction. Consequently, the African tropical flora is depauperate compared to that of South America, Madagascar, and southeast Asia. Early Miocene microfossils from the Cape region of Africa show the presence of several families that are now extinct on the continent, i.e., Winteraceae, Casuarinaceae, Chloranthaceae (Coetzee and Muller, 1984).

Mammals

Among the ungulates, the perissodactyls underwent their main radiation in the Eocene and thereafter began to decline. By the mid-Miocene, only four families remained. These were the Equidae, Tapiridae, Rhinocerotidae, and Calicotheriidae. As noted, the fossil record of the family Equidae provides an excellent account of long-term evolutionary change. Horses evidently evolved from a stem genus called *Hyracotherium* which is known from the early Eocene of both North America and Europe. This form gave rise to the Equidae in the former and the Palaeotheriidae in the latter. The palaeotheriids became extinct early in the Oligocene (Carroll, 1988).

In North America, beginning in the Eocene, one type of browsing horse succeeded another (MacFadden, 1992). These are usually recognized as a series of distinct genera: *Orohippus*, *Epihippus*, *Mesohippus*, and *Miohippus*. Toward the end of the Oligocene, a dichotomy took place which produced the genera *Parahippus* and *Anchitherium*. The latter gave rise to four succeeding genera in the early Miocene, two of which migrated across Beringia to the Old World (Fig. 46). These browsing horses continued in both North America and Asia, until their extinction in the late Miocene.

A significant transition in horse evolution took place in the early Miocene when *Parahippus* gave rise to *Merychippus*. This meant a change from a generalized or browsing feeding habit to grazing on grasses. A new hypsodont tooth structure permitted entry to a new (prairie) habitat and provided an evolutionary impetus. This resulted in the relatively sudden (middle-to late-Miocene) production of about 12 genera and a multitude of species. Two of the genera reached Asia by way of Beringia. All eventually became extinct except *Dinohippus* which evolved into *Equus* and *Hippidion*. In the late Pliocene, *Equus* migrated to South America and to Asia while *Hippidion* moved to South America. *Equus* then became extinct in North America.

The Proboscidea, which includes the modern elephants, were confined to Africa until the early Miocene. The earliest known proboscideans are from the early Eocene of southern Algeria. The Gomphotheriidae is considered an ancestral stock which gave rise to a succession of other groups (Carroll, 1988). By the Miocene, all the major lineages had emerged. Descendants of the early gomphotheriids reached all continents except Australia and Antarctica. Gomphotheriids and mammutids arrived in North America from Asia about 15 Ma ago and remained in the New World until the end of Pleistocene time (Savage and Russell, 1983).

Most modern artiodactyls (about 65 genera) belong to a single group, the Pecora. This includes the deer, giraffes, cattle, sheep, goats, and antelopes. Their major radiation occurred later than most of the perissodactyls. Deer first appeared in the early Miocene of Europe, Asia, and Africa. The North American antilocaprids also appeared in the early Miocene. Bovids occurred in the late/early Miocene of Europe but their major radiation took place in Africa. Cattle, sheep, goats, and Old World antelopes all belong to this family. The Giraffidae has been an Old World group occurring in Asia, Africa, and Europe.

The movement of the Arabian Peninsula against Eurasia in the early Miocene provided a migratory corridor for the terrestrial biota. This resulted in a major turnover in the European mammalian fauna. Some 57% of the genera and 23% of the families were apparently new immigrants (Savage and Russell, 1983). The newcomers included hyaenids, pliopithecids, giraffids, bovids, equids, and three families of proboscidians. The equids came from North America but the remainder from Africa and Asia. This changeover has been called the "Coupure miocène."

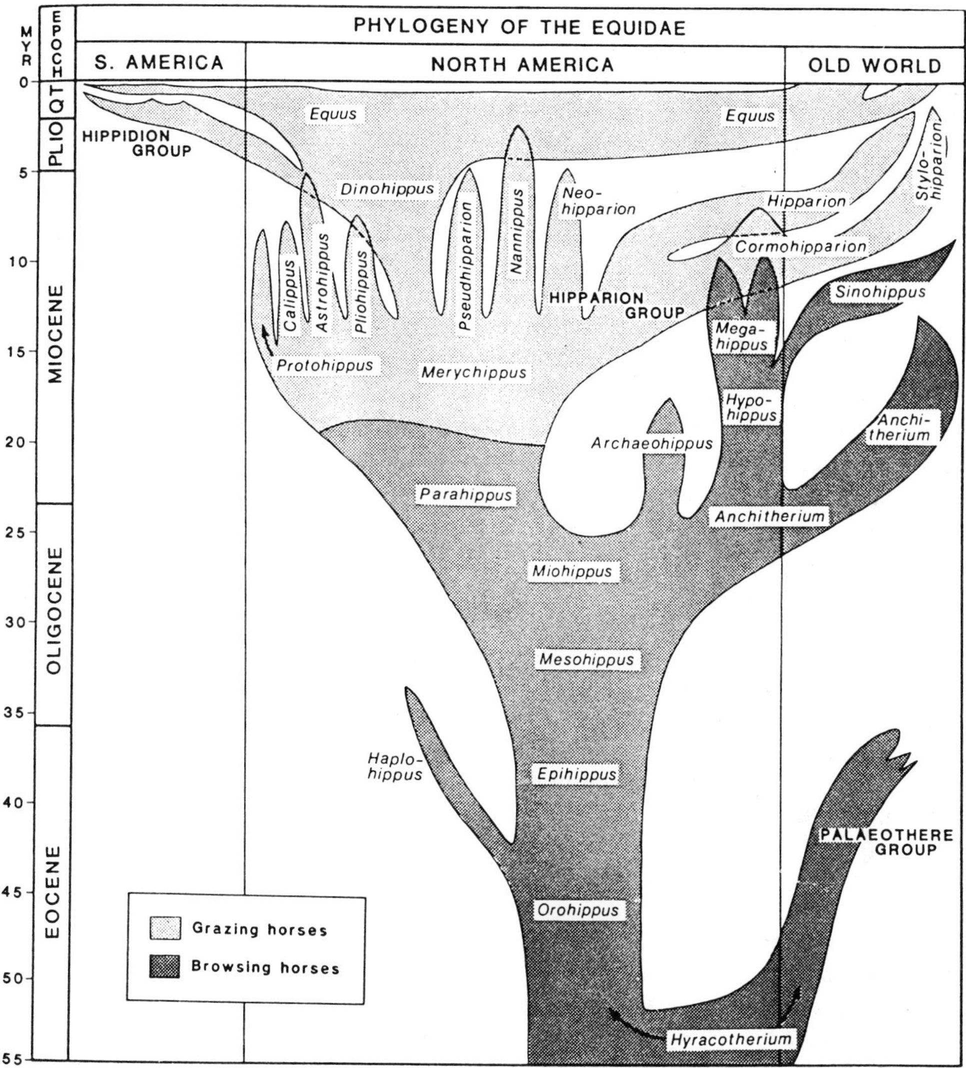

Fig. 46. A phylogeny of the Equidae indicating dispersal of several primitive genera to the Old World via Beringia. After MacFadden (1985).

The beginnings of the advanced, anthropoid primates took place in the Oligocene in both the New and Old Worlds. The first platyrrhine fossils have been found in the middle to late Oligocene of Bolivia and the first catarrhine fossils in the early Oligocene of North Africa. The old world monkeys, family Cercopithecidae, are a highly successful primate group that apparently arose in the early Miocene. The family ranges widely from Africa to southeast Asia and as far north as Japan. Molecular evidence appears to indicate that the division between humans (Hominidae) and the great apes, occurred between 6 and 10 Ma ago, but fossil evidence is lacking.

By the Miocene, the presence of mammalian fossils of North American affinity in the Panama Canal area (the Cucaracha Formation) indicate that the isthmian region had considerable contact with North America (Savage and Russell, 1983). In the late Miocene, North America received the first mammalian immigrants from South America since the late Paleocene (Webb, 1985b). These were two genera representing two different families of ground sloths. Presumably, they crossed considerable water gaps for they came alone and were not succeeded by other South American genera for another 4 Ma. At about the same time, a genus of the raccoon family showed up in South America (Webb, 1985b).

In North America, five different mammal ages are recognized for the Miocene. All five are characterized by some immigration from Asia (Webb, 1985a). In the earliest Miocene, new arrivals were an amphicyonid, a rhinocerotid, a chalicothere, a zapodid, and a eomyid. In the next age (the Hemingfordian) some 16 genera arrived. Among them were large ungulates such as antilocaprids, dromomerycids, moschids, and amphibious rhinocerotids. There were also many carnivore taxa including early genera of cats, bears, and several smaller forms. Rodent immigrants included fossil genera of beavers, an eomyid, and a flying squirrel.

Somewhat later in the Miocene, immigrants to North America from Asia included two proboscidean genera and cricetid rodents. These were followed by another cat, a small ruminant, an amphicyonid, and a shovel-tusked gomphothere. All these additions, plus the evolution of endemics, produced the largest standing crop of genera in the entire Tertiary record. In particular, the rich ungulate fauna was comparable to that of the recent African savanna (Webb, 1985a). In the Equidae alone, there co-existed nine grazing genera and three browsing genera. In the late Miocene, there were more immigrants including several carnivores, the first deer, an Old World antelope, modern beavers, mictotid rodents, and ochotonid hares.

Birds

The advanced or neognathous birds may be divided into two large assemblages, those that live primarily on or over land and those that live habitually in or over the water. Among the water birds, the Charadriiformes includes about 20 families with such familiar forms as the sandpipers, gulls, terns, and auks. They began their development in the late Cretaceous. Another very old order is the Procellariformes including the albatrosses, shearwaters, and petrels. The penguins (Sphenisciformes) are another primitive group dating from the late Eocene. Penguins were more numerous in the Miocene than at present. Their fossil history shows that they have been continually characteristic of the circum-Antarctic region (Carroll, 1988).

In the early Tertiary, before there were any large mammalian carnivores, there were two families of gigantic, flightless predaceous birds. The Phorusrachidae of South America included a dozen or more species, some of which reached nearly 3 m tall. The Diatrymidae reached more than 2 m in height and ranged throughout the northern continents. Within the modern land-bird assemblage, the Cuculiformes, including the hoatzin, cuckoos, and roadrunners, may be the most primitive (Olson, 1985). This order is widespread throughout the tropics of the world in a pattern that suggests an early Tertiary radiation. A similar pattern may be noted for the Galliformes (grouse, quail, pheasants, turkeys). Here, there is a worldwide pattern of specialized genera and species in southeast Asia with the more primitive members in peripheral areas (Briggs, 1984). This suggests a possible center of origin in the Asian tropics.

The parrots (Psittaciformes) are known from the early Miocene. They are primarily distributed in the southern hemisphere and reach their greatest generic diversity in the Australian region. It has been suggested that their early history took place on Gondwana (Cracraft, 1973) but the order does not appear to be nearly that old. The existence of several peculiar subfamilies in the Australian-New Zealand area indicates a long evolutionary history in that part of the world. Parrots either originated in the Australian region or else managed to get there early enough to undergo a significant part of their evolution.

The pigeons (Columbiformes) are first known from the late Eocene. They also have, primarily, a southern distribution. A related family, the Pteroclididae, apparently developed in Africa from an early columbiform stock (Cracraft, 1973). The dodos and solitaires, which belong to the pigeon family (Columbidae), existed only on the Mascarine Islands in the Western Indian Ocean. Based on our present knowledge of the two groups, it seems likely that the Australian region served as a center of origin for both the parrots and the pigeons. It seems apparent that the early evolution of the bird tribe Corvida (crows, ravens, magpies, and relatives) took place in Australia-New Guinea (Sibley and

Fig. 47. Early evolution of the bird tribe Corvini (crows, magpies, ravens and their relatives) was confined to Australia-New Guinea. Then, as that continent approached the Malay Archipelago in Oligocene/Miocene times, ancestral forms reached southeast Asia where a secondary burst of evolutionary radiation took place. Modern crows and ravens reinvaded Australia from Asia. Redrawn after Sibley and Ahlquist (1986).

Ahlquist, 1986) (Fig. 47). As Australia approached the Malay Archipelago in Miocene times, some forms reached southeast Asia where a secondary burst of evolutionary radiation took place. Modern crows and ravens dispersed from southeast Asia and also reinvaded Australia.

The most advanced birds belong to the order Passeriformes which comprise a huge group of more than 5000 species. They may have developed in the early Oligocene but their fossil record is very sparse (Carroll, 1988). The order may be divided into two large groups based upon the complexity of the syrinx and the structure of the stapes. The suboscines are considered to be the most primitive. They are the dominant passerine birds of South America where they total about 1000 species. The oscines or songbirds are the dominant passerines in the rest of the world. They appear to have originated in the Old World tropics and then spread into North America by the middle to late Miocene. They may not have reached South America until the Pliocene.

Herpetofauna

The Miocene herpetofauna of North America demonstrates some Asian and South American relationships as well as considerable increases in endemic diversity (Estes and Báez, 1985). The cryptobranchid salamander *Andrias* has living representatives in east Asia and a fossil record in Europe. Among the frogs, the cosmopolitan genera *Bufo* and *Rana* made their first North American appearance; also, first occurrences of the families Pelobatidae, Pelodytidae, Microhylidae, and Leptodactylidae took place. Dispersal via Beringia is likely for all except the leptodactylids which came from South America.

Among the lizards, the genus *Anolis* first appears in the Miocene of Florida. It probably invaded from the West Indies. The teiid lizard *Cnemidophorus* shows up in North America. It may have migrated directly from South America or could have resulted from a Central American radiation (Estes and Báez, 1985). There is also a first appearance of the helodermatid *Heloderma*. This genus has been recorded from the late Eocene or early Oligocene of France and probably represents a Beringian immigration. The same route was undoubtedly utilized by the ancestors of the viperid snakes *Agkistrodon* and *Crotalus* and the elapid genus *Micrurus*.

In South America, the living turtle genus *Geochelone* made its appearance in the Miocene of Argentina. And the first colubrid snake fossils were found in the Miocene of Columbia (Estes and Báez, 1985). Both probably represent immigrations from North America.

The Miocene approach of Australia may have permitted the immigration from Asia of four lizard groups, the agamids, geckos, skinks, and varanids. The ancestors of the pygopodids, an endemic family, may have gotten there earlier. Judging from their extensive radiation in Australia, the geckos may also have gotten there earlier. The Miocene invasion probably included the frog families Ranidae and Microhylidae; the turtle families Carettochelyidae and Trionychidae; and the snake families Elapidae, Colubridae, Acrochordidae, Uropeltidae, Typhlopidae, and Boidae. Of the snake families, the radiation of the elapids (25 genera and 63 species in Australia) indicates that they may have been the first arrivals (Briggs, 1987).

When it became connected with Asia in the early Miocene, Africa was probably invaded by two turtle families, the Testudinidae and the Trionychidae. Several families of frogs, lizards and snakes probably originated in Africa and subsequently spread to Asia and other places. These include the frog families Microhylidae and Ranidae, the lizard families Chamaeleontidae and Scincidae (and possibly the Lacertidae), and the snake family Viperidae.

Freshwater fauna

It has been noted that the ostariophysan fish (characoids, catfish, minnows), a primary freshwater group, have been able to dominate the freshwaters to which they have been able to gain access. The most successful family, in terms of its diversity, is the Cyprinidae (minnows, carps, daces, rasboras, etc.). There are at least 220 genera and some 1700 species. This family, and its close relatives, are generally considered the most specialized of the ostariophysans and they are probably the most recently evolved. While the characoids and catfish possibly developed in the Upper Jurassic (Briggs, 1987), the cyprinids arose, or at least radiated, much later.

The center of cyprinid diversity, and probably the center of origin for the family, is southeast Asia. In that general area, one may find all seven of the recognized subfamilies (Howes, 1991a) and a large portion of the genera. The initial radiation of the cyprinids apparently took place in the early Paleogene. They were in India by the Eocene (Hora, 1939), Western Europe by the Oligocene (Bănărescu, 1992), North America by the Oligocene (Cavender, 1986), and Africa by the Miocene (Van Couvering, 1977). Unlike the characoids and catfish, their radiation came too late for them to reach South America by means of freshwater connections.

The theory of a cyprinid origin in, and a subsequent dispersal from, southeast Asia (Briggs, 1979) has not been entirely accepted. Because two of the subfamilies are now well represented in Africa, Howes (1991) suggested that the family may have had an origin in Gondwana and was thereafter introduced to Eurasia by India, when the latter completed its interhemispheric voyage. There are several facts which argue against the Gondwana hypothesis: (1) there are no cyprinids in South America and that continent certainly formed a large part of Gondwana; (2) in Africa, diversity at the generic and subfamily levels is far less than in southeast Asia; (3) many of the African species have generic-level relationships to Asia which suggest a fairly recent dichotomy; and (4) several other freshwater fish families show similar patterns.

Fossils of cyprinid fish from Kenya have been dated from the early Miocene and it has been suggested that these fish entered Africa about 16–18 Ma ago (Van Couvering, 1977). Judging from their diversity patterns and African relationships, several other freshwater fish families probably originated in southeast Asia and eventually made their way to Africa. These are the Cobitidae, Anabantidae, Channidae, and Mastacembelidae. An interesting facet of the cyprinid invasion is that these fish do not seem to be compatible with the characoids. In Africa, the former may be in the process of replacing the latter.

The distribution of the freshwater fauna in Europe and the Near East was profoundly affected by the terrestrial connection between Africa and Eurasia. This event not only permitted a faunal exchange, but also affected the dispersal of freshwater biota throughout southern Europe. During the middle Miocene, the orogeny of the Alps, and associated mountain ranges, separated the Paratethys from the Mediterranean (Bianco, 1990).

As a result of the emergence, most of the European stream drainages were diverted into the Paratethys. This sea, although still connected to the Mediterranean in two or three places, gradually lost its oceanic salinity. By the end of the Miocene, it was probably a huge freshwater lake. This gave freshwater organisms easy access from Asia across most of southern Europe. Within the Paratethys, considerable endemism developed.

It has been recognized for many years that the freshwater fauna of Europe is dominated by lineages of Siberian or East Asian origin. The Holarctic affinities of the European fauna have generally been emphasized. It has been stated that, "All southern areas of the continent, as well as northwestern Africa, Anatolia, and the Middle East have been populated from the north (Bănărescu, 1992)." Bianco (1990) has suggested that primary freshwater fish dispersed through low salinity basins of the Mediterranean as it began to recover from the Messinian crisis. The current peri-Mediterranean distribution patterns seem to be consistent with this explanation.

In North America, cyprinid fish are known from the Oligocene, but may have entered the continent earlier. Three different genera are known from Oligocene localities in Oregon, Washington, and California (Cavender, 1986). Numerous cyprinid fossils are known from the Miocene but all occur west of the continental divide. They certainly must have spread eastward by this time. In the west, their diversity, along with that of other freshwater groups, was adversely affected by tectonic events. In the Rocky Mountain region and in the far west, block faulting accompanied by volcanism took place. These events, plus increased erosion, subdivided populations into small segments and greatly increased their rate of extinction. Eastern North America remained relatively quiescent and diversity among the freshwater fish, especially in the cyprinids and percids, increased rapidly.

In South America, the two primary freshwater fish groups (characoids and catfish), which had probably reached that continent in the late Jurassic or early Cretaceous, found a fertile area for their expansion. South America now has 13 families of catfish, all of them endemic. There are 15 families of characoids, 14 of which are endemic and one shared with Africa. It has become increasingly apparent that this enormous evolutionary expansion must have taken place in the early Tertiary. A fossil characoid from the Miocene of Columbia has been identified as a living species (Lundberg, 1986). And a catfish from the late Miocene of Venezuela has also been identified as a living species (Lundberg, 1988). These might be indications that the present level of diversification may have been reached by the Miocene and, thereafter, a kind of evolutionary stasis prevailed.

A Miocene extinction?

At one time, it was thought that a compilation of data on genera in the marine fossil record (Raup and Sepkoski, 1984) provided evidence of a middle- Miocene mass extinction. It is true that when the extinction data are plotted on a graph with a highly compressed time scale, a sharp peak appears in the mid-Miocene. However, the reliability of the data base has been seriously questioned (Smith and Patterson, 1988). In addition, recent works on Miocene planktic and benthic organisms indicate that faunal turnovers were associated with gradual changes in oceanic conditions, rather than being caused by some catastrophic event. An increase in the Antarctic ice pack, closure of the Indonesian Seaway, and restriction of the Panamerican Seaway south of Panama, affected ocean temperatures and currents. These changes probably raised the extinction rate, but they took place throughout the middle and late Miocene, not all at once.

In the North American Tertiary, the highest rate of mammalian generic extinctions took place in the early Miocene, about 20–24 Ma ago (Stucky, 1990). But, this interval also marked the time of the greatest number of Tertiary originations. This resulted in a historic peak of mammalian diversity. This turnover occurred much earlier than the supposed mid-Miocene marine extinctions.

PLIOCENE

Although the Pliocene Epoch was a relatively brief 3.5 Ma period at the end of the Tertiary, several events of biogeographical and evolutionary importance took place. There now exists a large volume of literature on the Great American Biotic Interchange. While this was an exceedingly important event, it was probably no more significant than another intermigration episode that took place some 0.5 Ma earlier. This was the flooding of Beringia which, for the first time in about 60 Ma, allowed the marine biotas of the North Pacific and Arctic-North Atlantic to intermix. This marine interchange has attracted less attention than the terrestrial one, probably because few mammals were involved. A comparison of the two gives the biologist considerable insight into the evolutionary consequences of the invasion process.

Another event of importance in the marine environment was the high-latitude cooling associated with the onset of northern hemisphere glaciation about 3.0 Ma ago. This caused the extinction of the boreal marine biota in the Arctic area and the beginning of a new cold-water biogeographic region. It was also the beginning of a series of cooling cycles in the North Atlantic which had a detrimental effect on its species diversity. In comparison, the North Pacific has not suffered such severe cooling and has been able to maintain a much higher level of diversity.

The Panamanian isthmus, when it became completed about 3.0 Ma ago, provided a dispersal corridor for terrestrial and freshwater organisms. Tetrapod fossils, particularly mammals, have allowed an unusually clear picture of past dispersal events. When the isthmus provided a land connection between the Americas, it simultaneously divided the tropical marine biota. The corridor in one environment became a barrier in the other, so an important dispersal event was at the same time an important vicariant event. This is an illustration of the fact that the tectonic or eustatic erection of a barrier in one environment will almost always provide dispersal opportunities in another.

Exchange of terrestrial animals between Asia and North America and between Africa and Eurasia continued through most of the Pliocene. The early Pliocene Paratethys basin was occupied by the Sarmatian Sea. This body gradually lost its salinity and became the Pontian Sea, identified by a peculiar biota. Tectonic movements then separated the Pontian Sea into two main parts, the Caspian and the Black Sea. The many shared species between these seas attest to their historic unity.

Marine patterns

Marine biogeographic patterns during the Pliocene were affected by two important tectonic events, the opening of the Bering Strait between Alaska and Siberia about 3.5 Ma ago and the emergence of the Panamanian land bridge about 3.0 Ma ago. When one considers the diversity and general distribution of the marine biotas that occupy the cold waters of the world, it becomes clear that the North Pacific has functioned as a center of evolutionary radiation (Briggs, 1974a). It has supplied species to, and to a large extent has controlled, the biotic characteristics of the Arctic and North Atlantic oceans; and it has made a major contribution to the diversity of the Antarctic seas.

Although the cool-water biotas of the North Pacific and the Arctic-North Atlantic began to evolve about the same time, they were completely isolated from one another until the mid-Pliocene. So, the opening of the Bering Strait permitted the mingling of biotas that had been separated for more than 60 Ma. The immediate consequences of the mixing are difficult to deduce from the fossil evidence, but considerable intermigration must have taken place. At that time, the North Pacific probably had, as it does today, a much more diverse biota. An hypothesis, applicable to biogeographic barriers in general, states that the area which develops the greatest species diversity will supply species to adjoining areas of lesser species diversity, but will accept few or no species in return (Briggs, 1974b). As noted in Chapter 9, empirical and theoretical evidence continues to indicate a positive relationship between species diversity and resistance to invasion.

The event of 3.5 Ma ago may be called the "Great Transarctic Biotic Interchange." Its biogeographic consequences have been evaluated, with an emphasis on the molluscan faunas, by Vermeij (1991). He identified 295 molluscan species that either took part in the interchange or had descended from taxa that did. Of these, 261 were determined to be of Pacific origin compared to 34 of Arctic-Atlantic origin. This gives a ratio of almost 8:1 in favor of the Pacific. The modern molluscan species diversity in the North Pacific is approximately twice as great as that of the Arctic-Atlantic. Although many northern molluscan species seem to have had early Pleistocene origins, Vermeij determined that the vast majority arose by anagenesis (without lineage splitting) so that it is reasonable to suppose that there has been little diversification since the early Pliocene. This means that the asymmetry of the invasions cannot be accounted for by the 2:1 ratio in species diversity.

There are two viable hypotheses that might account for the predominate success of the invaders from the Pacific: (1) the Pacific species, having come from a more diverse ecosystem, are competitively superior, or (2) an extinction event eliminated much of the Arctic-Atlantic fauna so that it was easy for the Pacific species to occupy the vacated niches. Vermeij (1991) emphasized the importance of the latter cause which he called an "Hypothesis of Ecological Opportunity". This is the same idea as the "Incumbent Replacement Model" presented by Rosenzweig and McCord (1991). Both are dependent on extinction to free ecological niches so that they can be occupied by an invader. Neither can explain how replacement occurs without the help of extinctions.

At the time of the transarctic interchange about 3.5 Ma ago, the Arctic Ocean was ice-free and boreal (cold-temperate) conditions prevailed (Golikov and Scarlato, 1989). So, a comparatively rich, North Pacific boreal fauna entered the Arctic-North Atlantic in large numbers. About 3.0 Ma ago, the cooling intensified and icebergs calved into the Arctic Ocean. Most of the boreal species were eliminated and the modern Arctic fauna began to develop.

Once established, the colder temperature of the Arctic waters prevented further penetration by boreal species, with the exception of some eurythermic arctic-boreal forms. This means that the Atlantic boreal species of Pacific origin were already in the Arctic-North Atlantic several hundred thousand years before the mid-Pliocene cooling episode. Therefore, it would be difficult to ascribe the success of the Pacific invaders to an extinction in the North Atlantic. It is more likely that the Pacific species, the products of a more highly diverse ecosystem, were the better competitors. The competition need not have been behavioral, but could have involved such factors as reproductive rate, individual size, or vulnerability to predators or parasites.

One must bear in mind that some invading species can successfully establish themselves by insinuation. This is applicable to species that have evolved so that the niche they occupy is unusual to the extent that the invader will not directly compete with a native species. This may be suspected when the invader succeeds in colonizing an area with a more diverse ecosystem. It might explain the success of the 34 molluscan species that dispersed from the Arctic-North Atlantic to the North Pacific. Although the molluscan movements are the best known due to the detailed analysis by Vermeij (1991), almost all groups of North Atlantic macroinvertebrates and fish possess some species of North Pacific ancestry.

Among the fish, the families Salmonidae, Osmeridae, Zoarcidae, Hexagrammidae, Cottidae, Agonidae, Liparididae, Stichaeidae, and Pholididae, probably all originated in the North Pacific but, during the transarctic interchange, contributed one or more species to the North Atlantic. The cod family Gadidae, on the other hand, developed primarily in the North Atlantic and contributed two species to the North Pacific. Among the marine mammals, fur seals and sealions of the family Otariidae and the walruses of the family Odobenidae originated in the North Pacific, while the seals of the family Phocidae are of Atlantic-Mediterranean origin (Carroll, 1988). The eelgrass *Zostera* (Hartog, 1970) and the kelp genus *Laminaria* (Estes and Steinberg, 1988) originated in the North Pacific, then spread to the North Atlantic.

Among the mollusca, it has been noted that the trans-Arctic invaders in the Atlantic have generally broader ranges than do native species with pre-Pliocene Atlantic histories (Vermeij, 1991). The apparent evolutionary consequences are also of interest. Of the trans-Arctic species that extend into the Atlantic, 48% are derived (speciated) forms. In the Pacific, 29% are derived. This indicates a remarkably low level of speciation for the past 3.5 Ma. Vermeij suggested that speciation among marine organisms appears to be much less frequent than assumed by evolutionary biologists. However, there is probably a difference in species longevity between cold-temperate and tropical habitats. Data from studies on the biota on each side of the Panamanian isthmus, where the separation has been in effect for about 3.0 Ma, show an exceedingly high level of speciation.

Molluscan fossils have shown that about 20–40% of early Pliocene boreal species in the North Pacific have become extinct; but, in the boreal North Atlantic more than 50% of the early Pliocene species have become extinct (Vermeij, 1989). Cooling episodes associated with northern hemisphere glaciation are generally recognized as the probable cause. However, Vermeij has maintained that cooling does not fully explain why the North Atlantic extinctions were so much greater, and that reduction in primary productivity must have played a part. We should note, however, that there are reasons why the North Atlantic, during glacial periods, undergoes more severe temperature declines than the North Pacific. It is a smaller ocean with a correspondingly smaller heat budget, and it is wide open to the inflow of ice from the Arctic basin. The North Pacific is protected from ice and cold water inflow by the Bering land bridge which has always been in place during glacial periods.

Another important effect of the cooling of the northern oceans and the establishment of the Arctic Biogeographic Region was the separation of the boreal regions. We have noted that, prior to the first Pliocene cooling episode of about 3.0 Ma ago, a boreal biota existed throughout the North Pacific and Arctic-North Atlantic. Boreal organisms, with the exception of some wide ranging arctic-boreal forms, were extirpated, by the temperature drop, from the Arctic Ocean as well as from the far northern parts of the Atlantic and Pacific. In the Atlantic, an Arctic biota now extends southward to the Strait of Belle Isle in the west and to the Kola Fjord at the base of the Murmansk Peninsula in the east. Included is all of Greenland and the northern half of Iceland (Briggs, 1974a). In the Pacific, an Arctic biota extends southward to Cape Olyutorsky in the west and Nunivak Island in the east. In each ocean, these southern extensions mean that the original Pliocene boreal regions were divided into two, one to the east and one to the west. Typical boreal species were no longer able to maintain amphipacific and amphiatlantic distributions and evolutionary change began to take place separately in each region. This is why we are now able to define a boreal region on each side of each ocean in terms of its endemic species.

Vermeij (1989) felt that the restriction of boreal species to one side or another of the Pacific or Atlantic was influenced by extinctions caused by reductions in primary productivity. Certainly, the larger sizes of individual organisms, such as those found in the northwestern Pacific, are probably a reflection of higher productivity. Yet, it is difficult to link higher productivity with the maintenance of a higher species diversity. In fact, it is clear that on a global scale, the areas of highest primary productivity generally have the lowest species diversity. It is the oligotrophic parts of the oceans that have the longer food chains and hence the greater diversity (Hallock, 1987).

The cooling episode of about 3.0 Ma ago, which decimated the cold-temperate fauna of the Arctic Region, also had a detrimental effect on the molluscan fauna of the tropical and warm-temperate Atlantic. This extinction, which may have continued with a succeeding cooling episode, eliminated as many as 65% of the early Pliocene bivalve fauna along the coast of the southeastern United States (Stanley, 1986). In comparison, the early Pliocene faunas of California and Japan lost only about 30% of their species. Stanley concluded that all of the endemic tropical species along the southeastern United States were lost, leaving only eurythermic forms that had extended into higher Latitudes. He suggested that perhaps two cooling episodes were involved, one at about 3.2 Ma ago and the other about 2.5 Ma ago. Sarnthein and Fenner (1988) have referred to a large-scale climatic deterioration that affected the North Atlantic 3.2–2.4 Ma ago.

At the same time, the huge early Pliocene bivalve fauna of the Mediterranean and North Sea basins suffered a heavy extinction (Raffi et al., 1986). This is indicated by the low survival of early Pliocene species to Recent times. Species richness was reduced from 323, for the two basins, to 198 living species. The authors noted that the heavy extinction of about 3.0–3.2 Ma ago coincided with the earliest deposition of glacial tills in Iceland and with low temperatures indicated by oxygen isotope data from planktonic foraminiferans. They detected an added extinction pulse about 2.4–2.5 Ma ago which continued into the early Pleistocene. There were few extinctions in the middle or late Pleistocene, probably because only eurythermic species had survived.

During most of the Miocene, the molluscan fauna of the western Atlantic was divisible into four provinces (Petuch, 1988). The endemic forms of the two highest latitude provinces, the Transmarian to the north and the Pernambucan to the south, became extinct at the end of the Miocene. This left for the Pliocene, an expanded Caloosahatchian Province ranging from Nova Scotia to the Texas coast and a Gatunian Province extending southward from Texas through the Caribbean and along the eastern coast of South America to Uruguay. It also included the Pacific coast from California to Chile (Fig. 48). The closing of the Isthmus of Panama bisected the Gatunian Province and altered the flow of the Gulf Stream system. Upwelling in the western Atlantic may have been reduced, leading to a drop in productivity and a starvation of the rich Gatunian fauna. While the Gatunian fauna on the Atlantic side of the isthmus collapsed, that of the Pacific side survived to become the predecessor of the modern Panamanian Province.

With the close of the Bolivar Gap in northern Columbia, the Panamanian land bridge was completed about 3.0 Ma ago. The isthmus suddenly separated a homogeneous, tropical New World biota. The effects of this vicariant event have been followed with great interest, for it constitutes a natural experiment that has been carried on sufficiently long to produce evolutionary changes. Although much work still needs to be done on the groups that are separated by this New World Land Barrier, it is evident that morphological differences have developed within the population on each side. There are about 1000 fish species on both sides of the barrier but only about 12 (aside from circumtropical species and a few euryhaline forms) are still considered to be identical. The great majority of the shallow-water invertebrates on each side are now recognized as separate species. Large numbers of geminate or twin species, separated by the isthmus, provide evidence of common ancestral species.

In the Mediterranean area, Sorbini (1988) found a rich marine-fish fauna in the Lower and Middle Pliocene deposits of east-central Italy. About 20% of the genera were tropical forms but the remainder were typical of the present-day Atlantic-Mediterranean region. He concluded that the presence of this fauna indicated a considerable survival of normal seawater fish through the Messinian crisis. This deposit was compared to material from an Upper Pliocene locality where many of the same taxa were present but all the tropical genera were missing. This suggests that, by the Upper Pliocene, the Mediterranean had attained its present warm-temperate regime.

As noted, formation of the present cold (as opposed to boreal or cold-temperate) Arctic Region began with a sharp decline in temperature at about 3.0–3.2 Ma ago. This was a time of glaciers which surrounded the Arctic basin. Most of the present endemic Arctic species probably began their evolution at that time. In the southern hemisphere, however, there are signs of increased Antarctic glaciation and high-latitude cooling between the early and middle Miocene (Kennett et al., 1985). If cold water temperatures (+2° to –2°C.) surrounded the Antarctic continent at that time, it means that the Antarctic marine

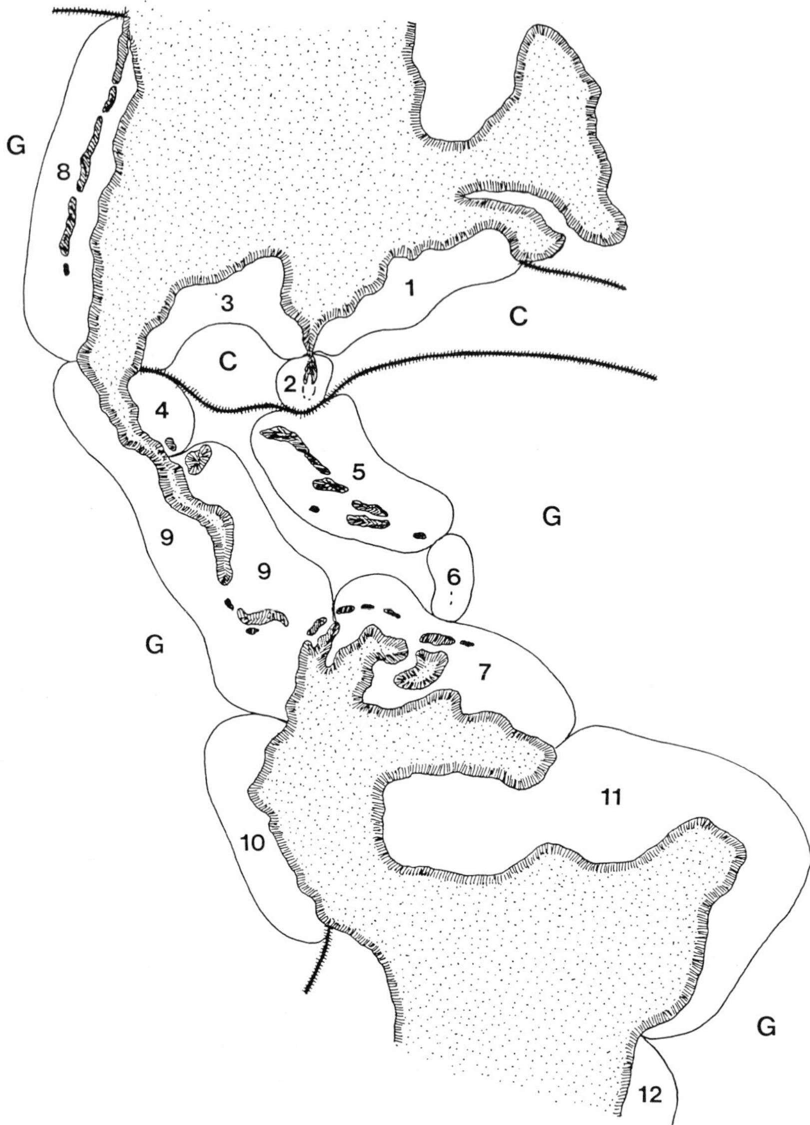

Fig. 48. Provincial arrangement of Pliocene molluscan faunas in the Americas. Provincial boundaries (crosshatched lines) are shown in relation to Pliocene continental configurations. C, Pliocene Caloosahatchian Province; G, Pliocene Gatunian Province. Subprovinces include: (Caloosahatchian) 1, Yorktownian; 2, Buckinghamian; 3, Jacksonbluffian (Gatunian); 4, Agueguexitean; 5, Guraban; 6, Carriacouan; 7, Puntagavilanian; 8, Imperialian; 9, Limonian; 10, Esmeraldan; 11, Piraban; 12, Camachoan. From Petuch (1988).

fauna got an evolutionary head start over that of the Arctic. It has been suggested that the Antarctic Convergence first developed about 20 Ma ago (Anderson, 1990). Warnke et al. (1992) found evidence that ice-rafted debris had reached 51°32′S at about 23.5 Ma ago. This would allow sufficient time for the development of the present highly diverse Antarctic fauna with its many endemic genera. In contrast, the Arctic fauna is much less diverse, with endemism expressed mainly at the species level.

Terrestrial patterns

Plants

After about 13 Ma ago, a decline in temperature from the Miocene through the Pliocene is associated with a retreat of paleotropical flora in the northern hemisphere and expansion of temperate deciduous trees, grasses, composites and other herbaceous dicots (Potts and Behrensmeyer, 1992). The cold-temperate broad-leaved deciduous forest was gradually replaced by coniferous forest, taiga, and birch forest. Angiosperm floras exhibited two important responses to cooling and increased seasonability in the mid-latitudes. First, trees and shrubs with deciduous habits became dominant. Second, herbaceous, probably annual, monocots and dicots evolved and diversified; this is thought to correspond to the expansion of grasslands and semiarid to arid conditions (Tiffney, 1984).

The spread of low-biomass vegetation during the late Neogene is linked to two parallel trends in climate: a consistent decline in mean annual temperature at high latitudes and an increase in mean temperature at low latitudes. Consequently, latitudinal temperature gradients were enhanced and subtropical high-pressure systems were intensified, causing summer drought along some continental margins (Wolfe, 1985). During the Pliocene, steppe vegetation appeared in the highlands of Asia and western North America. Tundra first appeared at high latitudes about 2–3 Ma ago. True desert vegetation may not have appeared before the Quaternary.

Webb (1991) investigated the floral changes that took place in the New World tropics in order to explain why certain mammals took part in the interamerican exchange and why others did not. It seems that during the cold or glacial phases of the Pliocene and Pleistocene, savanna or savanna-woodland habitats were predominant from the Gulf Coast of the United States, southward through Central America, the isthmus region, and the greater part of South America. The moist forest biotas were supposedly confined to relatively small refugia in areas now characterized by excessively high rainfall. Rain forests were also said to track river systems in relatively narrow bands of gallery forest.

The interchange of various plants in particular habitats was discussed by Simpson and Neff (1985). They noted that a few lowland tropical elements, such as members of the Aristolochiaceae and the Vitaceae, moved from North America into South America. But that many species now found in Panama and countries to the north belong to the genera that are "centered" in South America. In the montane habitat, there are some widespread Holarctic elements that arrived in South America by mountain hopping. Several important plant groups have disjunct distributions between the deserts of North and South America. Examples are genera in the Cactaceae, Celastraceae, Fabaceae, and Verbenaceae. The creosote bush, *Larrea*, evidently originated in South America and migrated northward while other genera, such as *Agave*, *Parthenium*, and *Zaluzania*, apparently traveled in the opposite direction. The Mediterranean scrub vegetation of the two continents also demonstrates some relationships. Genera of presumed northern origin that are now found in the Chilean matorral include *Myrica*, *Salix*, *Alnus*, *Berberis* and *Ribes*. Almost all such relationships are probably attributable to long distance dispersal.

Mammals

Although considerable intercontinental migration had its effect on the North American mammal fauna in the Miocene, the greatest change took place during the Pliocene. The whole of the mammalian fauna underwent an extensive turnover between 3 and 5 Ma ago (Webb, 1985a). More than two-thirds of all the genera were new appearances, about half of which were probably new immigrants. This was by far the highest rate of generic immigration during the entire Tertiary. The majority of the invaders came from Asia across Beringia.

The first record of South American animals that walked across the newly emerged land bridge occurs in rocks of late Blancan age that date from about 2.5 Ma ago (Marshall, 1988). Seven genera of land mammals and one large ground bird appear almost simultaneously in faunas of this age in Florida, Texas, New Mexico, Arizona, and California. These immigrants consisted of two armadillos, a giant glypodont, two ground sloths, a porcupine, a large capybara, and a phororhacoid ground bird.

A second major contingent of immigrants from South America appeared in rocks dating from about 1.9 Ma ago, mostly from localities across the southern United States. These included a giant armadillo, another ground sloth, a giant anteater, and a capybara. A bit later came an opossum, still another ground sloth, and a rhino-like toxodont. The toxodont and the latest ground sloth are known only from southern Central America, and the anteater from a single site in northern Mexico. Records of the last two have been established within the last six years (Marshall, 1988).

The first records of North American mammals traveling to South America appear in Argentinean rocks dating from about 2.5 to 2.8 Ma ago. These include a skunk, a peccary, and a horse. The main contingent of North American dispersants show up in Argentinean deposits that are about 2 Ma old. Sixteen genera representing nine families have been found including dogs (two genera), cats (a saber-tooth and a smaller form), skunks (two), a bear, an elephant-like gomphothere, a second horse, a tapir, camels (three), and deer (three). Rabbits and squirrels appeared somewhat later. North American shrews, pocket gophers, and kangaroo rats live in South America now but no one knows when they arrived. Field mice, known in North America as early as 9 Ma ago, have been found as fossils in Argentina dating from 2.8 to 3.0 Ma ago. Because of their extensive evolutionary radiation in South America, it has been suggested that these cricetid mice may have arrived in the early Miocene, or that they might have undergone some of their radiation in Central America.

The interamerican exchanges have apparently taken place under two contrasting climatic conditions (Webb, 1991). An interglacial phase was characterized by an unbroken rain forest that extended from northern South America, through Central America, to southern Mexico. This enabled a rich, moist tropical biota to spread northward from South America (Simpson and Neff, 1985). In the drier, glacial phase, the rain forest was largely supplanted by savannas or open woodlands. This change established an open-country highway which led through the tropics from one temperate zone to the other. There may have been a "low road", which led eastward through the llanos and then southward along the Atlantic coast of Brazil, and a "high road" leading directly south following the slope of the Andes (Webb, 1991). Probably all of the North American invaders dispersed southward during a glacial phase, since they were not adapted to rain forest conditions.

For the South American contingent in North America, one cannot make a good case for competition. Animals like the ground sloths, armadillos, porcupine, and the opossum, occupied peculiar niches that had no North American equivalents. For the reciprocal question, there seems to be considerable evidence of competition which eventually led to extinction of many of the South American endemics. The history of the ungulates in South America appears to reflect competition for, as the North American ungulates increased in diversity, the native species declined. A complicating factor may have been the placental carnivores, such as the mountain lion and sabre-tooth cat. The South American ungulates may have been especially vulnerable to such predators (Marshall, 1988).

Although the mammals that dispersed northward from South America have shown little subsequent evolutionary change, the same is not true for nine of the families that came from North America southward. For example, South America has seven living genera of the dog family Canidae, six living genera of deer (Cervidae), and 47 living genera of cricetid rodents (Cricetidae). At one time, these radiations were assumed to have taken place in South America. However, recent research indicates some of them probably took place in the tropics of North and Central America before the invasion of South America (Webb, 1991). Whether these evolutionary expansions took place in route, or in South America itself, they demonstrate a capacity for rapid adaptive change on the part of the invaders. Such a capacity was apparently lacking in the immigrants to North America.

How can one account for such a large evolutionary potential in one group of mammals while the other seems to lack it almost entirely? It has been noted that the primary reservoirs that contributed open-country biota to the interchange were in the temperate latitudes of North and South America. The land area of North America above the Isthmus of Tehuantepec is about six times larger than the area enclosed by Chile, Argentina and Uruguay. This gives an advantage to North America which possesses about twice as many mammalian species. In addition, there has been so much interchange between North America and Asia that one could argue that the northern mammals were actually the evolutionary products of the temperate Holarctic. In general, animals from a more diverse ecosystem will outcompete those in a less diverse system. It has been argued that habitat alteration, rather than competition, was responsible for the contrast in immigrant evolutionary success (Vrba, 1993). However, this cause is difficult to envision, for the glacial-interglacial climatic changes affected both hemispheres. Over the long run, the most important element in competition is probably the ability to adapt rapidly to a changing environment.

The impact of the interchange on the composition of the South American mammalian fauna has been summarized by Marshall (1988). When the interchange began, the fauna consisted of old families and genera that had been in South America since the Paleocene or had arrived there in the middle Tertiary. Today, the interchange invaders comprise about 44% of the families and 54% of the genera. So nearly half of the families and genera now on the South American continent belong to groups that emigrated from North America during the past 3 Ma.

Elsewhere in the world, the geographic relationships of Pliocene mammals have been studied in China and in Africa. The fauna of the Yushe Basin in eastern China spans the interval of about 6–2 Ma ago (Flynn et al., 1991). Both large and small mammals show a peak in species richness in the middle Pliocene but a relative faunal stability throughout the Epoch. Early Pliocene exchange between Asia and North America appears to have been balanced in both directions and involved a small proportion of the fauna. Immigration was probably opportunistic and contributed to faunal enrichment. Overall, the fauna was considered to be stable from about 5 to 2.5 Ma ago and changed mainly due to additions and congeneric species substitutions.

From the late Miocene through the Pliocene, African faunas experienced their greatest change since the Oligocene to Miocene transition (Potts and Behrensmeyer, 1992). At the generic level, 76% of the land mammals evident in the Pliocene were new, and about 53% of them were endemics. New arrivals from Eurasia included a camel, a calicothere, the modern giraffe, the horse genus *Equus*, and a sabre-tooth cat. The felid cats *Felis* and *Panthera* joined the canids and viverrids to form a significant carnivore component. There were dramatic turnovers and radiations in carnivores, ungulates, and ground-dwelling primates after about 5 Ma ago. The faunal changes corresponded with the development of more open, savanna-like communities. By about 2.5 Ma ago, desert conditions in the Sahara were apparently well established.

In Australia, the 28 fossil and living species of kangaroos of the genus *Macropus* probably evolved within the last 5 Ma. The shift from browse plants to grasses probably set the stage for their rapid expansion. With the radiation of the kangaroos, one can see the development of a set of correlated traits evident in placental mammals in other parts of the world: larger body size, greater mobility, greater reliance on grasslands, and larger aggregations of individuals. At about the same time, the wombats developed specializations parallel to the rodents in other continents.

Birds

The completion of the Panamanian isthmus had a profound effect on the bird faunas of the Americas. One of the most interesting of the invaders that came northward from South America is the phororachoid genus *Titanus* (Fig. 49). The phororachoids were giant, flightless predators that reached over 3 m in height (Marshall, 1988). Their remains have been found in the late Blancan and early Irvingtonian deposits of Florida. The Cenozoic fossil record of birds is not nearly as complete as that of the mammals due to the scarcity of well-preserved specimens. Leaving aside the passeriform and shore birds, Vuilleumier (1985) analyzed the relationships within the Americas. The fossils seem to show quite clearly that more North American birds have invaded South America than the reverse. Some 12 to 13 families dispersed southward while only four went in the opposite direction.

In the Eocene-Miocene, prior to the interamerican connection, 99% of the North American avian genera were endemic and 97% of the South American genera were endemic to that area. In the Pliocene-Pleistocene, however, the proportion of endemic genera in each continent falls to about 57%. The difference, according to Vuilleumier

Fig. 49. The flightless phororhachoid bird *Titanis* was the only large carnivore from South America to take part in the interchange. Scale is indicated by the human figure on the left. After Marshall (1988).

(1985), is due to the faunal interchange which began in the Pliocene and continued through the Pleistocene into the Recent. The fossil data on birds are not complete enough to tell us when the interchange started or when most of it took place.

It is remarkable that North American birds have apparently migrated southward in such superior numbers. Considering the Neotropical Region (South plus Central America) as a whole, it contains an enormous number of species – about one-third of all known species of birds. On the basis of comparative species diversity, one might expect most successful invasions to take place in the opposite direction. The reason probably lies in the evolutionary contrast between the two faunas. The Neotropical fauna is composed, to a large extent, of primitive families such as the Tinamidae (tinamous), Rheidae (rheas), Cracidae (curassows), Aramidae (limkpins), Psophidae (trumpters), Ceriamidae (seriemas), Steatornithidae (oilbirds), Nyctibiidae (potoos), Galbulidae (jacamars), Bucconidae (puffbirds), Ramphastidae (toucans), and nine families of the more primitive passeriforms. These groups give the Neotropics a decidedly plesiomorphic complex. The higher passeriform families (the oscines or songbirds) are predominant in North America and in most other parts of the world.

Herpetofauna

The effects of the Pliocene interchange on the American herpetofauna have been analyzed most recently by Vanzolini and Heyer (1985). The salamander family Plethodontidae evolved in North America and then dispersed south along the upland areas of Central and northern South America (Wake and Lynch, 1976). The South American species belong to the genus *Bolitoglossa* and are derived from a Central American radiation of that group. Since two of the species groups are endemic to South America, it has been suggested that the genus entered South America prior to the establishment of the isthmian link (Vanzolini and Heyer, 1985). The frog family Ranidae, represented in the New World by the genus *Rana*, entered North America from Asia via Beringia in the Miocene (p.) and underwent a modest radiation in both North and Central America. One species, a member of the Central American radiation, probably crossed the Pliocene isthmus to enter South America.

Among the reptiles, the turtle genus *Pseudemys* (Emydidae) entered South America from Central America long enough ago for differentiation at the species level. The data are most consistent with an entry via the Pliocene land bridge (Vanzolini and Heyer, 1985). The rattlesnake genus *Crotalus* probably arose in North America and has become widespread in South America. It and the colubrid snake *Drymarchon* probably reached South America via the isthmian bridge. Aside from these few examples, it is evident that the great majority of herpetofaunal interchanges between the Americas took place well before the formation of the isthmian link. These older exchanges were highly asymmetrical, with the fauna originating in South America being much more strongly represented in the present North American fauna than the converse.

Freshwater fauna

Formation of the complete isthmian link had a notable effect on freshwater fish distribution. Bussing (1985) recognized three distinct assemblages in Central America: an Old Southern, a New Southern, and Northern. The history of the Old Southern began in the late Cretaceous when an earlier isthmus permitted an entry for some South American fish. These were mainly secondary freshwater fish, but a few primary freshwater genera (one catfish and a few characoids) were apparently included. A new influx of South American primary freshwater fish began with the Pliocene development of freshwater stream systems along the new isthmus. Since that time, some 14 families and 31 genera of characoids and catfish have been able to move northward various distances.

Fishes of the New Southern fauna moved northward very slowly compared to the rapid pace of the mammals and herpetofauna. They do not presently extend north of the Golfo de Mosquitos on the Atlantic coast of Panama and only as far as Punta Mala on the Pacific coast of Costa Rica. Some species of secondary freshwater fish, descendants of the Old Southern group, have succeeded in moving southward into northern South America. These euryhaline forms probably moved along the estuaries, avoiding direct competition with the primary species. The Northern fauna consists of four genera that have been able to move southward along the Gulf of Mexico coast as far as northern Guatemala. The great bulk of the Central American freshwater fish fauna is the result of radiations that took place in the Poeciliidae and Cichlidae, members of the Old Southern fauna.

Although little information is available about the movements of aquatic insects between North and South America, Edmunds (1982) studied the mayflies and found a highly asymmetrical dispersal. Apparently, only a single genus invaded South America from the North, but no less than 21 genera dispersed in the opposite direction. Fourteen of the latter moved as far north as the United States and six extended into Canada.

In southern Europe and in the Near East, modern patterns, displayed by much of the freshwater biota, are related to the history of the Paratethys Sea. Seawater refilled the Mediterranean and Paratethys basins at the end of the Miocene. This formed a high-salinity Sarmatian Sea (Map 15). Over time, this sea gradually became smaller, contact with the Mediterranean became less, and the salinity dropped. By the late Pliocene, the former Paratethys was called the Pontian Sea in recognition of its peculiar low-salinity fauna.

In the late Pliocene, tectonic movements separated the Pontian Sea in two parts, the Caspian and Black seas. The Pontian fauna continued its development in the Caspian but, in the Black Sea, much of it was lost when the latter became reconnected to the Mediterranean. But a number of Pontic species survived in the Sea of Azov which maintained its low salinity. Today, about 75 out of a total of 305 Azov species are shared with the Caspian (Zenkevitch, 1963) and are called "Caspian relicts". In the modern Caspian Sea, about 47% of the species are endemics and most of the remainder are shared with the Black-Azov Sea.

PLEISTOCENE

Until the last few years, scientists have had great difficulty in determining Pleistocene climatic conditions, even though the changes have been dramatic and have occurred relatively recently. We used to think that there were only four major glaciations, but we now know that there were at least a dozen major and many minor ones (Crowley and North, 1991). This improvement in our knowledge has come about primarily through isotope analyses of deep-sea cores, examination of ice cores from Antarctica, study of uplifted coral reef terraces, and examination of windblown loess deposits.

As noted in the Pliocene account, the onset of significant glaciation in the northern hemisphere took place about 3 Ma ago. Since that time the succession of glacial and interglacial stages has profoundly affected the distribution and evolution of animals and plants. Compared to earlier epochs in the earth's history, there has been an enormous interest in the Pleistocene. Reasons are that Pleistocene fossils are relatively numerous and well preserved, the glaciers caused many spectacular geographical changes, and extensive migrations among humans and other mammals took place. The result has been a huge production of research and literature devoted to this brief period that began only 1.6 Ma ago, probably more than has been done on all the other intervals of earth history.

Each time the ice sheets built up, the sea level dropped and the shallow waters of the Bering Strait receded. The exposure of the land bridge presented, for the terrestrial biota, opportunities for intercontinental migration. At the same time, the land barrier separated the marine biota of the Arctic Ocean from that of the Bering Sea. These sea-level changes had important effects in other parts of the world as well. As the level dropped, the large islands off southeast Asia became connected to the mainland. This enhanced terrestrial dispersals but inhibited movement of tropical marine animals between the western Pacific and Indian oceans. Also, New Guinea became attached to Australia, Tasmania to Australia, and the Red Sea was cut off from the Indian Ocean. Fossil materials and distributions of the modern biota reflect these Pleistocene events.

Almost ever since the "Ice Age" was proposed as a theory by Louis Agassiz in 1840, scientists have argued about possible causes of glaciation. Once the stage was set by the current state of orogenesis and the present positions of the continents, there had to be some reason for the continued oscillation between glacial and interglacial stages. As Imbrie and Imbrie (1979) observed in their review, an astronomical theory was first proposed in 1842 by the French mathematician Joseph Adhémar, but it was best developed in 1930 and 1941 by the Serbian astronomer Milutin Milankovitch. His theory was based on an analysis of planetary perturbations on the receipt of solar isolation by the earth. The three variations of orbital eccentricity, obliquity, and precession are now thought to be responsible for the long term climatic cycles that control the ice ages.

Inspection of oxygen isotope records for the past 2.5 Ma indicates that, in addition to numerous oscillations of glaciers, there have been important long-term trends (Crowley and North, 1991). Distinct cycles of 100 000 years duration are present only for the last 700 000 years. Interglacials as warm as the present have occurred for only about 10% of the time in the late Quaternary. Before the late Pleistocene, glacial fluctuations had a periodicity of about 40 000 years. Distribution of glacial moraines on North America and Europe indicates that ice extent during some of the earlier advances was greater than the last glacial maximum.

Marine patterns

It has been noted that the cooling episode of about 3 Ma ago caused a wave of extinction in the early Pliocene fauna of the tropical and warm-temperate Atlantic. Among the mollusca of the western Atlantic, extinctions continued to take place during the early glacial periods of the Pleistocene (Petuch, 1988). It seems that the tropics of this side of the Atlantic were particularly vulnerable. Apparently, the steepening of the north-south temperature gradients strengthened the trade winds of the northern hemisphere. This may have caused the winds to blow more strongly along their diagonal paths towards the equator. As a result, the westward flowing South Equatorial Current may have been diverted entirely southward, instead of contributing part of its flow northward as it does now. This possibly deprived the western Atlantic, above the hump of South America, of some of its warm water supply and allowed it to cool down (Stanley, 1989).

Despite the widespread Plio-Pleistocene extinction of molluscs in the western Atlantic, some of the old Miocene and early Pliocene genera have survived in certain areas called "Neogene Relict Pockets" (Petuch, 1988). A Tampan pocket is located along the west coast of Florida, a Yucatanean on the north shore of the Yucatan Peninsula, a Honduran along the shore of that country, a Venezuelan in northern South America, and a Barbadan for the island of Barbados. The survival of relict genera is of considerable interest. So far, it is not known if these are pockets of only molluscan survival. It seems likely that other phyla would also be involved.

There are other indications of Plio-Pleistocene extinctions that affected the Atlantic Ocean to a far greater extent that the Pacific. We know that evolutionary change tends to occur very rapidly in the small populations of marine shore animals that become isolated around oceanic islands. As a result, those islands that are relatively old should possess faunas that show a high degree of evolutionary divergence. The best indication of the extent of such divergence is the amount of endemism. If the amount of endemism at such islands is very low, one may suspect that a major alteration of the environment has occurred resulting in an extinction followed by a repopulation by migrants from other areas.

When one examines the shore fish faunas of the old and well-isolated (500 km or more from the nearest mainland) oceanic islands of the North Atlantic and North Pacific, a very interesting pattern of endemism is revealed (Briggs, 1966). In the North Atlantic, the Azores, a group of nine islands of probable Miocene age at 36° to 39°N, has no endemism at all; Madeira, two islands of probable Miocene age at 32° to 33°N, shows about a 3% endemism; Bermuda, consisting of about 360 small islands of Eocene or Oligocene origin at 32°N, has about a 5% endemism; and the Cape Verde Islands, with 10 main islands of lower Cretaceous age at 14° to 17°N, has an endemic level of about 4%. In the North Pacific, the only old and well-isolated island group that has a well known fish fauna is Hawaii. This group of 20 islands, of a late Miocene or earlier origin at 18° to 23°N, has about a 25% endemism.

So, we find a remarkable contrast between the northern parts of the two oceans. Both the Cape Verde Islands and Bermuda occupy positions comparable to that of Hawaii in that each is within the tropical zone but lies close to its northern boundary, but these islands shown endemic rates of only 4% and 5% while that of Hawaii is 25%.

The classic study of coccolith distribution in the North Atlantic, that was carried out by McIntyre (1967), showed that the maximum cooling of the Wisconsin glacial period resulted in about a 15° southward shift in latitude for the planktonic populations. Comparable evidence for the North Pacific does not show such latitudinal shifts. These indications have been substantiated by the results of the CLIMAP (1981) project. While the North Atlantic underwent a substantial decline in surface temperature during the most recent glaciation, the North Pacific experienced very little change.

The reasons for the greater temperature fluctuation in the North Atlantic were given in the Pliocene account (p. 162). One result has apparently been a repeated extinction of the tropical marine faunas of the North Atlantic oceanic islands. With each glaciation the temperature dropped and the tropical species were eliminated. In the approximate 11 000 years since the last glaciation, these islands have each accumulated diverse tropical biotas, but this amount of time has not been sufficient for a significant amount of speciation. In the North Pacific, on the other hand, the Hawaiian islands suffered little temperature decline and were able to maintain their old tropical organisms with many peculiar species. Another probable result of the history of fluctuating temperatures in the North Atlantic, is the generally depauperate state of the fauna. In most biotic groups, there are only about one-third to one-half as many species as are found in the North Pacific (Briggs, 1970; Vermeij, 1991).

Another comparison can be made in which the characteristics of the present faunas reflect different histories. In the western North Atlantic, the Florida peninsula extends far enough south so that its distal end lies in tropical waters. This has the effect of dividing the Carolina Warm-Temperate Region into an Atlantic coast portion extending from about Cape Hatteras to Cape Canaveral and a Gulf coast portion that occupies the northern Gulf of Mexico. Despite their geographic separation, the faunas of the two parts are very similar. Almost all of the warm-temperate species found in the Atlantic coast segment also occur in the northern Gulf of Mexico.

In the eastern North Pacific, the Baja California peninsula also extends south so that its tip lies in tropical waters. This divides the California Warm-Temperate Region into an outer coast portion extending from Point Conception to about Magdalena Bay and an inner portion that is confined to the Gulf of California. Here, geographical separation has had a profound effect on the species that comprise the two faunas. Each is characterized by a high degree of endemism and only a few of the warm-temperate species of the outer coast are also found in the Gulf of California. So, we find that on the Atlantic coast of North America, the tip of the Florida peninsula has not functioned as a significant zoogeographic barrier while, on the Pacific coast, the end of the Baja California peninsula has been a highly effective barrier (Briggs, 1970). It appears obvious that Pleistocene temperature declines in the Atlantic have allowed warm-temperate species to circumvent the Florida peninsula, while similar declines did not take place in California.

Of the four tropical shelf regions of the world, it is the Eastern Atlantic (EA) Region that has the most depauperate fauna. Its species diversity is far less than might be expected from its geographic size. A count of the species of fish and in some of the major invertebrate groups (echinoderms, molluscs, stomatopods, brachyurans, reef corals) indicated only about half the number found in the Eastern Pacific (EP) Region (Briggs, 1985). The EA shelf appears to occupy about 400 000 km^2 while the EP shelf covers some 380 000 km^2. So, the EA Region should be able to support a diversity close to that of the EP.

Another strange aspect of the EA Region is its relationship with two of the other regions, particularly the tropical western Atlantic. An examination of the shore-fish fauna, has revealed the presence of a large number of trans-Atlantic species. Most of these wide-ranging species are clearly representatives of American genera, but none of them belong to genera that are typically eastern Atlantic. This appears to indicate a migratory movement across the mid-Atlantic deep-water barrier from west to east. These trans-Atlantic species comprise about 25% of the shore-fish fauna of the tropical EA Region. Works on the West African invertebrate groups also indicate appreciable numbers of trans-Atlantic species. Finally, there is a group of about 24 species of shore fish that have invaded the EA Region by rounding the Cape of Good Hope (Briggs, 1967). Such relationships suggest that the EA Region is still in the process of recovering from a Pleistocene extinction, so that species from the other regions are finding room for colonization.

Another way by which the Pleistocene glaciations affected distribution and speciation was through variations in eustatic sea level. When the large islands of the Indo-Australian Archipelago were coalesced into a peninsula during glacial stages, it left only a few narrow passages to connect the tropical waters of the western Pacific and the eastern Indian Ocean. Today, there are related species divided by the Malay Peninsula and/or the archipelago that probably owe their origin to restrictions of gene flow that took place during the Pleistocene regressions.

The modern fauna of the Red Sea also reflects Pleistocene regression. The southern end of the sea is blocked by a shallow sill that lies at about 125 m. During the major glaciations, the sea level probably dropped that far or close to it. Today, some 17% of the Red Sea fish are endemics (Ormond and Edwards, 1987), 14% of the echinoderms (Campbell, 1987), and significant numbers of other invertebrate groups. Since the mouth of the Red Sea is wide open to the Indian Ocean, it does not appear to be likely that these levels of endemism could have developed unless passage between the two oceans had been cut off or severely restricted.

In addition, there is the puzzle of the endemism that is apparent along the northwestern coast of Australia. Although a tropical fauna occupies the area between Cape York and Shark Bay, it demonstrates, among those groups that have been well-studied, a significant level of endemism (Wilson and Allen, 1988). Also, a number of species typical of the Great Barrier Reef do not extend to the northwestern coast. Although there is no present block to dispersal between northwestern and northeastern Australia, there was one during the glacial stages. The Torres Strait between Australia and New Guinea is a shallow sea that became dry land during the low sea-level excursions. This separated the two parts of northern Australia for apparently sufficient amounts of time to permit evolutionary changes. While the northern coast is considered to be part of the widespread Indo-Polynesian Province, northwestern Australia has been placed in a separate Northwestern (or Dampierian) Province (Briggs, 1974a).

Finally, there is the question of the origin of the corals that form the reefs in the tropical eastern Pacific. It has been argued that the coral genera concerned were remnants of an old, widespread pan-Tethyan coral biota that was disrupted by tectonic events (Heck and McCoy, 1978). However, it must be noted that the eastern Pacific genera have a strong affinity with the Indo-West Pacific (100%) but only a weak affinity with the western Atlantic (17%). These corals have planktonic larvae but no one knows exactly how long the larvae live before they have to settle. Another, and perhaps more important, factor is that small coral colonies are often found attached to pumice and other kinds of floating debris. Transport across the Pacific could have been provided by the Equatorial Countercurrent.

It is known that reef-coral assemblages are relatively intolerant of harsh environmental conditions. Glynn and Wellington (1983) have proposed that the eastern Pacific community was periodically exterminated during the glacial stages and then reconstituted by recruitment from the central Pacific during the interglacial periods. They suggested that low temperatures may have caused the extinctions, but data from the CLIMAP (1981) project shows no glacial sea-surface temperature decline in that area. Glacial period strengthening of the thermohaline circulation could, however, have caused enhanced upwelling which, in turn, could have resulted in a destructive eutrophication (Hallock and Schlager, 1986).

Although the Pleistocene was very brief compared to previous epochs, the constant fluctuation between glacial and interglacial stages had noticeable effects on the marine biota. At the higher latitudes, the temperature swings were detrimental to diversity, especially in areas where the water temperature was greatly influenced by the atmospheric temperature. Smaller seas with lesser heat-budget capacities suffered the most. In general, the tropics were not much affected, except for the eastern Atlantic where the latitudinal spread was apparently decreased to the extent that many species were lost. On the other hand, in some places the fluctuations in sea level probably stimulated diversity. In areas such as the East Indies, with many large continental islands, the Pleistocene making and breaking of land and sea barriers may be one reason for the present high level of species diversity.

Terrestrial patterns

Plants

The late-Quaternary fossil record shows massive geographic shifts that demonstrate the temporary nature of floral communities in an environment characterized by major climatic changes (Potts and Behrensmeyer, 1992). In North America, boreal ecosystems were largely overridden by continental ice sheets; those that remained supported tundra vegetation rather than forest. At mid-latitudes, temperate ecosystems gave way to boreal communities that, in terms of species composition, were unlike those of today. Significant changes also occurred at more southern latitudes (Spaulding, 1990).

The dominant theme for the North American Pleistocene record is that species rather than communities moved in response to climatic change (Potts and Behrensmeyer, 1992). If this process was the same for the vegetation of the earlier epochs, then what appears to have been coordinated expansion and contraction of communities may only reflect our inability to resolve the separate histories of the individual species. A second theme for the Pleistocene is that, despite the radical shifts in climate, few plant taxa became extinct. Some, such as the *Metasequoia*, which was common over western North America, became locally extinct but survived elsewhere.

The major period of plant extinctions in North America apparently occurred between 5 and 0.7 Ma ago. By the late Pleistocene, the vegetation had evidently become sufficiently adapted to the recurrent glacial stages so that few species were lost. Pollen data show that between 18 000 and 12 000 years ago vegetation in the northcentral United States was dominated by communities rich in spruce and sedges. The general picture is that of a spruce parkland, perhaps similar to that found in the southern part of the Ungava Peninsula in northern Quebec (Webb and Bartlein, 1992).

The Quaternary record of South America demonstrates marked expansions and contractions of forest and other types of vegetation. At lower altitudes, during the glacial periods, savanna vegetation expanded while forests contracted; as many as 20 cycles have been postulated (Van der Hammen, 1982). When the forests contracted, they presumably retreated to small, high-rainfall areas and along the courses of the great rivers. These forest refuges are said to have played a role as generators of species richness and taxon pulses (Erwin and Adis, 1982). But, there is some controversy about the placement and importance of the refuges (Endler, 1982). Considering the very high species diversity of the rain forest, the glacial-stage contractions must have caused considerable extinction. Although the fractionation process may have stimulated some speciation, the intervals between glacial periods were too short to permit significant recoveries of diversity.

In Africa, desert conditions in the Saraha were apparently well established by about 2.5 Ma ago (Potts and Behrensmeyer, 1992). There is abundant evidence for Pleistocene climatic conditions that alternated between being more humid and drier than today. Although there is no one point when savanna replaced forest, the overall trend has been towards restriction of forests and the development of continuous tracts of grass. In the Serengeti, the shift to virtually 100% grasslands may not have occurred until about 0.6 Ma ago (Cerling et al., 1989).

The main vegetational changes during the Quaternary in China include the following: warm-temperate deciduous, broad-leaved forest moved southward; Siberian cold-temperate conifers moved south to form Taiga in northern China; and Xeromorphic vegetation expanded in north and southwest China. During the late Pleistocene, herbaceous communities, simpler in taxonomic composition, became dominant in north China. Alpine meadows and desert formed on the Tibetan plateau. In eastern China, alpine forests expanded downward during the glacial phases (Wang, 1984).

In Australia, the late-Tertiary and Quaternary radiation and dominance of *Eucalyptus* was a notable event. In some places, considerable climatic change has occurred over the past 100 000 years. In Queensland, a complex rain forest gave way to a drier forest with *Podocarpus* and other gymnosperms after about 80 000 years ago. This changed to a sclerophyllous woodland during the last glacial stage. The present rain forest then became reestablished about 9500–7000 years ago. For the continent overall, the climate 30 000 years ago was much wetter than today. Expansion of Australia's arid center began by about 26 000 years ago, and by about 18 000 years ago, the large interior lakes had almost disappeared (Potts and Behrensmeyer, 1992).

Mammals

In North America, the peak of land mammal immigration occurred in the Blancan North American Land Mammal Age (NALMA) about 3 Ma ago, when at least 18 genera arrived from Asia and from South America (Webb and Barnovsky, 1989). The Irvingtonian NALMA fauna of the early to middle Pleistocene is characterized by the immigration of the mammoth, the bovid *Soergelia*, *Canis* (wolf), mustelids (ermine), *Alces* (elk), *Rangifer* (reindeer), *Panthera* (jaguar), and the cricetid genus *Clethrionomys* (Potts and Behrensmeyer, 1992). Common North American mammals included advanced *Equus*, *Smilodon* (saber-tooth cat), *Lepus* (hare), and *Microtus* (vole). There were five families of artiodactylids. Xenarthrans were a conspicuous element with glyptodonts, several species of armadillo, and three families of ground sloths. Proboscidea included a gamphothere, a mastodon, and several species of mammoth.

The cricetid rodents included at least five lineages that independently developed microtine teeth as an adaptation for eating seeds and leaves in grassland habitats (Repenning, 1987). Their explosive radiation in North America (and Europe) corresponded to the transition from savanna to grassland and steppe environments. The Irvingtonian fauna as a whole inhabited the extensive grasslands, deciduous woodlands, deserts, boreal forests, and tundra that characterized North America during the ice ages. The provinciality of the faunas was considerably greater than it had been during the later Tertiary.

The Rancholabrean NALMA began about 0.3 Ma ago. Rodents continued to diversify, while proboscideans, perissodactyls, xenarthrans, and camelids experienced progressive losses (Potts and Behrensmeyer, 1992). The immigration of *Bison* into North America provides a fossil marker for the beginning of the Rancholabrean. Other notable immigrants included *Ovis* (mountain sheep), *Ovibos* (musk ox), *Alces* (moose), *Panthera* (lion), and *Homo sapiens*. The pace of immigration and the blending of Holarctic faunas continued throughout the remainder of the Pleistocene. The driving force was the climate with its effects on the exposure and ecology of Beringia (Webb and Barnovsky, 1989). The types of animals that were able to cross the land bridge changed towards more cold-tolerant and steppe-adapted forms as the vegetation shifted to steppe and tundra.

Hoffman (1985) noted that Holarctic distributions are very high among Recent tundra mammals and birds, somewhat lower for taiga and alpine species, and less for deciduous forest and other habitats. From these and present systematic relationships, he concluded that the Beringian environment was tundra or cold steppe in the late Pleistocene, taiga (in part) in the middle or early Pleistocene, and steppe (in part) in the early Pleistocene or late Pliocene. He also concluded that isolating barriers between the Nearctic and Palearctic were competitive as well as ecological and physical.

In addition to the Asian immigrants, the Rancholabrean fauna was typified by *Smilodon* (saber-tooth cat), *Canis* (dire wolf), *Equus*, and *Mammuthus* (mammoth), and a diverse assemblage of sciurids and cricetids. Bovids were the dominant family of artiodactyls with 10 genera. By the end of the Rancholabrean, about 10 000 years ago, 43 genera of mammals had disappeared. These included the mammoth, mastodon, gomphothere, saber-tooth cat, dire wolf, horse, sloths, camels, and a number of species of bear, deer, antilocaprids, and others. This megafaunal extinction apparently took place over a period of less than 5000 years, mainly from 12 000 to 10 000 years ago.

The history of the land-mammal dispersal across Beringia from the late Miocene through the late Pleistocene has been reviewed by Webb and Barnovsky (1989). Eurasia, with its much larger area and greater faunal diversity, exported more genera than it received. These exchanges are interesting for, during the Pleistocene, they were more one-sided than would be suggested by the contrast between diversity or total land area. In the late Pleistocene, the successful (colonizing) traffic became entirely one way, with 21 genera going from Asia to North America but none in the opposite direction. This pattern suggests a competitive superiority for the Asian mammals. It is reminiscent of the successful invasion of South America from the north at the end of the Pliocene, and of the invasion of Europe during the Grande Coupure of the Oligocene.

In South America, the Pleistocene was a time of continued diversification of the mammalian invaders from North America. On the other hand, the large native mammals, such as the litopterns and notoungulates, became greatly reduced. The large marsupial carnivores may have become extinct prior to the invasion of carnivorous forms from North America. In contrast to the large endemic species, the southward invasion of cricetid rodents does not seem to have been accompanied by extinctions of the small native mammals. This implies that numerous ecological opportunities, that had not been exploited by the native fauna, were available in the southern ecosystems. The megafauna that disappeared at the end of the Pleistocene (proboscideans, edentates, equids) did so at about the same time as in North America.

In Eurasia, at least 17 glacial-interglacial cycles are known over the past 1.7 Ma. Eight of these occurred just in the past 0.8 Ma (Potts and Behrensmeyer, 1992). Climatic oscillations during the latter period were apparently associated with temperate forests during the interglacials, interrupted by invasions of grasses and herbs to form a steppe vegetation during the glacial stages. By about 0.9 Ma ago, a modernization of the European faunas took place with the appearance of larger mammals such as *Bos*, *Bison*, *Ovibos*, and *Megaceros*.

Extinctions in Eurasia did not reach the dramatic peak seen in the Americas or Australia. In Asia as in Africa, no late-Pleistocene families were lost. Out of 37 genera of large mammals that became extinct in Europe over the past 3 Ma, 13 became extinct at the end of the Pleistocene (Martin, 1984). The relatively gradual disappearance of large-mammal genera from Eurasia during the Quaternary does not seem to require special explanation, although both climatic change and the effect of human hunters have been proposed as responsible factors (Potts and Behrensmeyer, 1992).

In Australia, megafaunal extinctions during the late Pleistocene appear to have been of the same magnitude as those in North and South America. Thirteen genera of large marsupials met their demise, including three of the six families (Martin, 1984). Some giant birds, a giant varanid lizard, a large snake, and a horned tortoise also became extinct. The survivors tended to be smaller, more mobile, or nocturnal in habit. Human occupation, which to varying degrees has been implicated in these extinctions, occurred at least by 35 000 and possibly more than 40 000 years ago.

In Africa, the fauna of the first half of the Quaternary showed considerable continuity with that of the Pliocene, although new taxa appeared and disappeared (Potts and Behrensmeyer, 1992). Bovids were the most abundant of the large mammals. Archaic genera of bovids declined and were replaced by living genera. Cervids appeared in north Africa but not south of the Sahara. Suids diversified and were characterized through the middle Pleistocene by hypsodonty and increases in body size. In contrast, the hippopotamids were reduced in diversity; the restricted terrestrial habitat of the pigmy hippo and the aquatic specialization of the larger form, were possibly responses to increased competition from the bovids. By the mid-Pleistocene, the diversity of giraffids had decreased with the loss of *Sivatherium* and several species of *Giraffa*. The horse *Hipparion* became extinct but *Equus* persisted, probably because it was better suited to the grassland habitat. The specialized browsing chalicotheres also became extinct. Among the proboscideans, the deinotheres and gomphotheres disappeared leaving only the modern genus of elephants. During the latter half of the Pleistocene, relatively little faunal change occurred at the generic level, but there were minor changes as modern species replaced their older counterparts.

Members of the superfamily Hominoidea are known from the Miocene of east Africa, some 22 Ma ago (Carroll, 1988). This superfamily is the most extensively studied of all mammalian groups, yet the nature and interrelationships of the fossil species remain subject to continuing debate. During the Miocene, there was an extensive radiation of genera similar to *Proconsul*. At the end of the early Miocene, when a connection was made between Africa and Asia via the Arabian Peninsula, primitive hominoids migrated into Europe and Asia. The genus *Kenyapithecus*, known from strata approximately 16 Ma old in east Africa, may be a member of a stem group that was ancestral to both African apes and humans. Molecular evidence, based upon a variety of techniques, sug-

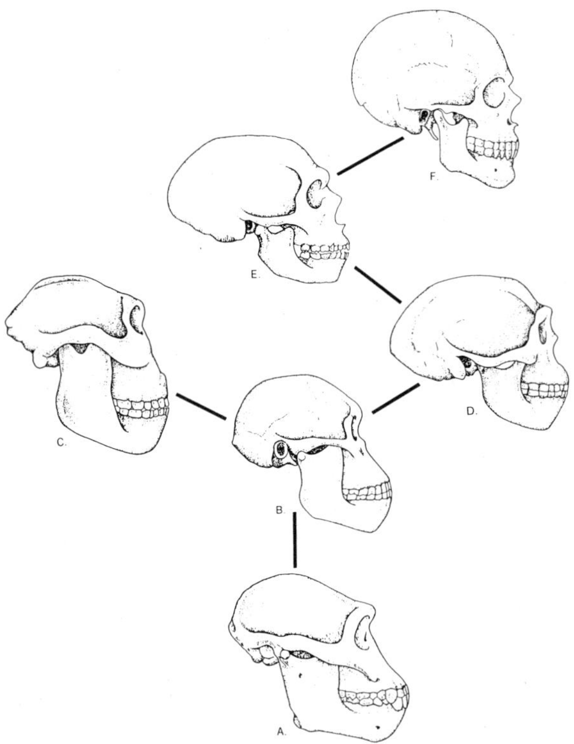

Fig. 50. Hominid skulls. (A) *Australopithecus afarensis*, (B) *Australopithecus africanus*, (C) *Australopithecus robustus*, (D) *Homo erectus*, (E) *Homo sapiens* (neanderthal), (F) *Homo sapiens* (modern human). After Stearn and Carroll (1989).

gests that the dichotomy between humans and the African great apes took place between 5 and 14 Ma ago.

The fossil record of the family Hominidae began in Africa about 4 Ma ago (Carroll, 1988) (Fig. 50). The earliest known species is *Australopithecus afarensis*. The emergence of hominids from the earlier apes may have been associated with a shift from life in the wet forest to dryer savanna and grassland. The necessity for rapid movement in the open may have provided the selective advantage for the development of bipedality. The fossil record shows that bipedality was achieved while the brain was little advanced over that of the great apes.

It is known that there were several species of hominids living at the same time in southern and eastern Africa (Carroll, 1988). Between 2 and 3 Ma ago, there appears to have been at least three coexisting lineages. The first member of the modern genus *Homo* (*H. habilis*) appeared about 2 Ma ago. The species *H. habilis* is not recognized after about 1.75 Ma ago, but it appears to have been replaced by *H. erectus*. A nearly complete skeleton of *H. erectus* has been discovered in east Africa from beds about 1.6 Ma old.

About 1 Ma ago, *Homo erectus* spread out of Africa into eastern and southern Asia. As Carroll (1988) has noted, there is considerable debate about the pattern and rate of evolutionary change between *H. erectus* and *H. sapiens*. However, by about 300 000 years ago, populations with an anatomy that is considered typical of *H. erectus* were replaced by primitive members of our own species. The Neanderthals were archaic *H. sapiens* that lived from about 300 000 to about 40 000 years ago. *Homo sapiens* of a more modern aspect are known in Africa as long as 100 000 years ago. From Africa, they probably spread first to Asia, then Europe (Aiello, 1993). They apparently invaded Europe between 30 000 and 40 000 years ago, and may have replaced the Neanderthals with a minimum of interbreeding.

The great apes and humans comprise a related group of the higher primates. While it seems clear that the family Hominidae originated in Africa, this is possibly true for the great apes (Pongidae) as well. The more advanced apes, *Pan* (chimpanzees) and *Gorilla* (gorillas), are confined to Africa while the more primitive *Hylobates* (gibbons) and *Pongo* (orangutan) are restricted to southeast Asia. This common geographic pattern, where the more advanced taxa occupy the putative center of origin, together with the African fossils of the possibly ancestral *Kenyapithecus*, suggests that Africa was the center of evolutionary radiation for all of the higher primates (Briggs, 1984).

The "overkill hypothesis", of which Martin (1984) has been the chief proponent, invokes human predation as the principal cause of the late-Pleistocene extinctions of large mammals and birds. There is an apparent connection between the timing of the megafaunal extinctions on the various continents and the dispersal of human hunters. Late-Pleistocene extinctions in Africa, where hominids have the longest history, were minor possibly due to the coevolution of predator and prey. Certain very large mammals, and primates that were competitors of humans, became extinct during the early Pleistocene. Megafaunal extinctions in North and South America, on the other hand, were concentrated between 15 000 and 8000 years ago and coincide with rise of human activity. In Australia, they began earlier, perhaps around 30 000 years ago; perhaps not long after human occupation. Europe experienced fewer extinctions during the late Pleistocene, but certain large animals such as the mammoth, woolly rhino, and musk ox did disappear.

The human overkill hypothesis has been supported by data from certain oceanic islands (Diamond, 1992). The Mediterranean islands of Cyprus, Corsica, and Sardinia are surrounded by such deep water that they remained unconnected to Europe even during the glacial stages of low sea level. Fossils appear to indicate that the first humans to reach Cyprus about 8500 years ago quickly exterminated its populations of pigmy hippos, pigmy elephants, and large birds. For Corsica and Sardinia, it also seems that megafaunal species were alive when humans arrived and vanished soon afterwards. On Madagascar, giant lemurs, elephant birds, and semi-pigmy hippos began their decline when humans arrived sometime between AD 0 and 300.

The argument against the overkill hypothesis is that late-Cenozoic extinction episodes among the mammals have been correlated with intervals of rapid climatic change (Webb and Barnovsky, 1989). In North America, the largest episode took place about 6 Ma ago when 62 genera were lost. Another 35 genera were lost about 1.9 Ma ago and in the latest extinction, ending about 10 000 years ago, some 43 genera became extinct. Although rapid climatic change can be postulated for the first two events, the third presents some difficulties. By 10 000 years ago, the ice age, with its rapid glacial-interglacial fluctuations, was well established. The mammalian fauna should have become adapted to such conditions, so that extinctions would occur at only the normal background rate. There appears to be no reason why the end of the Wisconsin glaciation should have provided any more stress than the terminations of the preceding ice advances. This makes it difficult to invoke the climatic hypothesis.

Freshwater fauna

In the higher latitudes of the earth, particularly in the northern hemisphere, the ice sheets of the Pleistocene had a profound effect on the distribution of fish and other aquatic organisms. The glaciers eliminated all aquatic life from the areas they occupied, changed the climate, altered drainage basins, and created lakes. As the ice sheets melted, they produced enormous amounts of water that resulted in a pluvial period characterized by the existence of torrential rivers and large lakes. As the ice retreated, the sea level rose covering the lowland streams and lakes that had existed on the continental shelves.

In North America, the Wisconsin glaciation reached its maximum about 17 000–18 000 years ago. Although most of the northern parts of the continent were covered with ice, certain areas such as most of Beringia and parts of Alaska and northwestern Canada remained ice free (Fig. 51). These areas, and others to the south of the ice sheets, functioned as refugia for many species that had once been widespread in the northern part of

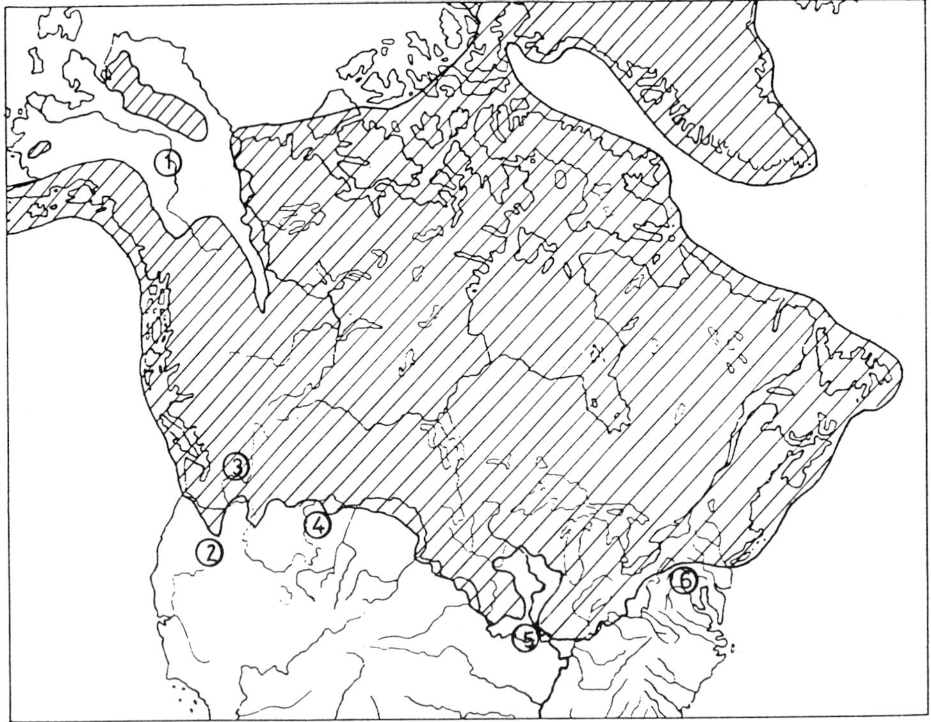

Fig. 51. Maximum extent of Wisconsin glaciation in North America. Recognized refugia: 1, Beringia; 2, Pacific; 3, Banff-Jasper; 4, Missouri; 5, Mississippi; 6, Atlantic. After Crossman and McAllister (1986).

the continent. As the ice retreated, some groups of animals and plants were able to reinvade the glaciated areas.

For many aquatic forms, the postglacial reinvasion was facilitated by the large quantities of melt water that initially occupied much of the land surface. The large lakes and streams of the pluvial period supported a variety of aquatic life. Some of the most notable lakes were formed in the Great Basin region. The largest were Lake Bonneville in Utah and Lake Lahonten in western Nevada. Each of these lakes covered up to 50 000 km^2 (Dott and Batten, 1988). Farther north, preglacial drainage valleys were gouged by the ice to leave large depressions that were subsequently filled with water. The Finger Lakes of New York are good examples.

The Great Lakes basins were formed by glacial action. They had a complex development and changed shape during various retreats and advances of the ice. The huge, postglacial Lake Agassiz, which was about four times the size of Lake Superior, formed in the southern Canadian plains. Large lakes formed several times in the northern Rocky Mountains. One such lake, called Lake Missoula, was probably about 300 m deep (Dott and Batten, 1988). About 18 000 years ago, its moraine dam in northern Idaho suddenly broke and a wall of water rushed through eastern Washington scouring channels and depositing gravel bars over large parts of the Columbian Plateau. The affected area is called the channeled scablands.

The Wisconsin glacial period began more than 70 000 years ago and terminated in most areas about 10 000 years ago. It not only had a disruptive effect on the distribution patterns of aquatic animals, but there were also evolutionary effects. Large numbers of small, isolated populations apparently occurred at two stages in Pleistocene history. One took place during the height of the glaciations when many populations were restricted to small refugia, and the second occurred when the postglacial lakes and streams began to dry up. In the Great Basin today, one may find small populations in isolated springs and streams. These are remnants of huge populations that once existed in the pluvial lakes about 10 000 years ago. Many of them have undergone sufficient evolutionary change in their restricted environments to be considered unique species.

The highest species diversity among the aquatic animals of North America, is found in the Mississippi River Basin. This fauna remained free of extinction, partly because of its large size and stability and partly due to the lack of major barriers between the northern tributaries and faunal refuges to the south (Smith, 1981). During the glaciations, the fauna was able to move south and then north again as the ice sheets receded. In the west, small, climatically unstable basins contained relatively small populations with high probabilities of extinction. Barriers prevented dispersal and recolonization, and the lack of long-term stability may have further depressed diversity by interfering with speciation.

The Pleistocene history of Beringia has been reviewed by Lindsey and McPhail (1986). They pointed out that, from the viewpoint of North American biogeography, the significant northwestern boundary is not at the Bering Strait; it is some 800 km farther west along the Siberian coast. They emphasized that Beringia is a biogeographic unit that has provided a refugium for the survival and evolution of a distinctive fauna, and that has alternately provided a barrier and access route for dispersal between North America and Asia.

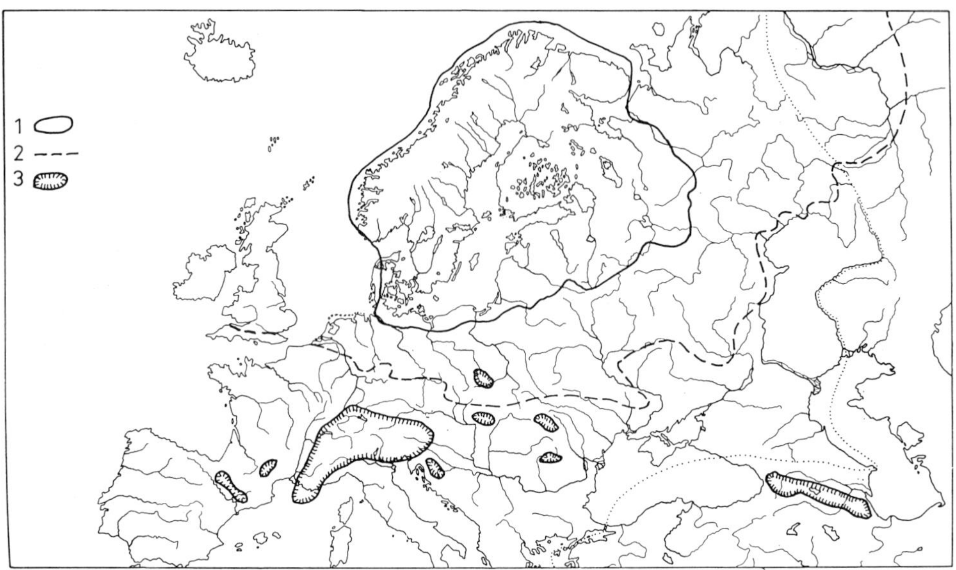

Fig. 52. European glaciation during the Pleistocene. 1, Boundary of ice during the last (Würm) glacial period; 2, maximum extension of Pleistocene ice cover; 3, high mountain glaciers. After Bănărescu (1992).

The Pleistocene history of the Euro-Mediterranean fauna has been reviewed by Bănărescu (1992). During the Würm Period, much of northern Europe was covered by a large ice cap (Fig. 52). Glaciers also developed in the high mountains from the Pyrenees to the Balkans and the Caucasus. Although aquatic life suffered in the vicinity of the ice sheets, a comparatively rich fauna survived in the southern parts of the continent. Considerable extinctions took place in the part of central Europe lying between the northern and alpine ice caps.

The European cold-water species that were affected by the glacier movements have been divided into two categories. The "northern ice cap marginal species" originated in the extreme north and were pushed southward by the advancing ice sheets. Most of this group returned north in postglacial times, but some of them remained in central Europe. Those that remained are recognized as disjunct boreo-alpine species (Bănărescu, 1992). Several fish, aquatic insects, and water mites belong in this group. The second category, called "southern ice cap marginal species", are believed to have been present in the Alps and other central European mountain ranges during preglacial times and to have moved south during the ice age.

The post-glacial warming and the melting of the northern and alpine ice caps were rapidly followed by recolonization of the formerly glaciated areas. For example, Zwick (1981b) was able to trace the migration of several species of European stoneflies from refuges in Iberia and northern Africa, Italy, and the Balkans. Some of the freshwater recolonizations may have taken place through the Black Sea since its salinity was probably lowered during the pluvial period. The melt water may have also allowed dispersals among the Black, Caspian, and Aral Seas.

There is a considerable European literature dealing with the occurrence of a group in the Baltic Sea called "glacial marine relicts" (Bănărescu, 1992). As the northern European icecap melted a large freshwater lake formed between the Scandinavian Peninsula and the mainland. The development of a connection to the North Sea created a brackish body called the Yoldia Sea. This was followed by a new freshwater stage, the Ancyclus Lake. Another marine connection produced a relatively high-salinity Litorina Sea, which gradually developed into the present Baltic Sea.

The marine relicts of the Baltic, which is essentially a freshwater sea, consist of a fish (*Myoxocephalus quadricornis*) and seven crustacean species. Supposedly, this fauna originally came from the Arctic Ocean and entered the Baltic from the White Sea during the Yoldia period. These relicts have survived in the northern Baltic Sea and have been able to penetrate into some of the surrounding lakes. Some authors would derive the relicts from a large Siberian lake rather than directly from the White Sea (Bănărescu, 1992).

Lake Baikal in Siberia is the world's oldest and deepest lake. Apparently, two or three basins, in the area of the present lake, began to expand and deepen during the Miocene. A subsidence of the depression lasted throughout the Neogene. The lake itself, in its present size and shape, is not older than the late Pliocene, but parts of it may have been in existence since the Mesozoic (Kozhov, 1963). Most of the Baikal fauna appears to be derived from the Tertiary freshwater fauna of Siberia. However, some groups such as the sponges, gammarid crustaceans, and possibly the fish families Cottocomephoridae and Comephoridae, are probably of an early Tertiary marine origin. Recent immigrants from the Arctic Ocean are two genera of gammarids, two fish genera, and the seal *Phoca sibirica*. All of these euryhaline marine derivatives probably reached Baikal by means of freshwater connections (Bănărescu, 1992).

Although the Pleistocene encompassed a brief time period of only 1.6 Ma, it had important evolutionary and biogeographic effects. In the terrestrial and freshwater habitats, the most noticeable changes took place in the northern hemisphere and were directly controlled by the waning and waxing of the great ice sheets. But, the ice-age removal of water from the hydrologic cycle and the recession of sea level brought on arid conditions over much of the remainder of the world; tropical forests receded, drier habitats expanded, and the land biota migrated on to the exposed continental shelves. In the marine environment, the fluctuating sea levels produced changes which affected distribution patterns all over the world. The width of the continental shelves varied widely, sea passages were created or broken, and the surface temperature changed. In both habitats, species diversity declined under the stress of the changes. The evolution and subsequent dispersal of *Homo sapiens* caused additional extinctions among the large terrestrial animals.

SUMMARY

1. During the Miocene, several plate tectonic movements had important biogeographic effects. These included a connection between Africa and Asia (which created the Mediterranean basin), orogenesis of the Alps, closure of the Strait of

Gibraltar, northward movement of the Australian Plate, and completion of a peninsula from Mexico to Panama.

2. The restriction of the Indonesian Seaway between the Western Pacific and Indian oceans and the shoaling of the Central American Seaway changed the current structure of the tropical Pacific Ocean. The countercurrent systems were strengthened retaining more warm water within the tropics and contributing to high-latitude cooling.
3. The Red Sea basin had formed by the Miocene and, at first, was connected to the Mediterranean. Towards the end of the Miocene, the northward connection was severed and separation of the African and Arabian plates formed a deep trough that opened into the Indian Ocean.
4. During the Messinian Stage, the Mediterranean was cut off from the Atlantic. The sea began to dry up, thick layers of evaporites were deposited, and much of the biota was extinguished. Connection with the Paratethys formed a freshwater "Lago Mare" stage, which terminated when the Gibraltar opening was reestablished.
5. The worldwide climatic deterioration of the Miocene stimulated a tremendous evolutionary radiation among the herbs and grasses. This, in turn, allowed a great increase in diversity among certain animal groups such as proboscideans, horses, bovids, and rodents.
6. The early Miocene influx of mammals from Africa and Asia caused a major turnover in the European fauna. Some 57% of the genera and 23% of the families were new immigrants. This changeover has been called the "Coupure miocéne."
7. In the late Miocene, North America received the first mammalian immigrants from South America since the late Paleocene. These were two genera representing two different families of ground sloths. About the same time, a genus of the raccoon family showed up in South America.
8. Continued mammalian immigration from Asia, plus the evolution of endemics, produced in North America the largest standing crop of genera in the entire Tertiary.
9. Evolutionary and distribution patterns among the bird order Galliformes suggest a possible center of origin in the Asian tropics. The same appears to be true for the songbirds (Passeriformes). The parrots (Psittaciformes) and pigeons (Columbiformes) probably originated in the Australian-New Zealand area. The early radiation of the bird tribe Corvida (crows, ravens, magpies) appears to have occurred in Australia.
10. In the herpetofauna, various groups made their way into North America from Asia and into Australia from Asia. The Middle American route between North and South America, even though not complete, provided for exchanges in both directions. Interchanges also took place between Africa and Asia.
11. No less than five freshwater fish families probably invaded Africa when the connection to Asia was achieved. Such fish also moved westward into southern Europe during the "Lago Mare" phase of the Mediterranean.
12. Evidence for a major extinction event at the end of the Miocene appears to be weak. Marine planktonic and benthic organisms show that turnovers were asso-

ciated with gradual changes in oceanographic conditions that took place throughout the middle and late Miocene.

13. During the 3.5 Ma of the Pliocene, three great events took place. These were: (1) completion of the Panamanian isthmus which permitted the Great American Biotic Interchange, (2) the flooding of Beringia which permitted the Great Transarctic Biotic Interchange, and (3) the beginning of glaciation in the northern hemisphere.
14. At the opening of the Bering Strait about 3.5 Ma ago, some 261 boreal marine, molluscan species entered the Arctic-Atlantic from the North Pacific. About 34 species migrated in the opposite direction. At the time, the North Pacific fauna was about twice as rich but this diversity difference does not explain the almost 8 to 1 ratio in favor of successful dispersals from the Pacific.
15. An "Hypothesis of Ecological Opportunity" was proposed to explain the difference. But, this idea is dependent on extinction in order to free ecological niches so that they can be occupied by an invader. But, the great extinctions in the Arctic-North Atlantic did not begin until about 0.5 Ma later, so the hypothesized effect came well before the cause.
16. The temperature drop of about 3.0 Ma ago, eliminated the boreal marine fauna from the Arctic basin and surrounding waters. This set the stage for the evolution of a new cold or Arctic biota. Because the new Arctic Region occupied the northern parts of the North Pacific and North Atlantic, the boreal biota in each ocean was separated in two parts, one to the east and one to the west.
17. The northern hemisphere glaciations had a much more severe effect in the North Atlantic than in the North Pacific. This because the former is a smaller ocean with a smaller heat budget and was invaded by ice and cold currents from the Arctic Ocean.
18. The rise of the Panamanian isthmus separated the tropical marine biota of the New World. Evolution during the past 3 Ma has proceeded to the extent that few species are now common to both sides of Panama.
19. On land, there was a general spread of low-biomass vegetation and, at high northern latitudes, steppe and tundra ecosystems appeared for the first time.
20. Interamerican exchanges took place under two contrasting conditions, glacial and interglacial. The interglacial phase was characterized by rain forest enabling the tropical biota to spread northward. During the glacial stages, a dry savanna existed which established a highway leading from one temperate zone to the other. The latter favored southward migrations.
21. The northern mammals that dispersed southward had a detrimental effect on the South American species, particularly the ungulates. On the other hand, the southern mammals that moved northward did not appear to encounter any direct competition.
22. The northward migrators have shown little subsequent evolutionary change. But many of the mammals that invaded South America underwent rapid and extensive changes. Exchanges continued between Asia and North America and also between Asia and Africa.
23. Even though the South American bird fauna is much the richer, some 12 to 13

families have dispersed from North America southward while only four have gone in the opposite direction.

24. Besides mammals and birds, many frogs, salamanders, lizards, snakes, and turtles have moved across the isthmian connection. The same is true for a variety of freshwater fish and other members of the aquatic biota.
25. The Pleistocene occupied the past 1.6 Ma of earth history. There were at least a dozen major glaciations and many smaller ones. The succession of glacial and interglacial stages profoundly affected the distribution and evolution of animals and plants.
26. The old, tropical oceanic islands of the North Atlantic (Bermuda and the Cape Verdes) evidently had their shore fauna extirpated by cold sea-surface temperature during the glacial periods. Their present faunas are probably the result of immigration since the last ice age. They have few endemic species.
27. The Hawaiian Islands in the North Pacific were evidently little disturbed by the glaciations. Their present fauna is older with many endemic species. During the last ice age, the sea-surface temperature of the North Pacific was only slightly depressed so that the Hawaiian shore fauna was not severely impacted.
28. The Florida peninsula separates the warm-temperate, western North Atlantic in two parts, one along the Atlantic coast and the other along the Gulf coast. The Baja California peninsula separates the warm-temperate eastern North Pacific in two parts, one along the outer coast and the other in the Gulf of California. The Florida peninsula is not an effective biogeographic barrier but the Baja California peninsula is. The difference is attributable to the contrasting temperature history of the two oceans.
29. Of the four tropical shelf regions of the world, the Eastern Atlantic Region has the most depauperate fauna. Its fauna is characterized by large numbers of trans-Atlantic species that are representatives of typical New World genera. These, plus a few species from the Indian Ocean that have rounded the Cape of Good Hope, appear to be filling in an Eastern Atlantic ecological vacuum. The space was probably created by ice age extinctions.
30. Ice age sea-level regressions restricted marine passage between the tropical Western Pacific and Indian oceans, isolated the Red Sea, separated northwestern from northeastern Australia, and eliminated the Bering Strait. The coral reefs of the Eastern Pacific were repeatedly decimated, probably by eutrophication due to increased upwelling. All of these events had important biogeographic consequences.
31. The floral fossil record of the Quaternary shows massive geographical shifts that took place, not only at high latitudes but in the tropics as well. It appears that species rather than communities moved in response to climatic change.
32. The North American-Asian mammalian exchange continued but with more Asian species coming eastward than vice versa. By the late Pleistocene, the successful (colonizing) traffic became entirely one way with 21 genera entering North America.
33. The fossil record of the family Hominidae began in Africa about 4 Ma ago. The first member of the genus *Homo* appeared about 2 Ma ago. This species, *Homo*

habilis, was evidently replaced by *H. erectus*. The latter then spread out of Africa into eastern and southern Asia. About 300 000 years ago, *H. erectus* was replaced by primitive members of our own species.

34. The Neanderthals were archaic *Homo sapiens* that lived from about 300 000 to 40 000 years ago. The modern *H. sapiens* has been known from African fossils as old as 100 000 years. They apparently invaded Europe some 30 000–40 000 years ago.
35. The great apes and humans comprise a related group of higher primates. It seems clear that the family Hominidae originated in Africa, but this is probably true for the great apes as well.
36. Towards the end of the Pleistocene, there was an extinction episode among the large mammals and birds. There is an apparent connection between the timing of these megafaunal extinctions on the various continents and the dispersal of human hunters. This connection has been reinforced by data showing that the arrival of humans on various oceanic islands coincided with the extinction of the large animals.
37. In the freshwater habitat, important biogeographic changes, that were directly controlled by the movement of the great ice sheets, took place in the northern hemisphere.

CHAPTER 8

Historic extinctions

One of the most pervasive effects during times of biotic crisis is the massive disruption of tropical and low-latitude ecosystems, and the relative non-disturbance of high-latitude and polar ecosystems.

George R. McGhee, Jr., *Catastrophes in the History of Life*, 1989

The terms "mass extinction", mass killing", "mass murder" and "catastrophic event" have all been used in the recent literature to describe historical extinction episodes. The use of such terms implies evidence of sudden disasters which wiped out significant portions of the earth's animal and plant species. They give the impression of a devastated globe littered with the remains of dead organisms. For example, Allaby and Lovelock (1983) in their book entitled "The Great Extinction" said that 65 Ma ago the earth had collided with a small planet. They described the impact as a conflagration when "volcanoes erupted, tidal waves swept the oceans, and earthquakes shook the continents. For years, a dust cloud shrouded the earth blocking out the sun. Three quarters of all species died including the mighty dinosaurs". In their *Scientific American* article, Alvarez and Asaro (1990) referred to a "sensational crime" that killed off half of all the life on earth and provided evidence which supposedly showed that a giant asteroid or comet had committed this "mass murder".

Are we, in light of our present knowledge, justified in using such lurid descriptions and terms? Aside from the matter of terminology, there are a number of important questions about historic extinctions that need to be addressed: (1) how did the mass extinction idea get started? (2) did the events occur so suddenly that they need to be called mass extinctions? (3) were they so widespread that they deserve to be termed global events? (4) did they really have drastic effects on the world's species diversity? (5) did they have a common cause? and (6) did they have important biogeographic and evolutionary effects?

HISTORICAL DEVELOPMENT

From a historical standpoint, it may be said that we are living in a time of neocatastrophism. This is a recapitulation of an era which began in the first part of the 19th Century and died out about 30–40 years later. The original idea of catastrophism began with the famous French anatomist and paleontologist Georges Cuvier. He became convinced that the earth had undergone a series of great catastrophes, the most recent being the biblical Deluge. After each catastrophe, the earth was repopulated by remnants that had somehow survived the crisis. The new species that subsequently appeared were supposed to have come from parts of the world previously unknown. Cuvier was followed by Alcide D'Orbigny who divided fossil bearing rocks into a series of stages. He believed each stage to represent an independent fauna made by a special act of creation. The early theory of catastrophism finally succumbed to the concept of uniformitarianism introduced by James Hutton and effectively championed by Charles Lyell and Charles Darwin.

Charles Lyell viewed life as a continuous fluctuation of living populations which expanded or contracted their boundaries as geological agents altered local topography and climates (Browne, 1983). Wallace (1855) observed that new species gradually arose to take the place of those that had become extinct. Darwin (1859) emphasized the imperfection of the geological record and noted that the sudden appearance or disappearance of fossil species was probably due to the fragmentary nature of that record. By the 20th Century, paleontologists had recognized two important times of change in the history of life, one at the end of the Permian and one at the end of the Cretaceous. But these changes were not, for many years, considered to be sudden catastrophes. For example, Dunbar (1960) described the Permian/Triassic change as "orderly and gradual, not cataclysmic".

More recently, changes in viewpoint about the rapidity of extinction began to be expressed, particularly in regard to the Cretaceous/Tertiary (K/T) boundary. Schindewolf (1962) announced his concept of "neocatastrophism" in reference to an essentially synchronous annihilation of major groups of Mesozoic organisms. Bramlette (1965) described "massive extinctions" in biota at the end of Mesozoic time; Newell (1967) referred to the boundary as marking a "mass extinction;" and Percival and Fischer (1977) discussed a "Cretaceous-Tertiary biotic crisis". These more dramatic terms, plus continued speculation in popular science articles about the fate of the dinosaurs, apparently set the stage for acceptance of theories involving catastrophic events.

A new era of speculation about historic extinctions began in 1979 when Luis Alvarez and coworkers discovered a level of enriched iridium in the clay of the K/T boundary. At first, they thought a supernova had been responsible but, in 1980, they changed their minds and decided that an asteroid had struck the earth about 65 Ma ago. The impact was supposedly followed by a cosmic winter lasting several years, which caused a biotic catastrophe. This startling news received extensive coverage in the scientific and popular press.

Several paleontologists, who had worked with fossils from the K/T boundary era, protested the widespread assumption that the extinctions came about with catastrophic suddenness (Kauffman, 1979a; Clemens et al., 1981; Hickey, 1981; Archibald and Clemens, 1984). But they were paid little heed and Alvarez (1983) expressed his frustration at his inability to convince them that an asteroid had done the job. General acceptance of the impact theory was aided by the fact that, at first, there seemed to be no other reasonable explanation for the iridium enrichment in the boundary clay.

The next important event was the publication of the periodicity theory of Raup and Sepkoski (1984). They analyzed the entire marine fossil record and found evidence of a 26 Ma cycle, which suggested that most historical extinctions were caused by extraterrestrial impacts. This, plus the advent of volcanism as an alternative high-energy theory (Officer and Drake, 1985) put us firmly into an era of neocatastrophism.

TEMPO OF THE EXTINCTIONS

Recent information on the time span over which each of the major extinctions occurred may be obtained from the review by Briggs (1990) and from two volumes that contain useful articles (edited by Briggs and Crowther (1990) and by Kauffman and Walliser, (1990)). Although it has been suggested that the middle part of the Vendian, about 650 Ma ago, was the time of a mass extinction (McMenamin, 1990a), so little information exists about the timing or severity that its recognition may be premature. The end of the early Cambrian and the end of the Cambrian are sometimes termed extinction events (Boucot, 1990b), but the evidence is not clear cut, although extensive evolutionary and community changes did take place.

The first widely recognized extinction episode took place during the Ashgillian Stage near the end of the Ordovician. The tempo has been estimated at 1–2 Ma (Brenchley, 1989, 1990) or "several million years or so" (Boucot, 1990b). The general marine-invertebrate data on the Frasnian-Famennian extinction, towards the end of the Devonian, were analyzed by McGhee (1988, 1990a, 1990b), who observed that extinction rates were elevated for perhaps 3–4 Ma. Schindler (1990), who studied the event in Europe, found a stepwise sequence of extinctions covering a period of more than 1 Ma.

The Permian-Triassic (P/T) extinction was the most severe of the Phanerozoic. High extinction rates took place through the final three stages of the Permian (Erwin, 1990), an extended period of perhaps 3–8 Ma. The late Triassic has generally been recognized as a time of great extinction, but the analysis by Benton (1990b) indicates that, during the Triassic, three large extinction events took place, each of them apparently lasting several Ma. In the Jurassic, the Pliensbachian Stage has been recognized as a minor global event (Sepkoski, 1986b), but Hallam (1986) pointed out that it was focused in central Europe. Some attention has also been called to the Tithonian Stage, but it also is probably best characterized as a regional rather than a global event.

The Cenomanian Stage of the mid-Cretaceous may or may not have been the time of a global extinction. It is considered to be a minor event by Boucot (1990b). It has been referred to as a stepwise extinction that probably spanned about 2.5 Ma (Shoemaker and Wolfe, 1986). The Cretaceous-Tertiary (K/T) extinction was a major event that was examined in Chapter 5. As far as duration is concerned, it appears that some large terrestrial animals, such as the dinosaurs and pterodactyls, had been declining for several million years prior to the K/T boundary. Smaller land animals seem to have been scarcely affected. Considerable destruction among higher plants took place in North America but not elsewhere. Most marine benthic animals that were affected disappeared over about 2.5–2.7 Ma. Some planktonic forms declined over about 0.8–1.0 Ma. In general, the K/T boundary information points to a gradual, noncatastrophic change.

The one major extinction of the Tertiary supposedly took place in the Priabonian Stage towards the close of the Eocene. However, as noted in Chapter 6, a variety of recent works have demonstrated that there was not a single massive extinction episode, but an extended series of minor events. In the marine environment, these faunal turnovers began in the middle/late Eocene and lasted until the early/late Oligocene, a period of some 14 Ma (Keller et al., 1992).

Impacts by giant, extraterrestrial bodies are still being proposed as the primary cause for almost all of the extinction episodes (McLaren and Goodfellow, 1990). While impacts and volcanic eruptions may have occurred during or near times of biotic decrease, the tempo of the extinctions is inconsistent with hypotheses involving single high-energy events. These were gradual deteriorations in organic diversity which developed in response to environmental changes. Terms such as "mass extinction", "mass killing", or "catastrophic event" are misleading when applied to extinction episodes in the geologic record.

SCOPE OF THE EXTINCTIONS

Of the seven episodes in Phanerozoic history which have been identified as times of great extinctions, only the Permian/Triassic event deserves to be considered a truly global phenomenon. This is the only extinction in which there was a drastic diversity decrease in *both* marine and terrestrial environments. For the K/T boundary, which has often been described as a global catastrophe, the overall effect on the terrestrial biota was small compared to the reduction of the tropical marine life (Briggs, 1990). Certainly, the relatively few remaining dinosaurs eventually died off but most other vertebrate and invertebrate groups survived quite well. The angiosperm plants received only a temporary setback in biomass which had no discernible effect on their long-term expanding diversity (Niklas, 1986).

EFFECTS ON GLOBAL SPECIES DIVERSITY

Estimates of the severity of so-called global extinctions have often been biased because of their dependence on fossils from the shallow waters of the marine tropics. Samples from high latitudes or the deep sea almost always show a better survival during times of environmental crisis. Another problem is that the fossil record as a whole may not be dependable to the extent that it can be used for an accurate assessment of the magnitude of extinction events (Smith and Patterson, 1988). For some years still, information about the magnitude of widespread extinctions probably should come from detailed studies of individual groups where the histories of discrete phyletic lines have been determined.

There are indications that the effects of widespread extinctions on the global species diversity have been greatly exaggerated. For example, estimates of species extinguished at the K/T boundary range from 50 to 80% of all species (Hsü, 1986; Raup, 1988; Alvarez and Asaro, 1990; Courtillot, 1990). However, as noted in Chapter 5, the decline in global species diversity may have been less than one percent.

A COMMON CAUSE?

There were nine episodes in Phanerozoic history at which the rate of extinction might have been extraordinarily high. In chronological order, they are the late Ordovician (Ashgillian), late Devonian (Frasnian), Permian/Triassic (P/T), late Triassic (Norian), early Jurassic (Pliensbachian), late Jurassic (Tithonian), mid-Cretaceous (Cenomanian), Cretaceous-Tertiary (K/T), and the late Eocene (Priabonian). If we eliminate, on the basis of their apparent regional rather than global significance, the Pliensbachian and Tithonian events, this leaves seven possibly significant extinctions. An important question is, are these apparent evolutionary setbacks attributable to a single common cause or a combination of causes?

When one compares the extinction events with the eustatic curve of sea-level change (Hallam, 1992), it is apparent that the extinctions generally coincide with notable regressions (Fig. 13). This was first pointed out by Newell (1967). As observed earlier, such regressions affect both the marine and terrestrial environments. In the sea, they reduce the continental shelf area where most of the marine biota lives and they probably decrease primary production in such areas. On land, the enlargement and higher stand of the continents produces dryness and seasonally more severe weather patterns. These changes, in turn, have a detrimental effect on primary production. In winter, the offshore movement of air masses from enlarged mid- or high-latitude continents will in some areas lower the sea-surface temperature and the diversity of the marine biota.

It has been argued that regression should not have much effect on diversity in the sea because small areas such as oceanic islands often have rich biotas (Stanley, 1987). There is little doubt that, on land, the size of the geographic area has an important effect on species diversity (richness). This has been reviewed for terrestrial island populations by Williamson (1981) and for larger areas by Darlington (1959) and Preston (1962). For the marine environment, Flessa and Sepkoski (1978) felt that the species-area relationship would be a productive tool in the evaluation of fossil diversity but its use is limited by the lack of documentation about the nature of species-area effects in the recent marine benthos. Abele (1982) compared the number of crustacean species in the four great tropical shelf regions of the world and plotted those numbers against the area of continental shelf in each region. The result showed an almost linear relationship. Briggs (1985) accumulated data on the echinoderms (crinoids, asteroids, ophiuroids, echinoids), molluscs, crustaceans (brachyurans, stomatopods), hermatypic corals, and fish. These were the groups that had been subjected to the most thorough systematic work. The total number of species for each region was tabulated and then compared to the area of continental shelf. The result showed an almost linear relationship among the tropical eastern Pacific, western Atlantic, and Indo-West Pacific. The diversity of the eastern Atlantic proved to be markedly lower. It was suggested that its relative poverty was due to the lowered sea-surface temperatures during the Pleistocene glaciations, which had a more severe effect on the eastern Atlantic than on other parts of the tropics.

Valentine and Jablonski (1991) investigated the effects of sea-level change on California Pleistocene and Recent molluscs. They found that extinctions were minimal, well below predicted species-area effects. However, they were dealing with faunas that had already passed through a dozen or more glacial/interglacial stages. The initial regressions, beginning about 3.0 Ma ago, could have sharply reduced the local species diversity. The periods of sea-level rise, during the succeeding interglacial stages, may have been too short for the diversity to rebound. Aside from the latter work, the general indication is that the species-area relationship is indeed applicable to large areas of the continental shelf. It means that a significant reduction in shelf area will be followed by a reduction in species diversity.

There are evidently three important causes of sea-level regression. First, plate-tectonic movement may slow down, permitting the mid-ocean ridges to cool and shrink. Second, ice accumulation in the form of ice sheets or glaciers removes water from the hydrological cycle and prevents it from reentering the sea. Third, increased volcanic activity or continental collision may cause the uplift of sections of continents and hence provide more space in the ocean basins.

Another possible common cause of extinctions is temperature change. Marine organisms are, in general, very sensitive to the temperature of their environment. The oceans can be divided into four latitudinal temperature zones, each with its own characteristic biota (Briggs, 1974a). The same zones can be recognized in the terrestrial habitat but the separations are not as sharp because the biota is physiologically acclimated to large excursions in diurnal and seasonal temperatures. Also, many terrestrial animals are endothermic, making then even more independent of environmental temperature.

In the late Ordovician, there was evidently some glaciation in North Africa which, at that time, was situated over the South Pole (Stanley, 1987) and brachiopod distribution (Sheehan, 1982) may indicate that a temperature decline accompanied the sea-level regression. For the late-Devonian extinction, there seems to be no direct evidence of glaciation and global cooling. However, the reef and other shallow-water communities were hit very hard and this may indirectly indicate that a temperature drop was involved.

In regard to the P/T extinction, it would be difficult to involve a temperature decrease as a major factor. The earth was apparently in a warming phase following a mid-Permian glaciation. As already mentioned, the huge drop in sea level, the loss of marine shoreline due to continental amalgamation, and a possible reduction in salinity would all have severe effects on marine diversity. While there may have been some ice formation towards the poles (Stanley, 1987), it probably did not have much effect on the rest of the world. In a like manner, there seems to be no direct evidence of a temperature decline during the late-Triassic and mid-Cretaceous extinctions but the losses were high among the major reef-building organisms and this may be an indication of a temperature effect.

In the K/T extinction, the climate was generally warm with no indication of ice formation at the poles. However, it seems likely that the intense volcanic activity at that time produced a global drop in temperature (or a series of them) for possibly several years. This may have been followed by a sustained temperature rise due to excess atmospheric CO_2. There is little doubt that the late-Eocene extinction took place at a time of global cooling when there was glaciation in the Antarctic. This introduced the modern era of cold water in the deep-ocean basins.

The most powerful argument for temperature reduction, as an effective agent in extinction episodes in the sea, is that in each case, it was the tropical or Tethyan organisms that appear to have suffered the greatest losses. We know, from the study of living forms, that they are the most stenothermic of all marine organisms. In discussing extinction causes, there has been a tendency to use the historic occurrence of polar-ice formation as an indication of global temperature deterioration which would then account for the destruction of tropical organisms (Stanley, 1987). The difficulty with this approach is that the world can stand a lot of polar glaciation and still maintain extensive low-latitude regions that remain warm enough to support a tropical biota. Even during the most recent Pleistocene glaciation, the tropical regions, although reduced in size, maintained their integrity.

Crowley and North (1988), in their review of the effects of abrupt climate change on extinction events in earth history, suggested that terrestrially induced climate instability was a viable mechanism for causing rapid environmental change and biotic turnover. They observed that sometimes abrupt transitions can occur in climate models involving atmospheric circulation, general ocean circulation, and thermohaline deep-water circulation. They concluded that such transitions cause climatic changes which affect the biotas. The best examples of abrupt transitions involved changes between polar-ice and ice-free conditions.

Volcanic aerosols and dust reflect sunlight and can markedly reduce the surface temperature. If a major eruption occurs at a low latitude, it will probably affect the major parts of both hemispheres. The same effect could be caused by the impact of a large comet or asteroid, which may inject fine debris into the upper atmosphere. As noted, one effect of regression is cooler continental temperatures but these alone would probably not eliminate large segments of the tropical marine biota. Another point is that polar cooling will stimulate the oceanic thermohaline circulation. The increased upwelling may bring up excess nutrients that would have a destructive effect upon the shallow oligotrophic system in the tropics (Hallock, 1987). There is some information which suggests that the nutrient supply to major parts of the low- and mid-latitude oceans is controlled by upwelling (Sarnthein and Winn, 1988). This means that the eutrophication caused by increased upwelling may be the primary cause of species extinction over large areas.

It seems that the only readily identifiable common cause for most extinctions is sea-level regression. And, it is perhaps significant that the greatest regression, at the P/T boundary, occurred at the time of the greatest extinction. There is also some evidence that sea-level transgressions, when they introduce anoxic water to epicontinental seas, may cause significant extinctions in the marine environment. This may have been the primary cause of the Cenomanian extinction (Hallam, 1989b). Judging from the differential effects on the marine biota, it seems likely that a temperature decrease was also involved. Other factors, such as acid rains, a drop in oceanic salinity, the competition engendered by the coming together of previously separated continents, and the loss of shoreline habitat, were probably important at particular times. These all come under the category of direct effects which had immediate consequences for the vulnerable populations. The extinction of the largest terrestrial and marine animals was probably due to the indirect effect of reduced primary productivity. Short food chains, with very large animals at the top, are the evolutionary result of high primary production that has been sustained for millions of years. When that production declines, the large animals are the first to go.

The periodicity theory was challenged on the basis of its statistical procedure (Hoffman, 1985), but the theory was then strengthened by refining the data and utilizing information on generic as well as family extinctions (Sepkoski, 1986b). Another such challenge was issued by Stigler and Wagner (1987). These and other criticisms were answered by Sepkoski (1989) in a detailed defense of the theory. However, the most serious criticisms have to do, not with the statistics, but with the dependability of the data base itself.

Patterson and Smith (1987) and Smith and Patterson (1988) conducted an examination of the fossil record in their areas of expertise (echinoderms and fish). They found that, at first, their subset of the record appeared to show the same extinction peaks that Raup and Sepkoski had discovered for the whole. However, when the evidence for each extinction was reviewed, an incredible amount of error and misleading data were found. Only 25% of the entries represented correctly dated last occurrences of monophyletic groups, which constituted valid extinction data. The other 75% comprised entries that for various reasons could be considered as noise. These were monophyletic groups whose last occurrences were wrongly dated, monotypic families, exclusively freshwater fish, and nonmonophyletic groups.

Smith and Patterson (1988) pointed out that the nonmonophyletic groups were taxonomic artifacts that did not represent real phyletic lines. Therefore, their disappearance from the fossil record constituted pseudoextinctions. When they constructed a plot of all last occurrences of fish and echinoderm monophyletic families and genera, they found no statistically significant peaks of mass extinction since the start of the Triassic. They concluded that there was no unambiguous evidence that either fish or echinoderms were affected by periods of mass extinction on the scale currently envisaged for the marine biota during the Mesozoic and Tertiary.

If the fish and echinoderm data are representative of the fossil record as a whole, and there is no reason to think they are not, the periodicity hypothesis becomes nonviable. It also means that the paleontological record, in its present state, cannot be used for broad analyses that attempt to define global events in the marine environment. It suggests that, for some years in the future, information about widespread extinctions and their evolutionary consequences should come from detailed studies of individual groups where the histories of discrete phyletic lines have been determined.

The evidence now before us indicates that all of the major extinction events took place over extended periods of time ranging from about 1 to 10 Ma or more. Furthermore, it seems likely that all may have occurred in the form of a sequential series of minor episodes that, only collectively, comprise a significant extinction. Hoffman (1989b) concluded that the major extinction peaks may in fact be clusters of separate events more or less accidentally aggregated in time. However, sea-level regression has almost always been a major factor. Regressions have direct and indirect effects detrimental to biotic diversity. Impacts by extraterrestrial bodies could have added to the magnitude of some of the extinctions. But, none of the major extinctions can possibly be attributed to a single impact event.

BIOGEOGRAPHY AND EVOLUTION

There has been, in recent years, considerable speculation about the evolutionary effects of extinction episodes. These provide a consistent message which gives the impression that a "wiping out of the old forms to make way for the new" has long-term evolutionary benefits. Gould (1984) suggested that extinctions might be the primary and indispensable seed of major changes and shifts in life's history; that if environments did not undergo such changes, evolution might well grind to a virtual halt. Raup (1986) referred to the common belief that, during an extinction event, the better adapted types would survive and that this would lead to an improvement in the entire biota.

Eldredge (1987) expressed the firm belief that, without extinctions to free up the ecological niches, life would still be confined to a primitive state somewhere on the sea bottom. Stanley (1987) stated, "Had the dinosaurs survived, there is no question that we would not walk the earth today. Mammals would still remain small and unobtrusive, not unlike the rodents of the modern world". The idea that major extinctions convey evolutionary benefits has been developed into a new theory of evolution by Hsü (1986). He would substitute the concept of evolution by means of global extinctions for Darwin's mechanism of natural selection. In this "new catastrophism", evolutionary advances would take place as the survivors adapt to spaces created by extinction events. In his view, "it is time to awaken to the absurdity of the idea of natural selection".

The available data on the widespread extinction patterns do not support the common assumption of evolutionary benefits. A number of authors have observed that, in the groups they studied, the warm-water or tropical marine organisms have been more prone to extinction during crises than those in cold or deep waters (Hsü et al., 1982; Aubry, 1983; Gerstel et al., 1986; Anstey, 1987; Fagerstrom, 1987; Brasier, 1988). Vermeij (1987) noted that this pattern has been found in late-Cambrian trilobites, late-Ordovician and early-Silurian brachiopods, late-Devonian brachiopods and corals, and late-Cretaceous gastropods and planktonic foraminiferans. In regard to all large extinctions, Jablonski (1986) stated that low-latitude taxa, particularly in reef communities, were invariably more severely affected than temperate, polar, and cosmopolitan taxa.

Gerstel et al. (1986) found, in regard to the K/T crisis, that cool-adapted species of planktonic foraminiferans migrated to low latitudes replacing warm-adapted species that had become extinct. Lipps (1986) called attention to the fact that several groups of the pelagic plankton exhibit a much simpler structure in cooler, higher-latitude waters and, following the K/T and E/O extinctions, it was these species that survived while the more complex forms became extinct. It seems probable that preferential removal of the complex tropical species by extinction episodes and their replacement by simpler forms would ordinarily involve movement by the latter from higher to lower latitudes. Relatively primitive taxa of many groups also tend to inhabit deeper, cooler waters or special refuges such as caves and the interstitial environment. It is possible that some of the replacement forms could have come from such places as well as from higher latitudes.

For some groups of marine animals that possess good fossil records, such as the hermatypic corals (Stehli and Wells, 1971), bivalve molluscs (Hecht and Agan, 1972), and benthic Foraminifera (Durazzi and Stehli, 1973), the younger, more advanced genera are found in the tropics while, the higher the latitude, the more ancient they become. In other groups, whose fossil history is not as well known, the higher-latitude representatives often have a simpler, more primitive morphology.

Many, if not most, geographically widespread families and higher taxa of shallow-water marine organisms demonstrate their greatest diversity in tropical waters. Recent systematic work has shown that the tropics, and especially certain centers of origin within them, tend to be inhabited by the youngest and most specialized genera and species (Briggs, 1984). As such taxa slowly disperse to higher latitudes, they become older, thus accounting for the patterns of increasing age with increasing latitude (Kafanov, 1987). In the meantime, still younger and more advanced species continue to evolve in the tropics. This general process may be the rule in times of non-crises or background evolution. When a widespread tropical or circumtropical extinction occurs it has its most drastic effect on the youngest, most advanced taxa.

It is apparent that most, perhaps all, of the great extinctions preferentially eliminated the tropical shelf and epipelagic organisms. This created an ecological vacuum which was gradually reoccupied by simpler, eurythermic organisms from higher latitudes, and possibly from deeper water and other refuges. This process of succession by older, more primitive forms may set back the evolutionary clock to an earlier time. Furthermore, the damage to the tropical centers of origin must have interrupted, for some time, the production of successful new species and genera. The concept of evolutionary setbacks is supported by the model of Kitchell and Carr (1985) which predicts that taxonomic turnover is generally delayed by extinction episodes, rather than being precipitated by them.

There is another way in which extinctions caused evolutionary setbacks. Jablonski (1986) has related how the shell-drilling habit in gastropods first appeared in the late Triassic, only to be lost in the end-Triassic extinction. It took another 120 Ma for it to re-originate. Bivalves first developed the ability to bore into rocks in the late Ordovician, but this adaptation was lost in the Ashgillian extinction. Such bivalves did not reappear until more than 100 Ma later. If these evolutionary innovations had persisted and developed through time, what would our molluscan assemblages look like today?

There are data that indicate, in general, the amount of time that it takes a community to recover from a severe extinction. In the P/T event, all of the existing reef communities completely collapsed (Fagerstrom, 1987). It took approximately 8 Ma until such communities began to reestablish themselves. The Norian extinction affected the scleractinian corals so severely that reef construction was halted and did not get underway again until some 15 Ma later. Hansen (1988) studied the long-term effects of the K/T extinction on marine molluscs. Along the Gulf Coast of the United States, the extinction reduced the diversity from about 500 species to a few over 100. Afterward, the diversity slowly returned in a series of steps to about 400 species in the mid-Eocene. So it took nearly 25 Ma for the diversity to build back to almost its late- Cretaceous level. Hansen estimated that, for bivalves and gastropods as a whole, it takes about 10 Ma for a doubling of species diversity.

As part of their argument for the evolutionary advantage of extinctions, Gould and Calloway (1980) gave the example of the brachiopods versus the pelecypods. They felt that the pelecypods gained the advantage because the brachiopods were hit harder by the P/T extinction. But this idea overlooks the fact that the pelecypods are clearly the superior competitors. Pelecypods can actively move about instead of being attached to the substrate, they pump water through their gills far more efficiently, and they grow faster (Vermeij, 1987). The present dominance of pelecypods in the warm, shallow habitats of the world, compared to brachiopods, which are mainly relegated to colder and deeper waters, is a tribute to the superior competitive ability of the former.

Similar evolutionary replacements have taken place in other groups. The stalked crinoids, once common on the continental shelves, became restricted after the Jurassic to waters deeper than 100 m (Meyer and Macurda, 1977). Their shelf habitat was taken over by their relatives, the mobile, active feather stars. Glypheoid crustaceans, once widespread in shallow waters during the early Mesozoic, are now represented by a single deep-water species (Forest et al., 1976). Its modern descendants, the lobsters and crabs, now occupy the shallow waters. There are many other examples. These represent gradual replacements that have taken place over very long periods of time. It would be difficult to find a positive role for sudden extinction episodes in these processes.

In the terrestrial environment, the large-scale changes in plant communities (Fig. 53) have been summarized by Niklas (1986). The gymnosperms began their development in the mid-Paleozoic and, for about 100 Ma, increased their diversity along with the pteridophytes. In the latter part of the Paleozoic, gymnosperms became somewhat the more diverse. Both groups suffered about equally in the P/T extinction but afterward the gymnosperms continued their expansion while the pteridophytes played an increasingly minor role. The angiosperms arose in the beginning of the Cretaceous and quickly developed into the dominant plant group. In each case, the more primitive group that was succeeded continued to do well for about 100 Ma beyond the time of origin of its successor. Both pteridophytes and gymnosperms eventually began to decline, but these changes had become evident prior to the P/T and K/T boundaries, respectively. The angiosperm increase continued despite a brief interruption at the K/T boundary.

The biogeographic consequences of the changes in the major plant groups are that the angiosperms now dominate the tropical and warm-temperate zones of the terrestrial world where edaphic and rainfall conditions are favorable. Angiosperms, in the form of

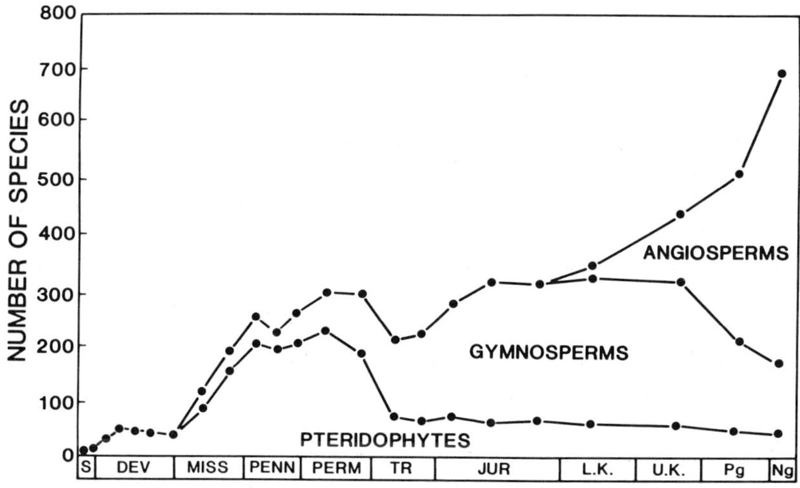

Fig. 53. Historical changes in the species diversity of the major groups of vascular plants during the Phanero zoic. After Niklas (1986).

sea grasses, have also invaded the warm, shallow marine waters and make an important contribution to primary production in that habitat. In contrast, the gymnosperms predominate at higher latitudes and in less favorable locations, while the pteridophytes tend to be interspersed in marginal habitats. Neither the origination of the major plant groups nor the relative decline of two of them is clearly attributable to large-scale extinction events.

The popular belief that the evolutionary success of the mammals was due to the extinction of the large reptiles at the K/T boundary is still prevalent (Stanley, 1987). The general view among vertebrate paleontologists is that the large dinosaurs had been declining in diversity for several Ma prior to the boundary (Dodson, 1991). Of greater importance is the fact that mammalian evolution did make considerable progress in the late Cretaceous when the large dinosaurs were still prevalent. In northwestern North America, there existed ten genera of multituberculates, five genera of marsupials, and seven genera of placentals (Savage, 1988). These data provide evidence that a mammalian radiation was already well underway before the large dinosaurs left the scene. Sloan et al. (1986) described fossils from the Hell Creek Formation which suggested to them that dinosaur extinction and ungulate (hooved-mammal) radiation were simultaneous events. This may imply that a competition existed in that region.

As in the case of the flora, the great evolutionary replacements among the major groups of terrestrial vertebrate animals do not seem to have been caused by or greatly affected by extinction episodes. As Colbert (1986) has shown, several orders of tetrapods continued from the Paleozoic well into the Mesozoic. They were gradually replaced by newer orders beginning in the Triassic. In most cases, there was an extended overlap lasting several Ma in which the older group coexisted with its newer counterpart. Much of this temporal overlap occurred during the late Triassic. Feathered dinosaurs (birds) came on the scene in the late Jurassic, but the older pterosaurs were able to carry on, in diminishing diversity, for another 70–80 Ma. Another example concerns the Mesozoic amphichelyd turtles (Rosenzweig and McCord, 1991). Beginning in the Upper Jurassic, they began to be replaced by more advanced groups. Although, in North America, the replacement process seems to have been somewhat accelerated across the K/T boundary, the worldwide process went slowly. It was not until the late Pleistocene that the amphichelyds finally became extinct. The replacement by the two successor groups took about 150 Ma. These exceedingly slow, evolutionary replacements of major groups appear to be the almost invariable rule in both the terrestrial and marine environments.

In the terrestrial environment, when one compares the biota of the high-latitude versus the low-latitude parts of the world, it becomes at once apparent that the former usually contains the older and more primitive organisms (Briggs, 1987). In this habitat also, it is probable that the tropical organisms were more vulnerable to extinction episodes. Higher-latitude forms are better adapted to extremes of temperature and humidity. Unusual fluctuations in these physical factors are likely to have taken place in conjunction with regressions of the sea and consequent formation of larger and higher continents. With higher rates of extinction in the tropics, we might expect an invasion of more primitive forms from higher latitudes. The end result would be a setting back of the evolutionary clock on land as well as in the sea.

Data that might indicate a setback to evolution on land are scarce because fossil localities are few and widely scattered so that it is almost impossible to correlate extinction events on a global scale. The bed of the ocean, on the other hand, is a great fossil repository which can be sampled in many places and often correlated chronologically. At the present state of our knowledge, it seems worthwhile to hypothesize that the great historical extinctions were disasters which interfered with rather than enhanced the course of evolution. As Vrba (1987) has concluded for African mammals, "It is the specialists that are shaved off wholesale from the Tree of Life by mass extinction. The system is reset, leaving a few hardy generalists to start all over again".

Although the fossil evidence does not support the concept of historical mass extinctions or mass killings, there is a catastrophic extinction event occurring in contemporary time. Raven (1990) has estimated that, by the first quarter of the next century, the world will have lost two million out of a minimal world total of 10 million animal species and about 65 000 out of 300 000 species of vascular plants. These losses, due to habitat destruction by humans, are occurring with a rapidity that is unprecedented in Phanerozoic time. Historic extinction episodes were so gradual that many lineages were able to accommodate in an evolutionary and ecological sense. The tempo of the current extinctions precludes any such adjustments.

CONCLUSIONS

Our present era of neocatastrophism is a recapitulation of the original theory of catastrophism which was popular in the early 19th Century. Neocatastrophism is predicated on the assumption that there occurred an historic series of global catastrophes called mass extinctions or mass killings. However, study of the tempo of the extinction episodes reveals that they took place over extended periods of time ranging from about 1 Ma to 10 Ma or more. The deliberate pace of the extinctions was, in reality, the antithesis of catastrophic. They took place over evolutionary, not contemporary time.

The scope of most of the extinctions was far less than has usually been described. Only the Permian/Triassic event deserves to be considered a global phenomenon. This is the only extinction in which there was a drastic diversity decrease in both marine and terrestrial environments. There are indications that the effects of widespread extinctions on the global species diversity have been greatly exaggerated. At the K/T boundary for example, estimates of species extinguished ranged from 50 to 80% of the world total. Yet, when the evidence is examined, it appears that the actual demise could have been less than one percent. Certainly, many important groups of marine animals disappeared, and some terrestrial as well, but their relative diversity was very small. The only event which deserves the title of mass extinction is that which is going on right now. It appears likely that, within the next 35 years, the world will lose about 20% of its total species diversity.

It is proposed that, contrary to conventional wisdom, the historical extinction episodes were disastrous interruptions to evolutionary progress. They set back the clock of evolutionary time, destroyed communities that took millions of years to reassemble, and eliminated new ecological inventions that did not re-originate for 100 million years or more.

During the past 15 years, two extraordinary theories have had a galvanizing effect on research pertaining to the extinction episodes. First came the extraterrestrial impact hypothesis which was primarily developed and defended by Alvarez et al. (1980). Then came the periodicity hypothesis of Raup and Sepkoski (1984). The fact that neither of these two ideas now appears to be entirely correct is of little importance. They were responsible for inspiring an enormous amount of thought and research. The real value of a hypothesis lies in its utility to scientific progress, not whether it turns out to be right or wrong.

There is an ongoing debate about what caused historical mass extinctions – asteroid impacts or volcanic eruptions? But this debate is based on the false assumption that mass extinctions or mass killings have taken place. Aside from the present human destruction of the earth's tropical ecosystems, there is no evidence that global mass extinctions (defined as short-term, catastrophic events) ever took place. Instead of using such misleading terminology, paleobiologists might well consider the recommendations of Walliser (1990). For exceptional happenings that took place in a relatively short time span, we could use the term "extinction events". For events in which a large portion of the biota disappeared, we might use "extinction crises".

PART B – Contemporary biogeography

CHAPTER 9

Marine patterns, Part I

> Who can explain why one species ranges widely and is very numerous, and why another allied species has a narrow range and is rare? Yet these relations are of the highest importance, for they determine the present welfare, and, as I believe, the future success and modification of every inhabitant of this world.
>
> Charles Darwin, *On the Origin of Species*, 1859.

As viewed from outer space, the planet we inhabit is a beautiful, blue and white sphere. The blue is reflected from the oceans which cover 71% of the surface, and the white comes from clouds which float above both oceans and land.

In our solar system, only earth has liquid water. The other planets are either too hot or too cold. Our total volume of water has been estimated at 1500 km^3, about 97% of it forming the oceans. If all the water were liquid and the surface of the earth perfectly smooth, our planet would be covered by a layer of water about 3000 m deep (Penman, 1970).

Life on earth is confined to the globe's fluid envelope, or covering. This envelope is divided into two parts: a gaseous portion called the atmosphere and a liquid portion called the hydrosphere. The atmosphere is so thin that very few living things have been able to achieve a purely aerial existence. Almost all atmospheric organisms are obliged to depend on the physical support of the substrate that forms the lower boundary of the atmosphere. So, from the standpoint of their overall distribution, we can see that such organisms occupy a thin, two-dimensional layer on the earth's surface.

In contrast, the liquid of the hydrosphere is viscous enough to provide physical support and contains sufficient nutrients in suspension so that almost all its volume is continually occupied by a vast array of living things. Here, we find that life exists on a truly three-dimensional scale with its vertical component extending from the surface to the greatest depths, almost 11 000 m. If we call that portion of the earth inhabited by life the biosphere, it is clear that the oceans comprise by far the greatest part of it. Thorson (1971, p. 20) estimated that the total oceanic living space is roughly 300 times larger than that available for life on land.

Although the first metazoan animals appeared in the sea about 800 Ma ago (Stanley, 1989), they did not invade the land until the late Silurian (Selden, 1990), somewhat more than 400 Ma ago. Most of the basic differentiation in the animal kingdom took place in the sea, and many of these marine groups never developed the ability to invade freshwater or the land. No less than 34 of the 37 phyla of living animals are found in the sea, while only 17 occur in freshwater, and 15 on land; at the class level, 73 occur in the seas,

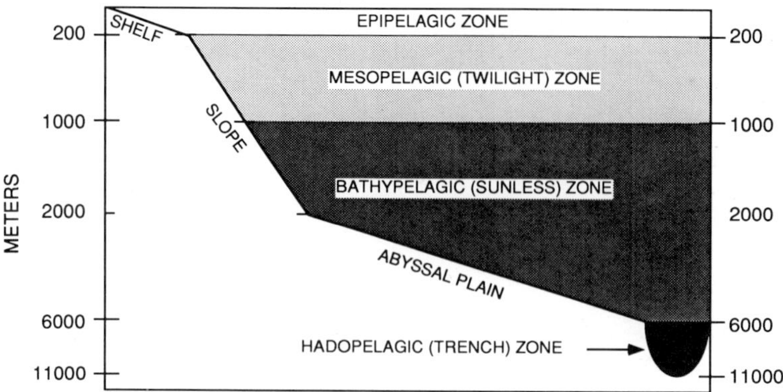

Fig. 54. Diagrammatic representation of the vertical divisions of the world ocean based on the distribution of the biota.

35 in freshwater, and 33 on land (Nicol, 1971). One might expect this superior diversity at the higher taxonomic levels to be accompanied by a greater species diversity, but this is evidently not true. There are probably less than 200 000 multicellular species in the sea while those on land probably number around 12 million (Briggs, 1994). Why so few species in the sea? This question is pursued later (p. 375).

Biogeographers traditionally subdivide the ocean into a number of parts which may be defined by the distributions of the organisms they contain. The two fundamental divisions of the oceanic biosphere are the Benthic Realm and the Pelagic Realm. Each realm may, in turn, be subdivided into a number of depth zones (Fig. 54) as well as a series of horizontal regions and provinces. While the shallow (less than 200 m) areas have been generally well defined and recognized, the horizontal patterns of the deeper zones have not been well established. The major deep-sea trenches are an exception, since most of their organisms appear to be endemic to individual trenches.

In the 20-year period between 1974 and 1994, a vast amount of literature, containing significant information about the distribution of various marine species, has appeared. There is, in fact, so much that one could no longer do justice to it within the covers of one book. In the case of the present work, which attempts to include historical biogeography as well as to outline contemporary regions and provinces, there is not enough space to discuss the evidence for each biogeographic subdivision. Most of the recent literature does substantiate the decisions about the existence and locations of the various subdivisions that were made earlier (Briggs, 1974a). Where more research has indicated the necessity for changes, they have been made.

LATITUDINAL ZONES

Any biologist who studies the distribution of species in the sea, soon becomes aware of the fact that marine organisms are highly sensitive to the temperature of their environ-

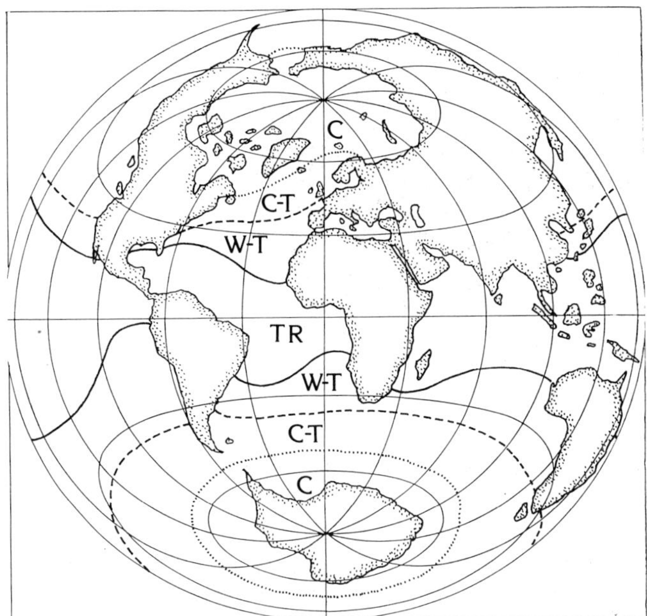

Fig. 55. The four major temperate zones of the ocean surface. The Tropical Zone (TR) is delimited by the 20°C isotherm for the coldest month. It is bordered by a Warm-Temperate Zone (W-T), a Cold-Temperate Zone (C-T) and a Cold or Polar Zone (C).

ment. Distribution patterns in local situations may be affected by such factors as food, shelter, salinity, oxygen, etc. But, on an oceanic or global scale, it is temperature that primarily controls distribution. The widespread patterns demonstrated by many species tell us that the surface of the ocean is subdivided into four zones, each with its own distinctive complex of species (Fig. 55).

The tropical biota generally occupies the lower latitudes, but the shape of the Tropical Zone is affected by the flow of the major oceanic currents. On the western sides of the Pacific and Atlantic oceans, the North and South Equatorial currents turn towards higher latitudes and bring with them warm water and tropical organisms. On the eastern sides of the two oceans, the major currents transport cool water towards the tropics. This allows the tropics to occupy a broad latitudinal area to the west but only a relatively narrow one to the east. Since there is a positive relationship between species diversity and geographic area, it means that the diversity of the western regions will exceed that of the eastern ones. The northern Indian Ocean is entirely tropical so, when its diversity is added to that of the western Pacific, to recognize an Indo-West Pacific Region, we find more species there than in the other three regions combined.

The shallow marine biota on the continental shelves and in the upper 200 m of the pelagic environment, also exhibits a greater diversity than the equivalent biota at higher latitudes. This is due to the latitudinal gradient in diversity, generally attributed to the greater solar radiation received by the earth at lower latitudes. A contributing factor may be that the tropics of today have a longer history than the cooler zones. During the Mesozoic and early Cenozoic, the tropics occupied almost all the globe, except for warm-temperate waters at the highest latitudes. The tropics did not become significantly reduced until a global temperature decline began in the middle Eocene.

As James D. Dana (1853) observed, "The cause which limits the distribution of species northward or southward from the equator is the cold of winter rather than the heat of summer or even the mean temperature of the year." His remarks were accompanied by a chart relating his "isocrymal lines" to the distribution of marine animals. The 20°C isocryme for the coldest month of the year (Fig. 55) is most important, since it appears to limit the tropical fauna regardless of whether it is rich (Indo-West Pacific) or poor (Eastern Atlantic) or whether the boundary is located to the north or south (Briggs, 1974a).

Although most tropical species are restrained within the 20°C winter boundaries, many of the more mobile forms migrate to higher latitudes during the summer when the surface waters become warmer. Another factor, which may obscure the location of the tropical boundary, is the presence of species called "eurythermic tropicals." This group, although it ranges broadly in the tropics, also habitually occupies the warm-temperate zones that lie next to the tropics.

A Warm-Temperate Zone, with a suite of endemic species, borders the tropics everywhere in the world. A boundary between the two may be said to exist where the greatest number of range terminations take place. Boundaries, although they may be indicated by a line on a map, are actually places of considerable overlap. Sometimes, the depth of the water will make a difference. In places where tropical currents move along the shelf to higher latitudes, the surface will cool more rapidly than depths of 100–200 m. Consequently, tropical species on the outer shelf will extend to much higher latitudes than those on the inner shelf. Latitudinal or temperature boundaries are usually indicated at the places of greatest change in the inshore biota. Warm-temperate zones, with winter surface temperatures usually from 20°C down to about 12°C, also have a long history. They were located around the poles in the Mesozoic, then moved to lower latitudes in the mid-Eocene.

The climatic deterioration, which began in the mid-Eocene, caused the formation of colder surface water at the highest latitudes. Winter surface temperatures of 12°C down to 2°C created conditions favorable to the formation of a new Cold-Temperate Zone. The colder waters absorbed more atmospheric oxygen and their increased density stimulated the thermohaline circulation. A result was increased upwelling which brought more nutrients to the surface and enhanced primary production. In the northern hemisphere, cold-temperate waters occupied the Arctic Basin and the northern parts of the Atlantic and Pacific oceans. In the south, they occupied the circumantarctic region including the southern part of Australia. There is some evidence of glacier formation in Antarctica as early as 42 Ma ago (Keller et al., 1992).

Eventually, as the global climate continued its decline, new cold-water zones were established with surface temperatures between +2°C and –2°C. This took place first in the Antarctic. The Antarctic Cold Zone may have been established as early as 20–23 Ma ago. The presence of many endemic genera and some families indicate an extensive period of evolution under cold-temperature conditions. In comparison, the Arctic Zone, while having many endemic species, has few endemics at the higher taxonomic levels.

Although most of the decisions about zonation were made on the basis of animal distribution, it is important to note that the same four zones, with almost identical shapes, have been recognized for marine plants (Michanek, 1979). The only noticeable difference in the botanical scheme is the recognition of a separate Subarctic Zone.

About 3.0 Ma ago, northern hemisphere cooling intensified and icebergs began to calve into the Arctic Ocean. Most of the cold-temperate (boreal) species were eliminated and the modern Arctic biota began to evolve. Because the Arctic Zone soon extended into the North Pacific and North Atlantic, the boreal biotas in each ocean were forced southward. The southward displacement resulted in the formation of separate east and west boreal regions in each ocean. Although the four present boreal regions still show some interrelationship, they are each distinguishable by a significant complement of endemic species.

The continental shelves are considered to be part of the Benthic Realm, even though some of the species exist in the upper part of the water column. As on land, species diversity on the continental shelves, and in the shallow waters around oceanic islands, appears to be amenable to the Energy-Stability-Area Theory of Biodiversity, or ESA theory for short (Wilson, 1992). That is, the more solar energy, the greater the diversity; the more stable the climate, the greater the diversity; and the larger the area, the greater the diversity. Another factor, that of evolutionary time, needs to be added (Briggs, 1994).

The richest marine biota is found in the shallow waters of the tropical shelves at depths generally less than 200 m. Four great biogeographic regions may be identified: The Indo-West Pacific, the Eastern Pacific, the Western Atlantic, and the Eastern Atlantic. Each region may, in turn, be subdivided into provinces. Longitudinally, the tropical regions are separated from one another by barriers that are very effective, since each region possesses, at the species level, a biota that is highly endemic.

INDO-WEST PACIFIC REGION

The shelf waters of the Indo-West Pacific Region (IWP) are spread out over an enormous geographic area extending longitudinally more than halfway around the world and through more than 60° of latitude. Furthermore, its biota is incredibly rich with a species diversity that exceeds the total of the other three regions. This huge area includes approximately 6 570 000 km^2 of continental shelf. The other tropical regions total slightly more than 2 000 000 km^2 (Briggs, 1985). The IWP has more than 6000 species of molluscs, 800 species of echinoderms, 500 species of hermatypic corals, and 4000 fish species.

One of the most interesting features of the IWP is that, despite a basic homogeneity caused by the occurrence of many wide-ranging species, there are great differences in species diversity among the various parts of the region. Over the years, several authors who studied widespread tropical groups, have noted a concentration of species in the comparatively small triangle formed by the Philippines, the Malay Peninsula, and New Guinea (Fig. 56).

As one leaves the East Indies Triangle and considers the biotas of the peripheral areas, there is a notable decrease in species diversity that appears to be correlated with distance.

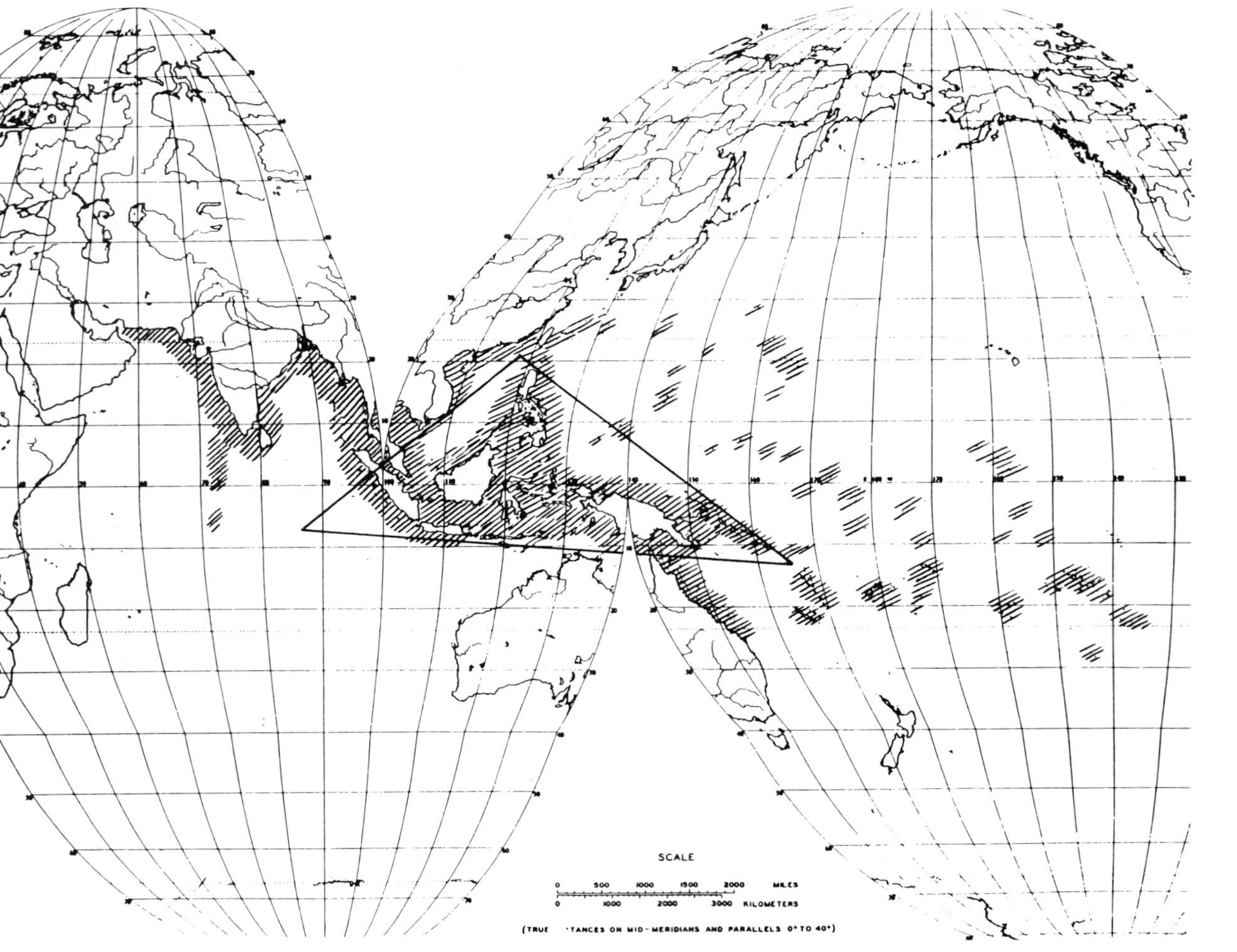

Fig. 56. The East Indies center of evolutionary radiation shown as a triangle within the Indo-Polynesian Province (hatched) of the Indo-West Pacific.

This is true of almost every tropical family whose systematics are well known. Examples are the damselfish family Pomacentridae (Allen, 1975a) (Fig. 57) and the molluscan family Strombidae (Abbott, 1960) (Fig. 58). The same pattern can often be seen at the generic level. In the hermatypic corals, there are 80 genera in the East Indies, 66 in the western Indian Ocean, 45 in the outer islands of Polynesia, and only 8 in the eastern Pacific (Veron, 1986). Similar diversity declines take place to the north and south of the East Indies.

The present East Indies center of high diversity owes its origin to the Tethys Sea, the Cretaceous to early Miocene ocean which was situated between the northern and southern continents. At first, the area of greatest species diversity, at least for molluscs, appeared to have extended from Europe and north Africa to northern India (Piccoli et al., 1987). Beginning in the middle Eocene, fossil materials from Java indicated the start of a diversity buildup in the East Indies. By the early Miocene, there was a substantial increase in Java, with many of the species showing up as far east as the Fiji Islands. By the late Miocene, the tropical diversity center in the East Indies was evidently well established.

Two events were primarily responsible for the eastward migration of the diversity center. First, was the global temperature deterioration which began in the mid-Eocene. Second, was the early Miocene collision between Africa and Eurasia which eliminated the Tethys and established the Mediterranean. The tropical biota that was trapped in the Mediterranean became gradually decimated as the climate got colder. A contributing factor was probably the fusion of northern India to Asia, with the formation of the Arabian Sea and the Bay of Bengal. The northern parts of the latter seas, with mud and sand bottoms, do not support coral-reef communities although high infaunal diversities may be found.

THE EAST INDIES: A CENTER OF ORIGIN?

What is the nature of the high-diversity center in the East Indies? Is it simply an area of accumulation where peripheral species have been deposited by ocean currents (Jokiel and Martinelli, 1992)? Or is it a center of origin where successful new species are produced (Briggs, 1984, 1992)? These questions are of fundamental biogeographic and evolutionary importance.

The controversy began with Ekman (1953) and Ladd (1960). Ekman referred to the Indo-Malayan region as a faunistic center from which other subdivisions of the Indo-West Pacific (IWP) recruited their faunas. But Ladd, who studied recent and fossil molluscs, thought that the original faunistic center was located in the islands of the central Pacific and that prevailing winds and currents had carried species into the East Indies. In the meantime, studies of relationships within the individual families and genera began to make it possible to compare phylogenetic with geographic patterns. Briggs (1955) found that in the fish family Gobiesocidae, the most specialized genera and species occurred in the central, tropical area of the IWP while the more generalized forms existed on the geographic periphery. Abbott (1960), in his monograph of the genus *Strombus* (family Strombidae), considered the central part of the Western Pacific Arc (Philippines to Indo-

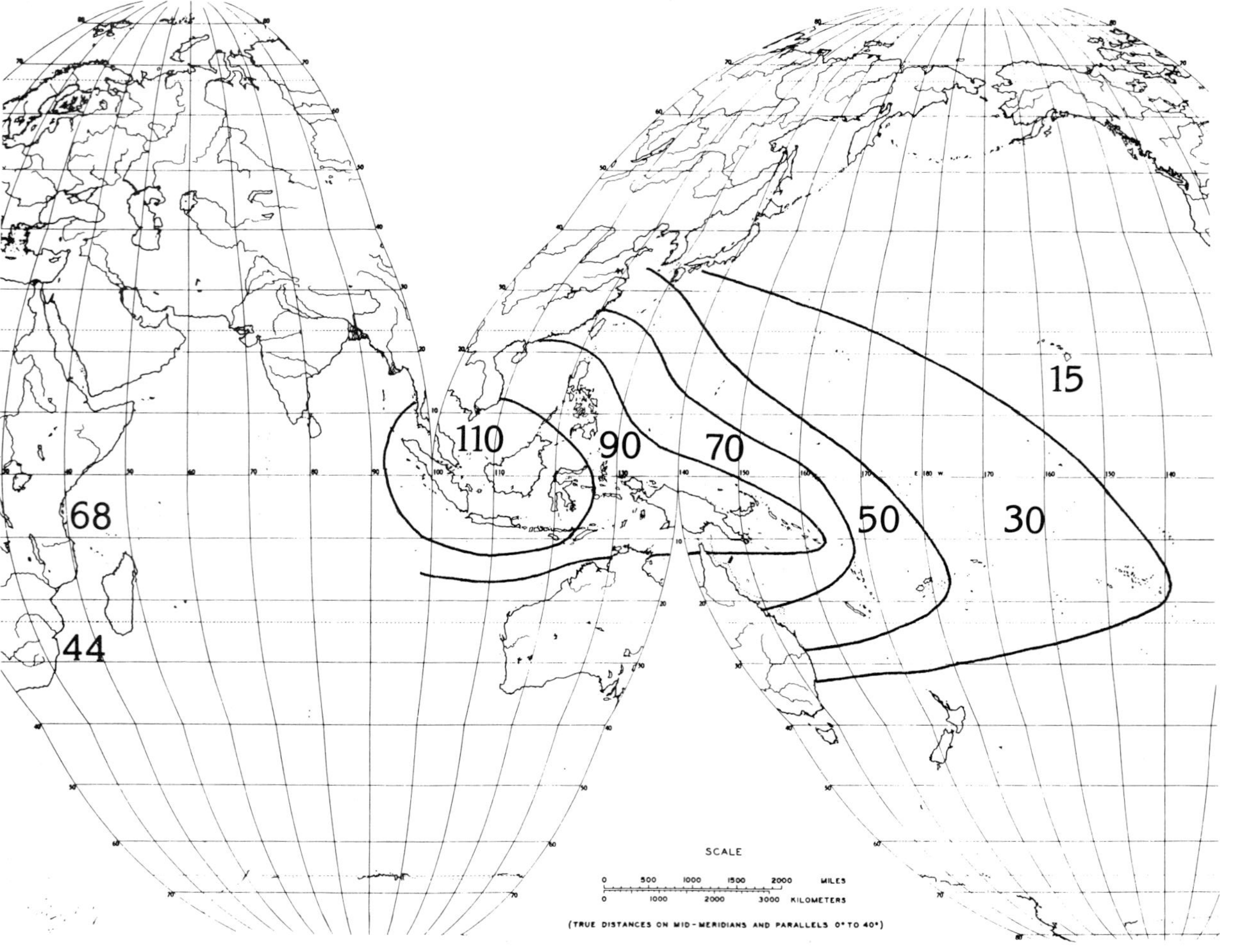

Fig. 57. Species diversity of damselfish (Pomacentridae) in the Indo-West Pacific. After Allen (1975a).

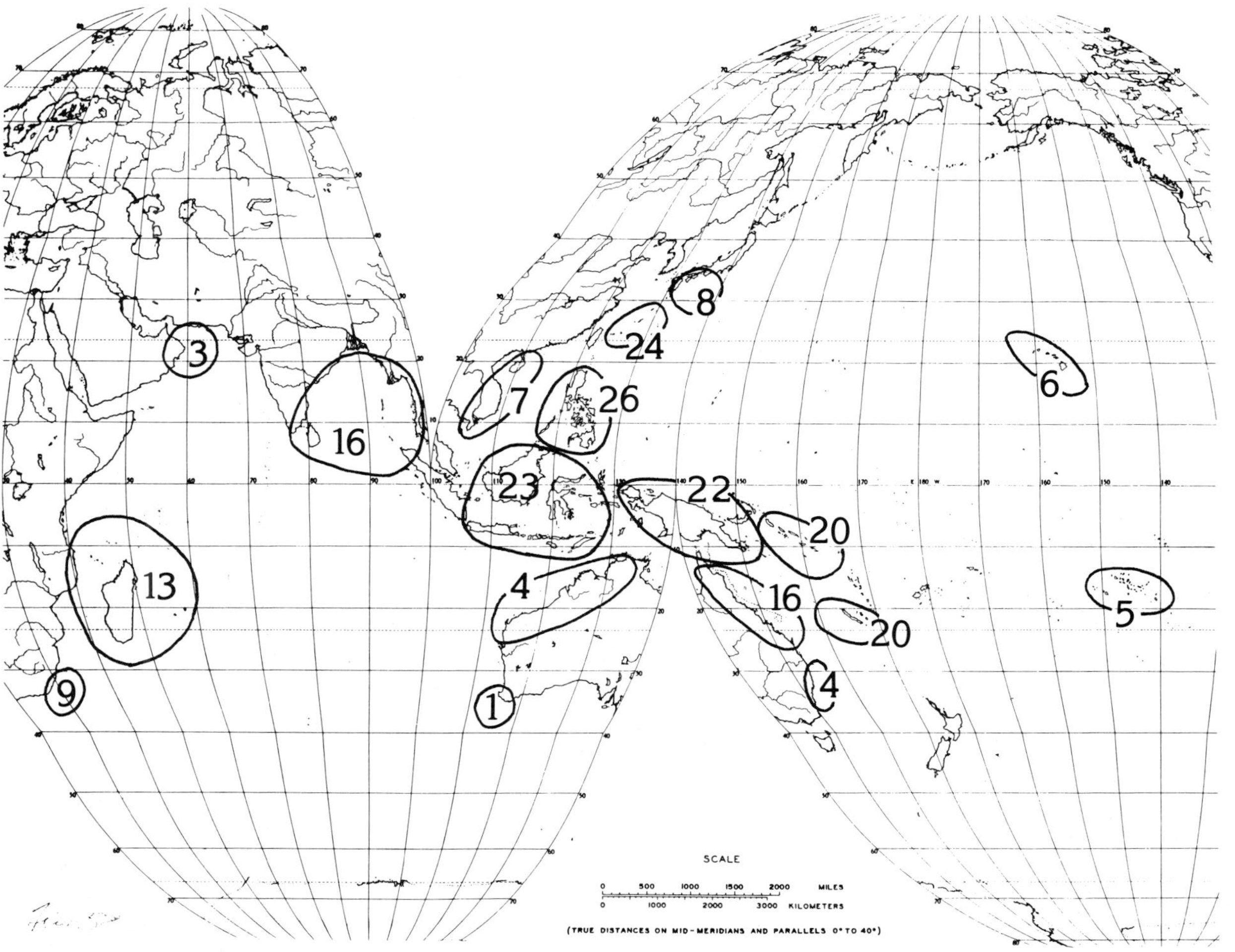

Fig. 58. Species diversity of the family Strombidae in the Indo-West Pacific. Modified after Abbott (1960).

nesia) to be the center of origin for most of the Pacific "races" and for some of those in the Indian Ocean.

In his book on the damselfish (Pomacentridae) of the South Seas, Allen (1975a) published a map illustrating a center of origin in the East Indies and the principal routes of dispersal from that area (Fig. 59). Similarly, Myers (1989) depicted the likely routes of colonization, taken by the inshore fish fauna as a whole, in order to reach the outer parts of the Western Pacific from the East Indies. Fricke (1988), who did a cladistic analysis of the evolution and distribution of fish family Callionymidae, identified a primary center of specialization in the East Indies (Philippines) and traced their dispersal from there into various parts of the Indian Ocean. Foin (1976), in his study on the biogeography of the Cypraeidae, included data on the marginal occurrence of the more generalized species and indicated paths along which dispersal from the East Indies may have taken place. East Indian centers of origin have been postulated for a variety of other groups such as the fish genus *Cynoglossus* (Menon, 1977), the hermatypic corals (Stehli and Wells, 1971), mangroves (Chapman, 1976; Ricklefs and Latham, 1993), sea grasses (Hartog, 1970; Specht, 1981), and sea grass-associated Foraminifera (Brasier, 1975).

The age gradient

If, as suggested by some of the foregoing studies, the more specialized forms do inhabit the East Indies, they should generally be younger than those found in the peripheral areas. The first indication that this may be true was provided by Stehli and Wells (1971) who showed that the central East Indies was inhabited by the youngest hermatypic coral genera and that the average age of the genera increased with distance from that area. This was followed by the discovery that coral-inhabiting balanomorph barnacles also exhibited an increasing trend of average generic age from the East Indies outward (Newman et al., 1976). Then, the same trend was found apparently to hold for shore barnacles (Newman, 1986). This may also be true for the stalked crinoids; Roux (1987) found that genera from the East Indies were mostly Cenozoic and Recent, but those from outer Polynesia were late Jurassic to Upper Cretaceous. Among the ostracods, Whatley (1987) and Titterton and Whatley (1988) found that, since the Miocene, species have been migrating outward from the East Indian-Southwest Pacific region. By studying the ages and geography of various species, they identified dispersal flows westward through the Indian Ocean to east Africa, northward, southward, and to a limited extent, eastward.

In addition to age data bearing specifically on the East Indies, it has been observed that, in general, the ages of genera seem to increase with the increase in latitude. This has been found in such diverse groups as the cetaceans (Gaskin, 1982), hermatypic corals (Stehli and Wells, 1971), bivalve molluscs (Hecht and Agan, 1972), and benthic Foraminifera (Durazzi and Stehli, 1973). It has been suggested that this pattern reflects an historical process of physiological adaptation to cooler latitudes by forms that were once tropical (Briggs, 1984). However, some fossil data have been interpreted to indicate that certain families and genera originated at relatively high latitudes and only later moved into the tropics. Taylor et al. (1980) argued that this was true for certain neogastropod families, but Jablonski et al. (1985) noted that this apparent pattern may be an artifact

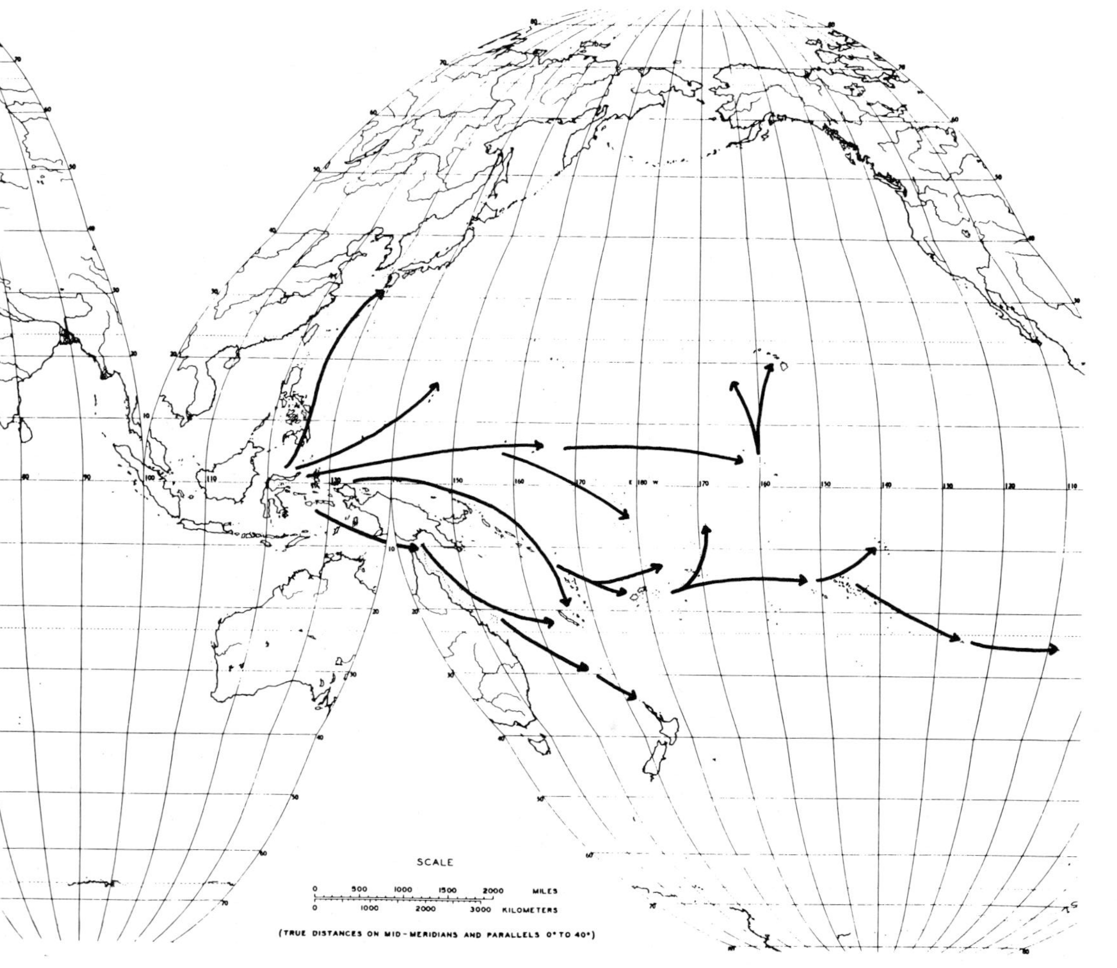

Fig. 59. Principal routes of dispersal for damselfish from their center of origin. After Allen (1975a).

due to a much better preservation of North American late-Cretaceous faunas compared to Tethyan deposits.

In their studies on an Upper Eocene fauna on Seymour Island, Antarctica, Zinsmeister and Feldmann (1984) discovered 11 genera that had been known only from lower latitudes in the late Cenozoic. The Seymour Island deposit apparently took place in shallow water but several of the genera survive today in deep waters below 100 m. Zinsmeister and Feldmann concluded that the high-latitude region of the southern hemisphere had acted as a center of origin for a broad spectrum of taxa.

Feldmann (1986) discussed the distribution of two decapod crustacean taxa. The family Aegilidae had previously been known as a living group inhabiting the fresh and brackish waters of southern South America. The discovery of fossils from the late-Cretaceous marine rocks of New Zealand suggested to him that the family originated in New Zealand and had spread from there to South America. The genus *Lyreidus* is broadly distributed, living primarily in the outer shelf and slope habitats of the IWP. Fossils are known from the Eocene of the Antarctic and New Zealand. Feldmann suggested that the genus originated in the New Zealand-Antarctic area. Crame (1987) described the presence of a distinct, circum-Gondwana, cool-temperate bivalve fauna from the late Mesozoic. He noted that competition, at that time, may have been greater around the poles and that organisms could have dispersed from high latitudes to the tropics.

It is important to recognize that these examples of apparent "high latitude heterochroneity" are based on the findings of fossils of certain groups older than fossils of the same groups from lower latitudes. The earliest fossils of a given taxon do not necessarily indicate the earliest evolutionary stage nor its center of origin. Whenever possible, the fossil data need to be correlated with the biogeography of living species. High-latitude centers of origin in the North Pacific and Antarctic have been recognized, through the relationships of living groups, for some years (Briggs, 1974a).

Onshore-offshore gradients

In 1983, Jablonski et al. published a significant article which called attention to onshore-offshore patterns in the evolution of Phanerozoic shelf communities. As evidence, they presented data on faunal changes that took place in the Cambrian and Ordovician periods of the early Paleozoic and the late-Cretaceous period of the Mesozoic. Their analysis indicated that the major new community types appeared first in nearshore settings and then expanded into offshore settings. This occurred despite higher rates of species evolution in the offshore habitats. By the end of the Ordovician, the three major evolutionary faunas of the Phanerozoic were arrayed in distinct community associations across the continental shelf and slope. The remnants of the Cambrian fauna were on the slope, the Paleozoic fauna was on the mid-to outer shelf, and the early members of the modern fauna on the inner shelf. A parallel pattern of change was found for the late Cretaceous.

Sepkoski and Miller (1985) provided more evidence that an onshore to offshore replacement had taken place. Bambach (1986) published a review of Phanerozoic marine communities. He corroborated the previous observations and emphasized that replacements were caused by the origination of future faunal dominant groups in nearshore environments and their expansion through time to offshore settings. Some of these paleontological data may not be directly applicable to the East Indies, for they often refer to times when the alignment of continents and oceans was much different. But they do illustrate a process that apparently is still going on and that does involve the East Indies. The age and distribution data on ostracods (Whatley, 1987) indicate the importance of the southwest Pacific (including the East Indies) as a locus of introduction into slope and abyssal environments during the Miocene to Quaternary interval. Once having reached deep water, many species migrated to other parts of the world's oceans.

The concept of a continuous evolution of dominant species in shallow waters is compatible with the center of origin hypothesis: successful evolution, which results in the formation of new phyletic lines, takes place in the nearshore waters of centers of origin. From such centers new species, which may eventually give rise to new higher taxa, spread out both vertically and horizontally. As they spread vertically, they replace older forms from the mid- and outer shelf and eventually from the slope. As they spread horizontally, they replace their more generalized relatives, leaving as evidence the phylogenetic and age patterns already described.

Barrier effects

Judging from relationships within the genera that inhabit the IWP, it appears to be relatively easy for species to disperse from the East Indies to other parts of that vast area. Many shelf species range westward to east Africa and eastward to outer Polynesia, a distance more than halfway around the world. But the IWP is bounded by two barriers separating it from the other tropical shelf regions. These are the deep-water East Pacific Barrier that lies between Polynesia and the New World, and the Old World Land Barrier comprised of Africa plus Eurasia. The East Pacific Barrier has been crossed by relatively small numbers of animals, most of which have pelagic larval stages. These include fish, molluscs, crustaceans, echinoderms, and hermatypic corals (Briggs, 1974a).

Almost all of the trans-Pacific species range broadly in the IWP and usually belong to genera that are best developed (more speciose) in that area. In contrast, they generally have a limited foothold in the eastern Pacific, most being confined to the offshore islands. There is not a single example of a shelf species belonging to a typical New World genus, having established itself in the western Pacific. This circumstantial evidence appears to indicate that the trans-Pacific species originated in the west and migrated eastward across the East Pacific Barrier (Briggs, 1974a). The route of dispersal must have been the North Equatorial Countercurrent or possibly the Equatorial Undercurrent. Although the westward flowing North and South Equatorial Currents are much stronger, eastern Pacific species carried by them have not been able to establish themselves in the IWP.

A vicariance hypothesis states that there has been little long-distance dispersal and that the trans-Pacific species are relicts of a former widespread Tethys biota (McCoy and Heck, 1976; Heck and McCoy, 1978, 1979). However, Rosenblatt and Waples (1986) examined 12 trans-Pacific fish species using starch gel electrophoresis and found populations from Hawaii and the eastern Pacific to be very closely related, much closer than amphi-American populations that had been separated by the Isthmus of Panama for about three million years. They concluded that only long-distance dispersal could account for the presence of trans-Pacific species in the eastern Pacific, and that the samples suggested the occurrence of ongoing gene flow across the ocean.

The Old World Land Barrier is also highly effective due to the fact that the southern tip of Africa extends into warm-temperate waters. This creates a temperature block that tropical species usually cannot pass. South of the mouth of the Kei River on the African southeast coast, the average temperature for the coldest month drops below 20°C so that purely-tropical species cannot survive. However, some tropical organisms, called eurythermic tropicals, do range into warm-temperate waters. Some of these have been able to make their way around the Cape of Good Hope to establish themselves in the eastern Atlantic region. A few of the latter, mainly circumtropical species, range across the Atlantic to the western side. Most of the species that have circumvented the Old World Land Barrier via the Cape of Good Hope belong to well-developed IWP genera. None belong to well-developed Atlantic genera. This has led to the suggestion that dispersal has taken place in one direction only, from east to west (Briggs, 1974a).

An artificial disruption to the Old World Land Barrier took place in 1869 when the Suez Canal was opened. At first, the Bitter Lakes portion of the canal, with its high salinity of 68‰, appeared to provide an effective barrier to migrants. But, over time, the salinity gradually decreased to its present level of about 41‰. While there has been an enormous influx of IWP immigrants into the eastern Mediterranean, there has been virtually no successful (colonizing) movement in the opposite direction. Por (1978) documented the invasion of more than 200 species into the Levant Sea area where they now comprise about 10% of the biota. The only examples of successful dispersal in the opposite direction are two species of estuarine fish that have reached a lagoon some 200 km south of Suez.

These barrier effects show that successful (colonizing) dispersals across the biogeographic boundaries that delimit the IWP have almost always taken place in one direction only, outward into areas where the fauna is less diverse. This demonstrated ability of IWP species to colonize adjacent areas is a reflection of their dominance. Whether this dominance is achieved by direct competition or by better resistance to predation, disease, or other ecological factors, no one knows. But the dispersal ability of the IWP species and their consequent influence over other parts of the tropical marine world are an indication of the importance of the East Indies, the evolutionary center in which most of them were probably produced.

Disjunct patterns

When one examines the geographic patterns shown by the better-known IWP disjunct species, it becomes evident that they may be divided in two main categories. There are those that have an antitropical distribution in the western Pacific and those that, in addition, have relict populations in the western Indian Ocean. Similarly, one may divide most non-disjunct IWP species into two large groups: one ranges broadly through the IWP and the other is primarily confined to the western Pacific. It has been suggested that these two prominent disjunct patterns were developed by extinction (Briggs, 1987). The scenario, based on Théel's (1886) relict theory, is that the extinction process for a given species usually starts in the East Indies.

The high diversity of the East Indian Triangle may suggest a high level of species competition. But there is also a high diversity of predators, parasites, and diseases. Any of these factors, or a combination thereof, could initiate the extinction process. Once an extinction begins, it might be expected to spread outward from the East Indies. As this happens, the range of a species that originally extended broadly in the western Pacific could be converted to an antitropical distribution in that ocean. In the same manner, a species that originally ranged throughout the IWP might now be represented by antitropical populations in the western Pacific and a remnant in the western Indian Ocean. Examples of both patterns have been delineated (Briggs, 1987).

There are examples of antitropical distribution in which the original populations apparently had a circumtropical distribution. In some of these instances, the extinction process at first left antitropical populations entirely across the Pacific and Atlantic Oceans. This was apparently followed by extinction in the North Pacific, then extinction in the North Atlantic, which left relict populations in the southern hemisphere. This sequence, indicated by the distributions of fossil and living barnacles (Newman and Foster, 1987), may be applicable to other groups. Another kind of disjunct pattern is produced when species or higher taxa become extinct on the shelf but manage to survive as relicts on the slopes in certain areas where the habitat is suitable.

Center of origin alternatives

As noted in the introduction to this section, the common alternative to the East Indian center of origin was proposed by Ladd (1960) who thought that the East Indies passively accumulated species that were formed elsewhere. This idea was reinforced by McCoy and Heck (1976) and in subsequent works (Heck and McCoy, 1978, 1979). The latter authors suggested that genera as well as species have been accumulating in the East Indies instead of originating there and spreading outward. Another explanation for the high level of species diversity was that of Woodland (1983). He suggested that it was due to the fact that the East Indies represented an area of overlap between the biotas of the Indian and Pacific Oceans. Rosen (1984) conceived the existence of a "diversity pump" in which new species were formed on the periphery and then fed into the central region.

More recently, the idea that the East Indies had accumulated species that had evolved in peripheral areas was supported by Jokiel and Martinelli (1992) and Kay (1990). The former devised a model which showed that ocean currents could theoretically transport species into the East Indies. Kay observed that the earliest Indo-Pacific cowries (Cypraeidae) are recorded from the Paleocene of India and Pakistan, and that speciation had occurred in the central Pacific and Indian Ocean. One would expect most tropical Paleocene molluscs to come from the Tethys Sea before the East Indies center was established in the late Miocene. The concept of the East Indies as an evolutionary center does not preclude speciation in peripheral areas. It does predict that peripheral speciation is less successful in the long run.

The foregoing theories all suggest ways in which species could accumulate in the East Indies without being formed there, and so they are used as arguments against the presence of a center of origin. However, the evidence for such a center does not rest on the presence of high diversity alone. Other, perhaps more pertinent, data leading to that conclusion are those provided by phylogenetic patterns in certain groups, age gradients, onshore-offshore gradients, barrier effects, and the geography of disjunct distributions. It is all of these factors put together that provide the modern case for the center of origin hypothesis.

Conclusions

Is the East Indies a place where older genera and species, that have been formed elsewhere, have accumulated or is it an active evolutionary center where new taxa are constantly being formed? The pertinent information may be summarized as follows:

1. The relationship within several families and genera suggest a phylogenetic pattern in which the more generalized genera and/or species are located in the peripheral areas of the IWP and the more specialized forms are found in the East Indies area.
2. The average generic ages of hermatypic corals, coral inhabiting barnacles, shore barnacles, and stalked crinoids, increase with distance from the East Indies. The ages and distribution patterns of various ostracod species appear to indicate dispersal flows outward from the East Indian-southwest Pacific region.
3. The age data are possibly counteracted by evidence of high-latitude heterochroneity in which fossils of certain groups first occur at high latitudes. However, such "earliest fossils" do not necessarily represent the area of origin for the group to which they belong.
4. The paleontological data demonstrate an historical onshore-offshore gradient in which faunal dominant groups originate in nearshore environments and expand, through time, to offshore settings. This is compatible with the center of origin concept in which newly evolved forms gradually extend vertically into deeper waters as well as horizontally into peripheral shelf waters.
5. The East Indies, located in the central part of the IWP, theoretically produces species that can eventually achieve broad ranges in that ocean. Some of these widespread species can occasionally surmount biogeographic barriers to colonize adjacent regions. But species from adjacent oceans have not been able to colonize the IWP. This appears to indicate that species from the IWP possess a dominance that enables them to overcome the resistance of other, less-diverse ecosystems.
6. Antitropical and other disjunct patterns in the IWP suggest that the extinction process first begins in the East Indies and then spreads outward. These kinds of patterns are also compatible with the center of origin concept. Areas in which successful new species are constantly being produced might be expected to also have high rates of extinction. This does not necessarily mean that the extinctions are the direct result of competition. Other factors such as predation, disease, and parasitism could be important.

There is an insistence, in some of the recent literature, that information bearing on scientific theories must be falsifiable in order to be legitimate. In biogeography this is often difficult if not impossible. However, in each of the above categories the evidence is potentially falsifiable depending on the results of further research. The center hypothesis could be weakened and the accumulation hypothesis strengthened if: (1) phylogenetic patterns are found where the more specialized species do not occur towards the center, (2) age gradients in species or genera are found that do not increase with distance from the center, (3) relict patterns are discovered that do not array in the peripheral areas, (4) onshore-offshore gradients are identified that do not lead to more primitive groups, and (5) numerous species are found crossing the biogeographic barriers towards instead of away from the East Indies. All of these factors have a bearing on the significance of the East Indies as a center of origin. Either it is an area of considerable evolutionary and distributional importance or the information so far gathered is misleading.

"In the zero-sum game of the Red Queen each species competes, directly or indirectly, with all others and no species ever wins; new adversaries replace the losers" (Van Valen, 1976). In the biogeographic game, the winners produce continuing phyletic lines while the losers do not. The winners are developed in the crucible of high-diversity centers while the losers arise under more benign circumstances.

MODES OF SPECIATION

Since the central East Indies apparently operates as a center of evolutionary radiation, it means that it must be a site of speciation that is successful in terms of its products being able to spread outward into adjacent areas. This idea has been criticized on the basis that it would require a high degree of sympatric speciation (Jokiel and Martinelli, 1992). This criticism overlooks the fact that, although the East Indian Triangle is a relatively small part of the IWP, it still encompasses a large and varied geographical area. In addition to the Philippines and the Malay Peninsula, it covers all the large islands of the Indo-Australian Archipelago, the northern tip of the Great Barrier Reef, and extends beyond New Guinea to include the Solomon Islands and the Bismarck Archipelago.

The present geographic complexity of the area appears to offer many opportunities for the isolation of populations and subsequent allopatric or parapatric speciation. For the past 3 Ma, these opportunities have been multiplied by the making and breaking of barriers caused by glacio-eustatic changes in sea level. In addition to these kinds of speciation, there are some indications that the sympatric mode may be involved.

A distinction has been made between intrinsic and extrinsic barriers to gene flow (Diehl and Bush, 1989). Intrinsic barriers are inherent in the organisms themselves and include both postmating reproductive incompatibilities and premating barriers due to behavioral or ecological characteristics. Allopatric isolation requires extrinsic barriers that are external to the organisms involved, such as isolation by distance or by some physical or chemical obstruction. In the absence of extrinsic barriers, populations are regarded as sympatric when individuals can readily move between habitats. In almost all cases of sibling sympatry, one may find habitat differences on a local scale. In the past, there has been a tendency to refer to these as indicating "microallopatric" or "microgeographic" speciation but, as Diehl and Bush have pointed out, these terms are misleading because extrinsic barriers to gene exchange do not exist.

The modes of speciation in vertebrate animals were investigated by Lynch (1989) using data from three genera of cyprinodontiform fish, three genera of frogs, and one genus of passeriform bird. These are groups on which detailed systematic work had been done to determine relationships within the genera and where sufficient information on geographic range was available. If the distribution pattern of one daughter species entirely enclosed the other, his first-order explanation was that sympatric speciation had occurred. The alternative explanation would be that the sympatry was due to dispersal following allopatric or parapatric speciation. But it is unlikely that newly formed sibling species would be able to immediately achieve complete sympatry.

Distribution patterns

In the marine East Indies, recent systematic work on fish has revealed some remarkable distribution patterns. Within the genus *Lutjanus* (Lutjanidae), there exist three closely related species. All three species are euryhaline and occur in both freshwater and marine habitats (Allen and Talbot, 1985; McDowall, 1988). One, *L. goldiei*, has a restricted range along the south coast of New Guinea. It is sympatric with the more widely distributed *L. fuscescens*. Those two occur within the range of the very widespread *L. argentimaculatus* (Fig. 60). In the pipefish (Syngnathidae) genus *Phoxocampus*, there are three known species (Dawson, 1985). *P. tetrophthalmus* is found only in the East Indies and occurs entirely within the range of *P. diacanthus*. *P. belcheri* is very widespread and is sympatric with and almost completely encompasses the ranges of its two sibling species (Fig. 61).

The butterflyfish of the family Chaetodontidae comprise a group of conspicuous, well-studied reef fish (Burgess, 1978; Blum, 1989). Within this group, there are two pairs of sibling species that have sympatric distributions. *Chaetodon oxycephalus* occurs entirely within the range of its sibling C. *lineolatus* (Fig. 62) and *C. ocellicaudus* is contained within the range of *C. melannotus* (Fig. 63). In each case, the species with the smaller territory occurs primarily in the East Indies while its more widespread relative ranges broadly in the Indo-West Pacific. Similarly, in the family Pseudochromidae (Winterbottom, 1985) *Congrogadus hierichthys* occupies a restricted territory in the East Indies but is sympatric with and occurs entirely within the range of its very close relative *C. subducens* (Fig. 64).

Not all sibling species in the East Indies are sympatric. In his study of the anemonefish of the genus *Amphiprion*, Allen (1975b) remarked on the presence of three pairs of closely related, geminate species. One of each pair occurs to the west of the Indo-Australian Archipelago and the other to the east. Allen suggested that their evolution was due to the restriction of contact between the tropical parts of the Eastern Indian Ocean and the Western Pacific that occurred during the glacio-eustatic drops in sea level. Other examples were given by Springer and Williams (1990). Woodland (1986) noted the distribution of two pairs of sibling species in the fish family Siganidae. Each pair was found to have a shared border in the vicinity of Wallace's Line.

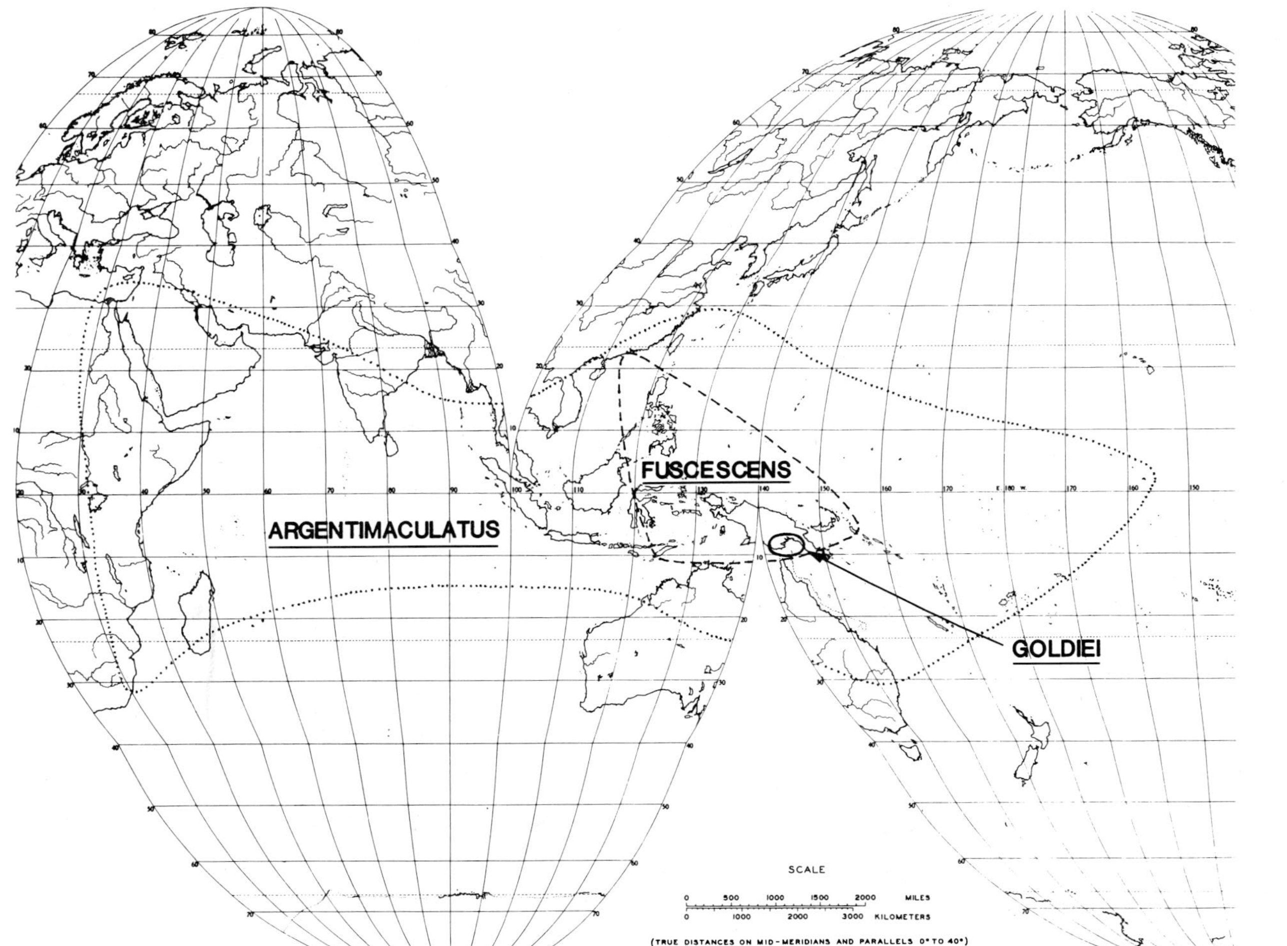

Fig. 60. Distribution patterns of three closely related species in the genus *Lutjanus* (family Lutjanidae). After Allen and Talbot (1985) and McDowell (1988).

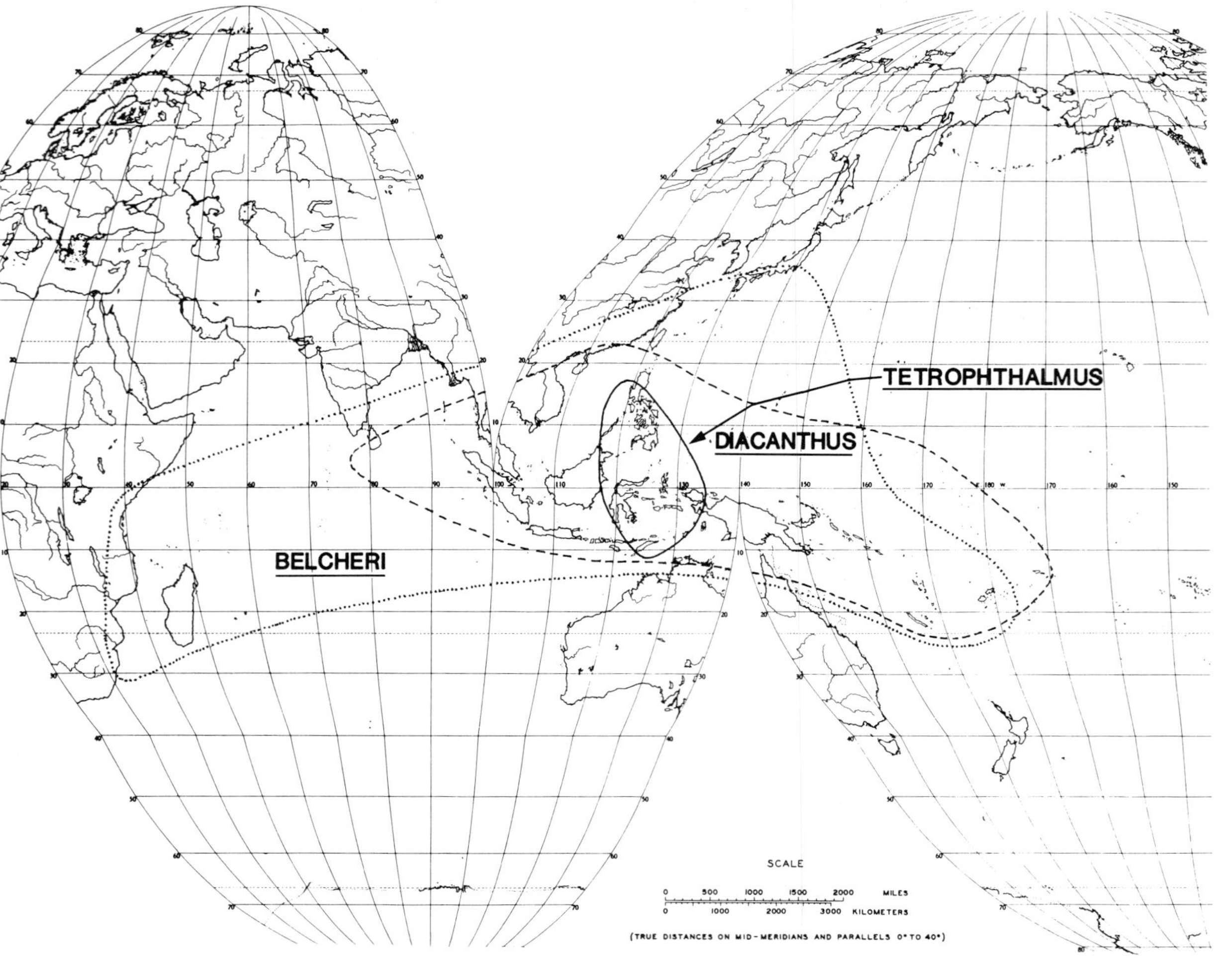

Fig. 61. Distribution patterns of the three known species in the genus *Phoxocampus* (family Syngnathidae). After Dawson (1985).

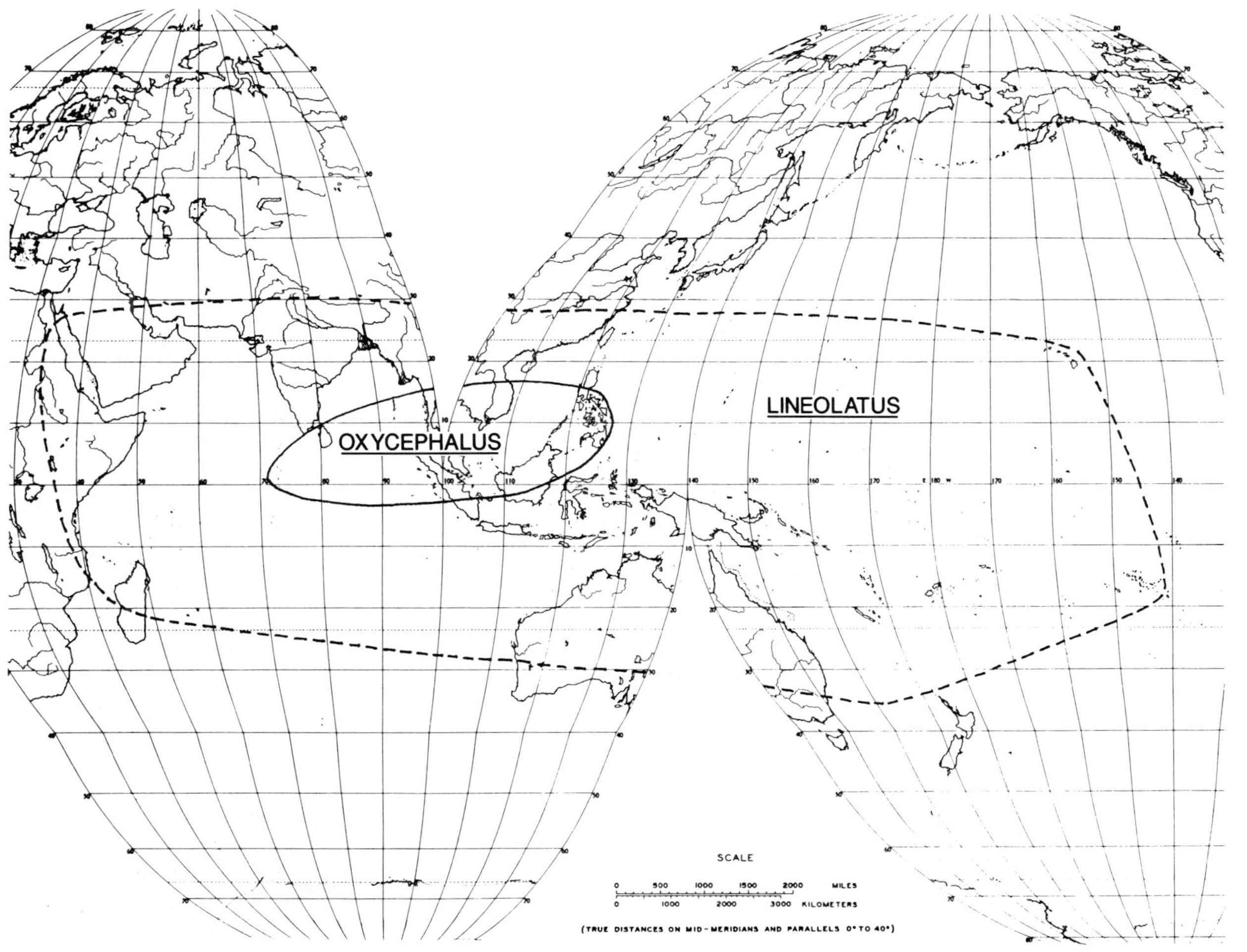

Fig. 62. Distribution patterns of a sibling pair of butterfly fish (family Chaetodontidae). After Burgess (1978) and Blum (1989).

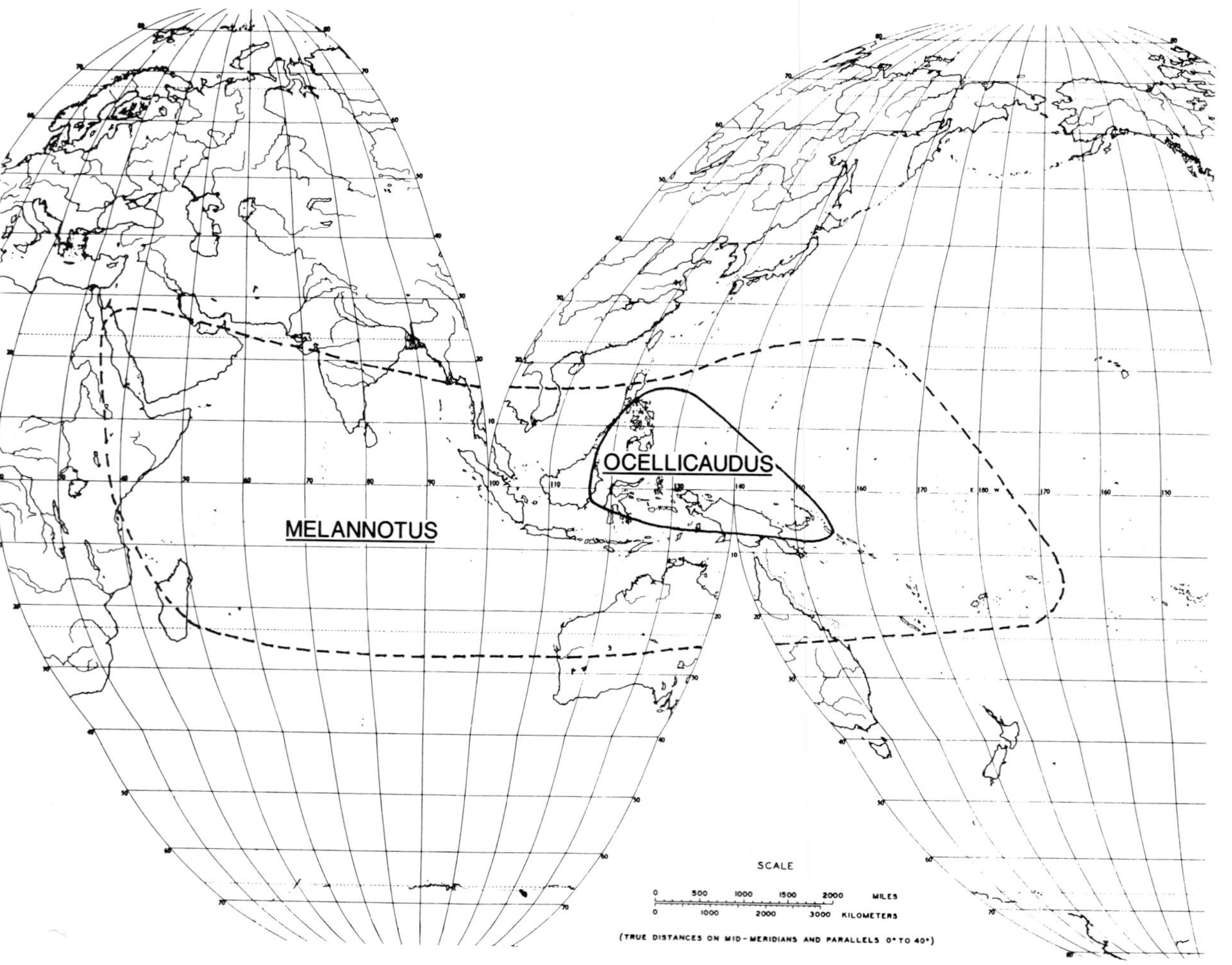

Fig. 63. A second sibling pair of butterfly fish. After Burgess (1978) and Blum (1989).

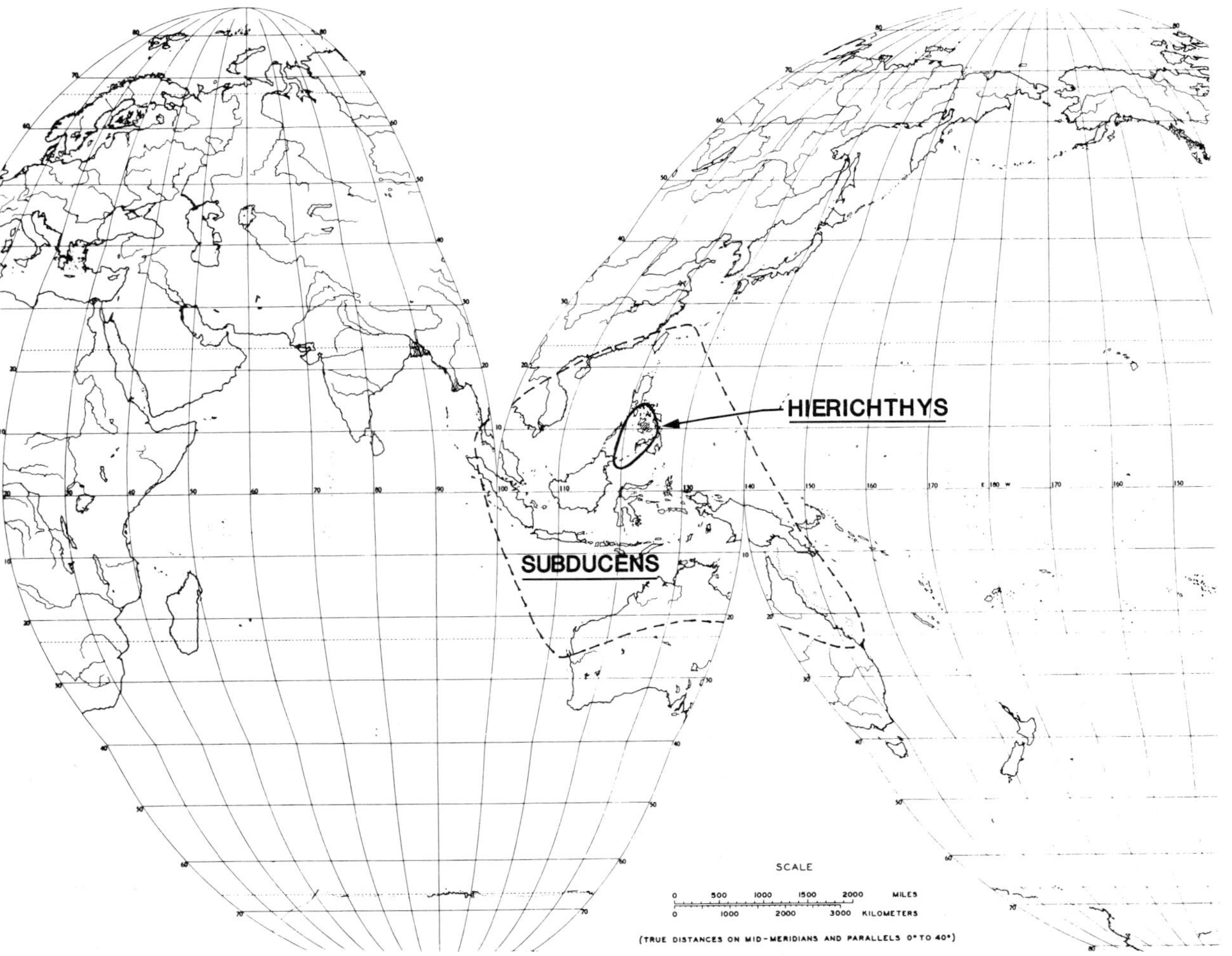

Fig. 64. Distribution patterns of two closely related species in the family Pseudochromidae. After Winterbottom (1985).

Discussion

In his discussion of butterflyfish evolution, Blum (1989) assumed that the sympatry between the two pairs of sibling species was achieved by the dispersal of the more widely distributed species into the original range of the narrowly distributed species. In the case of the *Congrogadus* siblings, Winterbottom (1985) also suspected that the wide-ranging species dispersed into the range of its more restricted relative. In each of these situations, it is difficult to imagine a widespread species leaving a gap in the center of its range, so that a sibling could evolve by allopatry, and then closing the gap to achieve sympatry.

Although sympatric speciation has not previously been identified in the marine environment, the foregoing patterns are highly suggestive. The geographic concordance of the patterns is virtually complete and there are no apparent extrinsic barriers separating the sibling species. Supposedly, before the completion of speciation, the populations concerned were at the subspecies level of development. Yet in this region, no sympatric subspecies have been reported. The reason may be that most systematists are quite willing to describe, as subspecies, closely related populations that are geographically separated. Conversely, if the closely related populations occupy the same geographic area, there is a tendency to describe them as separate species even though they may be difficult to distinguish.

In his study on the frequency of the modes of speciation in vertebrates, Lynch (1989) found that nearly 50% of all cases of sympatric occurrences in his data set involved "trivial" sympatry. In those cases, the vast majority of the distributional area of both daughter species was allopatric, which made the argument for postspeciational dispersal seem obvious. But in 6% of the 66 cases of speciation the sympatry was complete or nearly so. In such cases, Lynch advocated sympatric speciation as the most parsimonious conclusion. An unstated but inherent factor in this conclusion is that the younger the species, the closer its geographic pattern is to the condition at the time of speciation. Thus, without fossil material (which is the case for most families of marine vertebrates), perhaps the closest we can come to determining the mode of speciation is to study the geographic patterns of sibling species.

In each example of sibling sympatry, it may be noted that the most restricted member of the related group is confined primarily to the central East Indies. In the two instances where three siblings are involved, the patterns are remarkably similar; one species has a very small East Indies range, the next encompasses a larger portion of the area, and the third has become very widespread. There is the possibility that these are evolutionary patterns with the youngest sibling occupying the smallest area, the next eldest the intermediate size territory, and the oldest the larger. The existence of this mode of speciation, in addition to the allopatric and parapatric processes, might help to explain the very high species diversity in the East Indies Triangle.

A possible criticism of the hypothesis that the sibling patterns are evidence of recurring successful evolution, is that they might instead represent extinction patterns. Thus it could be argued that the smaller the territory occupied by a sibling, the closer it is to extinction. However, this argument does not appear to be valid in view of the evidence that the East Indies apparently harbors relatively young taxa that become older as they gradually spread out to occupy other parts of the tropical marine world.

Conclusions

In the past, the concept of sympatric speciation has either been completely rejected or it has been suggested only for exceptional situations such as host switching by phytophagous insects. It has only recently been recognized in some terrestrial and freshwater vertebrates (Lynch, 1989). The fact that this mode is evidently occurring in the marine East Indies means that it may be of considerable evolutionary importance. It seems clear that this area functions as a center of evolutionary radiation. Does the sympatric process produce species that can become numerous and widespread? Judging from the patterns produced by the sibling species of *Lutjanus*, *Phoxocampus*, *Chaetodon*, and *Congrogadus*, this may be true.

A neglected area of biogeographic research is the examination of the geographic patterns that are produced as the result of speciation. As species are formed and extend their ranges, and as others become scarce and finally extinct, they produce geographic patterns. We should examine such patterns and attempt to interpret them, for they will help to elucidate the underlying evolutionary process.

INDO-WEST PACIFIC SUBDIVISIONS

Within almost all of the biogeographic regions, minor barriers to gene flow exist that, over extensive periods of time, have resulted in the formation of local endemic species. In places where the endemism has developed to the extent that the fauna or flora has taken on a definite provincial character, we should consider some further biogeographic subdivision. In such instances, it seems reasonable to identify provinces, but the difficult question is, how much does a local biota have to differ from the parent in order to merit formal recognition as a province? In this work, an admittedly arbitrary decision has been made: if there is evidence that 10% or more of the species are endemic to a given area, it is designated as a separate province.

As already noted, the area with the greatest species diversity in the marine world is the East Indies Triangle. Can this area by itself be considered a distinct province? The answer must be "no" because the peripheral areas, although demonstrating notable reductions in diversity, do not as a rule possess species that are not also found in the central triangle. Considering that such distant places as the Tuamotu Archipelago, India and Sri Lanka, Taiwan (Formosa), and the Great Barrier Reef of Australia are populated primarily by wide-ranging species and possess few local endemics, all should be placed in the same province.

So, we find that the heart as well as the greater part of the Indo-West Pacific Region is occupied by one huge province that is larger than any of the regions in other parts of the world. The Indo-Polynesian Province extends all the way from the entrance to the Persian Gulf to the Tuamotu Archipelago, and from Sandy Cape on the east coast of Eastern Australia to the Amami Islands in southern Japan. All of Polynesia is included with the exception of the Hawaiian archipelago, the Marquesas, and Easter Island (Fig. 56).

Springer (1982) produced a detailed work in which he examined the western Pacific distribution of many families of fish and some invertebrates. His purpose was to demonstrate that the Pacific Plate comprised a biogeographic region of major significance. Although his data on the distributions of the various groups are accurate and useful, his conclusions are questionable. By including all of the endemic species known from individual islands and island groups (such as Hawaii, the Marquesas, and Easter Island), as well as the relatively few widespread endemics, he was able to predict an endemism rate for the shorefish of 22–25%.

The difficulty with the foregoing estimate is that it does not utilize the data on ende-

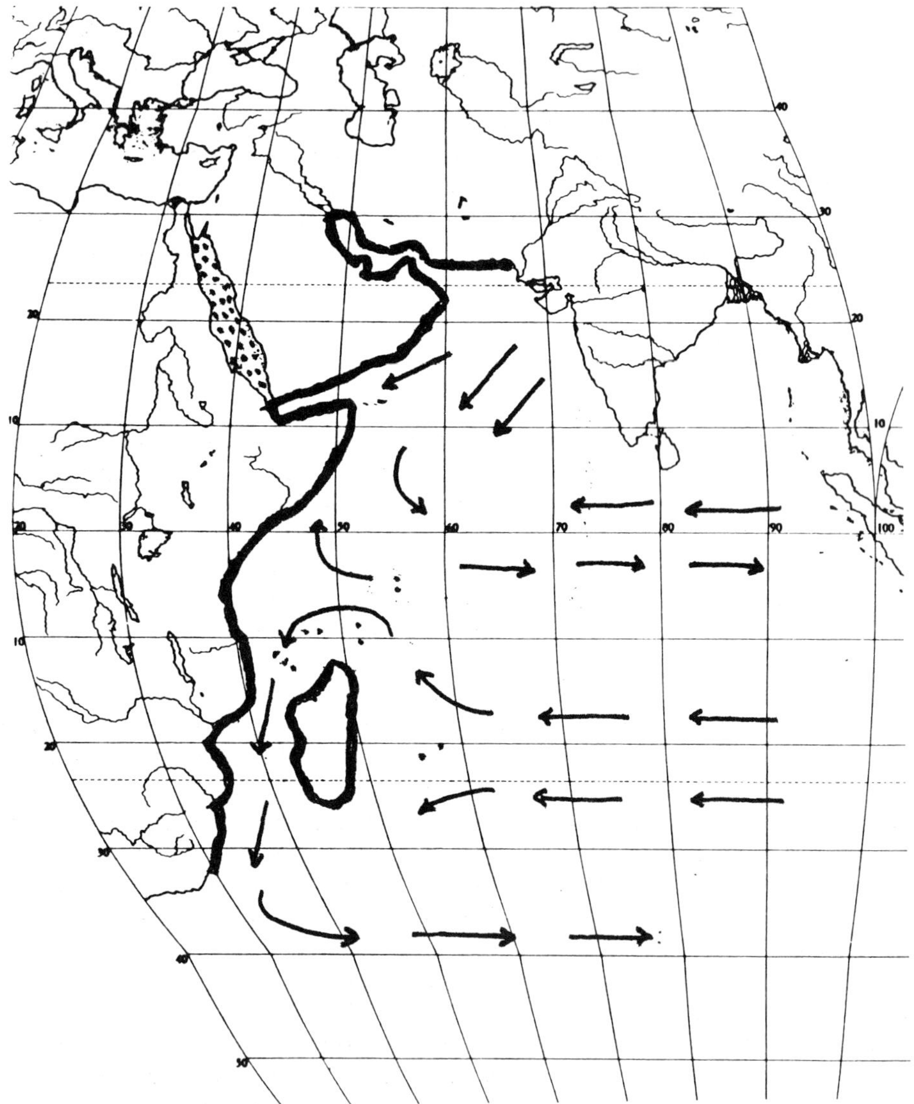

Fig. 65. Western Indian Ocean and Red Sea Provinces. Currents in Arabian Sea depict prevailing patterns under influence of the northeast tradewinds.

mism in a legitimate manner. Species that are endemic to the Hawaiian Islands are so characterized because they occur at Hawaii and nowhere else. They are the result of evolutionary changes that occurred only around that group of islands. They are not, at the same time, Pacific Plate endemics. The latter, by definition, need to be characteristic of and confined to the Pacific Plate. How many species have some kind of broad plate distribution yet are confined to it? Springer (1982) provided a list of "widely distributed Pacific Plate endemic fish". But, among those 48 species about half also exist on the Philippine, Nazca, or Australian Plates so are not really endemics. At best, there are about 25 widely distributed species that are truly confined to the Pacific Plate. Since a total of 1312 have been recorded from the plate, the rate of endemism may be calculated at less than 2%.

In addition to the Indo-Polynesian Province, seven other areas within the IWP have been recognized as separate provinces because their levels of endemism appeared to be greater than 10% (Briggs, 1974a). Recent literature indicates that all but one should still be so designated. Along the African east coast, the tropical Agulhas Current runs southward creating tropical conditions all the way to the mouth of the Kei River. To the north, there is evidently a faunal change at about the mouth of the Arabian Gulf. Between these two points, as well as offshore to include Madagascar and its associated islands, is the distinct biota of the Western Indian Ocean Province (Fig. 65).

Although the Red Sea is open to the Indian Ocean, its southern end is partially blocked by a sill that lies at about 125 m. During the glacial periods, the eustatic level of the world ocean probably dropped to the extent that the Red Sea was partly or completely isolated. These periods of gene flow interruption occurred often enough to produce significant amounts of endemism so that a Red Sea Province is recognized. Of the shore fish, about 17% are endemics (Ormond and Edwards, 1987).

The Northwestern Australian Province lies between Cape York to the north and Shark Bay to the southwest (Fig. 66). There is some controversy as to whether this area should continue to be designated as a province. Recent investigations continue to indicate high rates of endemism in groups such as the fish and echinoderms but little or none in other groups (Wilson and Allen, 1988). The northwestern section of the province has an impoverished biota due to discharge from large, muddy rivers. This, plus the closure of the passage between Cape York and New Guinea during the ice ages, may account for the apparent endemism.

Early collections of fish from Lord Howe Island, Norfolk Island, Middleton Reef, and Elizabeth Reef – all oceanic islands in the Tasman Sea – indicated endemism sufficient to recognize a Lord Howe – Norfolk Province (Briggs, 1974a). Although some endemism is still apparent, subsequent investigations suggest that the biota is not sufficiently distinct for provincial recognition.

The Hawaiian archipelago is far enough from the rest of Polynesia that the shallow marine fauna has become very distinct. Except for Johnson Island, which lies about 620 km to the south, the Hawaiian chain is separated from the other island groups by a deep-water gap of about 1240 km. Johnson Island probably served as a stepping stone for the migratory movement of marine organisms to Hawaii. The biota of Hawaii is now better known than any other IWP locality, so that its status as a distinct Hawaiian Province is well accepted (Fig. 67). About 25% of the shore fish are endemic (Randall, 1992).

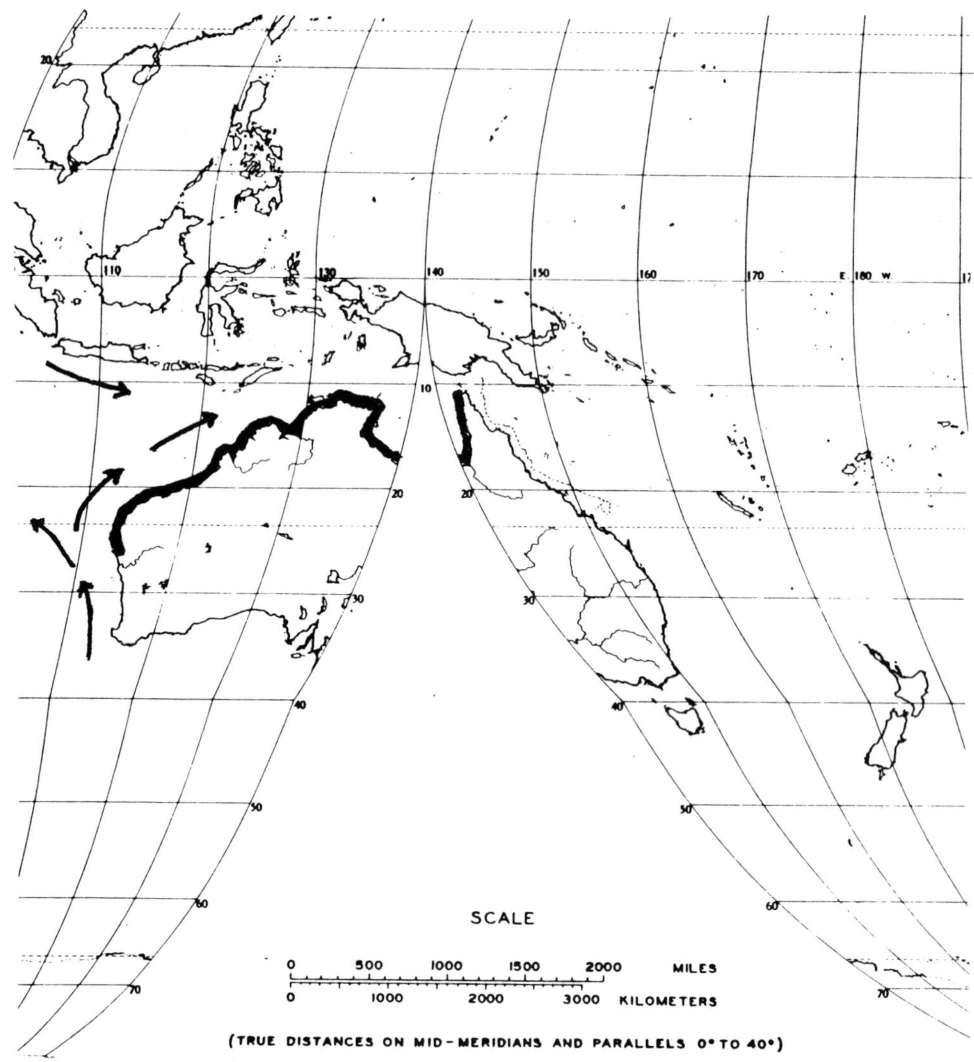

Fig. 66. Northwestern Australia Province.

Easter Island, together with Sala y Gómez about 400 km farther east, is the most geographically isolated part of the IWP. The many collections from Easter Island indicate an endemism about equivalent to that of Hawaii. Recent collections from the seamounts on ridges between Sala y Gómez and San Felix – San Ambrosio yielded an IWP fauna with many endemics (Parin, 1991). This indicates that the Easter Island Province needs to be expanded eastward to include the area between Easter Island and San Felix – San Ambrosio. This has the effect of extending the IWP biota to within about 700 km of the South American coast. Some 23.2% of the Easter Island shore fish are endemic (Randall, 1992).

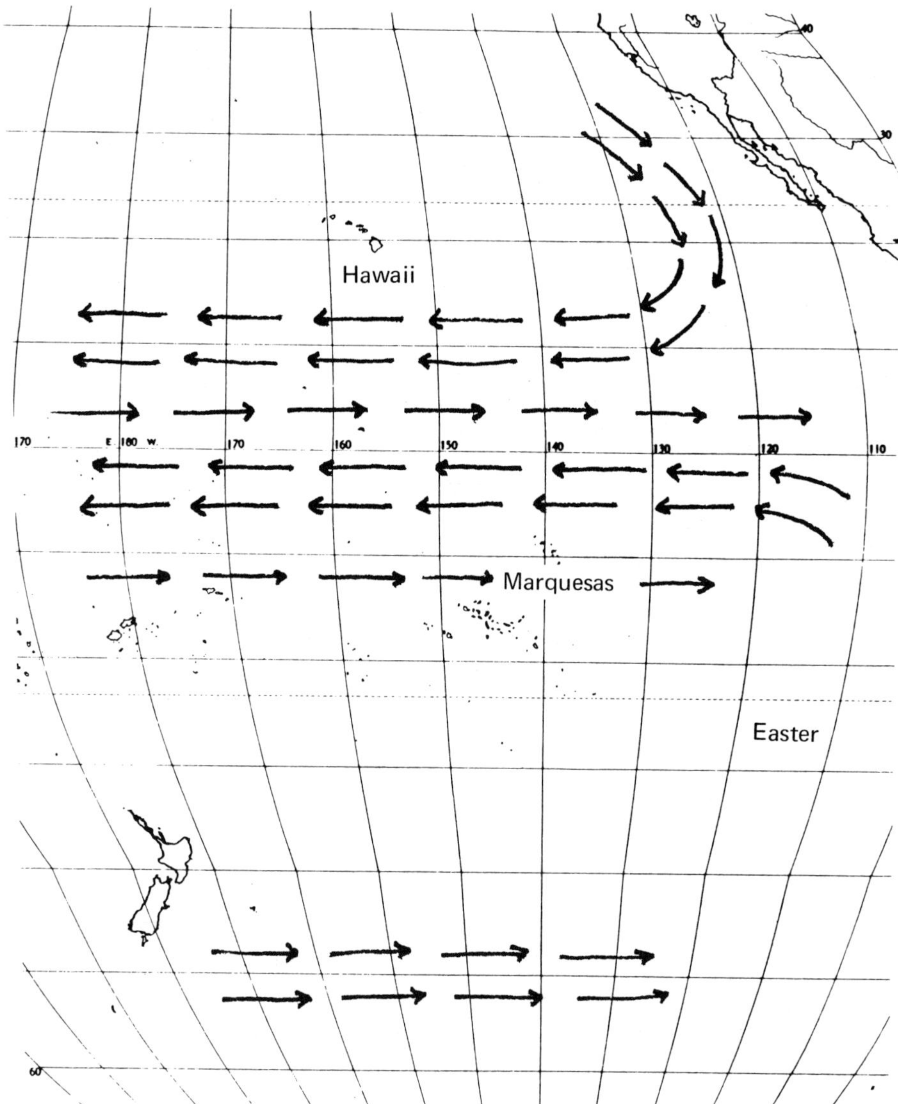

Fig. 67. Hawaiian, Marquesas and Easter Island Provinces.

The Marquesas Province pertains to a group of 13 islands that lie about 350 km north of the nearest islands in the Tuamotu Archipelago. Although 350 km is not a great distance for shore animals that have pelagic larvae, the prevailing surface current may contribute the isolating factor that is the cause of significant endemism. The South Equatorial Countercurrent apparently runs between the two island groups. About 10% of the shore fish are endemic (Randall, 1992).

EASTERN PACIFIC REGION

The Eastern Pacific Biogeographic Region includes the mainland coast from Magdalena Bay and the lower end of the Gulf of California to the southern Gulf of Guayaquil. The entire area covers about 28° of latitude and extends westward to include five offshore island localities. The most remote of these is Clipperton, an atoll lying off the coast of Costa Rica about 900 km from the mainland. The general topography of the region offers marked contrast to that of the Indo-West Pacific. Both the inshore and offshore islands and archipelagos are small, few in number, and widely scattered.

To the south, the cool Peru (Humboldt) Current comes far up the coast of Peru to prevent the tropical waters from ordinarily extending more than about 3° below the equator. However, to the north the corresponding California Current makes its main swing to the west off the southern part of Baja California. Apparently, the discrepancy in the behavior of the two main, cool, equatorial-bound currents of the Eastern Pacific is due to the difference in coastline topography. The South American coast and continental shelf line

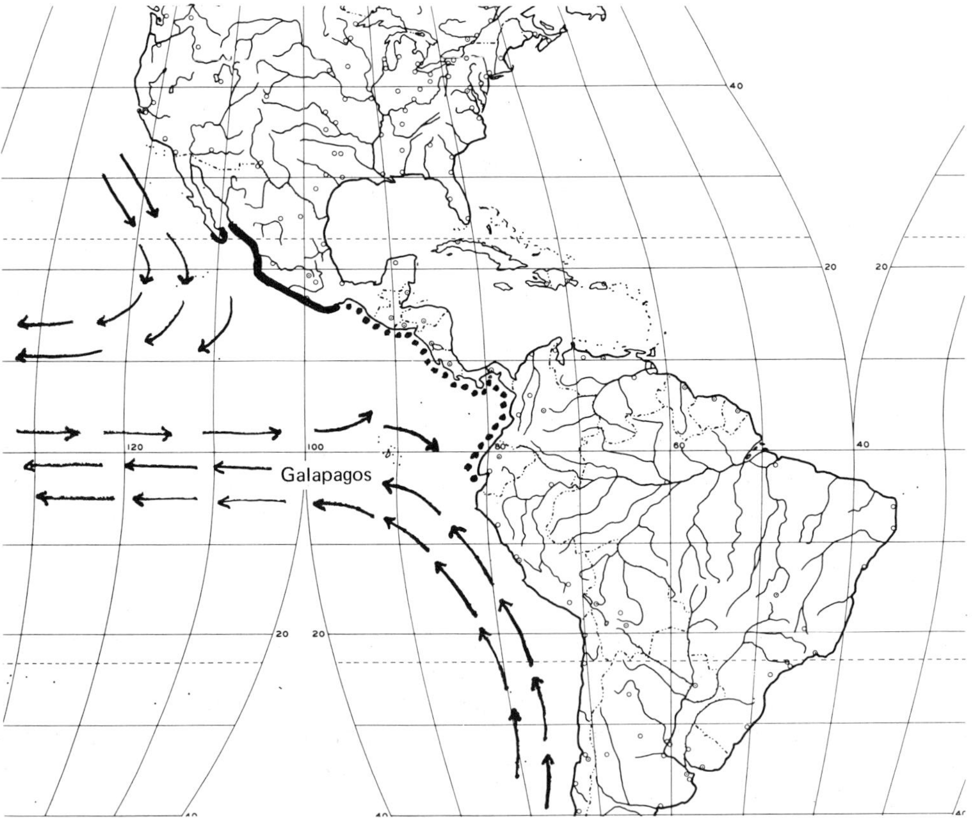

Fig. 68. Mexican, Panamanian and Galápagos Provinces.

extends almost due north from the Antarctic while that of North America – at least from northern Mexico southward – runs to the southeast. This leaves most of the Mexican and all of the Central American coast untouched by cool water from the north (Fig. 68). The result is that almost all of the tropical waters of the Eastern Pacific shores are located to the north of the equator.

A Mexican Province is recognized extending southward from Magdalena Bay and La Paz on the Baja California peninsula, and from Topolobampo on the eastern side of the gulf. The location of these northern boundaries means that the tip of the Baja California peninsula exists as a spatially isolated portion of the province. Dispersal of species from the mainland on the east is limited by a deep-water gap of about 140 km, and tropical animals cannot come down along the edges of the peninsula from the north because of the colder water temperatures in those areas. The fish fauna at the end of the peninsula demonstrates the effects of isolation. The number of species is reduced, there are several endemics, and there has been an invasion by some transpacific species.

Although the Mexican Province was originally recognized primarily on the basis of fish distribution (Briggs, 1974a), a general review of shallow-water zoogeography (Brusca and Wallerstein, 1979) has reinforced its status. There is a faunal change at about Tangola-Tangola Bay in the northern part of the Gulf of Tehuantepec at 16°N. From that point southward to the Gulf of Guayaquil at 3°S, one may recognize a Panamanian Province. Here, there are many short-ranged species, some of them apparently confined to the Gulf of Panama. As the biota becomes better known, a subdivision may be justified.

The Galápagos Islands (Archipiélago de Colón), a group consisting of 13 large and many smaller islands, is located approximately 830 km off the coast of Ecuador (Fig. 68). The islands are oceanic in origin, having been formed over a hotspot as the crustal plates moved along (Chavez and Brusca, 1991). The older islands are found in line with the crustal drift direction, to the southeast. The islands have never been closer to the mainland than they currently are. Although the Galápagos Platform is no more than 10–15 Ma old, the oldest exposed land is on the order of 2.5–4.0 Ma old.

The Galápagos have been of interest to biologists ever since Darwin wrote about them in 1839. There is now a large literature dealing with the marine biota. The most comprehensive work is that edited by James (1991). The islands may be recognized as a distinct Galápagos Province because there are many endemic species. In several animal groups, these amount to around 20% of the total species. The province is of special biogeographic interest due to its geographic position in relation to several ocean currents and the Caribbean Sea.

The Galápagos have derived most of their shallow-water species from the nearby Panamanian Province. However, the islands are also influenced by the Peru Oceanic Current, especially during the May–December period. So, a few Peruvian warm-temperate species have established themselves. A significant number of species, in some groups more than 10%, have emigrated from the Indo-West Pacific. Their mode of transport was either the North Equatorial Countercurrent or the Equatorial Undercurrent. Finally, there is a Caribbean relationship in that a few species are shared with that sea, either exclusively or also with the Panamanian Province. The latter are examples of slowly evolving species that have not changed since the formation of the Panamanian isthmus.

There are other offshore islands in the Eastern Pacific that have interesting biogeographic relationships. Clipperton Island is the only true atoll in the Eastern Pacific. It lies about 890 km southwest of Acapulco, Mexico. Its biota consists of relatively large numbers of Indo-West Pacific species, although the majority of the total are shared with the Panamanian Province. Other islands such as Cocos, Malpelo, and the Revillagigedos have also been invaded by the IWP species. All of these islands exhibit some endemism but not enough to qualify as separate provinces.

WESTERN ATLANTIC REGION

From Bermuda, southern Florida, and the southwestern Gulf of Mexico, a rich tropical biota extends down the continental coast and through the West Indies. This biota is found along the South American mainland all the way to Cape Frio near Rio de Janeiro and is continued offshore to Fernando de Noronha and Trindade islands and St. Paul's Rocks. As recently as 1953, Ekman regarded this region as "too little known in detail" to separate it into biogeographic subdivisions. There are now ample indications that the tropical western Atlantic should be divided into three parts, each with its own characteristic assemblage.

The eastward projection or "hump" of Brazil splits the warm South Equatorial Current into two streams. One runs southward (the Brazil Current) and remains within the South Atlantic Gyre system. The other runs northwest parallel to the shore and picks up speed as it becomes reinforced by flow from the North Equatorial Current. It runs strongly into the southern Caribbean and on through to the Yucatán Channel. From there, the main stream makes a rather abrupt turn to the east and, now named the Florida Current, surges rapidly through the Florida Straits.

The Florida Current is called the Gulf Stream as it leaves the coast of the eastern United States to head across the North Atlantic. The Gulf Stream has a profound effect upon the distribution of shore animals in the western Atlantic. Because of its tendency to transport larvae and occasional adults, many tropical forms are left stranded along the inhospitable shores of northeastern North America.

The position of Bermuda is unique, for nowhere else in the world does a tropical fauna occur at such a high latitude (32°15′). The swiftly flowing and often-erratic Gulf Stream is Bermuda's lifeline, carrying warmth, food, and genetic reinforcement for a thriving tropical community across some 1200 km of ocean. At the southern end of the region, the island of Fernando de Noronha is probably close enough to the continental shelf to account for its principal faunal relationship, but Trindade is well isolated and, apparently is dependent on the Brazil Current. St. Paul's Rocks are influenced primarily by the Atlantic Equatorial Undercurrent.

The Caribbean Province extends southward from about Cape Canaveral in eastern Florida, Cape Romano in western Florida, and Cape Rojo (just below Tampico), Mexico. From these points, it follows the mainland shore all the way to the northern edge of the Orinoco River delta in Venezuela (Fig. 69). The west coast of Florida supports a complex biotic assemblage. From north to south, warm-temperate species reach their range limits and tropical species make their appearance at various points. There is also considerable

Fig. 69. Caribbean, Brazilian and West Indian Provinces.

evidence of faunal change with depth, the tropical species being more numerous on the deeper parts of the shelf. In some places, there are considerable seasonal changes involving an inshore movement of tropical species during the warm months of the year.

The Brazilian Province occupies the area between the Orinoco delta and Cape Frio. The distinctiveness of its biota is probably due to an abrupt change in habitat. Beginning with the Orinoco River in Venezuela, almost the entire northeastern coast of South America as far as the vicinity of Fortaleza, Brazil, is virtually devoid of coral reefs. Instead, there are vast stretches of mud bottom, and near the mouths of the great rivers such as the Orinoco, Amazon, Tocantins, and Paranaiba, the shallow waters have greatly reduced salinities. Some reef species have been found on the outer shelf below the freshwater outflow, but the break in continuity has apparently been sufficient to produce a significant endemic component.

Although their biotas are still poorly known, the offshore islands of Fernando de Noronha and Trindade should probably be considered parts of the Brazilian Province. The marine zoology of St. Paul's Rocks has been investigated by Edwards and Lubbock (1983). Of 77 benthic species, they found 5% to be endemic, 15% shared only with Ascension/St. Helena, and the majority of the remainder with western Atlantic relationships. They concluded that St. Paul's Rocks was essentially an impoverished oceanic outpost of the Brazilian Province.

Although the West Indian Province includes an extensive geographic area, it consists entirely of islands. Bermuda is an isolated northern outpost while the main portion is an archipelago stretching from the Bahamas to Grenada – the southernmost of the Windward Island group. The southern Netherlands West Indies (Aruba, Curacao, Bonaire) are still a distributional enigma; a number of Antillean species have been found there as well as some that are otherwise restricted to the mainland coastline.

Recognition of a West Indian Province means that the Straits of Florida – only 80 km wide – are an important barrier to the dispersal of the marine shore biota. It seems clear that the barrier results not from distance, but from the fast flowing Florida Current. A close proximity to the Caribbean Province is found in two other places. One is the Yucatán Channel, about 180 km wide, and the other is the passage between Grenada and Trinidad (or Tobago), about 100 km across. As in the case of the Florida Straits, the effectiveness of these barriers depends primarily on the strong current movement.

The oceanic islands of St. Helena and Ascension lie towards the eastern part of the tropical Atlantic (Fig. 70) and were formerly placed as a province within the Eastern Atlantic Region (Briggs, 1974a). However, a detailed investigation of the shore fish of Ascension by Lubbock (1980) and of St. Helena by Edwards and Glass (1987) have clarified the biogeographic relationships of the islands to each other and to the mainland shores. St. Helena was found to have 81 species of shore fish, 15.2% of which were western Atlantic species with the island as their eastern limit, 13.9% were eastern Atlantic species, and 13.9% were endemics. St. Helena is the most remote of all tropical islands. It lies about 1870 km to the west of Angola, 1290 km from Ascension, and 800 km east of the mid-Atlantic Ridge. Its age is about 15 Ma.

Ascension Island lies northwest of St. Helena and about 1500 km from Africa. It is only about 120 km from the mid-Atlantic Ridge and, presumably, is a much younger island (Pleistocene?). Its shore-fish fauna consists of 71 species. Of these, 30% were found to be shared with the western Atlantic and only 7% with the eastern Atlantic. Twelve species or 17% were shared only with St. Helena and 15.7% were endemic to Ascension. As things now stand, each island exhibits enough endemism to be considered a separate province. However, in view of the evident relationship between the two islands, and the fact that the remainder of their biotas are not well known, it seems best to follow the suggestion of Edwards and Glass (1987) and recognize an independent St. Helena-Ascension Province. More recently, new records indicate that 16 neritic fish species, known from nowhere else, are shared by the two islands. These shared endemics

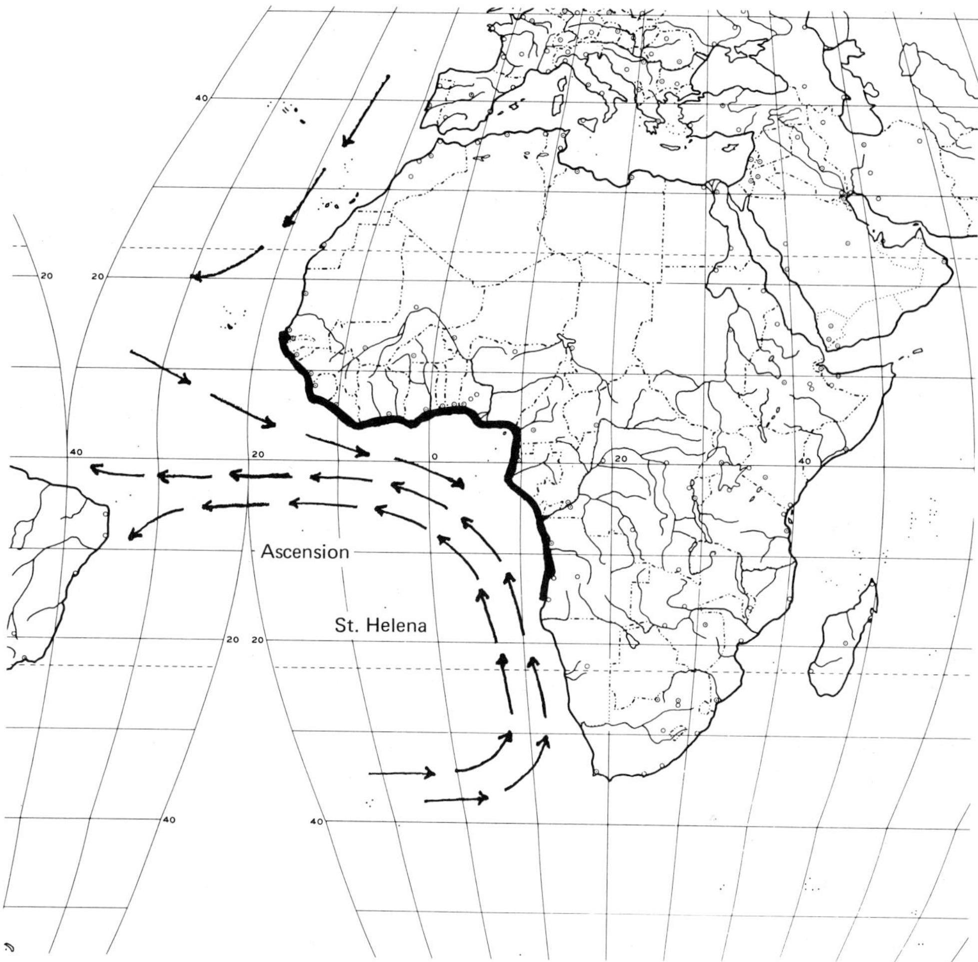

Fig. 70. The Eastern Atlantic Region and the oceanic St. Helena-Ascension Province.

comprise 20% of the St. Helena and 22.2% of the Ascension neritic fish fauna. In addition, Ascension has seven endemic species and St. Helena 10 (Edwards, 1993).

EASTERN ATLANTIC REGION

The tropics of the eastern Atlantic have been referred to as a comparatively poor faunal area because the number of species (and higher taxa) is less than those found in any of the other three regions. The reasons for this situation are the latitudinal restriction of the area (to about 30°), the thin surface layer of tropical water, and the virtual absence of coral reefs. These factors and especially severe ice-age climatic changes have considerably limited opportunity for speciation.

The Eastern Atlantic Region extends from the Cape Verde Islands and Cape Verde on the mainland probably as far south as Mossâmedes, Angola (Fig. 70). It is the only tropical region that is not divisible into provinces. The Cape Verde Islands comprise an archipelago consisting of 10 main islands located about 440 km west of Cape Verde, Sénégal. The islands are very old, apparently dating from the Cretaceous.

Despite its great age and isolation, the Cape Verde Islands do not display much endemism. Among the shore fish, only about 4% are endemics (Briggs, 1974a). This low endemic rate is probably due to the geographical location of the archipelago. As in the case of Bermuda, it is very close to the northern border of the tropics so that a marked drop in sea-surface temperature, such as apparently occurred during the most recent glaciation, would have eliminated its tropical biota. It follows that if the islands had to be essentially repopulated since the last ice age, very little endemism would have had time to develop.

RELATIONSHIPS OF THE TROPICAL SHELF REGIONS

We have observed that the richest (most diverse) marine biota is found in the shallow waters of the tropical oceans and that this environment may be divided into four great zoogeographic regions: the Indo-West Pacific, the Eastern Pacific, the Western Atlantic,

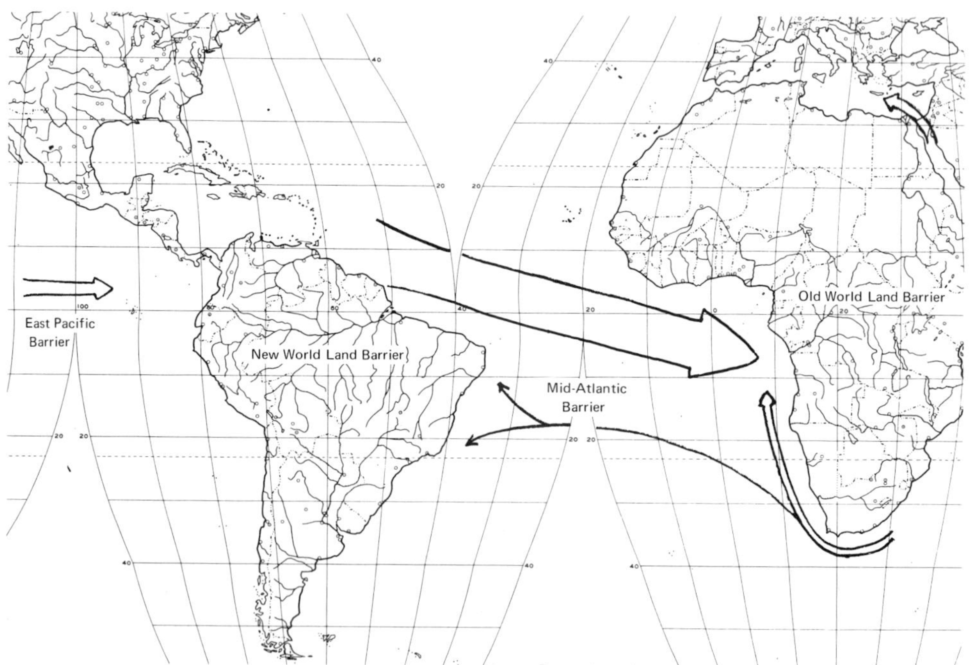

Fig. 71. Barriers separating the tropical shelf regions of the world. Arrows indicate the direction and approximate relative amount of successful (colonizing) migration that has taken place across the barrier.

and the Eastern Atlantic. Longitudinally, the tropical shelf regions are separated from one another by barriers that are very effective, since each region possesses, at the species level, a biota that is highly endemic. By studying the operation of these longitudinal barriers (Fig. 71), one can learn something about the interrelationship of the regions and, more important, obtain information leading to a better understanding of biogeography and evolution.

East Pacific Barrier

The East Pacific Barrier is the formidable stretch of deep water that lies between Polynesia and America. Although a variety of shallow-water fish and invertebrates, that are widely distributed in the IWP, have succeeded in crossing the barrier, particular attention has been paid to the fish, the molluscs, and the hermatypic corals. The most recent evaluation of the fish has been published by Leis (1984). He recognized a total of 55 species, all of them belonging to families whose larvae are often found several to many kilometers from shore. In addition, about 7% of the eastern Pacific endemic shore fish are very closely related to, and are probably derived from, trans-Pacific species.

Emerson (1991), and in a series of earlier papers, has kept track of the trans-Pacific molluscs. A total of 61 species of prosobranch gastropods with an IWP faunal affinity are known to inhabit eastern Pacific waters. Most of them occur only around the offshore islands, namely Clipperton (33 species), Revillagigedos (6), Cocos (18), Galápagos (10), and Guadalupe (1). Only 20 species occur on the west American shelf. This distribution is somewhat similar to that found for the fish, in that most of the trans-Pacific species also occur around the offshore islands. Also, the gastropods, as well as most of the fish, belong to groups that are known to have long-lived larval stages.

It is important to note that no shore species of apparent eastern Pacific origin, occur in Polynesia. This means that the successful migration (invasion followed by colonization) of shallow-water species takes place in one direction only, from west to east. Among the fish, there are probably about 650 species inhabiting the coastal waters between southern Mexico and Peru. The 55 trans-Pacific species comprise about 8% of that fauna. If one adds the 7% of eastern Pacific endemics that were probably derived from trans-Pacific species, it suggests that about 15% of the present eastern Pacific fish fauna is attributable to trans-Pacific migrations.

The origin of the corals that form the reefs in the eastern Pacific was discussed in the Pleistocene section (p. 175). Because of the strong IWP affinity of the corals, Glynn and Wellington (1983) proposed that they were periodically exterminated by cold temperature during the glacial stages and then reconstituted by recruitment from the Central Pacific during the interglacial periods. A possibly more-viable theory is that enhanced upwelling, due to a glacial-period strengthening of the thermohaline circulation, could have caused extinctions by means of eutrophication (Hallock and Schlager, 1986).

As noted, some of the shallow-water IWP fauna extends to the east of Easter Island along the seamounts of the Nazca Ridge. It has been surmised that migration along this route might have enabled IWP species to invade the eastern Pacific (Leis, 1984). However, tropical species would have been blocked by the cool Peru Current running parallel to the South American coastline. Although 173 fish species have been taken from the Nazca and Sala y Gomez ridges, apparently none of the 55 known trans-Pacific species were included (Parin, 1991). This makes it highly unlikely that the trans-Pacific invasion of shore species has been taking place along that route.

It is the geographic pattern shown by the trans-Pacific species that offers the best clue to the migration route. Most of them occur at four island locations between the equator and 20°N. Clipperton Atoll, which has the greatest number, lies about 10°N. The Equatorial Countercurrent is generally located about 5°N, but it is subject to seasonal variations. Another possibility is the Equatorial Undercurrent, a subsurface (50–300 m), eastward-flowing, stream located about on the equator. The Line Islands are the easternmost Polynesian group that lie along the course of the Equatorial Countercurrent. They are the most likely source of the trans-Pacific migrants.

New World Land Barrier

The New World Land Barrier, with the Isthmus of Panama forming its narrowest part, is virtually a complete block to the movement of tropical marine species between the eastern Pacific and the western Atlantic. This state of affairs has existed since the late Pliocene so that, at the species level, the two faunas are highly distinct. The present Panama Canal has not notably altered this relationship since, for most of its length, it is a freshwater passage forming an effective barrier for all but a few euryhaline species.

Despite the obvious physical effect of the isthmus in separating the marine populations of the American tropics, its evolutionary effects have not been fully investigated, and there is still an urgent need for research along this line. As more systematic work was done on the families and genera of amphi-American distribution, the species which could be recognized as common to the eastern Pacific and western Atlantic became fewer.

Considering that the total number of shore-fish species now estimated to be present on both sides of Central America (about 1000), the fact only about 12 (aside from the circumtropical species and a few euryhaline forms) can still be considered identical indicates that the New World Land Barrier is highly effective. Some of the invertebrate groups have more amphi-American species but seldom more than 10% of the total. This means that, for the great majority, some 3 Ma has been sufficient to produce distinct morphological evidence of evolutionary change.

For the past 75 years, almost since the opening of the Panama Canal, various proposals for the construction of a sea-level canal have been put forth. These have generated considerable enthusiasm among shipping interests and consternation among marine biologists. The prospect of a sea-level passage across Panama, to replace the present freshwater route, means the creation of a migratory avenue for marine species. Considering that 80 to 85% of all tropical benthic invertebrate species have planktonic larvae and that most fish larvae and adults are relatively mobile, it is probable that within a few years, after the opening of such a canal, most of the shelf species in the vicinity would be able to pass through.

What happens to the fauna in an area that is apparently ecologically saturated (as most mainland shore areas probably are) when additional species are introduced? It seems reasonable to expect, if we can profit by records of similar happenings in the terrestrial environment, that the faunal enrichment would be temporary and, by means of the elimination of species through competition, the area would eventually return to its original level of diversity.

The tropical faunas on each side of Central America are not only very rich but, in general, are so poorly known that marine biologists cannot say how many species are present. A rough estimate of more than 8000 species on the Atlantic side and nearly 6000 species on the Pacific side has been made (Briggs, 1974a). Considering its more diverse ecosystem, it has been suggested that many of the Atlantic species would be able to outcompete their Pacific equivalents. On the other hand, a small but significant portion of the eastern Pacific biota consists of IWP species that have made their way across the East Pacific Barrier.

Among the IWP species that have colonized the eastern Pacific are the crown-of-thorns starfish (*Acanthaster planci*) and the poisonous sea snake (*Pelamis platurus*). The starfish feeds on living coral and is responsible for widespread reef destruction in the western Pacific. The sea snake is highly poisonous and is capable of causing human fatalities. Should these species pass through a sea-level canal, it is likely that they would establish themselves, since they apparently have no Atlantic competitors. Considering that, in the Atlantic, the prey of these species would consist of organisms that have had no opportunity to evolve defense mechanisms, the result may well be a population explosion. This happened when the parasitic sea lamprey gained access to the western Great Lakes via the Welland Ship Canal and virtually destroyed the native populations of lake trout.

Aside from the possible economic and human welfare aspects of invasions by particular organisms, it is the prospect of the terrible loss of species that presents the real conservation problem. What is the value of a unique species? Do we have so many that we can afford to perform alterations to the earth's surface that are likely to result in large-scale extinctions? If one undertook the task of devising a simple construction project that would eliminate the greatest number of marine species from the face of the earth, none would be more effective than a sea-level passage through the New World Land Barrier – the most efficient of all the barriers separating the tropical faunas of the world.

Mid-Atlantic Barrier

The Mid-Atlantic Barrier is the broad, deep-water region that separates the western Atlantic tropics from those of the west African coast. For the shallow-water biota, the distance across the deep-water barrier is not as great as it is in the Pacific. For this reason, there is a greater proportion of trans-Atlantic species in most groups. In the shore fish, it appears that about 120 species have trans-Atlantic distribution (Briggs, 1974a). This may be related to a pool of about 900 species in the west and about 450 species in the east. Aside from a group of about 24 species that apparently make their way around the Cape of Good Hope and then westward across the Atlantic, the predominant migratory movement across the Mid-Atlantic Barrier seems to be from west to east.

Many of the trans-Atlantic fish species range broadly along the western Atlantic shelf but have attained only limited purchase in the east. Others that have achieved broad distributions in the east are clearly representatives of speciose American genera. None of the trans-Atlantic species belong to genera that are typically eastern Atlantic. Trans-Atlantic species comprise about 25% of the shore-fish fauna of tropical west Africa. Works on several groups of west African invertebrates appear to indicate the same kind of relationships.

Old World Land Barrier

The Old World Land Barrier was established in the early Miocene when the collision between Africa and Asia terminated the existence of the Tethys Sea. Passage around the southern tip of Africa probably became difficult for tropical shelf species with the onset of high-latitude cooling between the early and middle Miocene (Kennett et al., 1985).

Since that time, it appears that a few circumtropical shore species have been able to preserve their genetic continuity by migration around the Cape of Good Hope. Occasionally, the Agulhas Current of the western Indian Ocean will bud off rings of water that are carried around the cape to be picked up by the northward-flowing Benguela Current. In addition to the circumtropical species, a few others are common to the IWP and the tropical Atlantic. Among the fish, eight such species are found on both sides of the Atlantic and four are confined to the west African fauna.

It looks as if the successful migration of tropical shore species around the Cape of Good Hope may have been taking place in one direction only. Aside from several species that belong to monotypic genera, the rest represent genera that are best developed in the IWP. Similarly, a few tropical invertebrate species range from the IWP to the Atlantic. Their relationships also appear to lie with the IWP.

Although water-borne traffic between the Red Sea and the Mediterranean was possible in the days of the Pharaohs, it was accomplished by means of canals supplied with freshwater from the Nile, so there was little or no effect on the distribution of the marine biota. However, since the opening of the modern Suez Canal in 1869, some significant changes have taken place. At first, the Bitter Lakes portion of the canal had a very high salinity (68‰) which provided an effective barrier to migrants. But now, the salinity has dropped to about 41‰, not very different from that of the northern Red Sea or the southeastern Mediterranean.

A thorough analysis of migrations through the Suez Canal has been published by Por (1978). He discussed data pertaining to the presence of more than 200 species of Red Sea organisms in the Mediterranean. He suggested further that, due to our incomplete knowledge of the migrations, as many as 500 species of immigrants may actually be present in the Mediterranean. In contrast, a few Mediterranean species have moved south in the canal as far as Suez but have not colonized the open gulf. Exceptions are two species of Mediterranean euryhaline fish that have become established in lagoons of the Gulf of Suez. All other cases of southward migration involve species belonging to groups where passive transport on ships is common.

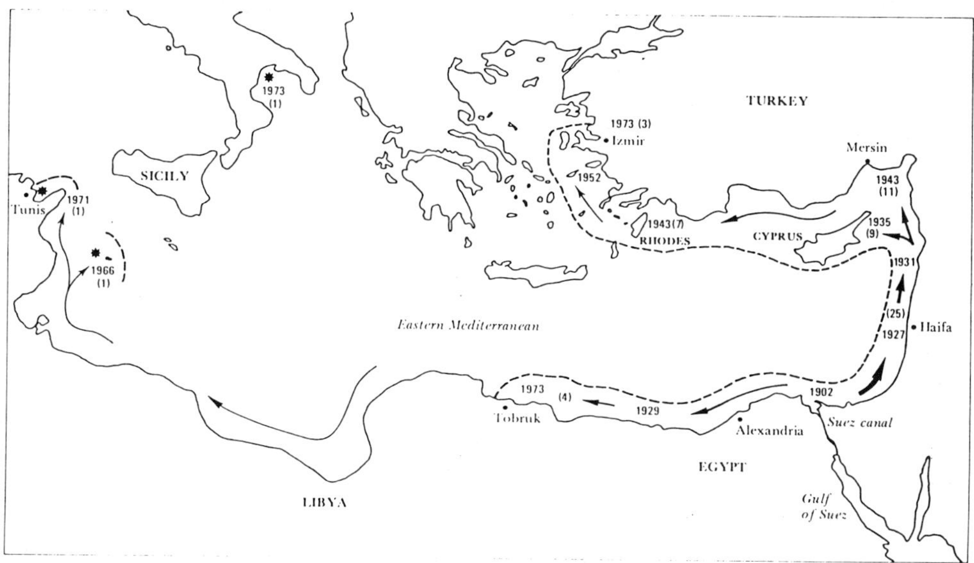

Fig. 72. Invasion of the eastern Mediterranean by Red Sea fish species. The dates indicate the advent of the first published records at the various localities, the figures in parentheses show the numbers of Red Sea species presently recorded at those localities and the dashed outline shows the limits of the area in which the invaders appear to be reasonably well established. After Ormond and Edwards (1987).

About 30 shore-fish species from the Red Sea have invaded the Mediterranean (Ormond and Edwards, 1987). For the most part, the invaders have been confined to the eastern Mediterranean (Fig. 72). Some of them support important commercial fisheries. There are a few examples of habitat displacement. The Red Sea goatfish *Upeneus moluccensis* and the lizard fish *Saurida undosquamis* proved to be competitors of the local red mullet *Mullus barbatus* and the hake *Merluccius merluccius*, respectively. In each case, the local species has been displaced into deeper, cooler water.

Conclusions

It may be said that the available information on the relationship of the tropical shelf regions permits the identification of three important facts: first, very few species have been able to migrate from one region to another and successfully establish themselves; second, the four biogeographic barriers differ considerably in their efficiency or effectiveness; and third, the successful migratory traffic that does overcome the barriers tends to be unidirectional.

The most effective of the four barriers is the New World Land Barrier since it has provided an almost complete block to migration for about 3 Ma. The few species that are still considered to be identical on each side of Central America consist of forms that have been slow to speciate, or they are circumtropical forms that probably continue to migrate around the Cape of Good Hope and across the East Pacific Barrier. We must realize that proposals for a sea-level canal will continue to be put forth. That project has the potential for causing a disastrous loss of species in the New World tropics.

Successful (colonizing) migrations across the biogeographic boundaries that delimit the Indo-West Pacific can apparently take place in one direction only – outward into areas where the biota is poorer and the competition is less. The realization that the East Pacific and Old World Land Barriers operate as one-way filters enables us to understand better how the Indo-West Pacific Region serves as *the evolutionary and distributional center* for much of the tropical shore biota of the world. We can see that competitively dominant species continue to migrate, as they probably have for millions of years, from the Indo-West Pacific eastward across the open ocean to America and westward around the Cape of Good Hope into the Atlantic; since 1869, some of them have also been able to pass northward through the Suez Canal into the Mediterranean.

The Western Atlantic Region may be considered a secondary center of evolutionary radiation. Many species evolved in this area have proved capable of migrating eastward to colonize the tropical eastern Atlantic. However, species originating in the eastern Atlantic are apparently incapable of successfully invading the western side. Again, the advantage seems to lie with the area that possesses the richer fauna and the higher level of competition.

It can be seen that the completely eastward direction of successful migratory movements across the East Pacific Barrier and the predominantly eastward movements across the Mid-Atlantic Barrier take place in a direction opposite to that of the main flow of the surface waters via the North and South Equatorial Currents. In contrast, the surface and subsurface countercurrents in the tropical Pacific and Atlantic are weakly developed, but these smaller currents are obviously the principal means by which successful transport is achieved.

Centers of origin, such as the tropical East Indies and the cold-temperate North Pacific, not only produce dominant species but demonstrate a high level of resistance to invasions by species from other regions. Both functions are apparently attributable to relatively high diversity. In the past, there has been considerable argument among ecologists about the relationship between diversity and ecosystem stability. If resistance to invasion is a measure of stability, then stability does increase with diversity. This empirical observation is substantiated by theoretical models which indicate that the probability of extinction for a resident species increases with community size, and the probability of successful colonization by an invader decreases (Case, 1990).

LATITUDINAL BARRIERS

In determining the latitudinal boundaries of each of the four tropical regions, an objective attempt was made to gather all possible evidence in order that a boundary could be located where the rate of species change was the greatest, that is, where the most tropical forms dropped out and/or where the most temperate species began. As a result, boundaries were proposed as follows:

1. Indo-West Pacific – Amami Islands, Taiwan, and Hong Kong in the north. In the south, Fraser Island and Shark Bay in Australia and the mouth of the Kei River in South Africa.

2. Eastern Pacific – Magdalena Bay, Baja California, and the mouth of the Gulf of California, to the southern Gulf of Guayaquil, Peru.
3. Western Atlantic – Cape Canaveral and Cape Romano, Florida, and Cape Rojo, Mexico, to Cape Frio, Brazil.
4. Eastern Atlantic – Cape Verde to Mossâmedes, Angola.

The important question is: What do these localities have in common that permits them to operate as major zoogeographic barriers? Biologists have been aware for a long time that the distribution patterns of marine organisms reflect a high degree of temperature sensitivity. Beyond this, however, there has been little agreement. The most influential of the modern writers (Ekman, 1953; Hedgpeth, 1957; Coomans, 1962) have utilized annual mean surface isotherms to help define the limits of biogeographical regions. Others (Stechell, 1920; Wells, 1963) have used isotheres (lines of mean maximum surface temperature), and still others (Dana, 1853; Forbes, 1859; Gill, 1884; Monod, 1957) have used isoscrymes (lines of mean minimum surface temperature).

It is possible to offer some solution to this problem, at least as far as the tropical biota is concerned, for when the above-listed places are located on the monthly sea-surface temperature charts it can be seen that they fall on or very close to the 20°C isocryme line (calculated from the mean for the month of February in the northern hemisphere and for August in the southern hemisphere). This serves to substantiate Dana's (1853) observation that tropical species are limited in their northward distribution by the cold of winter, rather than the heat of summer or the mean temperature of the year.

SUMMARY

1. The two fundamental divisions of the oceanic biosphere are the Benthic Realm and the Pelagic Realm. Each realm may be subdivided into a number of depth zones and a series of horizontal regions and provinces.
2. A series of four latitudinal surface zones is recognized. The Tropical Zone occupies the lowest latitudes and supports the greatest species diversity. This zone is contained within the 20°C isocryme for the coldest month of the year.
3. A Warm-Temperate Zone with its own suite of endemic species borders the tropics. Here, the winter surface temperatures generally range from 20°C down to about 12°C.
4. A Cold-Temperate Zone lies on the high-latitude side of each Warm-Temperate Zone. Winter surface temperatures range from 12°C down to about 2°C.
5. The Cold or Arctic and Antarctic Zones occupy the circumpolar areas. Their surface temperatures usually range between +2°C and −2°C.
6. The continental shelves of the tropics may be subdivided into four great biogeographic regions. These are the Indo-West Pacific, the Eastern Pacific, the Western Atlantic, and the Eastern Atlantic.
7. The Indo-West Pacific is the largest region. Its size and its species diversity is more than double that of the other three regions combined.
8. The greatest species diversity in the marine world is concentrated in a comparatively small triangle which includes the Philippines, the Malay Peninsula, New Guinea, and the northern Great Barrier Reef.

9. There is a current controversy about the cause and significance of the high species diversity in the East Indian Triangle. Is it an accumulation of species deposited by ocean currents or is it a center of origin where successful new species are being produced?
10. Data pertaining to phylogenetic patterns, generic ages of several animal groups, onshore-offshore gradients, migration across regional boundaries, and the locations of disjunct distributions all appear to provide support for the center of origin hypothesis.
11. Although allopatric and parapatric speciation may have been the dominant modes in the East Indies, geographic patterns exhibited by closely related (sibling) species suggest that sympatric speciation may also be important.
12. The Indo-West Pacific Region may be subdivided into several provinces, each of them having more than 10% or more of their biota comprised of endemic species. The largest is the Indo-Polynesian Province.
13. A large part of the Pacific Plate, here considered to be part of the Indo-Polynesian Province, was separated by one investigator into its own biogeographic region. However, the number of endemics characteristic of the Pacific Plate as a whole are so few that such a separation cannot be justified.
14. The remaining provinces within the IWP Region are the Western Indian Ocean, Red Sea, Northwestern Australian, Hawaiian, Easter Island, and the Marquesas.
15. The Eastern Pacific Region may be subdivided into Mexican, Panamanian, and Galápagos provinces.
16. The Western Atlantic Region includes three provinces: the Caribbean, West Indian, and Brazilian. The mid-Atlantic islands of St. Helena and Ascension are considered to form an independent province of their own.
17. The Eastern Atlantic Region is not divisible into provinces.
18. The tropical regions are separated from one another by four highly effective barriers: the East Pacific Barrier, New World Land Barrier, Mid-Atlantic Barrier, and the Old World Land Barrier.
19. The Indo-West Pacific Region (and within it the East Indies Triangle) represents an evolutionary and distributional center for much of the world's tropical shore biota. Species continue to migrate, as they may have done for several million years, eastward across the open ocean to the New World and westward around the Cape of Good Hope into the Atlantic. Since 1869, several hundred have been able to enter the Mediterranean via the Suez Canal.
20. The western Atlantic constitutes a secondary center of evolutionary radiation. Many species from there have migrated eastward to colonize the tropical eastern Atlantic. The completely eastward direction of successful (colonizing) migrations across the East Pacific Barrier and the predominately eastward movements across the Mid-Atlantic Barrier take place in a direction opposite to the main flow of surface waters.
21. Centers of origin, such as the tropical East Indies and the cold-temperate North Pacific, not only produce dominant species but demonstrate a high level of resistance to invasion by species from other regions. If resistance to invasion is a measure of ecosystem stability, then stability does increase with diversity.

CHAPTER 10

Marine patterns, Part 2

No matter how strange or rigorous the inanimate or noncompetitive environment may be, the evolutionary products of such environments are not as thoroughly refined by competition, and are thus not as well prepared for widespread success, as are the products of highly competitive associations.

Hobart M. Smith, *Evolution of Chordate Structure*, 1960

WARM-TEMPERATE REGIONS

Everywhere in the world the tropics are bordered by a warm-temperate biota. The warm-temperate regions and provinces contain many endemic species, but also have many species called eurythermic tropicals. The latter are generally wide ranging in the tropics but also occupy the warm-temperate regions. The tropical relationships of most of the warm-temperate endemics are obvious because they usually belong to tropical genera and families. In some places, such as southern Japan and the northeastern United States, the northward migration of tropical species in the summer tends to obscure the presence of the warm-temperate forms.

Southern hemisphere

The warm-temperate, and sometimes the cold-temperate, waters of the southern hemisphere are of considerable evolutionary interest because they harbor many relict families and genera that were once widespread in the tropics. Some of these groups are also represented by relicts in northern temperate regions. Among the fish, southern Australia has six families that are found nowhere else. In addition, it appears to have functioned as a distributional center for several other families.

It appears that a number of taxa, having become initially isolated in southern Australia, were then able to achieve circumglobal ranges by means of the West Wind Drift. Such a dispersal history has been suggested for the fish families Cheilodactylidae and Latridae (Briggs, 1974a). The spiny lobster genus *Jasus* also provides an example of such a distribution (Holthuis, 1991): while three species occur in the southern Australian-New Zealand area, single endemic species are found at Juan Fernandez (and nearby islands) in the southeastern Pacific, the Tristan da Cunha group in the south Atlantic, southern Africa, and the St. Paul-Amsterdam islands in the southern Indian Ocean. A

West Wind Drift dispersal for certain groups of echinoderms has been described by Fell (1962).

The Southern Australian Region comes under the influence of three important currents. To the west, the West Australian Current – the eastern arc of the main gyre of the south Indian Ocean – affects the coast from about Cape Leeuwin north as far as Shark Bay. The south coast from Cape Leeuwin to Victoria receives flow from the West Wind Drift. To the east, the East Australian Current comes close to the coast in the area between Brisbane and Cape Howe.

The Region may be divided in two Provinces that are separated from one another by the cold-temperate waters of Victoria and Tasmania (Fig. 73). The Southwestern Australia Province extends from the tropical boundary at Shark Bay down to about Robe, South Australia. On the western side of the continent, a Southeastern Australian Province extends from Sandy Cape on Fraser Island south to about Bermagui, New South Wales.

The Northern New Zealand Region comes under the influence of the East Australian Current as it heads northward from the Tasman Sea (Fig. 73). This Region may also be divided in two provinces. The Auckland Province includes the Auckland Peninsula and the shoreline from there to East Cape. A large number of its species are shared with southeastern Australia. The Kermadec Province pertains to the Kermadec Islands, a small volcanic group located about 690 km northeast of Auckland, New Zealand.

The Western South America Region (Fig. 74) includes the major part of the western coast of South America extending from the Gulf of Guayaquil to the northern end of

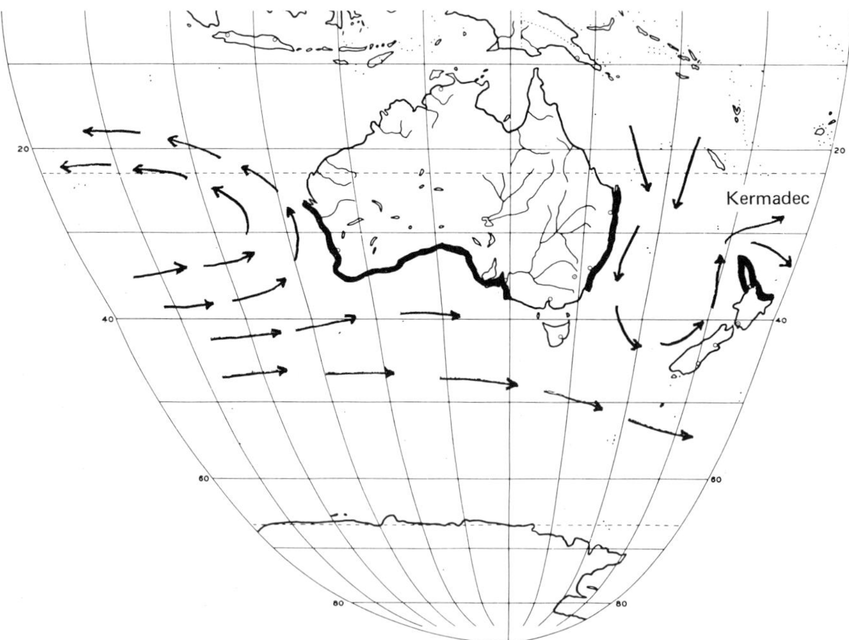

Fig. 73. Southwestern Australia, Southeastern Australia, Auckland and Kermadec Provinces.

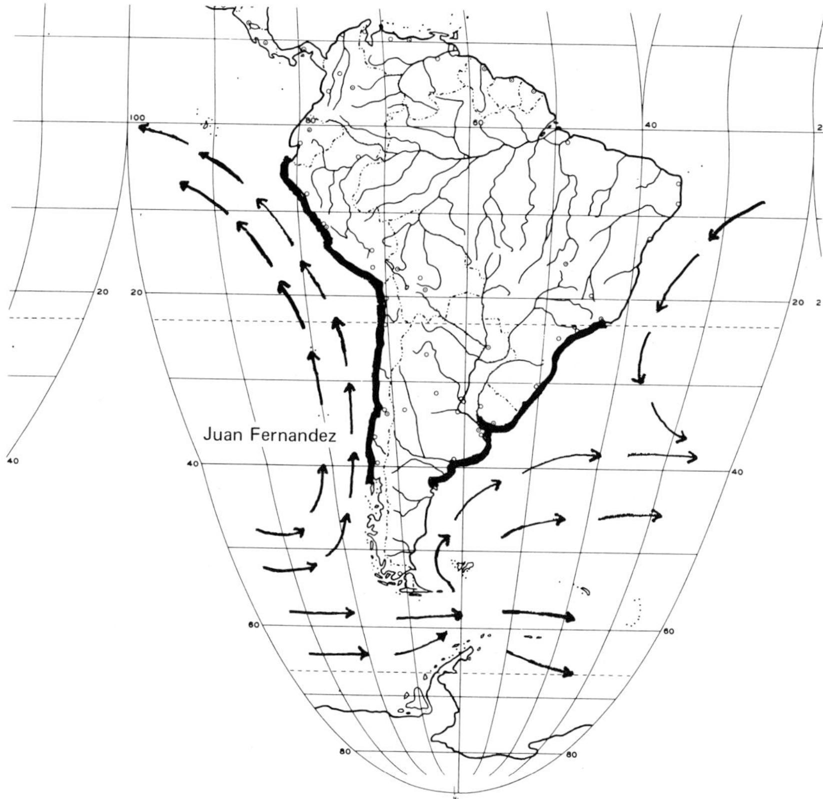

Fig. 74. Peru-Chilean and Juan Fernandez Provinces and Eastern South America Region.,

Chiloe Island. This mainland coast is considered to comprise the Peru-Chilean Province. It is influenced by the Peru Current which originates from the West Wind Drift. As this current flows northward along the Peruvian coast, it is affected by southwesterly winds which cause upwelling. The upwelling maintains relatively cool surface temperatures all the way to about 3°S. A Juan Fernandez Province is recognized for the islands of that name that lie about 550 km west of the Chilean coast.

The Eastern South America Region extends from the tropical boundary at Cape Frio southward all the way to the Valdés Peninsula in Argentina. The Region is influenced by the warm, southward-flowing Brazil Current. At one time, the southern boundary was placed at the mouth of the Rio de la Plata (Briggs, 1974a) but more recent work on the molluscs (Scarabino, 1977), the decapod crustaceans (Boschi, 1979), and the fish (Menni et al., 1984) indicates that it belongs farther south. While most of the inshore members of the cold-temperate biota appear to stop at the Valdés Peninsula, many of those on the outer shelf extend farther north. This Region forms a unit by itself that is not divisible into provinces.

The Southern Africa Region lies between the tropical boundaries at Mossamedes on the west coast and the Kei River mouth on the east coast. The Cape of Good Hope is the

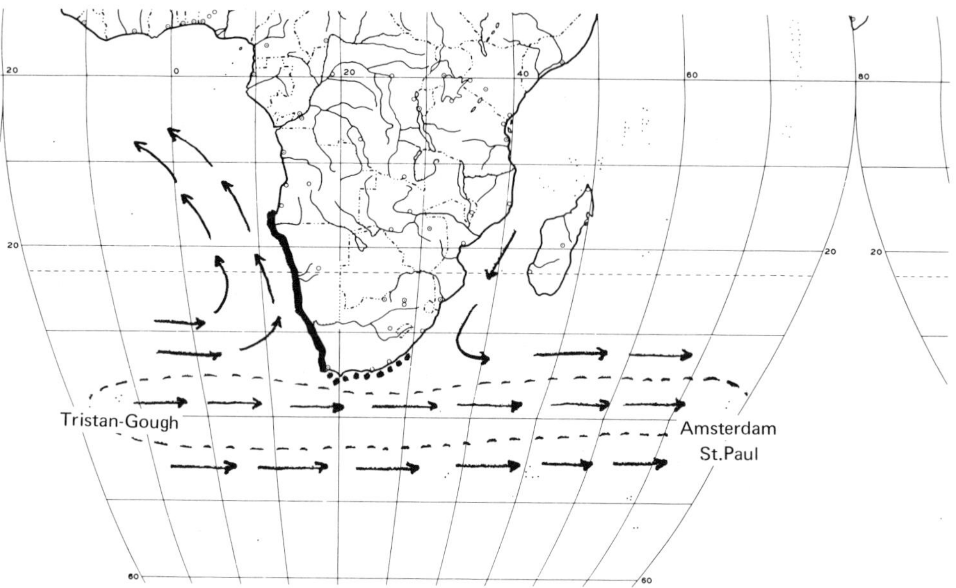

Fig. 75. Southwest Africa, Agulhas and West Wind Drift Island Provinces.

dividing line between the two provinces (Fig. 75). The Southwestern Africa Province is influenced by the Benguela Current which flows northward along that coast. The Agulhas Province, on the southeastern African coast, is affected by the Agulhas Current which flows southward from Madagascar. In the past (Briggs, 1974a), a third province was identified for the Amsterdam and St. Paul Islands about 3800 km to the east. But Collette and Parin (1991) have shown that the relationships of those islands are closer to other islands and seamounts to the west. They concluded, on the basis of a 30–40% endemism in the shallow-water fish fauna, that St. Paul-Amsterdam, Tristan-Gough Islands in the South Atlantic, and the intervening sea mounts (Walters Shoals, UN2, and Vema) should be incorporated into a West Wind Drift Islands Province (Fig. 75). This province can probably be included within the Southern African Region, although at present the relationship seems tenuous.

Northern hemisphere

Of the northern hemisphere warm-temperate regions, the Mediterranean-Atlantic Region is by far the most extensive. The Lusitania Province extends northward from the tropical boundary at Cape Verde on the African west coast to about the western entrance to the English Channel (Fig. 76). The Mediterranean Sea is an interesting and special part of the province. The biota of the latter is more diverse than that of the Atlantic coast and there is considerable endemism. However, almost all of the Atlantic coast species also occur in the Mediterranean so it has little endemism of its own. The Lusitania Province also includes the offshore islands of Azores and Madeira.

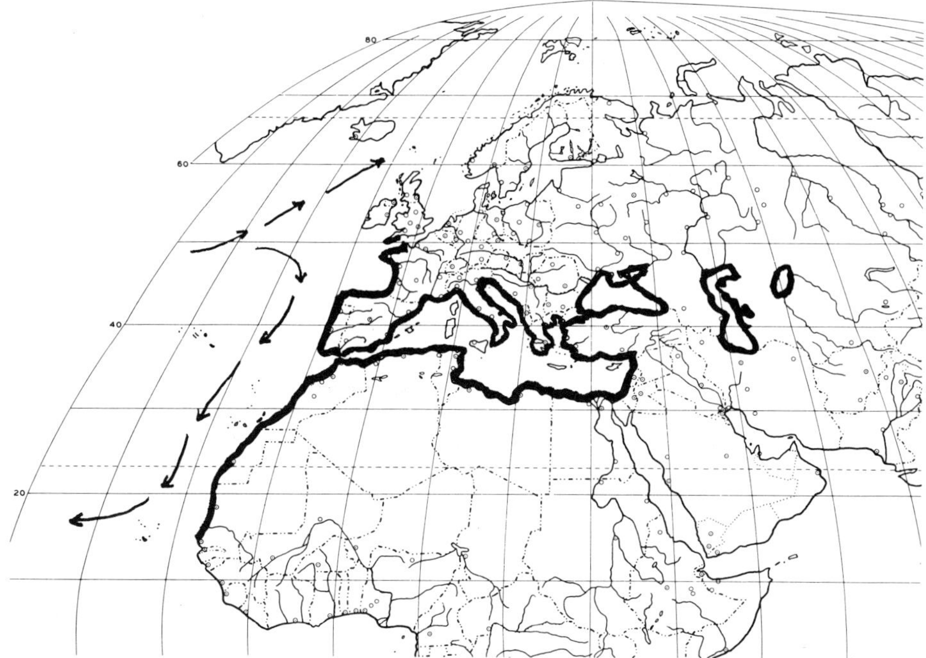

Fig. 76. Lusitania, Black Sea, Caspian and Aral Provinces.

Although the Black, Caspian, and Aral Seas have temperature regimes that are colder than the other warm-temperate provinces, they have derived their marine biotas from the Mediterranean so are included in the Mediterranean-Atlantic Region. The Black Sea Province has a low salinity of about 17–18‰ in the open sea and less in the Sea of Azov. Its species diversity is only about 20–25% as great as the Mediterranean. The Caspian Province – for the Caspian Sea – possesses a highly distinct, brackish and freshwater biota with many Pliocene relicts. The Aral Province has many species in common with the Caspian Sea but also a significant endemism. Most of the Aral Sea species are of freshwater rather than marine origin.

The Carolina Region is exposed to warm water carried northward by the Gulf Stream System. Currents from the Caribbean Sea enter the Gulf of Mexico via the Yucatán Channel. The Florida Current carries the outflow through the Straits of Florida and northward along the Florida east coast. The Region exists in two parts, one in the northern Gulf of Mexico and the other on the Atlantic coast (Fig. 77). Within the Gulf, the warm-temperate biota occupies the area north of the tropical boundaries at Cape Romano, Florida and Cape Rojo, Mexico. The Atlantic section is found between Cape Canaveral and Cape Hatteras. Although the northern Gulf section displays considerable endemism, almost all the warm-temperate species of the Atlantic section are also found in the Gulf of Mexico.

The California Region also exists in two separate parts, but here each part clearly comprises a distinct province. South of Point Conception, the central California shoreline

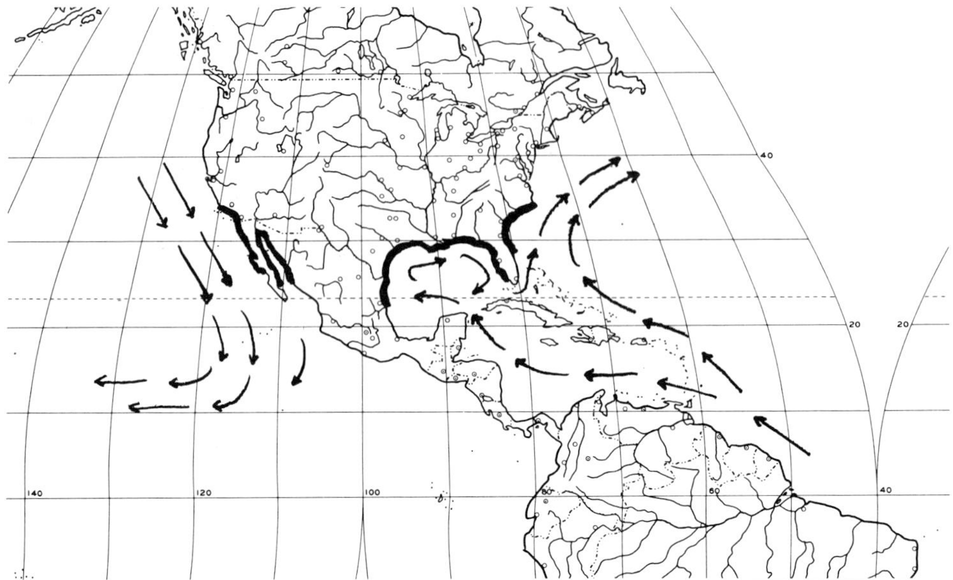

Fig. 77. Carolina Region and the Cortez and San Diego Provinces of the California Region.

takes an abrupt curve towards the east. This, and probably the presence of the Channel Islands, gives the inshore waters of the area some protection from the cold southbound California Current. These sheltering effects, plus the help of an onshore current movement from the south during the winter, permit the existence of a warm-temperate area along the southern California Coast. Aside from some local interruptions caused by upwelling, warm-temperate conditions extend south to the tropical boundary at Magdalena Bay, Mexico. This outer coast area is called the San Diego Province (Fig. 77).

Within the Gulf of California, north of about La Paz on the west and Topolobampo on the east, lies the Cortez Province. Although this province contains many eurythermic tropical species that range widely in the eastern Pacific, it exhibits a considerable endemism in the fish and most groups of invertebrates. There is, in addition, a small group of species that are shared with the San Diego Province. It is suspected that most of them gained access to the Gulf via isothermic submergence. The history of the New World warm-temperate regions is considered in the section on Pleistocene events (p. 174).

As the Kuroshio Current flows northward from the Philippines, a portion of it cuts through the Ryukyu chain of islands just east of Taiwan. A western branch, called the Tsushima Current, runs through the Korean Strait and into the Sea of Japan. The main branch of the Kuroshio Current passes to the east of Kyushu Island and remains close to the coast until it reaches Cape Inubo where it meets with the cold Oyashio Current from the north. The degree of exposure to the warm, northward-flowing currents has determined the geography of the Japan Warm-Temperate Region (Fig. 78).

The Japan Region biota extends, on the mainland coast of China from about Hong Kong north to Wenchou. It is also found on the west coast of Taiwan and the tip of the

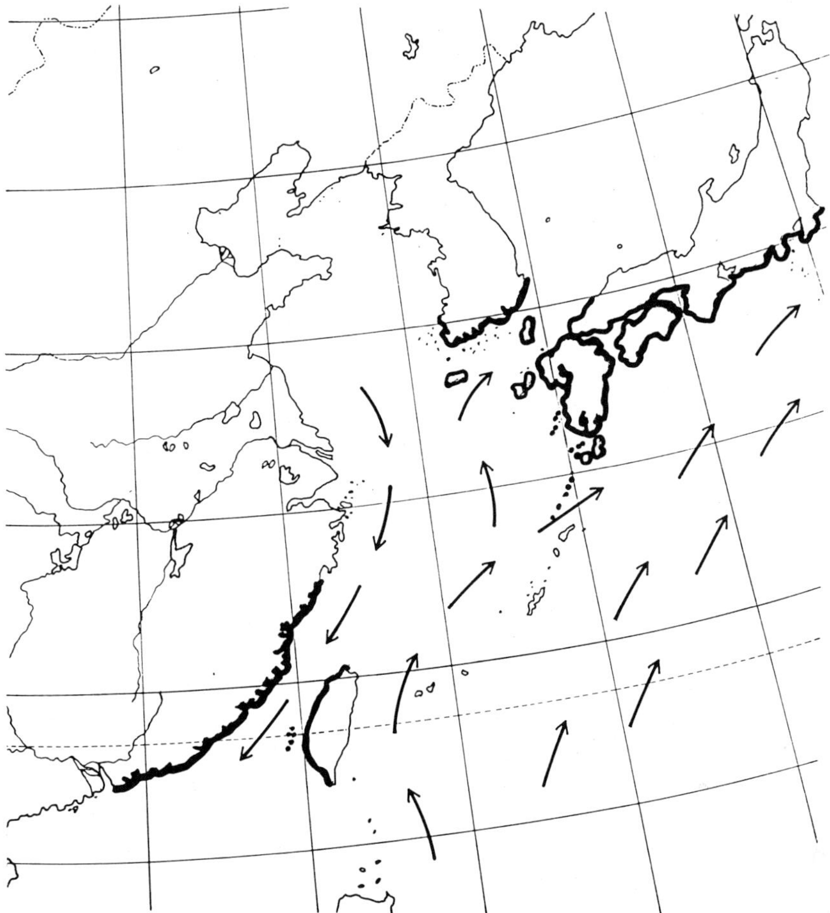

Fig. 78. Japan Region.

Korean Peninsula. In the Japanese Islands, it extends from just north of the Amami Islands to Cape Inubo on the Pacific Coast. In the Sea of Japan, it extends north to about Hamada in Shimane Prefecture. Although a number of cold-temperate species have been taken south of these northern boundaries, particularly on the Pacific coast, almost all have come from relatively deep water where the temperature is cooler.

COLD-TEMPERATE REGIONS

Although the cold-temperate regions lie adjacent to warm-temperate regions, the biotas are markedly different. Most of the cold-temperate families and genera have originated in evolutionary centers located in the North Pacific or in the Antarctic. Some of them arose in the North Atlantic. Although a few very-wide-ranging tropical species (called broad eurythermic tropicals) do penetrate cold-temperate waters – mainly in the pelagic environment – they have little effect on the main biotic composition. As noted in Chapter 7, global cooling events near the middle/late Eocene and the early/late Oligocene, were probably responsible for the establishment of cold-temperate conditions at the higher latitudes.

Southern hemisphere

A good example of the influence of the Antarctic evolutionary center may be found in the notothenioid fish. This group of five families is confined to the temperate waters of the southern hemisphere (Eastman, 1991). A cladogram illustrating their relationship was published by Iwami (1985). His phylogeny was quite similar to that of earlier workers. The most primitive of the families is the Bovichthyidae. Except for one species found on the Antarctic Peninsula, its distribution is subantarctic. Bovichthyid species are found in southern South America (plus the Juan Fernandez Islands), the West Wind Drift Islands Province, and southern New Zealand. The most primitive bovichthyid genus is confined to freshwater streams in Victoria and Tasmania. The family provides a good example of a group that has been pushed out of its center of origin by its more advanced descendants (Briggs, 1974a). Species belonging to the other four, more apomorphic, families comprise the majority of the fish fauna of the Antarctic continent.

Aside from the notothenoid group, much of the remaining cold-temperate fish fauna is comprised of species belonging to the families Liparididae, Zoarcidae, and Rajidae. These are groups of apparent North Pacific origin that migrated southward by means of isothermic submergence. In the cold-temperate areas of Tasmania-Victoria and New Zealand there has occurred a notable penetration of IWP tropical groups including the families Labridae, Syngnathidae, Tripterygiidae, Gobiidae, and Gobiesocidae. The cold-temperate species and genera belonging to these families appear to be phylogenetic relicts that may have accumulated over a long period of time.

It has been suggested that the dispersal of tropical fish into temperate seas in the southern hemisphere may be attributed to transport by southward flowing currents (Hutchins, 1991). However, this does not explain why tropical families are prominent in the cold-temperate waters of Australia-New Zealand but scarce in southern South America and in the sub-Antarctic regions. On the other hand, the families of North Pacific origin that are prominent in the latter areas are scarce in Australia-New Zealand. One may conclude that the cold-temperate fish fauna of the southern hemisphere has accumulated from three main sources: (1) an Antarctic center of origin, (2) invasion of boreal forms into southern South America and Antarctica-sub-Antarctica, and (3) invasion of tropical groups into Australia-New Zealand.

The tip of South America extends almost 20° farther south than any of the other continental masses reaching the circum-Antarctic region. A portion of the cold West Wind Drift is deflected northward and contributes to the formation of the Peru Current. By means of the latter, cold-temperate conditions are extended northward to the north end of Chiloe Island. In the deeper, offshore waters the cold-temperate biota ranges farther to the north. After the West Wind Drift flows through Drake's Passage, it gives rise to the

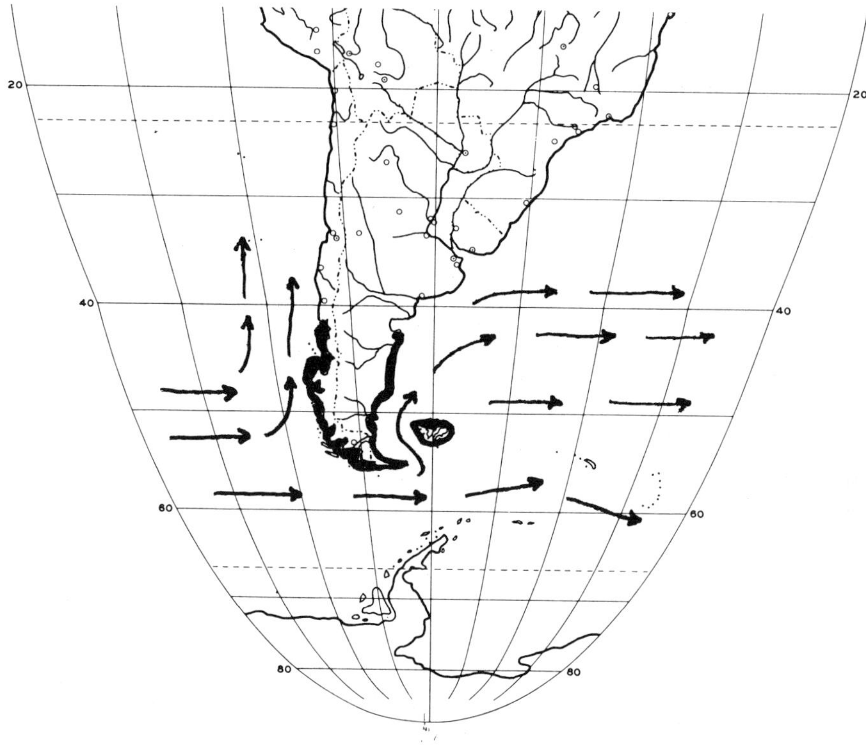

Fig. 79. South America (Magellan) Region.

Falkland Current which runs northward along the Argentine coast. As a result, a cold-temperate biota may be found northward to the Valdés Peninsula. Since the biota of both sides of southern South America seems to be fairly homogeneous, a single South America (or Magellan) Region is recognized (Fig. 79).

The Tasmanian Region (Fig. 80) includes all of Tasmania as well as the coast of Victoria in south Australia. Both areas are exposed to the West Wind Drift, although the circulation between the two is complex and subject to seasonal changes. Along the Victoria coast, there appears to be a considerable overlap of cold-temperate and warm-temperate species. Even so, the degree of endemism among the common intertidal species is significant. The giant brown algae *Durvillea potatorum* appears to be a useful indicator species.

The Southern New Zealand Region (Fig. 80) is also affected by the West Wind Drift. It impinges directly on the southern part of the South Island and the island groups to the south. It also gives rise to cold currents that flow northward along the east and west coasts of New Zealand. The Region consists of two provinces. The Cookian Province includes the whole of New Zealand south of the warm-temperate boundaries at Auckland and East Cape. Earlier (Briggs, 1974a) a Chatham Province was recognized for the Chatham Islands which lie about 550 km to the east of the North Island. However, the

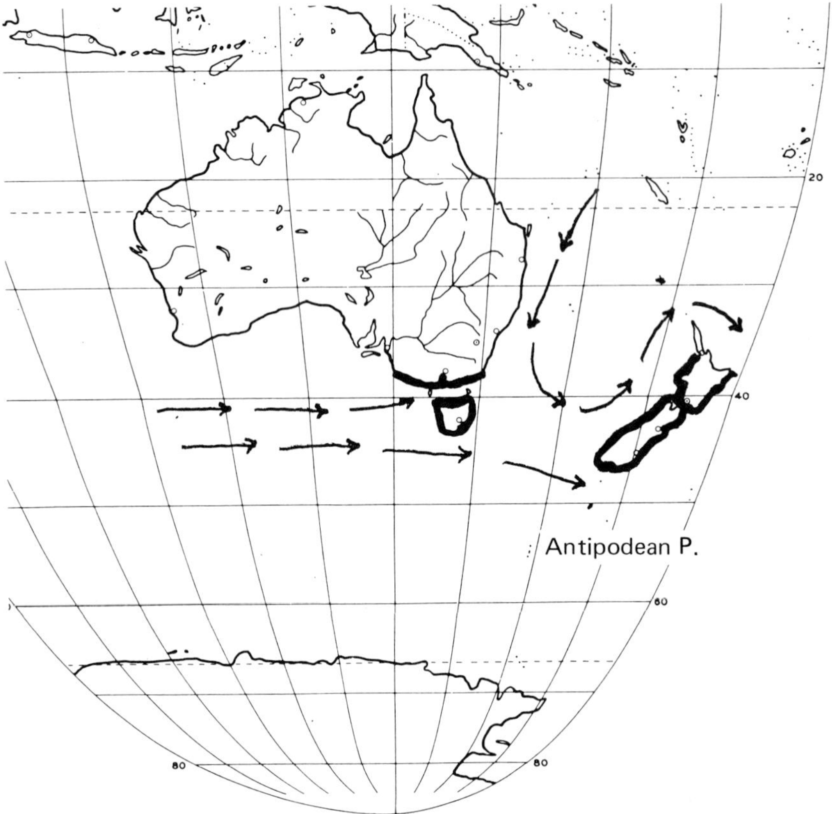

Fig. 80. Tasmanian Region and Cookian and Antipodean Provinces of the Southern New Zealand Region.

endemism now seems to be so low that the islands should be considered as belonging to the Cookian Province.

The Antipodean Province includes four groups of sub-Antarctic islands that lie to the south and east of New Zealand as follows: Auckland Islands, 260 km; Antipodes Islands, 67 km; Campbell Island, 440 km; and Bounty Islands, 670 km (Fig. 80). These islands project from an extensive submerged plateau that stretches from the southern tip of the mainland. The molluscan fauna is the best known and it appears to be highly endemic.

The Sub-Antarctic is the one biogeographic region in the world that is entirely made up of small, oceanic islands (Fig. 81). Although they are scattered over a broad expanse of southern ocean, the islands have many genera and species in common. The West Wind Drift affects the Region throughout the year and acts as an important dispersal agent. The surface temperatures are cold and show only a minor amount of seasonal fluctuation, from 2°C in winter to 5°C in summer.

The Kerguelen Province includes not only Kerguelen Island itself but the McDonald, Heard, Marion, Prince Edward, and Crozet Islands. The Kerguelen Islands lie in the southern Indian Ocean about 1800 km north of the Antarctic continent and directly on

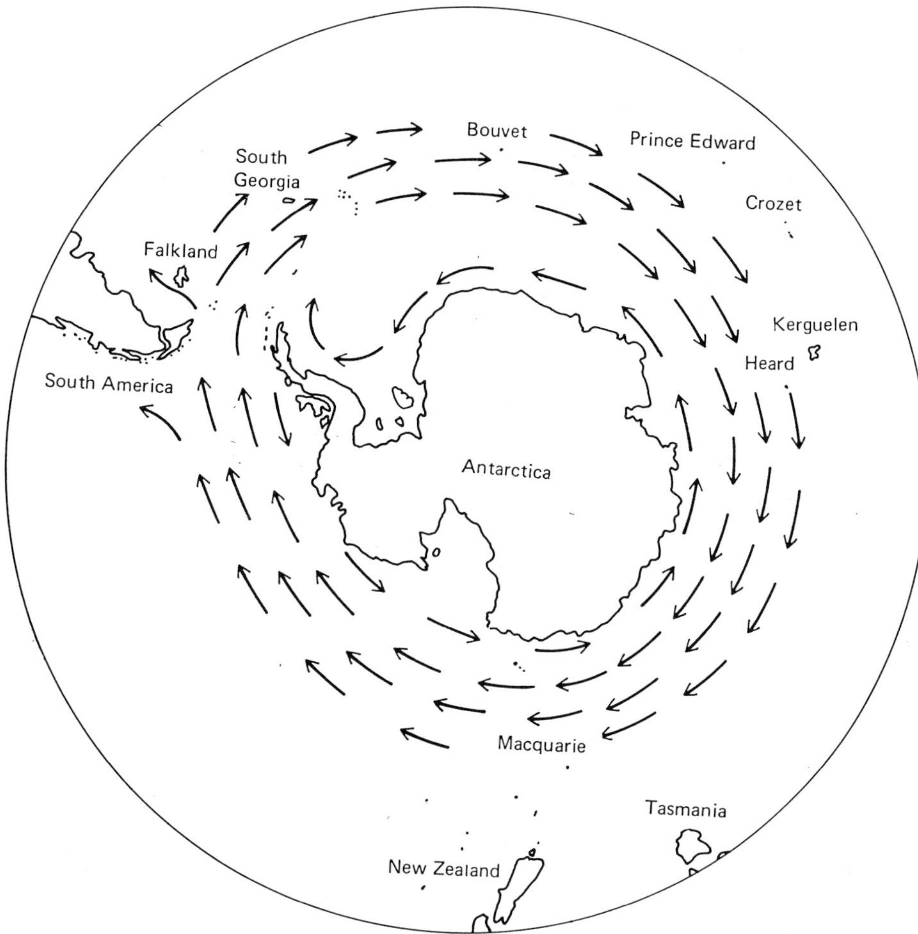

Fig. 81. South polar projection depicting geographic relationship of areas included in the Kerguelen, Macquarie, South Polar, South Georgia and Bouvet Provinces.

the mean position of the Antarctic Convergence. The McDonald-Heard Islands are located 400 km southwest of the Kerguelens and the two groups are linked by the submerged Kerguelen-Gaussberg Ridge. The Crozet Islands are about 1380 km west of Kerguelen and the Prince-Edward-Marion group is another 400 km farther west. Although its biota is still poorly known, it is possible that the Prince Edward-Marion group (with probably the Crozets) constitutes a separate province within the Region.

Macquarie Island is surrounded by deep water and is well isolated, being situated about 550 km southwest of the Auckland Islands and some 900 km from the New Zealand mainland. The molluscan fauna is the best known of the groups of shore animals. The existence of a high degree of endemism, along with notable external ties to the Kerguelen Province, indicate that a Macquarie Province should also be recognized within the Sub-Antarctic Region.

Northern hemisphere

Upon considering the general distribution of the marine biotas that occupy the cold waters of the world, it becomes clear that the North Pacific has functioned as a center of evolutionary radiation. It has supplied species to and has had a visible effect on the biotic composition of the Arctic and North Atlantic. Furthermore, North Pacific species have, by means of isothermic submergence, bypassed the equatorial region to invade the cold waters of the southern hemisphere. The rich, boreal biota of the North Pacific shelf may be subdivided into two regions, one on each side of the ocean. These regions may, in turn, be subdivided into five distinct provinces, three on the western side and two on the east.

On the western side of the North Pacific, an Oriental Province exists in three different segments. The first extends north from the warm-temperate boundary at Wenchou and continues through the Yellow Sea. Its continuity is broken by the tip of the Korean Peninsula (Fig. 82), but it then continues up the north side of the peninsula to about Chongjin. On the eastern side of the Sea of Japan, the Oriental Province extends from about Hamada to the Tsugaru Strait. From that point, it continues southward on the outer coast of Honshu Island to Cape Inubo.

It seems clear that a faunal break exists at about the location of the Tsugaru Strait between the islands of Honshu and Hokkaido. To the north of this point, both along the outer coast and within the Sea of Japan, one may find a different species complex that characterizes the Kurile Province. This province extends northward along the Kurile

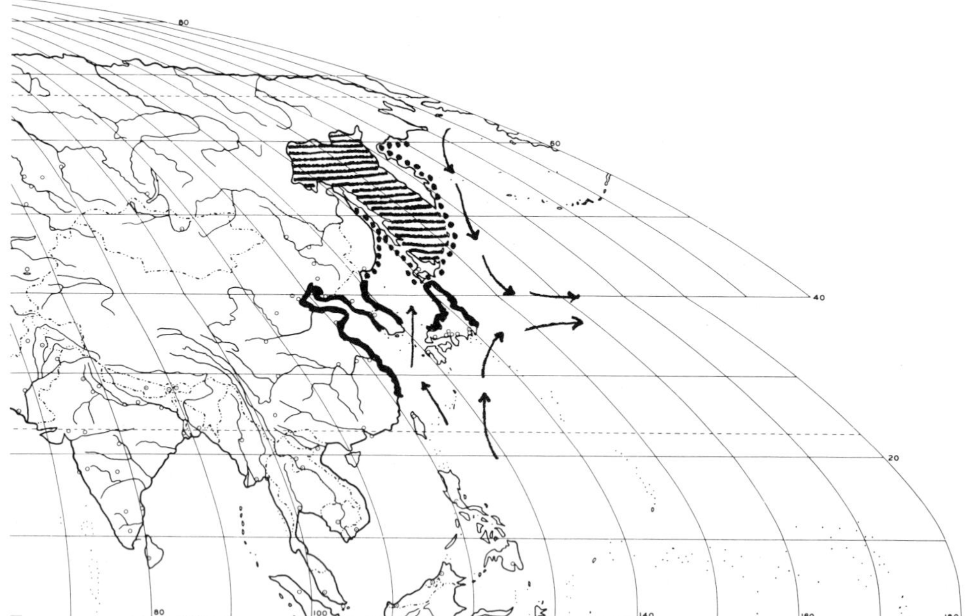

Fig. 82. Oriental, Okhotsk and Kurile Provinces.

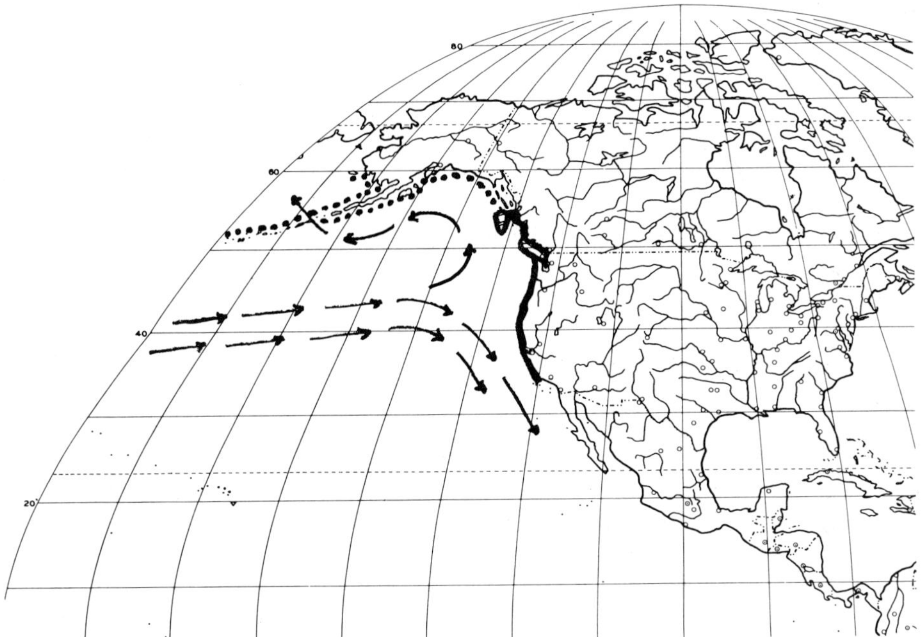

Fig. 83. Oregon and Aleutian Provinces.

chain of islands and the east coast of the Kamchatka Peninsula to about Cape Olyutorsky. The Okhotsk Province is confined to the Okhotsk Sea. Although this sea is now confluent with the North Pacific through the Kurile islands and with the Sea of Japan around Sakhalin Island, it was probably isolated during the glacial stages. Considerable endemism is apparent in a variety of groups such as the ascidians, pycnogonids, and fish.

In a paper on isopod biogeography, Kussakin (1990) recognized a single Beringian subregion extending entirely across the North Pacific from the Kamchatka Peninsula to the Gulf of Alaska and northward through the Bering Strait. Most of the Okhotsk Sea is placed in its own province. His Manchurian subregion is approximately equivalent to my Oriental Province. This means that the isopod fauna of eastern Kamchatka is about the same as that of the Gulf of Alaska. This kind of strong amphi-Pacific relationship at the species level, does not seem to occur in most other shallow-water groups.

The Eastern Pacific Boreal Region is influenced by the cool, North Pacific Current which flows eastward in the vicinity of 40°N. As it nears the North American coast, the current divides with one part turning southward to form the California Current while the other flows northward into the Gulf of Alaska (Fig. 83). The Region may be divided into two provinces. The southernmost of the two is called the Oregon Province. It extends from Point Conception on the California Coast north to about the Dixon Entrance, just north of the Queen Charlotte Islands at about 54–55°N.

Although Peden and Wilson (1977) identified concentrations of range end points at 50° and 57°N for the fish, the Dixon Entrance boundary appears to be best for the biota as a whole. The southern boundary at Point Conception has become well recognized.

Horn and Allen (1978) found that many California fish ranged only as far north as about 42°. This prompted them to recognize a separate "Montereyan Fauna" between that point and Point Conception.

The Aleutian Province lies between the Oregon Province and the cold waters of the Arctic Region. The northern boundary has been located at about Nunivak Island in the eastern Bering Sea. In this area, the distribution of the molluscs has been studied more closely than that of most other groups. This boundary was identified by Valentine (1966) and, so far, has stood the test of time. The mean southern limit of the pack ice in January-February lies at about Nunivak Island.

There is a group of eurythermic, arctic-boreal species that have continuous ranges through the northern Bering Sea and extend far to the south on each side of the boreal Pacific. The presence of such species accounts for most of the similarity found between the Oregon and Kurile provinces. Examples among the fish are the starry flounder (*Platichthys stellatus*) and four Pacific salmon species belonging to the genus *Oncorhynchus*. In echinoderms, there are *Strongylocentrotus franciscanus*, *Amphiura arcystata*, and *Amphipholis pugetana*. Stellar's sea lion (*Eumetopias jubatus*) is another example.

The Western Atlantic Boreal Region extends from Cape Hatteras to the Strait of Belle Isle (the northern entrance to the Gulf of St. Lawrence) (Fig. 84). As the Florida Current passes Cape Hatteras on its northeasterly course, it becomes the Gulf Stream. Previously, it had been flowing near or over the continental shelf, but beyond the Cape, it extends over deep water. It is composed of a series of meandering filaments, some of which may flow as fast as 5 knots, an extreme speed for an ocean current. For most of its course, the Gulf Stream stays well out to sea, but it approaches shallow water once more over the Grand Banks, which extend to the southeast of Newfoundland.

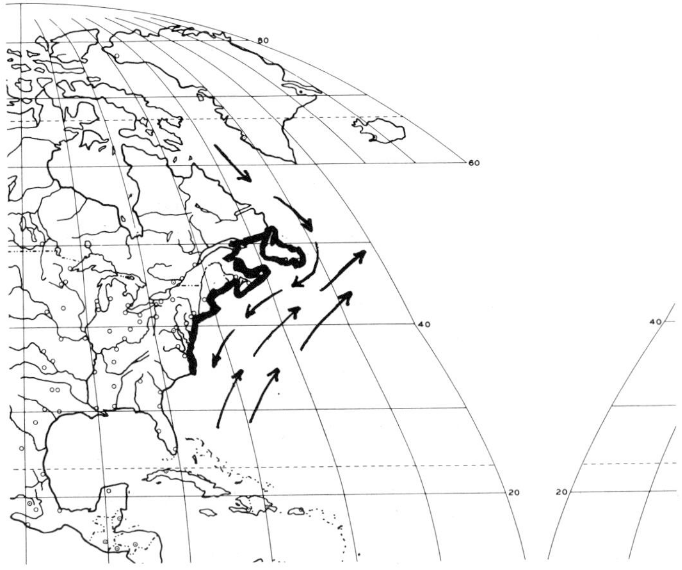

Fig. 84. Western Atlantic Boreal Region.

Biogeographically, the most important function of the Gulf Stream is that it forms a very effective barrier between a cold-water area to the northwest and the warm water of the Sargasso Sea (the western North Atlantic) on the southeast. Between the Gulf Stream and the shore of North America, there is a southwest-flowing, coastal current that tends to form an elongated, counterclockwise eddy. The coastal current is supplied with cold, low-salinity water from the Labrador Current, which flows south out of the Labrador Sea and around the tip of Newfoundland (Fig. 84). It is this coastal current that has in large part determined the nature of the fauna that occupies the shelf from the latter island south to Cape Hatteras.

In considering the geographic extent and distinctiveness of the cold-temperate, western North Atlantic, the biologist is confronted with a good deal of conflicting opinion. Most of the disagreement has been concerned with the relationship of the fauna that occupies the region between Cape Hatteras and Cape Cod, often called the Middle Atlantic Seaboard. The area is penetrated, during the summer months for the most part, by a huge number of tropical and warm-temperate organisms. This has often resulted in its being allied with the Carolina Region to the south of Cape Hatteras. However, the presence of large numbers of boreal species, together with very little endemism, shows that it clearly belongs to the Boreal Region.

Traditionally, Cape Cod was considered to mark the dividing line between the warm-temperate and cold-temperate biotas (Ekman, 1953; Hedgpeth, 1957). But since the latter is now known to be more or less continuous around the Cape, it can no longer be regarded as a major biogeographic boundary; yet, some of the reasons for considering it a boundary still hold. That is, a remarkable number of eurythermic tropical and warm-temperate organisms do occasionally migrate or are carried as far north as Cape Cod, and a good many wide-ranging arctic-boreal species manage to penetrate this far south.

The Eastern Atlantic Boreal Region extends from the western entrance to the English Channel north to the base of the Murmansk Peninsula in the vicinity of the Kola Fjord (Fig. 85). Beyond the Grand Bank of Newfoundland, the eastward continuation of the Gulf Stream becomes the North Atlantic Current. The latter forms the upper portion of the North Atlantic Gyre and its main flow is to the northeast towards the British Isles. In the middle of the North Atlantic, a major division takes place whereby one branch turns southward to remain within the gyre while the other extends northward to flow into the Norwegian Sea. Upon separation, the northern branch becomes the Norwegian Current.

By means of the Norwegian Current, relatively warm, saline water is carried into much higher latitudes than anywhere else in the world. A westerly branch affects the south and west shores of Iceland, another carries warm water into the Greenland Sea, and a third branch follows along the Norwegian coast to enter the Barents Sea.

Iceland possesses an interesting biotic mixture. There is a purely boreal component, pure arctic, arctic-boreal, and some eurythermic temperate forms. The boreal species, at least in the fish, appear to predominate. The biotic relationships are almost entirely with the eastern Atlantic. The only boreal species shared with North America are those that have an amphi-Atlantic distribution. Iceland may well have served as a way station in the westward migration of these amphi-Atlantic forms. The absence of any special American relationship and the almost complete absence of endemics (except for an occasional sub-

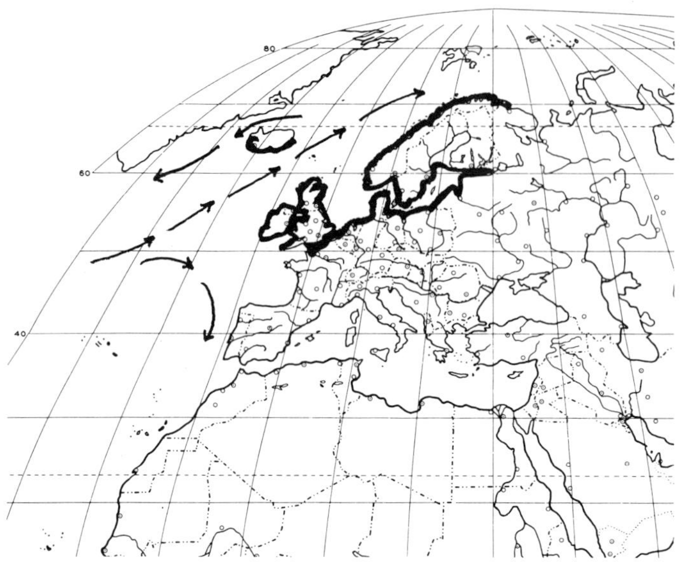

Fig. 85. Eastern Atlantic Boreal Region.

species) indicate that the island, or at least the south and east shores, should be included in the Eastern Atlantic Boreal Region.

The Faroes are a group of 21 volcanic islands located between Iceland and the Shetlands at about latitude 62°N and longitude 7°W. The primary source of our knowledge about the shore fauna and its relationships is the series entitled *The Zoology of the Faroes* which was published in Copenhagen (1928–1942). Stephensen (1937) contributed a series of articles on the various groups of Crustacea and, in general, found that the species were either boreal or arctic-boreal. Also, all were either Eastern Atlantic or transatlantic species, there being no endemism and no evidence of any special relationship to the western Atlantic.

The Baltic Sea is the largest estuarine area in the world. The salinity is relatively stable and decreases gradually towards the inner end of the long, narrow basin. At the east end of the Skagerrak the salinity is usually about 30‰, in the Kattegat it drops to 20‰, and in the Baltic proper it ranges from about 15‰ at the lower end down to about 2‰ at the extremities of the Gulfs of Bothnia and Finland. The distribution of species along this natural salinity gradient has interested many of the biologists from the countries that border the Baltic. The combined effects of low salinity and the extreme seasonal fluctuation in temperature make the environment a very rigorous one for the marine biota.

The relationship of the European boreal fauna to the American was apparently first investigated by the Swedish zoologist Lovén (1846). He found that, in the molluscs, about 75% of the arctic-boreal species were shared by the two sides of the Atlantic, but only about 8% of the purely boreal species were shared. Although more recent knowledge would alter these figures to some extent, the same general magnitude of difference continues to be evident.

THE COLD (POLAR) REGIONS

The Antarctic Region

Within the West Wind Drift area is a circumpolar zone in which the cold surface water tends to sink, causing the surface temperature to alter rapidly with a change in latitude. This is called the Antarctic Convergence. The Antarctic surface water, lying between the Convergence and the edge of the continent, is a cold, low-salinity layer that varies in thickness from 100 to 250 m. Much of it is covered by pack ice which forms a belt that expands and contracts with the seasons. The Antarctic continent and the South Shetland, South Orkney, and South Sandwich Islands all lie below the February 1°C isotherm and have winter temperatures as low as –2°C. The temperature regime at South Georgia is notably warmer, ranging from a winter mean of 1.3°C to a summer mean of 3.7°C. That of Bouvet Island is about –1° to +1°C.

The fauna of the Antarctic Region is relatively old compared to that of the Arctic Region. As noted (Chapter 6), ice sheets may have formed as early as 42 Ma ago. From that time, the temperature decline continued until it reached a maximum low at about 29 Ma ago. Drake Passage opened sufficiently to permit circulation of a deep current about 22–23 Ma ago. The isolation effect permitted the evolution of a highly distinctive biota with many endemic genera and species.

The South Polar Province (Fig. 81) includes the Antarctic continent itself plus the three small island groups already mentioned. The shore animals demonstrate an unusual ability to cope with variation in depth. A common nototheniid fish, *Trematomus bernacchii*, has been taken from the surface to over 600 m and many ophiuroid starfish range from very-shallow water down to 700 m. This bathymetric range is probably permitted by the sinking of cold water that takes place at the edge of the continent. Fairly uniform temperature and salinity conditions exist from the surface down to considerable depths.

The South Georgia Province (Fig. 81) contains only South Georgia Island. It lies about 400 km west of the South Sandwich Islands and 1100 km east of the Falkland Islands. It is surrounded by water more than 3000 m deep and lies within the mean limit of drift ice. Its very-high rate of endemism is probably due to its spatial isolation plus its warmer temperature compared to that of the Antarctic Province. Bouvet Island is situated about 1380 km from the Antarctic continent opposite Princess Astrid Land. It has been placed in a separate Bouvet Province (Briggs, 1974a), but its biota and external relationships are still poorly known.

The Arctic Region

As noted in Chapter 7, the present cold temperature regime of the Arctic probably began about 3.0–3.2 Ma ago. This means that the polar biota of the Arctic Region is much

Fig. 86. Arctic Region.

younger than that of the Antarctic. Consequently, although there are significant numbers of endemic species, there are very few endemics at the higher taxonomic levels. The Arctic seas have traditionally been subdivided into a number of separate zones and provinces; however, more recent works suggest an essentially homogeneous biota so that it is possible to recognize only one biogeographic region (Fig. 86). The polar cod, *Boreogadus saida*, may be considered an indicator species for it extends to all parts of the Region but no farther (Cohen et al., 1990).

Since the Arctic Region extends south into the northwestern Atlantic and also well into the Bering Sea, the general circulation scheme is quite complicated. A cold East Greenland Current, moving out of the Arctic Basin and through the Denmark Strait, is responsible for the consistently cold temperature of eastern Greenland and northeastern Iceland. When this current reaches the southern tip of Greenland, it turns westward into the Labrador Sea. There, it meets additional cold water from Baffin Bay to form the Labrador Current which flows to the southeast.

The main inflow of Atlantic water into the Arctic Ocean takes place just to the west of Spitsbergen. The direction of flow is towards the Laptev Sea. From there, the stream turns west to begin a broad, clockwise circulation in the Arctic Basin. In the Bering Sea, a northward current flows from about the middle of the Aleutian chain through the Bering Strait. Also cold currents have been observed to enter the Bering Sea from the north and to flow southward along the Asiatic side. The Arctic Region occupies all of the area north of the cold-temperate boundaries that have previously been identified.

THE PELAGIC REALM

As may be noted from the diagram (Fig. 54), the Pelagic Realm can be divided into four depth layers. It is apparent that the *epipelagic* zone is a distinct habitat. Although there is a good deal of migratory movement in and out of this layer, most of the zooplankton and the great majority of the phytoplankton species seem to be confined to it. Its lower boundary is most likely to occur at about 200 m, but it can vary from 100 to 500 m. This tends to be a mixed layer where the temperature is similar to that of the surface.

The *mesopelagic* or twilight zone extends down to about 1000 m. Here, there are many zooplankters and a variety of species that prey upon them. They live in a thermocline area where the temperature decreases rapidly with depth. The *bathypelagic* zone extends to 6000 m, and the decrease of temperature with depth is very slow. Sometimes an additional abyssopelagic zone is recognized for animals that swim just above the ocean floor. There is a poor but separate *hadopelagic* or trench fauna existing below 6000 m.

Epipelagic and mesopelagic zones

It is apparent from the high degree of correlation between the distributional patterns of the epipelagic and mesopelagic biota that, as far as horizontal dispersal is concerned, the two groups can be treated as one. This raises the question of why species that live in the 200–1000 m layer should show such a close correspondence to surface patterns. As one can see by examination of the temperature charts, the deeper layer is not only considerably colder, but the average latitudinal temperature changes are much less.

We know that the mesopelagic zone is populated by many species that undergo diurnal migrations. In the lower latitudes, such species are probably quiescent in the cold, mesopelagic waters but become very active upon migrating to the warm, upper layer. Compared to most marine animals, these species are unique in their ability to withstand rapid and extensive temperature changes. For this reason, they are usually considered to be highly eurythermic. However, it seems likely that, in terms of the temperature necessary for them to feed actively and efficiently, their requirements may be just as narrow as those of many forms that are confined to the surface.

The introduction of the T-S diagram (a temperature–salinity graph) by Helland-

Hansen in 1916 has had a significant influence on both descriptive and theoretical oceanographic research. Many biologists have used T-S diagrams, as water mass indicators, in order to explain the distribution of plankton. This research has been mainly confined to individual species but some of it has related water masses and the distribution of communities. As a result of this activity, much of which has been directed towards the goal of discovering indicator species for the various water masses, the older literature on the pelagic biota is strongly weighted towards the concept that each water mass has its own biological identity.

More recently (Briggs, 1974a; Haedrich and Judkins, 1979), it has been realized that the great majority of species in the better known planktonic groups are not confined to a single water mass. Many, in some groups most, of them extend through several water masses. As a general rule, longitudinal ranges, except for the neritic species, tend to be very broad while latitudes are very restricted. Such patterns emphasize the restraining influence of surface temperature. In the modern literature dealing with planktonic biogeography, one can still find an interesting dichotomy. Some authors still plot their distribution data in reference to water mass location, but others pay more attention to latitudinal changes based on temperature.

It has become apparent that the distribution patterns of plankton may be subdivided in two types, the neritic and oceanic. The neritic patterns have been outlined by Tokioka (1979) using chaetognaths as examples. He called attention to the fact that some neritic species are generally confined to an "inlet zone" of very-shallow, sometimes-estuarine,

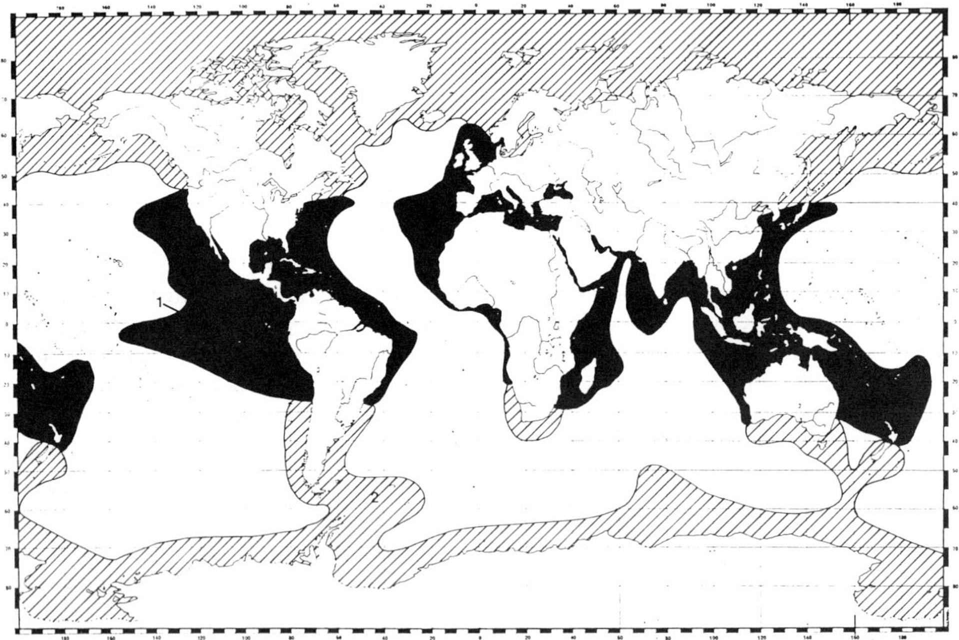

Fig. 87. The world distribution of distant-neritic plankton groups. After van der Spoel (1983). Warm-water communities in black, cold-water hatched.

waters. A second group called "open neritic" extend farther from shore. The latter have been termed "distant-neritic" by van der Spoel (1983) and he further subdivided that category into warm-water and cold-water types. He then provided a world map of general distant-neritic distributions (Fig. 87).

It is the oceanic species of plankton that have been difficult to analyze in a biogeographic sense. As Fleminger and Hulsemann (1973) pointed out, there are warm-water species that breed regularly up to mid-latitudes and are circumglobal in distribution; and there are warm-water species that breed regularly only at low latitudes and are provincial. The first group is able to maintain a gene flow around southern Africa while the second group does not. The first group generally ranges from about 40°N to 40°S and the second from 30°N to 30°S (Pierrot-Bults and Nair, 1991).

As noted, there has been a dichotomy in the presentation of distribution data on oceanic plankton. One group of research people emphasizes the effects of water masses and the other pays more attention to water temperature. Van der Spoel (1983) provided a solution to this problem when he recognized the existence of central-water patterns and belt-shaped patterns. There are five large central water masses, each outlined by a major subtropical gyre (two each for the Atlantic and Pacific oceans and one in the Indian Ocean). There are a few species that appear to be confined to such areas. On the other hand, there are large numbers of species that demonstrate belt-shaped patterns that either extend entirely across oceans or are circumglobal. Since the latter patterns are predominant in the open ocean, they need to be expressed when providing global distribution maps.

The planktonic biota of the Pacific Ocean is considerably richer than that of the Atlantic. When a world map of epipelagic-mesopelagic distributions was first attempted (Briggs, 1974a), the North Pacific was divided into four latitudinal belts, North Boreal, South Boreal, Warm-Temperate, and Tropical. But only two belts were identified in the North Atlantic, Temperate and Tropical. Since that time, additional work has identified belts in the North Atlantic that are equivalent to those in the Pacific. As these oceanic belts approach the shore lines, it may be noted that they tie in very closely to shelf regions of equivalent temperature (Figs. 88 and 89).

The scheme used here bears a general resemblance to that published by Rass (1986) except that the terminology for the belts is somewhat different. Both North Boreal and South Boreal are used but here the Russian Tropical = Warm-Temperate, Equatorial = Tropical, and Northern Notal and Southern Notal = Cold-Temperate. The original terminology is maintained so that a given temperature range, whether shelf or pelagic, can be referred to by the same name. Two cold-temperate belts in the southern hemisphere, as Rass has done, are not recognized here. Groups such as the fish and the chaetognaths (Pierrot-Bults and Nair, 1991) appear to be best represented by one circumglobal zone.

It has been observed that the process of parapatric speciation seems to be appropriate for oceanic populations where only a partial separation of populations can occur (Pierrot-Bults and Van der Spoel, 1979). The kinds of geographic variation shown by the five different forms of the pteropod *Limacina helicina* may demonstrate the early stages of the parapatric process. Many warm-temperate or cold-temperate populations demonstrate an antitropical (bipolar) distribution. Although migration across the tropics during glacial

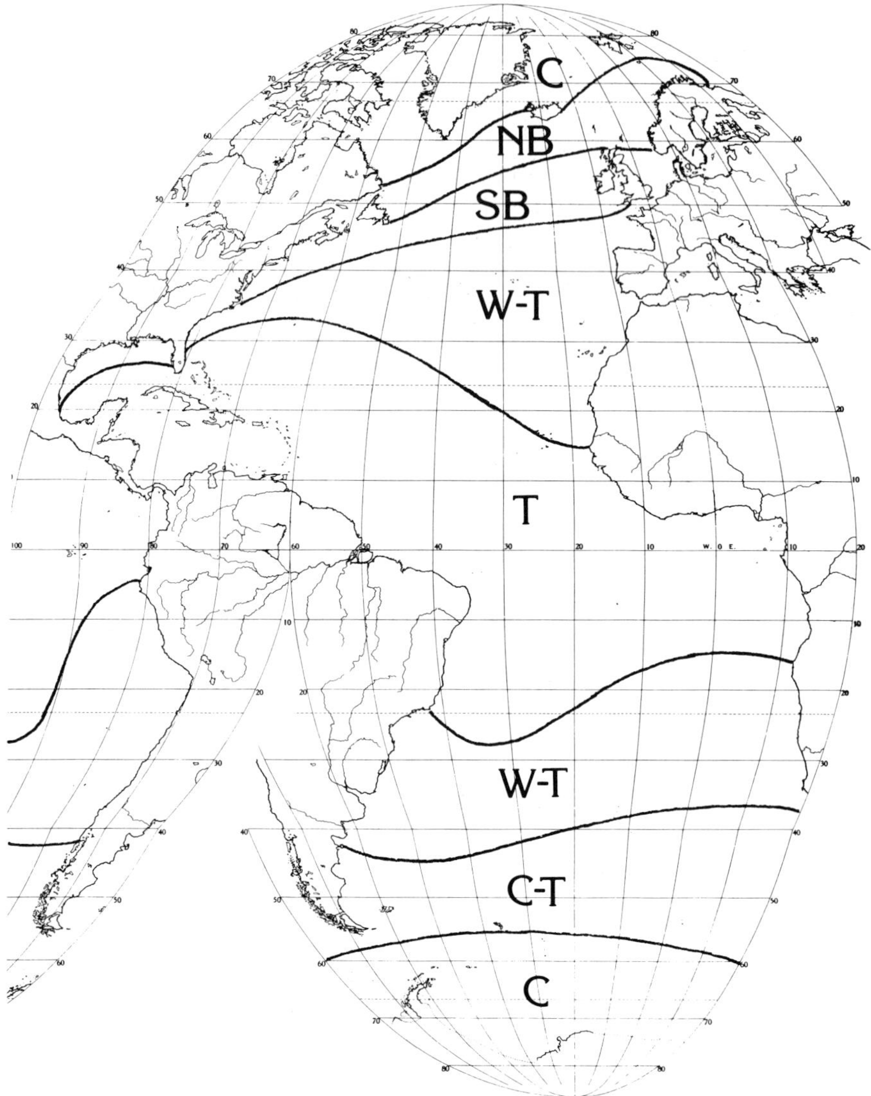

Fig. 88. Patterns of oceanic plankton groups in the Atlantic Ocean. These are Arctic or Antarctic (C), North Boreal (NB), South Boreal (SB), Cold-temperate (C-T), Warm-temperate (W-T) and Tropical (T).

stages is often mentioned as the probable cause, it is more likely that such patterns are the result of tropical extinctions (p. 221). Of course, several species that were initially thought to be antitropical were later found to have continuous populations through the deep waters of the tropics (isothermic submersion).

A hypothesis for the historical development of the major planktonic faunas was published by van der Spoel and Heyman (1983). They envisioned the initial presence of general warm-water taxa living between 40°N and 40°S. From these, more narrow 30°N to

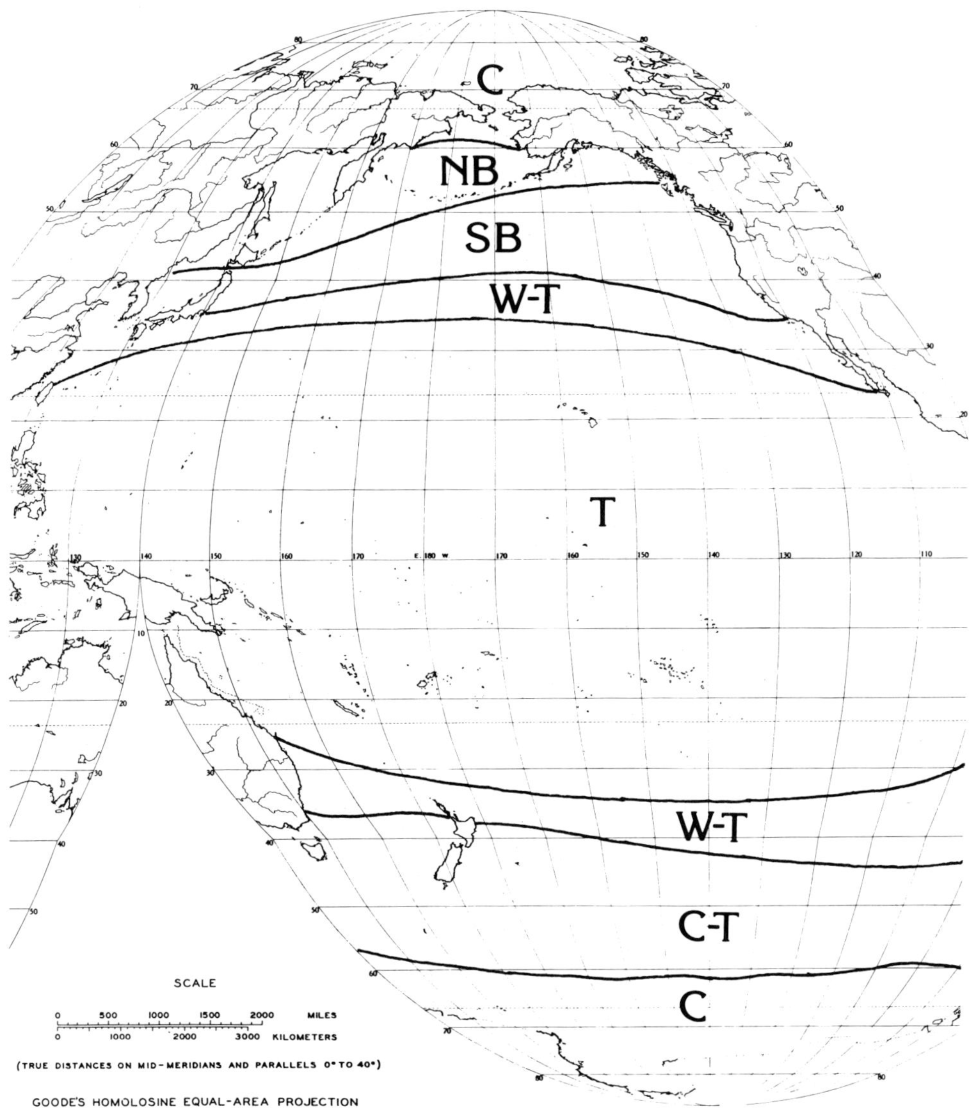

Fig. 89. Patterns of oceanic plankton groups in the Pacific Ocean. These are Arctic or Antarctic (C), North Boreal (NB), South Boreal (SB), Cold-temperate (C-T), Warm-temperate (W-T) and Tropical (T).

30°S taxa developed with the onset of cooler temperatures during the Eocene-Oligocene. The next stage was development of a temperate fauna which split into northern and southern hemisphere components. The cold-water fauna then gave rise to the bathypelagic species. Their scheme was illustrated by a geocladogram (Fig. 90). This sequence of events is apparently quite applicable to the Chaetognatha (Pierrot-Bults and Nair, 1991). The chaetognaths were assumed to have been oceanic in origin but the presence

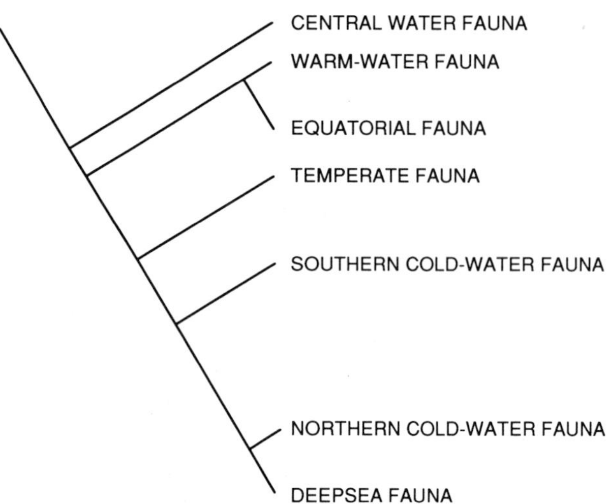

Fig. 90. Geocladogram depicting historic development of the major planktonic faunas. After van der Spoel and Heyman (1983).

of a large number of species in the shallow waters of the East Indies (Tokioka, 1979) may suggest an origin in that area.

Arctic and Antarctic

As noted in Chapter 7, formation of the present cold Arctic Region began with a sharp decline in temperature about 3.0–3.2 Ma ago. Most of the present endemic Arctic species probably began their evolution at that time. The greater part of the Arctic Ocean has an almost continuous cover of floating ice, occupying in summer about 60–80% of the surface area. Beneath the ice there is a poor epipelagic biota, but there are local situations where plankton may become very abundant. High concentrations, with phytoplankton species predominating, have been found in estuaries of tributary rivers and along the fringe of the pack ice. Considerable endemism has been found among the calanoid copepods and the amphipods. The Arctic pelagic waters include the Arctic Basin and extend south to encompass northern Iceland, southern Greenland, and southern Labrador (Fig. 89).

The Antarctic Region lies between the continental shelf and the Antarctic Convergence. Here, the surface temperature is between −1.9°C in winter and +4°C in summer. Among the conspicuous endemics are two kinds of krill (*Euphausia*), two hydromedusae, three polychaetes, two copepods, and a chaetognath. Also, there are about 12 species of lantern fish belonging to the family Myctophidae, as well as species belonging to several other fish families. Kennett (1982) found that the beginning of the Neogene, about 22 Ma ago, heralded the development of a distinctive siliceous microfossil assemblage. He pointed out that this transition marked a major change in the world's planktonic biogeography whereby a permanent high-diversity gradient was established between tropical and polar regions.

Bathypelagic zone

We know that the sea occupies about 71% of the earth's surface, but it is not generally realized that most of this water layer is relatively deep. In fact, about 88% of the world's oceanic area is more than 1000 m in depth. This means that bathypelagic animals are spread over about two-thirds of the earth's surface. Considering that the average depth of the ocean is about 3800 m and that the abyssal plain extends down to about 6000 m, it can be seen that the Bathypelagic Zone occupies an enormous volume of liquid space. If we consider that portion of our planet into which life has penetrated and call it the biosphere, then the bathypelagic habitat occupies the greater part of it.

How successfully has this major portion of our biosphere been occupied? We need to keep in mind that we are dealing with an area in which no food is manufactured by photosynthesis (since it is beyond the reach of sunlight) where the temperature is very cold (1–5°C), and where the pressure is high (100 atmospheres or more). Furthermore, the animals must suspend themselves in the water rather than rest on a firm substrate. Also, in general, bathypelagic animals are situated several links down a food chain that has its beginning in the sunlit, epipelagic zone at least 800 m above. Considering these factors, the variety of life is impressive.

Marshall (1954), in his classic work on deep sea biology, has related that when a net is towed horizontally at bathypelagic depths, the crustaceans are usually the most numerous in the catch with the copepods generally far outnumbering the other groups. Results from six hauls made off Bermuda between depths of 730 and 1650 m showed that 70% of the species were copepods, 15.5% were other crustaceans, 3.1% chaetognaths, 1.9% siphonophores, 1.7% fish, 1.2% radiolarian protozoans, and the remaining 0.6% comprising annelid worms, molluscs, tunicates, echinoderms, jellyfish, and ctenophores. Such information indicated that we are evaluating an area that could be entitled the "domain of the Crustacea", for that class has been by far the most successful in penetrating this vast but difficult environment. Since about three-fourths of all bathypelagic species are copepods, it is clearly this group that offers the greatest potential for biogeographic studies.

The major features of distribution in the bathypelagic zone have been reviewed by Marshall (1980). Although most species have exceedingly broad distributions and occur in all three oceans, they tend to be concentrated under the more productive regions. The most numerous bathypelagic fish, in numbers of individuals, belong to the genus *Cyclothone*. Six species are known; two of them are circumglobal between 40°N and 40°S, one has a broader distribution in the Atlantic (subarctic to subantarctic) and a narrow one (20°S to Antarctic waters) in the Indian and Pacific Oceans, and another is circumtropical (mainly 20°N to 20°S). The remaining two species are very restricted, one is found only off the West African coast and the other is endemic to the Mediterranean.

The dominant bathypelagic fish in number of species (about 230) are the barbeled dragonfish (family Stomatiidae). The second most speciose group (about 150) is the ceratioid anglerfish. Female anglerfish have been described as floating, baited traps.

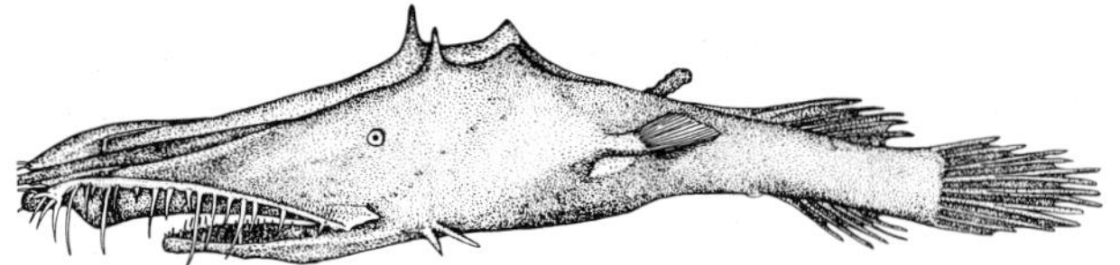

Fig. 91. *Lasiognathus saccostoma*. The ultimate angler fish. After Bertelsen (1951).

Most such species employ lighted, moveable lures that are marvels of evolutionary design. Some have "fishing poles" (illicia) that can be extended or contracted, and one, *Lasiognathus saccostoma* (Fig. 91), is completely outfitted with pole, line, "float", lure, and hooks! These fish tend to have very broad horizontal distributions but are captured most often beneath current margins, upwelling zones, and other areas of high productivity. This same tendency is found among the various invertebrate species. The predominance of very broad distribution patterns in all three oceans indicates that one huge Bathypelagic Region will suffice for all of the world except the Arctic Basin. The deep water of the latter appears to harbor many endemic zooplankton species.

Hadopelagic zone

So far, the pelagic fauna of the trenches, which occurs below 6000 m, is very poorly known. One of the Russian research vessels has operated closing nets in the Kurile-Kamchatka Trench and caught a variety of copepods, amphipods, and ostracods. All of these animals were reported to be completely unpigmented compared to their bathypelagic relatives where a dark red color predominates. It appears that the rate of endemism may be very high (Zenkevitch and Birstein, 1956).

DEEP BENTHIC REALM

Vertical distribution

Until 1962, our knowledge about the general pattern of vertical distribution in the deep-sea benthic fauna was extremely fragmentary. Ekman (1953) recognized only an "archibenthal" fauna extending from the outer edge of the shelf to about 1000 m and an "abyssal" fauna for the greater depths. Bruun (1956) termed the fauna from the shelf edge to 2000 m "bathyal," that from 2000 to 6000 m abyssal, and coined the term "hadal" for the fauna of the trenches (below 6000 m). In his chapter on the classification of marine environments, Hedgpeth (1957) depicted a bathyal fauna extending from the shelf to about 4000 m, an abyssal assemblage from about 4000 to 6000 m, and a hadal fauna for the trenches.

Vinogradova (1962) wrote a review article which analyzed data on the vertical zonation of the benthic animals that were taken mainly by the Russian vessel *Vitiaz*, the Danish *Galathea*, and the Swedish *Albatross*. The analysis included a total of 1144 species belonging to most of the major marine invertebrate groups. She found that the changes in the systematic composition of the bottom fauna were most pronounced at a depth of about 3000 m. Here, a large number of species, genera, and even families characteristic of the slope disappear and are replaced by groups peculiar to the greater depths. For this reason, she identified this depth as the upper limit of the abyssal zone.

Another, less abrupt, change in the fauna led Vinogradova (op. cit.) to subdivide the abyssal zone into two subzones, an "upper-abyssal" extending from 3000 m to about 4500 m and a "lower-abyssal" from the latter depth to about 6000 m. For the abyssal zone in its entirety, a preliminary estimate of the endemism indicated that 58.5% of the species were confined to that area. For depths greater than 6000 m, an "ultra-abyssal" (hadal) zone was recognized. Vinogradova stated further that these faunal changes were found to occur at similar depths in the Atlantic, Pacific, and Indian Oceans. More recently, our knowledge about vertical distribution patterns has been considerably augmented, so that some changes appear to be desirable.

At one time (Briggs, 1974a), there seemed to be a good correlation between the position of the permanent thermocline zone of 200–1000 m and the occurrence of a distinctive Upper Slope Fauna. Similarly, there seemed to be good evidence for the recognition of a Lower Slope Fauna from about 1000 to 3000 m. During the past 20 years, there has been a great increase in research on the benthic fauna of the deep sea and the depth patterns have proved to be much more complicated. Extensive trawling and photographic transects have taken place, especially along the continental margins of the United States. These show that the pattern of faunal change varies considerably from area to area and between canyons and normal slope conditions. In general, macroinvertebrate distributions tend to show major changes in species composition between 400 and 600 m, about 1000 m, and between 1400 and 1600 m. There is also a suggestion of an additional discontinuity at about 2000 m (Gage and Tyler, 1991).

On the other hand, it has been observed that specialists on the megafaunal groups have set the transition between the slope (bathyal) fauna and the abyssal fauna at about 2000 m (Marshall, 1980). Marshall also called attention to a diverse slope fish fauna that is found between 200 and 2000 m. The genus *Coelorhynchus* (Macrouridae), represented by more than 50 species, is a generic indicator of the slope fauna. Another view has been expressed by Haedrich and Merrett (1990), who had examined demersal fish data from nine deep-sea surveys in the North Atlantic. They concluded that they could find little evidence for faunal zonation or communities in their material.

In addition to typically benthic forms, there are fish species (called pseudoceanic or mesobenthopelagic) that are attracted to, and sometimes thrive in large numbers along the slopes (Merrett, 1986; Parin, 1986). Many of them are representatives of families that otherwise are predominately pelagic. They tend to be numerous near islands and ridges with steep slopes at depths ranging from about 200 to 2000 m. In general, it may be said that the foregoing information does not lend itself to the designation of sharp vertical boundaries, yet it can be seen that individual animal groups do undergo significant changes in species that appear to be correlated with depth.

That there are fundamental differences between the shelf, slope, and abyssal habitats seems to be clear. Sometimes a continental rise fauna, from about 2250 to 4000 m, is recognized but there seems to be little biogeographical justification for doing so. It appears that the most useful decision that can be made at this time is to define a single, Slope Zone that extends from about 200 to 2000 m. Beyond that, from 2000 to 6000 m, lies the Abyssal Zone which is much more thinly populated and contains many endemic species. The Hadal Zone of the trenches is generally recognized as extending from about 6000 to almost 11 000 m (the depth of the deepest trenches). The distinctive fauna of this zone has been reviewed by Belyaev (1989).

Horizontal distributions

In an earlier work (Briggs, 1974a), a number of characteristics of the upper slope were enumerated. With the benefit of our increase in knowledge during the past 20 years, these characteristics can now be seen as applicable to the Slope Zone in general. Although we can no longer recognize a clear demarcation between the upper and lower slope, it must be noted that many animals prefer relatively narrow depth ranges and that the upper region tends to have the greatest species diversity. In the earlier work, it was noted that, in some parts of the world, the horizontal distributions on the slope closely followed that of the shelf and that it was possible to recognize distinct tropical, warm-temperate, cold-temperate, and polar assemblages.

More recent information reinforces the idea that slope species tend to follow surface temperature zones. Perhaps the best demonstration of this is the comprehensive work on the gadiform fish of the world (Cohen et al., 1990). The largest family within that order is the Macrouridae (grenadiers) with more than 300 species. More than three-fourths of them are apparently confined to the slope habitat. In general, their horizontal distributions are highly restricted, with most of them being found beneath tropical or warm-temperate waters. Within such zones, the latitudinal ranges are limited and there are very few trans-oceanic distributions.

Endemic species of macrourids may be found on the slopes beneath many of the world's tropical, warm-temperate, and cold-temperate provinces. For example, *Coelorinchus argus* is confined to tropical East Indies, *C. kishinouyei* is found only off warm-temperate Japan, and *C. aspercephalus* only around cold-temperate New Zealand. There are, in addition, several eurythermic tropicals that extend from the tropics into warm-temperate areas. There are also some eurythermic temperate species that occupy both warm-temperate and cold-temperate zones. *Macrourus berglax* is an arctic-boreal species in the North Atlantic, *M. whitsoni* is Antarctic-subantarctic, and the other two species in the genus are cold-temperate in the southern hemisphere. These same patterns are common among the shelf species, so they are not peculiar to the slope.

Like the macrourids, many species of sharks belonging to the family Scyliorhinidae are obligate slope dwellers and demonstrate horizontal ranges that are similarly restricted (Compagno, 1984). Zenkevitch (1963) presented a summary of the Russian work on the slopes of the Arctic Basin. This fauna was found to be about four times as rich as that of the abyssal zone. More than 50% of the species were considered to be endemic to the Arctic Ocean. For the Antarctic, Vinogradova (1964) found that about 70% of the invertebrate species, occurring between 200 and 2000 m, were shared with other areas. This implies about a 30% endemism which is less than that found in the fish.

In contrast to the slope, the Abyssal Zone has many fewer species and a small fraction have exceedingly broad distributions. For example, the macrourid fish *Coryphaenoides armatus*, *Squalogadus modificatus*, and *Macrouroides inflaticeps* are widespread in the tropical and temperate waters of all three oceans (Cohen et al., 1990). Few abyssal invertebrates have such broad distributions. An analysis of the distribution of more than 2000 species of the deep-sea macrobenthos was conducted by Vinogradova (1979). This still remains the most comprehensive treatment of the distribution of the abyssal and hadal faunas of the world. She found that deep sea-animals may be separated in two types based on their vertical distribution. Eurybathic species, which extend upward on the slopes, tend to have the most widespread patterns. It is among this group, that the few circumglobal or cosmopolitan species are found. The stenobathic abyssal species that do not extend to the slopes, have generally restricted patterns.

The numbers of eurybathic versus stenobathic species were found to vary among the invertebrate groups (Vinogradova, 1979). Eurybathic distributions predominated among colenterates, urchins, ophiuroids, and decapods while most isopods, sponges, pogonophorans, holothurians and some asteroids were mainly stenobathic. On the whole, in the Pacific Ocean, eurybathic forms comprised about 60% of the species found at depths below 2500–3000 m. The isolation of the faunas in different parts of the ocean was found to increase with depth. For example, species found between 3000 and 4500 m consisted of 70.3 endemics; but species between 4500 and 6000 m had an endemic rate of 90.2%.

Biogeographic calculations for the World Ocean have shown that about 85% of the deep-sea fauna occurs in one ocean only, while only 4% are common to all three oceans. A map has been published by Vinogradova (1979) which divided the abyssal zone into three regions which, in turn, contain various subregions and provinces (Fig. 92). The map indicates latitudinal differences in species composition from north of 35°N, 35°N to 35–40°S, and south of 40°S. It may be noted that these changes correspond to major changes in the plankton of the upper layers. At the middle latitudes, it can be seen that eastern and western provinces are recognized with the northern Indian Ocean being separated from the western Pacific. Within the context of these larger divisions, we need to keep in mind that work on such groups as the isopods and molluscs suggests that each of the ocean basins may have a significant component of endemic species. This could mean that eventually all 47 of the basins will need to be considered as separate provinces (Briggs, 1974a).

Although the great bulk of the deep-sea fauna consists of widely scattered individuals dependent on food chains that begin in the surface layers, there are some interesting exceptions. Since the late 1970s, increasing numbers of hydrothermal vents and cold seeps have been discovered. They form small, highly productive islands where enormous numbers of chemoautotrophic bacteria are produced. The bacteria are preyed upon by, or have formed symbiotic relationships with, a variety of animals. Many hydrothermal vents are associated with the mid-ocean ridges. Fractures are formed in the ridge crest basalt as it cools, providing conduits for the circulation of cold bottom water into the underlying crust. The seawater reacts chemically with the hot basalt and emerges as hot springs. Sulfur-containing inorganic compounds serve as an energy source for the bacteria (Gage and Tyler, 1991).

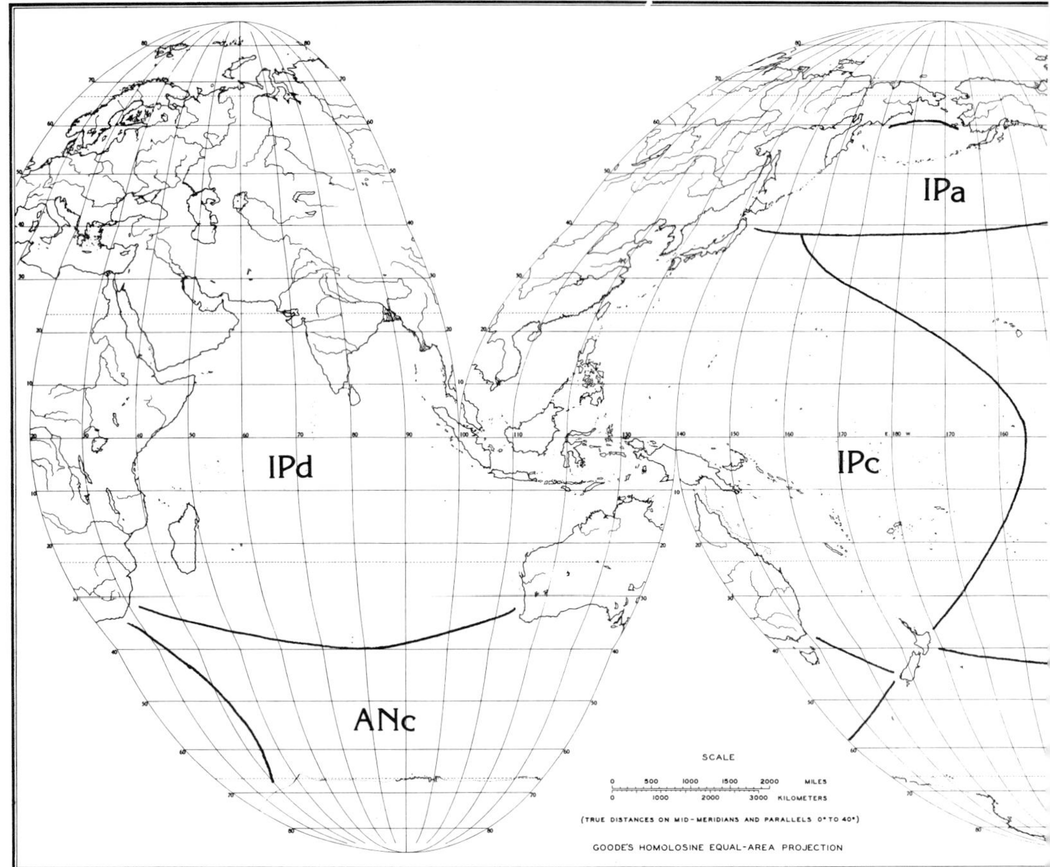

Fig. 92. General relationship of the abyssal fauna. The three major regions are the Indo-Pacific (IP), Atlantic (AT) and Antarctic (AN). Subregions are the North Pacific (IPa), East Pacific (IPb), West Pacific (IPc), Indian (IPd), Arctic (ATa), North Atlantic (ATb), East Atlantic (ATc), West Atlantic (ATd), Antarctic-Atlantic (ANa), Antarctic-Pacific (ANb) and Antarctic-Indian (ANc). Modified after Vinogradova (1979).

Various kinds of cool-water seeps have been discovered that also support colonies of autotrophic bacteria and higher animals. These may originate from groundwater, hydrocarbon deposits, or subduction zones. Vents or seeps that are geographically isolated from the main ocean ridges often support communities that are much different and are apt to contain endemic species. Many of the vent species are considered to be phylogenetic relicts of shallow-water groups. The nearest relatives of some of the limpets and a barnacle are known from the late Paleozoic and Mesozoic (Grassle, 1986).

The benthic fauna of the Hadal Zone is still poorly known. The trenches, which range in depth from a little over 6000 m to almost 11 000 m, occupy only about 1.5% of the benthic habitat. However, they are of considerable biogeographic interest. Most of them have high rates of endemism ranging from 50 to 75% (Belyaev, 1989). Colonization of

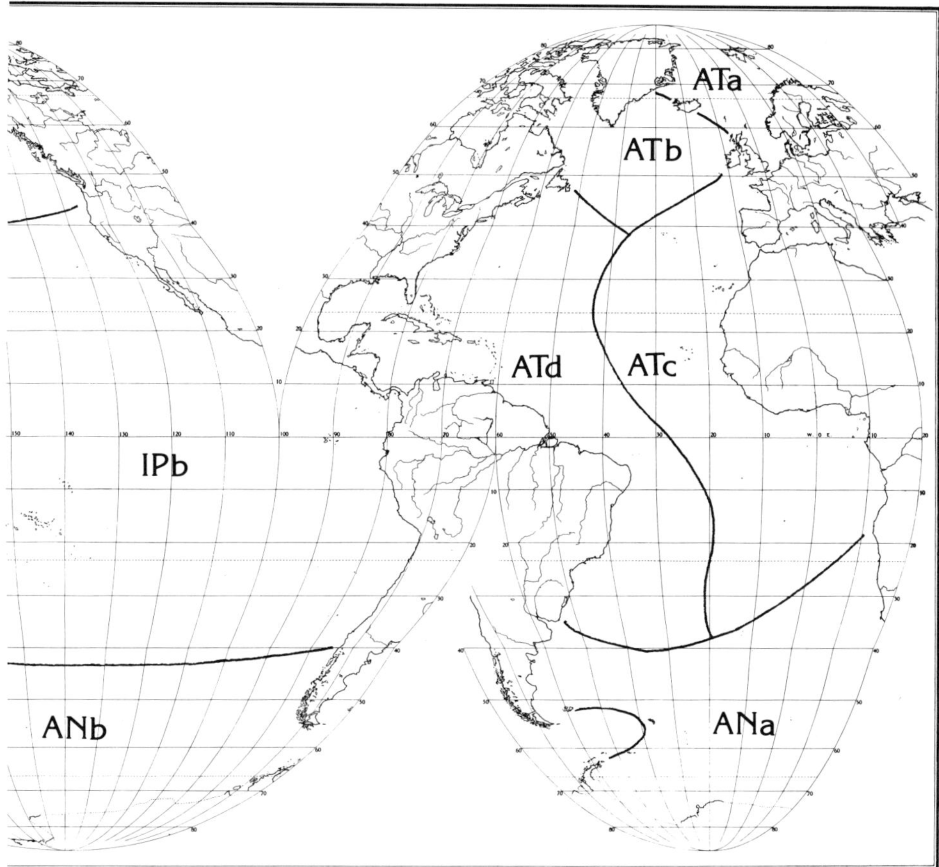

Fig. 92 continued. Legend on opposite page.

the trenches has generally taken place from the surrounding abyssal fauna. Some of the trenches exhibit connections to the abyssal and slope fauna of the Antarctic. The holothurian genus *Elpidia* occurs in most trenches, sometimes reaching as deep as 10 000 m. Vinogradova (1979) provided an hypothesis for its historical dispersal. This suggested that the genus originated in the Antarctic and dispersed northward with the spread of Antarctic bottom water.

Most of the collecting from the trenches has been accomplished by Russian research vessels. This material has been analyzed by Belyaev (1966, 1989). He found that hadal animals were surprisingly diverse with about 140 families, 200 genera, and 700 species. Within the boundaries of the zone, the fauna becomes impoverished with depth so that three subzones may be recognized: 6000–7000 m, 7000–8500 m, and below 8500 m.

Holothurians are dominant in numbers and biomass followed by bivalve molluscs and polychaetes. Eventually, each of the deeper trenches will need to be recognized as a separate province.

Diversity and origin

The questions of the relative species diversity in the deep sea below 1000 m, and whether or not there is a gradual decline with depth, are the basis for considerable disagreement. The pioneering studies of Sanders (1968) and Sanders and Hessler (1969) called attention to the fact that the species diversity of the small metazoans inhabiting the soft bottom of the continental slope in the northwestern Atlantic was much greater than previously realized. Despite the large numbers of species discovered, the diversity did not prove to be as great as that of the benthic habitat in tropical shallow waters. More recently, Grassle (1991) and Grassle and Maciolek (1992) extrapolated from 21 m^2 of box core samples from the northwestern Atlantic slope to reach an estimate of some 10 million species for the deep sea below 1000 m. This estimate was disputed by Briggs (1994) who presented data indicating a total of less than 200 000 species for the entire marine habitat (see p. 374).

The sampling transects that have been done in the northwestern Atlantic show conclusively that species diversity on the slope is much greater than on the continental shelf (Gage and Tyler, 1991). In that area, the diversity appears to gradually rise from the shelf edge down to about mid-slope depths and declines thereafter. This pattern is evidently not applicable to other parts of the world. In some places, macrofaunal diversity on the abyssal plain seems to be greater than on the slope. In the tropics, where the benthic fauna of the shelf is not exposed to the seasonal changes found in the boreal regions, the greatest diversity seems to occur in shallow waters. One may predict that, in the tropics, there will be a gradual decrease in diversity from the shelf down to the greatest depths.

The origin of the deep-sea fauna and the question of its relative antiquity are still being debated. It is clear that many ancient phylogenetic relicts have accumulated on the lower slopes and the upper part of the abyssal zone. These include many primitive molluscs, crinoids, starfish, decapods, isopods, and pogonophorans (Briggs, 1974a). Representatives of some of these groups once inhabited the shelf in Mesozoic and Paleozoic times. These modest depths seem to have offered a good refuge, since competition from advanced species is less, and the temperature has probably remained more stable than in the deeper waters.

For the deepest waters of the abyssal plain, phylogenetic relicts are fewer and the fossil record is generally poor. The foraminiferans from deep-sea cores indicate that taxa living there today mostly originated during the post-early Miocene period, so that almost all are less than 20 Ma old. Ostracods show a similar history (Lipps and Hickman, 1982). This accords with the general view that the cooling of the deep-ocean water mass, resulting in the formation of the psychrosphere, caused massive faunal extinctions. This means that a much colder, depauperate abyssal zone had to be repopulated.

Where did the modern deep-abyssal fauna come from? Two opposing ideas have been put forward: (1) evolution within the deep sea itself has been sufficient, and (2) migration from shallow water. The first theory comes mainly from systematic work on isopods where one suborder has its greatest representation in the deep sea. For this group, a theory of deep-sea evolution and high-latitude emergence has been proposed (Hessler and Wilson, 1983). It is based on greater deep-sea diversity, presence in deep water of the most primitive genera and species, and absence of eyes in both the deep and shallow forms. The latter is considered important because the blind forms must have lost their eyes in the deep sea before emerging at high latitudes.

Another theory of isopod biogeography says that the most primitive forms are principally confined to the tropical shelves while the cold water and the deep seas have the more advanced forms (Kussakin, 1973). This structure led to the hypothesis of the submergence of the deep-sea fauna from the Antarctic. It is this theory that has generally proven to be the most popular (Gage and Tyler, 1991). Due to similar temperature and salinity characteristics between Antarctic and deep-sea waters, a migratory pathway for eurybathic animals has probably existed for the past 20 Ma or more. On the other hand, Clarke (1962) in his monograph on the deep-sea molluscs, concluded that this group was derived by invasions from shallow waters all over the world. In their review, Lipps and Hickman (1982) decided that the Antarctic and deep-sea faunas have evolved primarily in place.

The age and distribution data on ostracods indicate the importance of the southwest Pacific, including the East Indies, as a locus of introduction into slope and abyssal environments during the Miocene to Quaternary interval (Whatley, 1987). What does one conclude in general about the origin of the deep-sea fauna? Certainly, the species that are there now, and probably most of the genera, have evolved in situ. Many of the families and higher taxa have probably been moving down the slopes from the shelf since Mesozoic or even Paleozoic times. The fauna of the deepest part of the abyss and the trenches is probably not very old since most of the preceding species would have been wiped out during the formation of the psychrosphere about 20–30 Ma ago. The replacements seem to have come from various places depending on the group concerned.

The marine system

The foregoing arrangement of regions and provinces is an attempt to recognize contemporary distribution patterns. We need to keep in mind that these are not fixed entities. Marine biogeography tells us that we are dealing with a dynamic system in which relationships are constantly changing. Geological and climatic changes may have immediate effects on distribution patterns, but they also have long-term effects that are expressed by evolutionary change. Even during times when physical conditions are relatively stable, evolution continues to be driven by the need to accommodate to the complexities of the biological environment.

A few of the changes presently underway include: (1) dispersal of tropical shelf species eastward from the Indo-West Pacific to the New World, around the Cape of Good Hope to the Atlantic, and through the Suez Canal to the Mediterranean; (2) migration of species from the western to the eastern Atlantic; (3) invasion of the southern hemisphere by cold-temperate species from the North Pacific; (4) the long-term movement of tropical forms towards higher latitudes; (5) the origin of new species in tropical onshore environments followed by their gradual horizontal and vertical dispersal; and (6) accommodation to the pelagic environment by species that were once benthic.

When all the ongoing changes are considered, together with reflection on the historic changes that were outlined in the first eight chapters, we can begin to appreciate the dynamics of the living system in the sea. In the distant past, there were some extinction episodes that were especially severe in the shallow waters of the tropics. Despite these setbacks, the system, after a few million years, recovered. The human caused extinction now underway is impacting the marine environment. Here, the effects of species extinction are magnified because the marine environment contains only about 2% of the world's species diversity.

SUMMARY

1. The warm-temperate regions border the tropics and have evident tropical relationships. In the southern hemisphere, especially around Australia and New Zealand, there are relict families and genera that were once widespread in the tropics.
2. The Southern Australia Region may be divided into Southwestern Australian and Southeastern Australian Provinces. The Northern New Zealand Region is divisible into Auckland and Kermadec Provinces.
3. An Eastern South American Region is now considered to extend southward to Valdés Peninsula in Argentina. The Western South America Region includes the Peru-Chilean and Juan Fernandez Provinces.
4. The Southern Africa Region is comprised of the Southwestern Africa and Agulhas Provinces. Tristan-Gough Islands in the South Atlantic, St. Paul-Amsterdam in the southern Indian Ocean, and some intervening sea mounts, are combined into a West Wind Drift Islands Province.
5. Among the northern hemisphere warm-temperate regions, the Mediterranean-Atlantic Region is by far the largest. It is subdivided into the Lusitania, Black Sea, Caspian, and Aral Provinces.
6. Although the Carolina Region exists in two parts that are separated by the tip of the Florida Peninsula, the biota is homogeneous to the extent that it is not divisible into provinces.
7. The California Region contains the San Diego and Cortez Provinces. They are separated by the tip of the Baja California Peninsula.
8. The Japan Region includes sections along the coast of mainland China, the west coast of Taiwan, the southern Sea of Japan, and the Pacific coast of Japan.
9. The cold-temperate regions have biotas that are markedly different than those of the warm-temperate regions. Many cold-temperate groups appear to have originated in the North Pacific or in the Antarctic.
10. Some old tropical families are prominent in the cold-temperate waters of Australia and New Zealand but are scarce elsewhere. It can be seen that the West Wind Drift has had a strong influence on southern hemisphere patterns in the cold-temperate zone.

11. The Tasmania Region includes all of Tasmania as well as the coast of Victoria in southern Australia. The Southern New Zealand Region includes the Cookian and Antipodean Provinces.
12. The Sub-Antarctic Region consists entirely of oceanic islands. The two provinces are the Kerguelen and the Macquarie.
13. In the northern hemisphere, it is the North Pacific that has the most diverse (richest) biota. The Western North Pacific Region is divided into three provinces, the Oriental, Okhotsk, and Kurile.
14. The Eastern North Pacific Region includes the Oregon and Aleutian Provinces. The Western Atlantic Region and the Eastern Atlantic Region are each single entities, not divisible into provinces.
15. The cold biota of the Antarctic Region is separable into South Polar, South Georgia, and Bouvet Provinces. The Arctic Region possesses a widespread biota that does not lend itself to provincial separation.
16. The Pelagic Realm is divided into four distinct depth zones: the Epipelagic (sunlit) Zone, the Mesopelagic (twilight) Zone, the Bathypelagic (sunless) Zone, and the Hadopelagic (trench) Zone.
17. Although the Epipelagic and Mesopelagic zones tend to be occupied by different species, the horizontal patterns at the two levels are virtually identical.
18. The horizontal patterns are of two types, neritic and oceanic. The oceanic distributions may correspond to the major subtropical gyres or they may exist as transoceanic, longitudinal belts. The latter appear to be the most common and dominate the Southern Ocean.
19. Tropical, warm-temperate, cold-temperate, and polar distributions can be identified in the upper two pelagic zones. It was suggested that parapatric speciation may be important.
20. Most Bathypelagic Zone species have exceedingly broad distributions and inhabit all three oceans. One huge Bathypelagic Region will suffice, with the exception of the Arctic Basin which displays considerable endemism.
21. The Hadopelagic fauna is poorly known but a variety of copepods, amphipods, and ostracods have been found.
22. The Deep Benthic Realm has three different depth zones, the slope, the abyss, and the trench. Although other, more restricted, depth zones have been recognized, this three part subdivision seems to fit the broadest spectrum of the fauna.
23. On the continental slopes and oceanic rises between about 200 and 2000 m, the animals do not generally have broad distributions. Many slope species appear to be confined beneath tropical, warm-temperate, or cold-temperate waters.
24. The Abyssal Zone has many fewer species than the slope. Some of the upper slope forms have broad distributions, but few of those taken below 4500 m are widespread. As the abyssal fauna becomes better known, all 47 of the ocean basins may need to have their own provincial designations.
25. Places such as hydrothermal vents and bottom seeps can support isolated communities consisting of large numbers of individuals.
26. The Hadal Zone designation belongs to the benthic fauna of the trenches. The fauna is depauperate, and that found in each trench is usually related to the sur-

rounding abyssal forms. Within the boundaries of the zone, the fauna become impoverished with depth. Each trench appears to have a high rate of endemism.

27. The fauna of the deep sea below 1000 m is not nearly as diverse as some researchers have speculated. The deep-sea fauna appears to have had multiple origins from shallow waters in various parts of the world. The seas around Antarctica may have supplied a conduit to the deep sea for several groups.
28. The marine system is dynamic. Its biogeographic patterns have changed in the past and they are currently undergoing changes. Extinction of species caused by human activities will further reduce an already low level of species diversity.

CHAPTER 11

Terrestrial patterns

The distinguishing of the aboriginal from the invading population, and the determination of the causes which have produced and directed the invasion, are among the problems which the investigator of the distribution of animated creatures, has to endeavour to solve.

Prof. Edward Forbes, *The Natural History of European Seas*, 1859

INTRODUCTION

Just as the dynamic nature of the marine biota was revealed by the comparison of past and present distribution patterns, the same procedure indicates that terrestrial and freshwater organisms have been, and continue to be, constantly moving. The ice-age climatic cycles of the past 3 Ma have induced extensive floral and faunal migrations in the northern hemisphere. Associated sea-level changes resulted in the making and breaking of terrestrial connections in such places as Beringia, Indo-Malaya, and Australia-New Guinea. At the same time, tectonic-plate movements and associated volcanism continued to create oceanic islands, which begin to accumulate their biotas by means of long-distance dispersal.

As Udvardy (1969) emphasized, present distribution is the result of processes by which the organisms have moved in space. Consequently, the distributional picture changes with the passage of time. As long as our inquiry probes into the reasons for the arrival and settling of a species in a certain area, we are within the field of dynamic biogeography. As soon as we ask how and why a species can live in a given area, we leave biogeography and the question is answered in ecological terms. Changes in distribution will result in the alteration of gene pools which will eventually be expressed by evolutionary change. The phenotypic consequences are of interest to the systematist and evolutionary biologist.

Distributional changes are driven not only by climatic events and by the migrations of the organisms themselves. Small populations, whether established by dispersal or vicarianism, are vulnerable to extinction; and extinctions may cause disjunctions. Extinctions are also frequent in or near evolutionary centers as the result of competition from newly formed species or other factors. As we have noted in the marine environment, some tropical extinction patterns appear to spread outward causing broad, antitropical patterns. The next step is often extinction in the northern hemisphere, leaving relict populations in southern latitudes (see examples in Chapter 12).

Aside from causing extinction patterns, the distributional and evolutionary effects of centers of origin are apparent in both terrestrial and freshwater habitats. Although the existence of such centers is denied by some vicarianists, one has only to look at the genesis of such well-known groups as the proboscidean, camelid, and equid mammals, or the cyprinid fish. These, and all other widespread groups, had their origin in one relatively restricted area, and became widely dispersed over time. There is no evidence that any species, or larger taxonomic group, came into being simultaneously over a large geographic area. We may conclude that present biogeographic patterns of species are the result of recent climatic changes, dispersals, extinctions, and originations. These took place on top of older changes, generally apparent at the higher taxonomic levels, which took place in earlier periods.

The terrestrial environment contains about 98% of the world's species (see p. 373) and a comparatively large amount of systematic work has been devoted to them and the groups to which they belong. A comprehensive terrestrial biogeography, or even one entirely devoted to the vertebrates, would require a large volume by itself. In order that the reader may achieve an appreciation for the major patterns, a few relatively well known groups have been selected as examples.

Despite the attempts of a group called panbiogeographers to locate the terrestrial biogeographic regions around the ocean basins (Grehan, 1991), the patterns of such disparate groups as the birds, mammals, insects, and freshwater fish all conform remarkably well to the regional system devised by P.L. Sclater (1858) and adopted by Wallace (1876). Although the floristic regions of the world are traditionally more finely divided, there is an overall resemblance to the zoogeographical pattern. Ecologists have divided the world into a large number of biomes which represent characteristic plant and/or animal communities. Biomes have been incorporated into world biogeographic provinces (Udvardy, 1975) and also illustrated within the framework of the six recognized zoogeographic realms (Berra, 1981) (Fig. 93).

ANIMALS

Freshwater habitat

Of all the animal groups that might be used to trace the earth's tectonic history, the freshwater fish are probably the most reliable. As Myers (1938) pointed out, there are certain families and higher taxonomic groups that have apparently been confined to freshwater throughout their histories. We have noted that these primary freshwater fish have been able to achieve intercontinental distributions only by means of terrestrial connections that were supplied with freshwater streams. Over time, such fish can spread by means of erosion events that result in headwater stream captures, or through stream diversions caused by tectonic events or glacier movements. By these means, they have been able to spread throughout continents and large islands to which they have been able to gain access, but they have not been able to cross saltwater barriers.

The ostariophysan fish are a monophyletic primary freshwater group comprised of three or four orders, 53 families, and about 6000 species. They probably arose in the Upper Jurassic, so that the more primitive orders (characoids and catfish) were able to travel

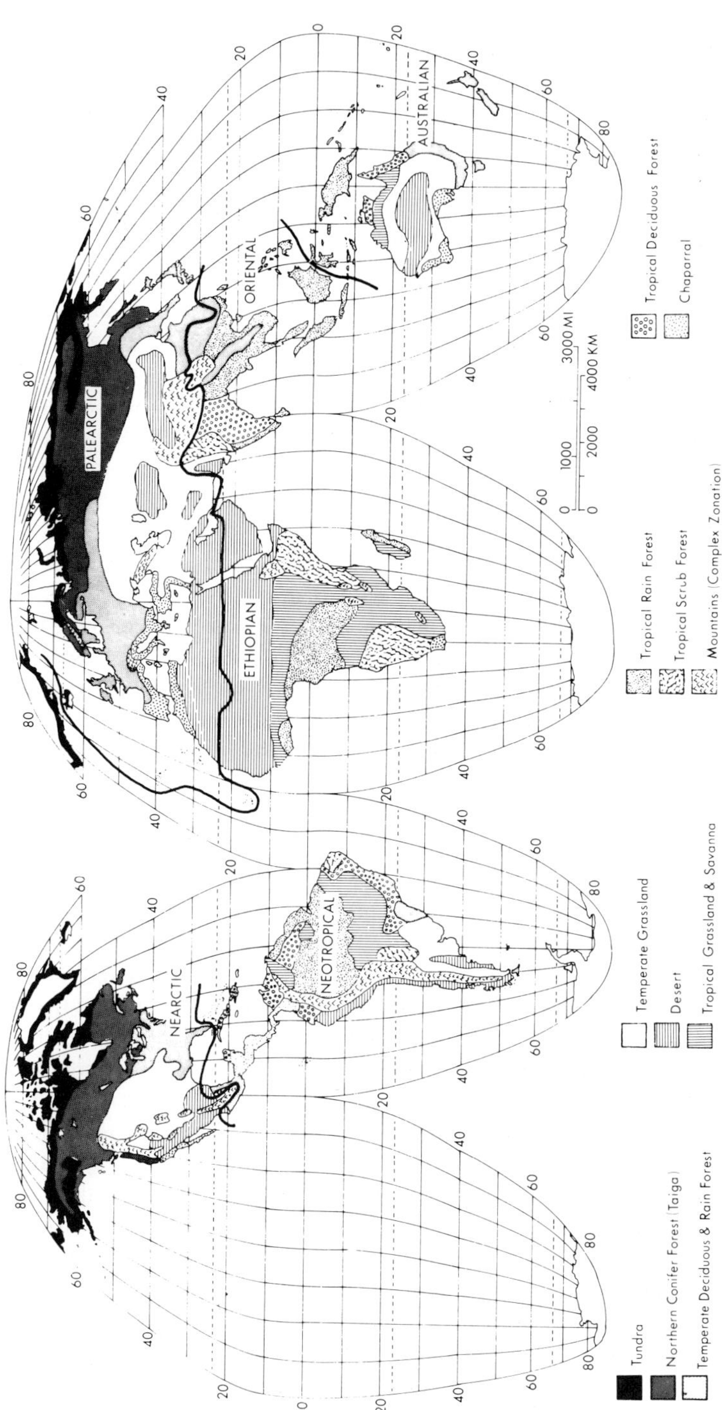

Fig. 93. The six Zoogeographic Realms and the major biomes. After Berra (1981).

among Asia, Africa, and South America before these continents were separated. The cypriniform ostariophysans probably arose in southeast Asia following the Mesozoic continental separations, and were subsequently able to migrate to Africa via the Arabian Peninsula and to the New World via Beringia (Fig. 94). Aside from two catfish families that became secondarily adapted to saltwater, ostariophysan fish have not been able to reach such places as Madagascar, Australia, New Zealand, and the West Indies. This permits us to predict that those land masses have not been connected to any of the larger continents since the Upper Jurassic.

Bănărescu (1992) recognized, primarily on the basis of fish distribution, eight regions for the freshwater fauna of the world (Fig. 95). While the general outline is similar to Sclater's regions, most areas are more finely divided. Madagascar, Australia, and New Zealand are each placed in a separate region. The reason for such designations is that each area has a highly endemic freshwater biota. Their fish faunas are comprised mainly of euryhaline species, belonging to marine families, that can successfully occupy streams where there is no competition from primary freshwater species. Some of them have diadromous life histories which are characterized by regular migrations between fresh and salt water. Many are capable of extensive migrations during the marine phase of their life cycles.

Fish that generally live in freshwater, but can tolerate high salinities to the extent that they can cross minor saltwater barriers, are called secondary freshwater fish. Various subcategories such as vicarious, complementary, and sporadic are often recognized, but it is sometimes difficult to assign species to them. Aside from the dominant ostariophysans, some other fish groups are often placed in the primary category. This has occasionally resulted in confusion about the history of land mass relationships. For example, the lungfish (Dipnoi) are sometimes placed in the primary division (Bănărescu, 1990). One family is found in South America and Africa and the other in Australia.

Does the fact that Australia has a lungfish mean that it must have arrived there via a freshwater connection? It has been assumed by some that an ancestral lungfish occupied all of Gondwana and that the present pattern is the result of continental drift. However, the Australian species belongs to a very old family (Ceratodontidae) with members that have been found in Triassic marine beds. The lungfish genera found in Africa and South America belong to a more recent family (Lepidosirenidae) that probably inhabited those two continents before they split apart. So, while lungfish distribution does not tell us that Australia had freshwater connections to the rest of Gondwana, it does signify an old relationship between Africa and South America.

Another ancient fish family often placed in the primary category is the bonytongues (Osteoglossidae). Various genera live in Australia, southeast Asia, Africa, and South America (Fig. 96). The same question arises about the significance of such a distribution. Does it reflect a Gondwana pattern of freshwater streams? Two species are found in Australia-New Guinea. Yet, Patterson (1975) indicated that at least one extinct osteoglossid was marine, indicating that the group may have had a marine origin. The Australian families Melanotaeniidae and Percichthyidae have also been considered primary freshwater (Bănărescu, 1990), but both are clearly related to marine families and have probably become secondarily adapted to freshwater. If one eliminates these suspicious

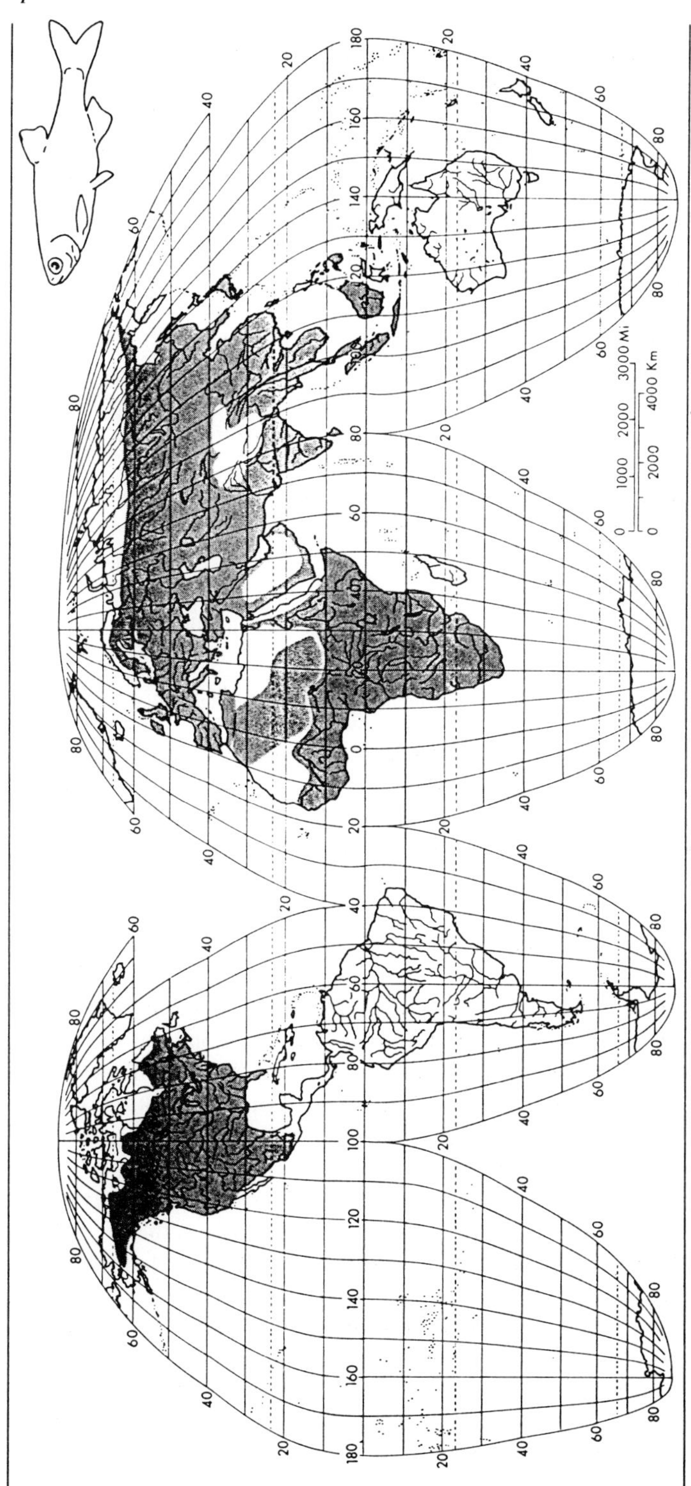

Fig. 94. Distribution of the cypriniform fish family Cyprinidae, a primary freshwater group. After Berra (1981).

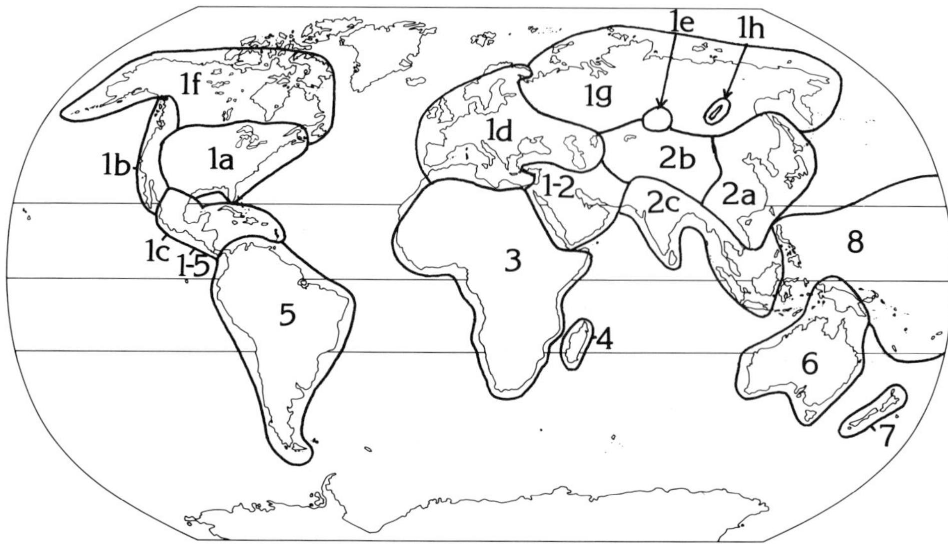

Fig. 95. Zoogeographical regions for the freshwater fauna. 1. Holarctic Region with eight subregions: (a) Eastern North America; (b) Western North America; (c) Central Mexican; (d) Euro-Mediterranean; (e) Western Mongolian; (f) Arctic North American; (g) Siberian; (h) Baikal; 2, Sino-Indian Region with three subregions: (a) East Asian; (b) High Asian; (c) South Asian; 3, Ethiopian Region; 4, Malagasy Region; 5, Neotropical Region; 6, Australian Region; 7, New Zealand Region; 8, Indo-West Pacific Region. The numbers 1–2 and 1–5 refer to transition zones between the indicated regions. Primary freshwater fish do not exist in Regions 4, 6, 7 and 8. After Bănărescu (1992).

groups, there is nothing in the present or fossil Australian fish fauna that requires freshwater connections to other parts of the southern hemisphere.

The leaf fish of the family Nandidae are sometimes (Bănărescu, 1990) placed in the primary category. They have a circumtropical distribution (Fig. 97) which has been related to the Gondwana assembly (Berra, 1981). One subfamily occurs over the entire range while the other two are confined to southern Asia. The worldwide distribution could not have been achieved in the Mesozoic for the family belongs to the suborder Percoidei which almost certainly did not evolve until the early Tertiary. They probably arose in southeast Asia and, like other freshwater families, could have entered Africa in the early Miocene. They would have had to reach South America via oceanic currents or some other means of transport. Nandids have been taken from brackish water.

There are six old freshwater-fish families (Denticipitidae, Mormyridae, Gymnarchidae, Pantodontidae, Phractolaemidae, Kneriidae), generally included in the primary category, that are confined to Africa. However, their relationships indicate that they are old enough to have been extant before Africa and South America were separated. South America, on the other hand is relatively depauperate in regard to such ancient families. That continent has only the three groups already mentioned, the lungfish, osteoglossids, and leaf fish (all three also occur in Africa). Fossils assigned to the African family Polypteridae have recently been found in the late Cretaceous of Bolivia (Gayet and Meunier, 1991). Of course, the younger ostariophysan characoids and catfish are shared

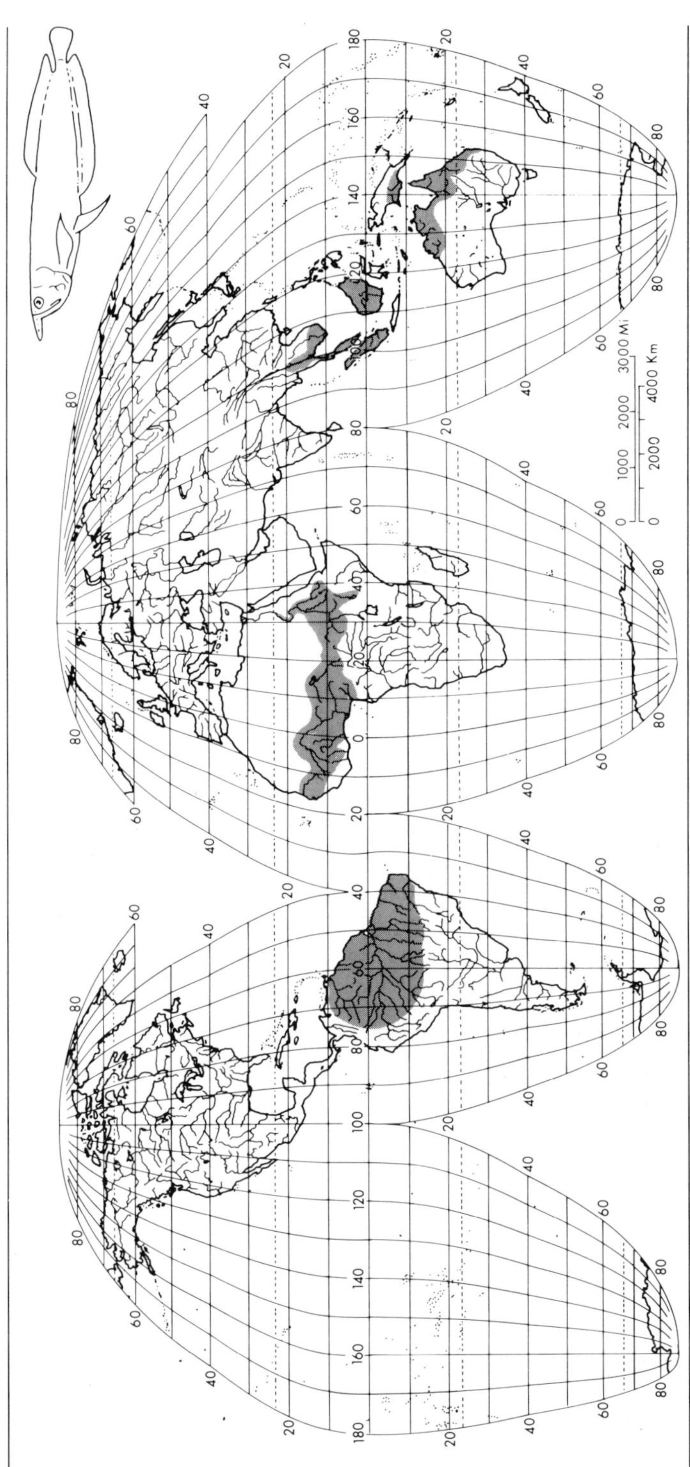

Fig. 96. Distribution of the fish family Osteoglossidae. After Berra (1981).

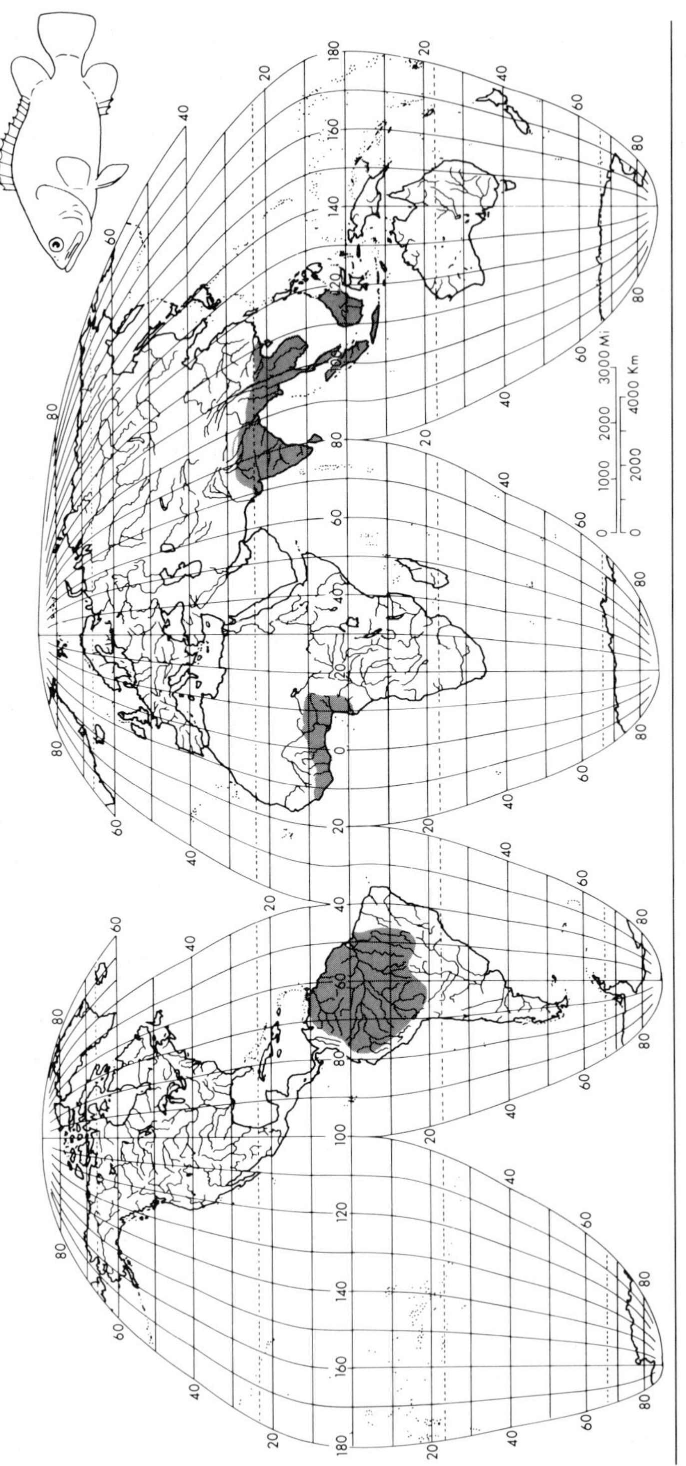

Fig. 97. Distribution of the fish family Nandidae. After Berra (1981).

between the two continents, but why not most of the older families, and why should they be concentrated in Africa? Some years ago, Darlington (1957) suggested that, due to its limitation and imbalance, the South American primary freshwater fish fauna is a derived one descended from a few immigrants that somehow reached South America from Africa. This indicates that from the mid-Mesozoic, and perhaps even earlier, freshwater connections between the two were not as common as is often assumed.

It has been proposed (Bănărescu, 1990) that the archaic fish families have been able to survive in Africa due to the lack of competition. This, because the characoid fauna includes only a few unspecialized genera, and cyprinids have entered so recently that they have not had time to become very diversified. But, there are no fossils to indicate that the old African families have ever been in South America. If South America received its major groups, the four archaic families plus the characoids and catfish, from Africa it could not have functioned as a center of origin as some have maintained. In the nematode family Camallanidae, which is parasitic on freshwater fish, the African fauna appears to have been derived from southeast Asia and the South American fauna, which is comparatively primitive and depauperate, was probably derived from Africa in the Mesozoic (Stromberg and Crites, 1974). As noted earlier, the same distributional history has been suggested for the characoids and catfish (Briggs, 1987).

If the early characoids and catfish did enter South America from Africa, they would have found a virtually empty continent as far as competitive species are concerned. This may explain why these groups have been able to expand to their enormous present diversity. It is assumed here that the gymnotid eels are a characoid offshoot, rather than representing a separate invasion from Africa. A biogeographic and phylogenetic pattern that may possibly contradict this hypothesis may be found in the catfish. The two most primitive (plesiomorphic) families are found at opposite ends of the New World. The family Diplomystidae, with two living species, exists in southern South America. Fossils of the primitive family Hypsidoridae have been taken from the lower Oligocene and Eocene of North America. This pattern could indicate that catfish arose in the South American tropics and, as they diversified, the most primitive families were displaced to the periphery. On the other hand, a primitive ancestor may have entered South America from Africa to bring about the same eventual result. Catfish are also well diversified in southeast Asia. It seems apparent that the progenitors of the two North American families, the fossil Hypsidoridae and the Ictaluridae, entered the New World by way of Beringia.

In the Miocene account we noted that the three principal groups of ostariophysan fish (characoids, minnows, catfish) have been able to dominate freshwaters to which they have been able to gain access. It seems clear that the minnows (cypriniforms) arose in southeast Asia, probably in the early Paleogene. But this group is part of a monophyletic triad which means that they had a common ancestor. The difficult question is, where did the ostariophysans themselves originate? Southeast Asia has been suggested (Briggs, 1979). This seems reasonable for the catfish, but what about the characoids which are presently confined to Africa and the neotropics (Fig. 98)?

In the discussion of Eocene events reference was made to the presence of characoid fossils on the Eocene and Oligocene of France. These fossils may represent the remnants of an earlier Eurasian distribution. Otherwise, they would have had to make their way from Africa across the Tethys Sea. A popular theory says that characoids, as

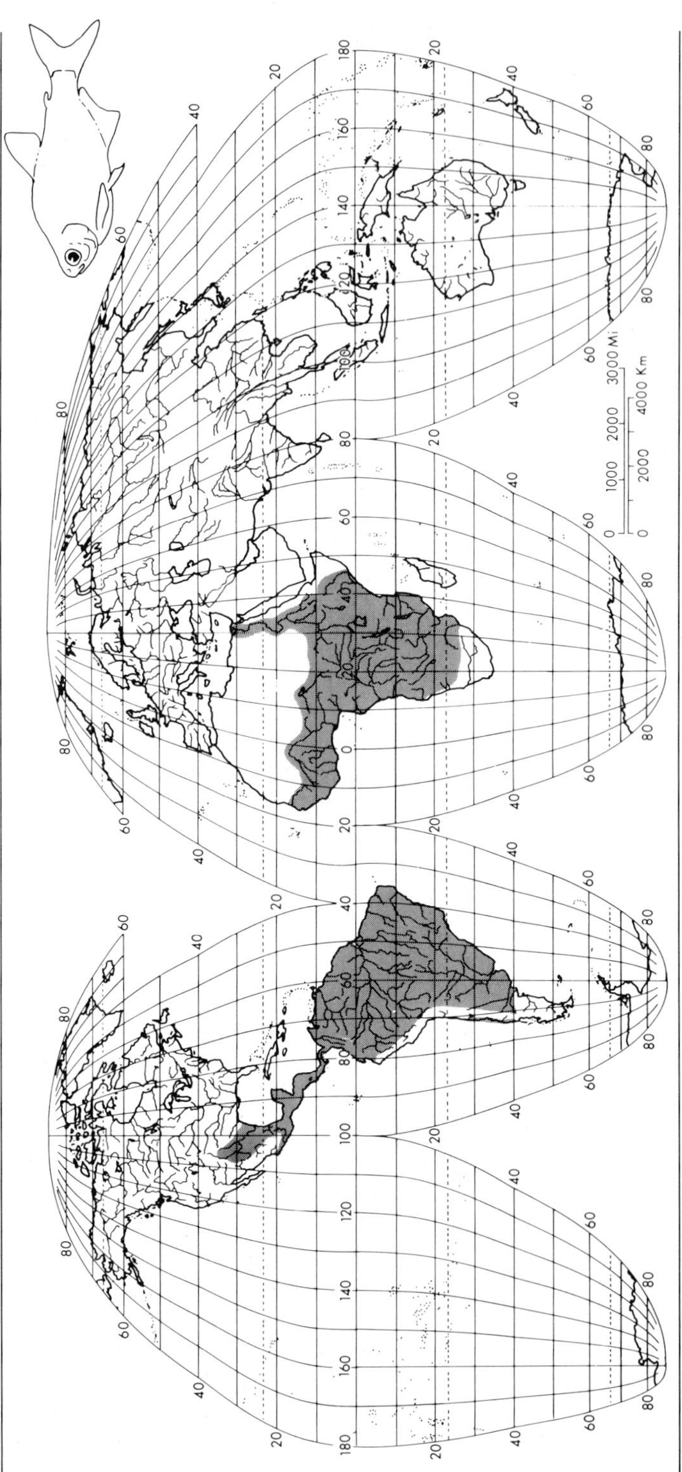

Fig. 98. Distribution of the characoid fish family Characidae, a primary freshwater group. After Berra (1981).

well as the minnows and catfish, arose in Gondwana and were transported to Asia when portions of that supercontinent were rafted northward (Bănărescu, 1990).

The difficulty is, there is nothing in the present or fossil fauna of India to suggest the transport of early ostariophysans. Portions of the Australian part of Gondwana became detached in the late Paleozoic and were annexed to southeast Asia by the late Triassic (Metcalfe, 1991), but the rifting of these fragments almost certainly took place prior to the origination of the ostariophysans. Supposedly, the gonorynchiform fish, which consist of two marine and two freshwater families, may be related to the earliest ostariophysans (Nelson, 1984), but the connection is tenuous. These fish do not have a functional Weberian apparatus which is common to all ostariophysan fish. Until the situation is clarified by additional fossil evidence, the theory of an ostariophysan origin in southeast Asia appears to be the most viable.

In the northern hemisphere, the primary freshwater fish comprise, at the higher taxonomic levels, a fairly homogeneous fauna. A connection between northern Europe and eastern North America lasted until the early Eocene. The connection between western North America and east Asia took place in the late Cretaceous and was in place during most of the Tertiary. The biogeographical effect of dispersals across these connections via their freshwater streams was the recognition by some ichthyologists of one huge, Holarctic Region (Bănărescu, 1992).

It is apparent that the North Atlantic Connection resulted in the intercontinental migrations of the ancient gars (Lepisosteidae) and bowfins (Amiidae). Eocene remains of both groups have been found on Ellesmere Island, northwest of Greenland (Patterson, 1981). A study of the systematics of the family Percidae by Collette and Bănărescu (1977), and more recently by Wiley (1992), indicated that the family probably originated in Europe and then dispersed over the land route sometime between the end of the Cretaceous and the early Eocene. After the interruption, the percid faunas of the two continents developed independently (Fig. 99).

Primary freshwater fish began to make their way across Beringia from Asia to North America as early as the late Cretaceous. The immigrations of the families Ictaluridae, Hypsidoridae, Cyprinidae, and Catostomidae have been described in the historical accounts. Also, the archaic families Polyodontidae and Hiodontidae evidently have Asiatic ties. Fossils from the early to middle-Eocene Green River Formation have yielded more fish with Asian relationships (Grande, 1985). The oldest-known North American members of the family Cyprinidae are Oligocene in age (Cavender, 1986). Several of the genera exhibit Eurasian relationships (Coburn and Cavender, 1992), which may indicate multiple crossings of Beringia. It seems that the major part of the present Holarctic relationships at the generic and family levels is attributable to historic migrations across Beringia.

Attention has been called to the contrast between the biogeographic patterns of primary versus secondary freshwater fish. Although the latter are not as useful in establishing past continental relationships, they are of interest particularly in parts of the world not dominated by the primary category. Many of the secondary families exploit the estuarine environment around continental shorelines. Some species have become well adapted to purely freshwater and, in the absence of primary species, can dominate the river systems. There are a large number of secondary families in the tropics. Notable

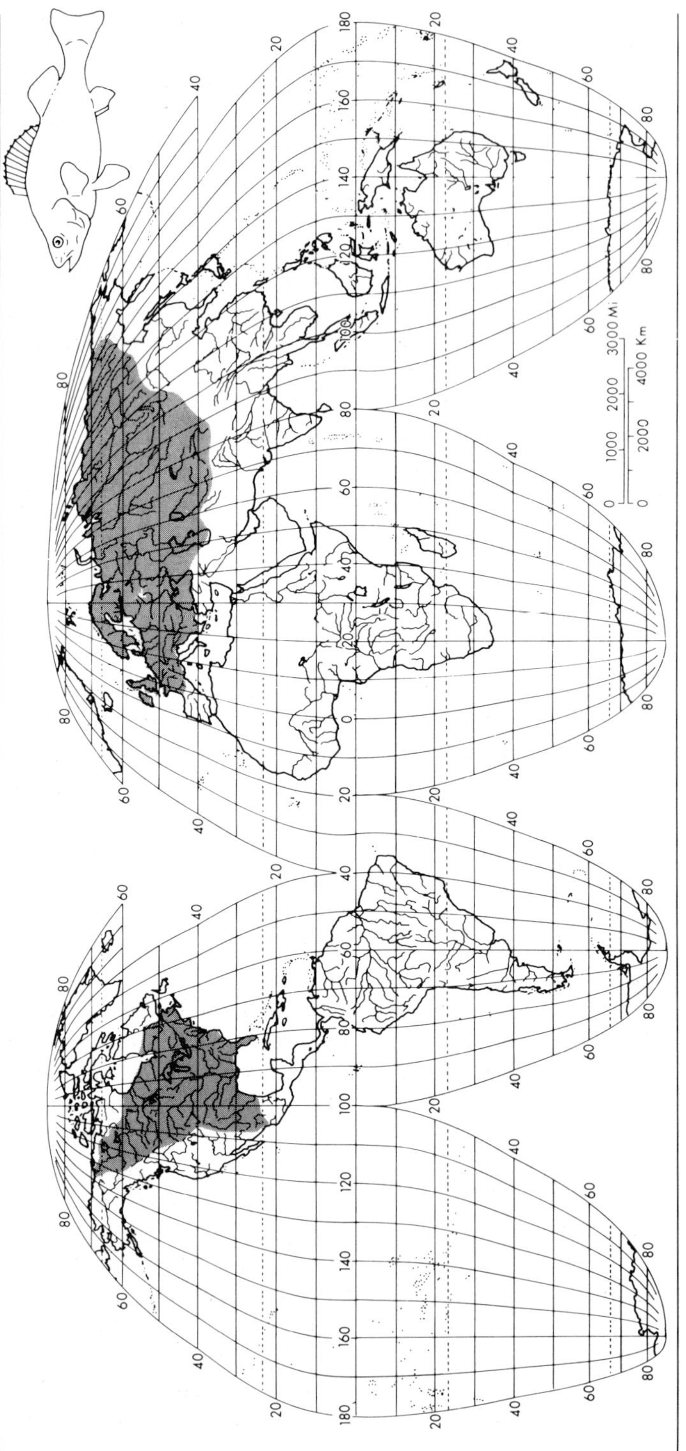

Fig. 99. Distribution of primary freshwater fish family Percidae. After Berra (1981).

examples are the killifish (Cyprinodontidae), cichlids (Cichlidae), silversides (Atherinidae), and the swamp eels (Synbranchidae).

The killifish have a circumtropical distribution (Fig. 100). It has been suggested that this family, and the order to which it belongs, arose in the New World in the late Cretaceous (Briggs, 1987). It may have been able to traverse the Atlantic when the distance between South America and Africa was much less than at present. An older, Pangean origin had been proposed by Parenti (1981). While the killifish are most diverse in the New World, the cichlids are best established in Africa and may have originally spread from that continent (Briggs, 1987). The most primitive cichlid genera are found in Madagascar (Stiassney, 1992). The latter group may also have had a late-Cretaceous origin. Both families have become established in Madagascar and the West Indies, places that primary freshwater fish have been unable to reach.

In northern high latitudes, the secondary families of sticklebacks (Gasterosteidae), cottids (Cottidae), smelts (Osmeridae), and sturgeons (Acipenseridae) are common. All of these families are worldwide in the far north. In the southern hemisphere, the family Galaxiidae may be found in southern South America, southern Africa, Australia, and New Zealand (Fig. 101). At one time, the galaxiids were thought to provide a good example of a group that had reached their present distribution by means of continental movement. But, some of the species are known to have marine life-history stages, so the dispersal of the family probably took place through the medium of ocean currents (McDowall, 1988).

As Bănărescu (1990, 1992) made clear in his comprehensive work on freshwater zoogeography, there are many other animal groups whose distributions have historical importance. Among the most useful are the crayfish belonging to the decapod infra-order Astacida. The family Astacidae is found in western Asia, Europe, and western North America, while the Cambaridae exists in northeast Asia, eastern North America and Cuba. Bănărescu has suggested that the Astacidae, which must have lived in Siberia, has been outcompeted and displaced by the Cambaridae. He noted that the cambarids, which have been introduced into Europe, proved to outcompete the native astacids. The astacids probably invaded western North America from Asia via Beringia.

The southern crayfish family Parastacidae has an expanded amphinotic distribution. The bulk of its diversity is in Australia where there are 10 genera. Single genera exist in South America, New Zealand, New Guinea, the Aru Islands, and Madagascar (Fig. 102). The distribution of this family has been used as a "strong argument" for a Gondwana origin (Bănărescu, 1990). However, its presence at New Zealand, Madagascar, and the Aru Islands are difficult to justify on the basis of an entirely freshwater history. Its extensive diversity in Australia may indicate that the family originated there and that its dispersal to the other areas was accomplished, at least partly, by marine routes. Similarly, the presence of a cambarid crayfish on Cuba, where no primary freshwater fish occur, makes one suspect a marine dispersal. Since there are marine families related to crayfish, the northern and southern hemisphere groups may have had independent marine origins.

Three groups of aquatic insects are among the most ancient of the insect orders. These are the stoneflies (Plecoptera), caddisflies (Trichoptera), and mayflies (Ephemeroptera). All three have lengthy larval lives in freshwater with short-lived flying adults. The stoneflies may be divided into two suborders, the Arctoperlaria are mainly Holarctic but

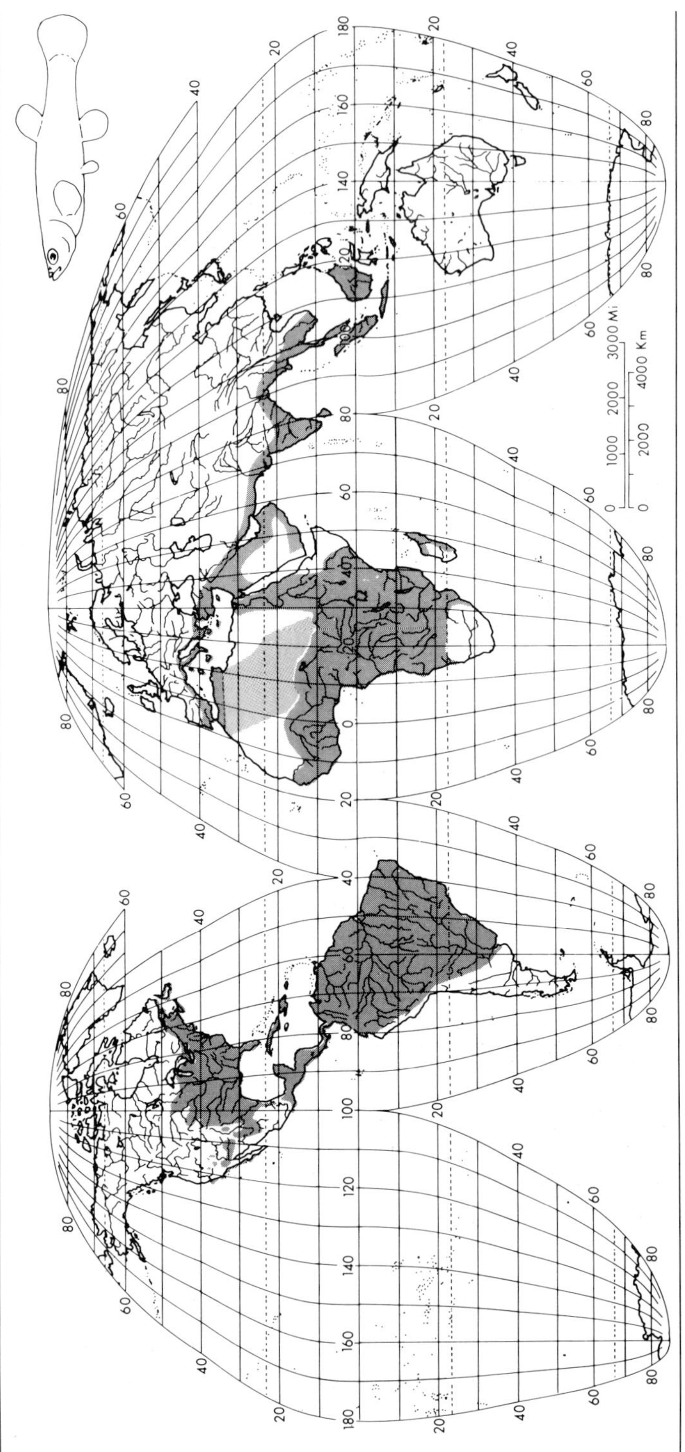

Fig. 100. Distribution of the secondary freshwater fish family Cyprinodontidae. After Berra (1981).

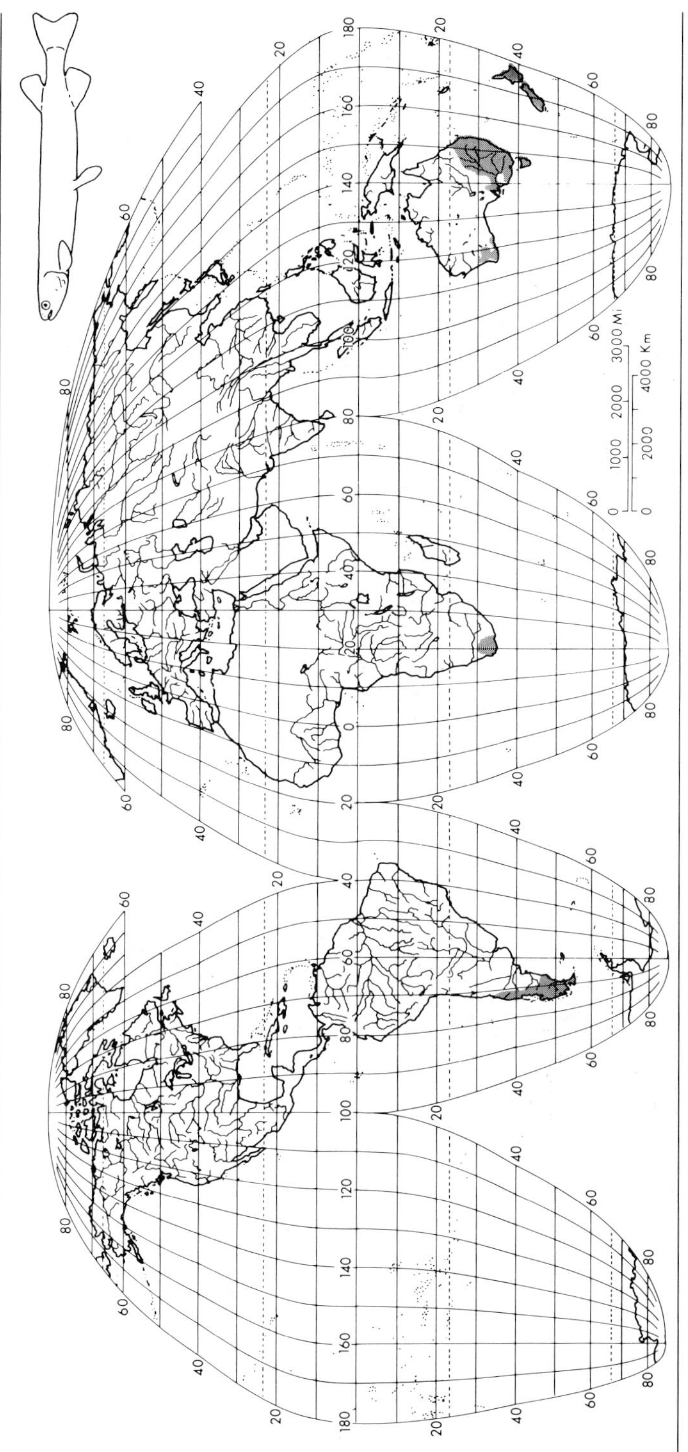

Fig. 101. Distribution of the secondary fish family Galaxiidae. After Berra (1981).

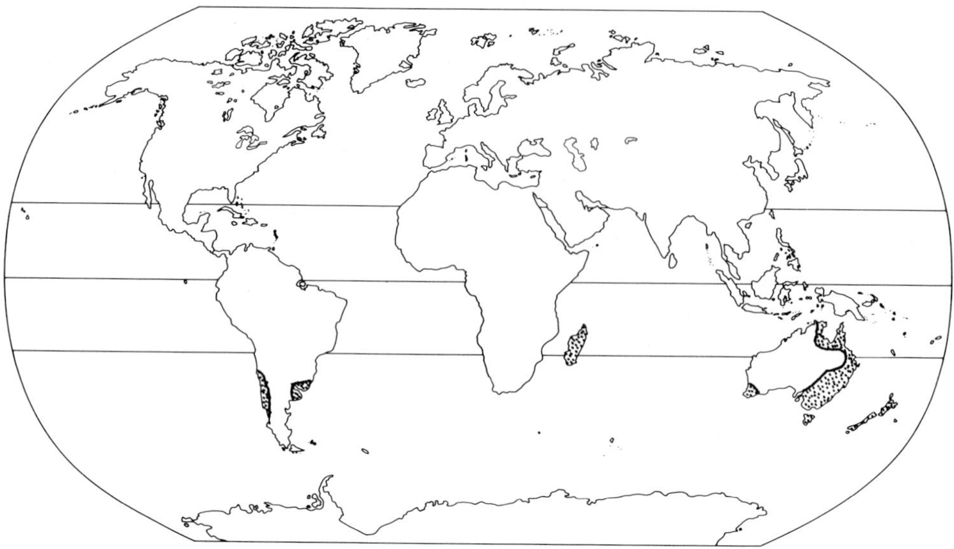

Fig. 102. Distribution of the southern crayfish family Parastacidae. After Bănărescu (1990).

have one family (the Notonemouridae) restricted to the southern hemisphere. The Antarctoperlia includes four families with a strictly amphinotic distribution. The phylogeny and distribution of the stoneflies has been analyzed by Zwick (1990).

Fossil caddisflies, known from the Permian and Upper Cretaceous of Siberia and Can-

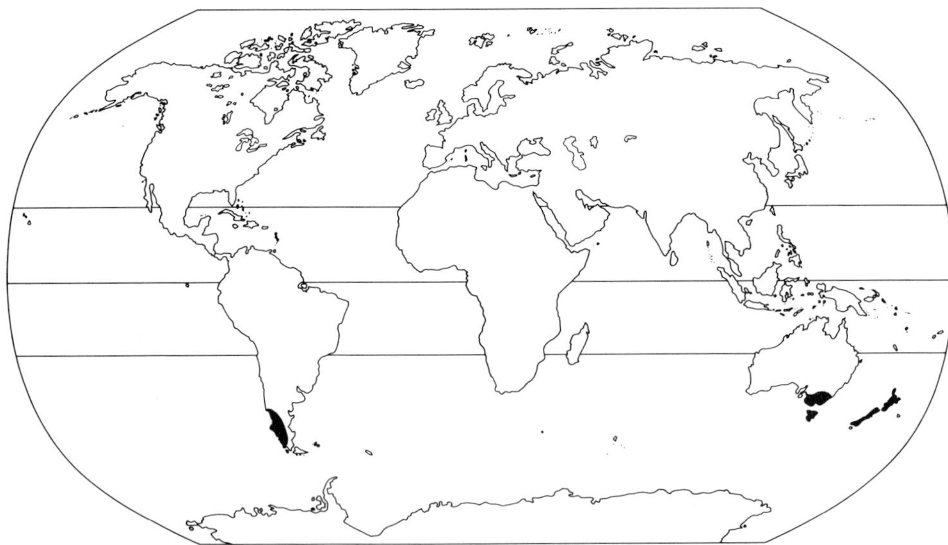

Fig. 103. The caddisfly family Kokiriidae. Two genera occur in Tasmania and one each in South America, southeastern Australia and New Zealand. After Bănărescu (1990).

ada, are placed in recent families or are closely related to them (Bănărescu, 1990). Some of the families probably have a Pangaean origin which may explain their very broad distributions. Of the 46 recent families, 15 are primarily Laurasian, 9 cosmopolitan, 3 antitropical, and 19 primarily Gondwanian. Of the last category, 12 are amphinotic (South America, Australia, New Zealand). Botosaneanu and Wichard (1984) have reported, from Upper Cretaceous Siberian fossils, three different groups of stoneflies that now have an exclusive or almost exclusive southern hemisphere pattern. Bănărescu has illustrated the distribution of one of the amphinotic families (Fig. 103).

The archaic mayfly family Lepidophlebiidae has a predominately southern, cool temperate distribution. Work on the South American genera indicates that they are most closely related to genera from Australia, followed by genera from New Zealand, Africa, and Madagascar, in that order (Pescador and Peters, 1980).

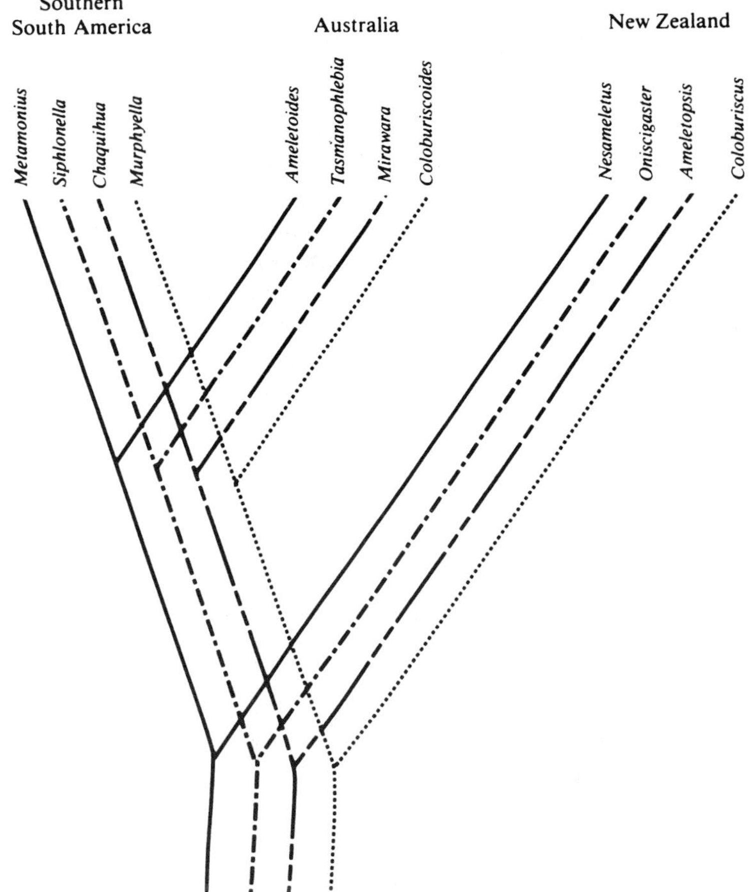

Fig. 104. Distributional relationships among four phyletic lines in southern hemisphere mayflies (Ephemeroptera). This cladogram indicates that the South American taxa are more closely related to those of Australia instead of New Zealand. From Brown and Gibson (1983), after Edmunds (1982).

In New Zealand, four leptophlebiid genera are most closely related to New Caledonian genera, two are related to Chilean genera, and the remainder are closer to other New Zealand genera (Towns and Peters, 1980). In the family Siphlonuridae, four of the five New Zealand genera belong to subfamilies that also exist in southern South America and Australia (McLellan, 1975). Edmunds (1972) gave evidence (Fig. 104) from mayfly genera that the closest relationships were between South America and Australia, with New Zealand next.

The dragonfly family Petaluridae, an ancient group that was dominant during the Jurassic, now contains only nine living species in five genera, *Petalura* (4 species) in Australia, *Uropetala* (1) in New Zealand, *Phenes* (1) in Chile, *Tachopteryx* (1) in eastern North America, and *Tanypteryx* in western North America and Japan. Thus, this family possesses a relict distribution that is both amphinotic and antitropical (Watson, 1981).

The water striders belonging to the family Gerridae are widely distributed, occurring on all major land masses except New Zealand and Antarctica. There are 57 genera; all except *Halobates* are typically freshwater. The group is concentrated in the tropics with the greatest diversity in the Oriental Region. In his zoogeographic analysis, Calabrese (1980) surmised that most of the genera attained their present distributions as their tropical Gondwana ancestors were split apart by the drifting continents. In Australia, there are 14 genera with two of them endemic. Of the 12 non-endemics, all are found also in southeast Asia and none of them extends farther south in Australia than northeast Queensland. This leads one to suspect that southeast Asia is operating as the important evolutionary center and that the species representing the 12 oriental genera may have invaded Australia in late-Tertiary times (Briggs, 1984).

The natural history of the blackfly family Simuliidae has been monographed by Crosskey (1990). The group is known from the Middle Jurassic and presumably was widespread in Pangaea. The most primitive of the recent groups is the one related to the genus *Prosimulium*. It has an antitropical distribution, and its southern hemisphere pattern was attributed to continental disbursement. Another archaic genus, *Austrosimulium*, confined to the southern hemisphere, was supposed to have evolved in Gondwana. The latter occurs in South America, Australia, and New Zealand. The rest of the family has a typical worldwide distribution in all six biogeographic regions, each of them with many endemic species and sometimes genera. Biogeographic conclusions for this group need to be tempered by the fact that almost all oceanic islands with freshwater streams have at least one blackfly species.

In a series of works, Brundin (1966, 1967, 1975) called attention to the distribution of the midges (Chironomidae). He decided that Western Antarctica, together with New Zealand and southern South America, comprised an austral center of evolution; this, because he found a concentration of related, primitive genera in the latter two areas. It seems odd that these places, which harbor geographical relicts from so many other insect groups, could have functioned as an evolutionary center for chironomids.

Another important group of Diptera is the net-winged midges of the family Blephariceridae. Although the family itself is widely distributed, the primitive subfamily Edwardsininae has a modified amphinotic range. It occurs in southern South America, Madagascar, and southeastern Australia (Zwick, 1981a). Within the genus *Edwardsina*, sister species are found in Australia and South America. Bănărescu (1990) assumed a

Gondwana origin, but the sister species are more likely late Tertiary or Pleistocene in origin.

An overview of the geographical relationships of the Australian aquatic insect fauna was provided by Williams (1981) who noted that the bulk of the fauna consisted of a southern element of primitive forms with austral (Gondwana) affinities and a younger, northern element of oriental affinities. The latter element was observed to have speciated greatly but not to have radiated much above the species level.

Freshwater mussels have a considerable potential for biogeographic work but their systematics at the levels of the higher taxa are in a state of flux. Within the larger unionacean group, the family Margaritiferidae is presumably the most primitive (Bănărescu, 1990). It has a disjunct distribution in the Holarctic and east to southeast Asia (Fig. 105). The family Hyriidae has a relict, southern hemisphere distribution. The subfamily Hyriinae is widespread in South America and has been there since the Jurassic; but fossils are known from the Jurassic and Cretaceous of North America. Two other subfamilies exist in Australia-New Guinea (including the Misool and Solomon Islands) and New Zealand. The family Muelleriidae has a remarkable disjunct range. One genus inhabits the Rio Magdalena drainage in South America and the other is found in Southern India.

Freshwater snails of the family Potamiopsidae present interesting patterns (Davis, 1979). An extraordinary species diversity involving the more advanced genera is found in the Mekong River of southeast Asia. The most primitive genera are located in South America, southern Africa and Australia. Davis' theory states that the family originated in Gondwana and were brought to Asia by the Indian Plate and that they were able to move from India to Malaya via stream captures during the Himalayan orogeny. However, there are no older potamiopsid genera in India today and Cretaceous fossils from there may or

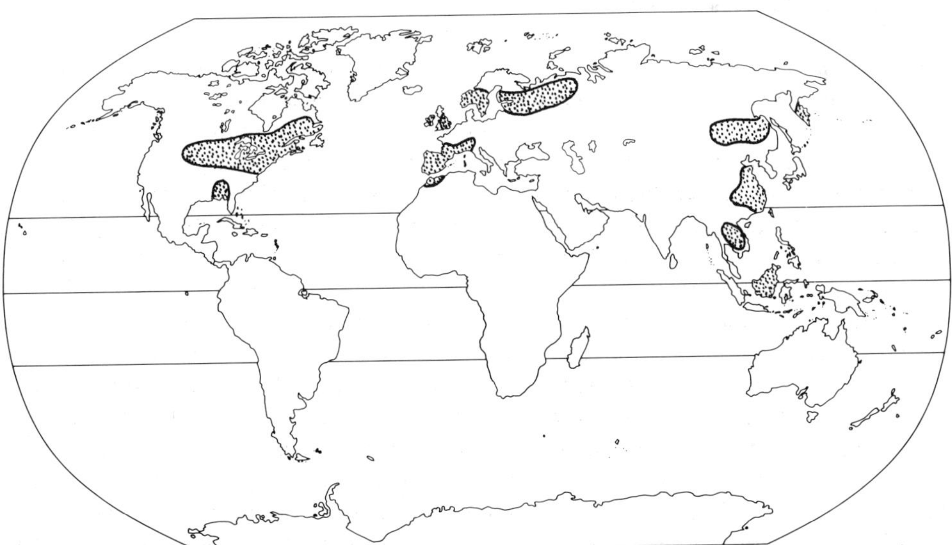

Fig. 105. The east-west disjunct distribution of the freshwater mussel family Margaretiferidae. After Bănărescu (1990).

may not represent that group (the characters that identify the family are found in the snail rather than its shell). An alternative theory states that the family originated in southeast Asia in the early Mesozoic and gradually dispersed southward along freshwater routes suggested for the ostariophysan fish (Briggs, 1984).

The aquatic earthworms have been considered as promising for the solving of biogeographic problems since they appear to be very unlikely candidates for overseas dispersal (Pielou, 1979). However, when one examines the distributional patterns of the various species it can be seen that many of them have proved to markedly adept at traveling from one place to another (Timm, 1980). Of the 12 families, the Glossoscolecidae appears to be the best known (Jamieson, 1971). Although the group is predominately tropical, four relatively primitive subfamilies have relict distributions in the Holarctic. Two relict subfamilies occur in the south, one in southern Africa and one in Madagascar. The overwhelming majority of the genera (23) and species (over 160) occur in the Neotropical Region. A Palearctic origin for this family and the related Lumbricidae, based on the position of the majority of the relict subfamilies, has been suggested by Jamieson. On the other hand, one might point out that South America has been the scene of most of the recent evolutionary action.

Attention has been called to the freshwater triclads as being ideal for determining past continental relationships. Supposedly, they do not possess resting stages, the adults are very fragile, and they spend their entire life cycle in the aquatic environment (Ball, 1974). The Dugesiidae is considered to be the most primitive family with a center of origin postulated to be in Gondwana. But the family is now very widespread, occurring on all continents plus New Zealand, Madagascar, the West Indies, Crozet Island, and Tahiti. Undescribed species apparently exist in Hawaii. A single species, *Cura pingius*, occurs in Australia, New Zealand, and New Caledonia. Patterns of this kind, which include oceanic islands, indicate overseas dispersals.

Mesozoic intercontinental relationships have been examined using the bathynellid crustaceans (Schminke, 1974). The family Parabathynellidae is widely distributed in Eurasia and in the southern hemisphere. Schminke considered the group to be historically freshwater except for one advanced genus that has become euryhaline. He postulated an origin in southeast Asia and illustrated the subsequent dispersals. The main difficulty with this hypothesis is that it assumes a Mesozoic connection between Asia and Australia. Of course, Australia's earlier continental relationships lie to the south, not the north. Another problem is that one species is common to Australia and New Zealand, so that a recent dispersal must be considered.

Although an attempt has been made to include the freshwater groups upon which the most notable systematic work has been done, there are many others that are of present and potential value. As may be seen from Bănărescu's (1990, 1992) volumes, other groups of interest are the freshwater crabs, mysids, isopods, amphipods, prosobranch and pulmonate snails, leeches, parasitic worms and crustaceans, fairy shrimps, clam shrimps, copepods, and water mites. Particular attention has been paid to the southern hemisphere, for this is where the greatest biogeographic problems exist. Freshwater patterns in the north are comparatively easy to interpret and a good correlation exists among the fish and invertebrates.

In the southern hemisphere, many of the groups we have examined, from secondary freshwater fish to a variety of invertebrates, exhibit amphinotic patterns. Southern South America is invariably included and usually Australia and New Zealand. Less often, the pattern extends to Madagascar, southern Africa, or New Caledonia. A striking feature of the amphinotic groups is that they generally represent the most primitive (plesiomorphic) genera or families within their respective larger groups. Their more advanced (apomorphic) relatives are located to the north, either in the tropics, or more commonly at boreal latitudes. Some demonstrate antitropical distributions while others, now confined to southern latitudes, once had northern relatives that are now extinct.

Many of the amphinotic groups have been used as examples to demonstrate the dispersal effects of the rifting of Gondwana. While many of the groups considered may be old enough to have been affected by such continental movements, we must ask if this was the primary reason for their present distribution? Prior to the mid-Miocene, before Antarctica was completely glaciated, the southern land masses were closer to one another. Migration across modest oceanic gaps was relatively easy. Strong westerly winds, characteristic of high southern latitudes, were probably a factor, especially for aquatic insects with flying adults. In cases where related species or related genera show amphinotic patterns, the chances are that the distribution took place after continental fractionation. Where subfamilies and families are different on each land area, a Gondwana history may be more plausible. There may be objections to this general rule, but we need to keep in mind that most living species are not more than 5 Ma old and that the great majority of modern genera are Tertiary in origin, making them less than 65 Ma old.

Terrestrial habitat

Invertebrates

Despite the impressive effort that has been and is being devoted to insect taxonomy in various parts of the world, there are only a few systematic works dealing with old, widespread groups where the author has taken pains to work out a detailed phylogeny. Without this step, one cannot do a logical biogeographic analysis. It is understandable that so many works are devoted entirely to alpha taxonomy, since most insect groups are very poorly known. But one cannot obtain useful information about the history of a group from taxonomic descriptions alone. It is up to the expert to complete the task of working out the relationships and biogeography. This cannot be done by anyone else. Although the lack of such complete systematic treatment is notable in entomology, it is true for other groups as well. The following are a few selected works that appear to be important for understanding insect evolution and distribution.

It is the beetles (Coleoptera), more than any other insect group, that have been the subject of biogeographic analysis. This is mainly due to the work of Philip Darlington and his students and disciples. The beetles are a very old insect group with a fossil record extending back to the Lower Permian (Crowson, 1981). The most primitive phylogenetic relicts are found in Australia, with a few also in New Zealand and Chile. The Australian genus *Omma* is considered to be the most archaic surviving type. Crowson noted that it is almost identical with fossils described from the Lower Jurassic of England. After working on the family Carabidae, Darlington (1971) suggested that the main area of the Old

World tropics appeared to be the most important center of dispersal. He concluded that the broad movement of Carabidae from Asia towards Australia was part of a world-wide pattern of evolution of successive dominant groups in Africa and tropical Asia and dispersal into smaller areas, with replacement of older by more recently dominant groups.

Early work on the family Carabidae by Darlington (1971 and other papers) has been augmented by several others. As a result, we now have a fairly good general outline of evolution and distribution in that family. The tribe Galeritini was analyzed by Ball (1985) who used cladistic methodology. He determined that western Gondwana was the principal area of differentiation. Only one genus is present in Australia and it may have reached there by dispersal from Asia. Ball suggested that Australia may have been too cold to support the Mesozoic genera. A biogeographical history of the Carabidae was presented by Noonan (1985). His work is a good illustration of the fact that both vicariance and dispersal are important in old, widespread families. The early Mesozoic fractionation of Pangaea, and almost all subsequent continental joinings and separations, can be followed by using carabid beetles as examples. Most of the Australian fauna is the result of dispersals from Asia.

A world revision of the subfamily Dynastinae was published by Endrödi (1985). The group is primarily tropical and southern hemisphere. The bulk of the diversity, together with the most specialized of the eight tribes, occurs in South America. The Alticinae is a widespread subfamily with about 500 genera and 8000–10 000 species (Scherer, 1988). It also is most diverse in South America where there are more than 200 endemic genera. With the completion of the Panamanian isthmus, species of 16 South American genera migrated northward and representatives of 4 northern genera entered South America. One Australian genus belongs to a South American tribe. The origin of the subfamily is supposedly Gondwanian, but the Australian fauna is sparse and at least some of its species have invaded from the Oriental Region.

The concept of repetitive "taxon pulses" was conceived by Erwin (1981) from his experience working with carabid beetles. He concluded that selection agents are repetitious through time and that they work on the basic groundplan of the organism. This leads to convergence and parallelism in structure and in geographic patterns. Taxa arise in equatorial latitudes and are derived through stages along pathways to extinction in three directions: towards the poles, up mountains, and into the forest canopy. Driving forces include more potent adversaries, climatic cycles, tectonic events, and microhabitat changes. The evolutionary results are of a repetitious nature as evidenced by disparate groups of carabids following the same paths over and over. This led Erwin to the conclusion that evolution has boundaries in which it operates, and that perhaps there are some rules.

The work by Wilson (1961, 1985) on the Melanesian ant fauna gave results similar to those cited by Darlington. Wilson found that the creation of "potent" new genera and higher taxa was confined to the large source areas of tropical Asia and New Guinea. Older taxa were forced to retreat before the new forms, but could shift only from larger to smaller land masses - not in the opposite direction. This information is consistent with Brown's (1973) analysis of the distribution of ant genera. He observed that, at least from mid-Tertiary times, evolution of world-dominating new taxa has proceeded mainly from the combined tropics of Africa and southeast Asia. Brown also noted that the genera or groups of genera now well represented in the Neotropical and Australian regions, and absent or very rare in Africa, are the peripheral relicts of older taxal waves that are now being replaced from the central Old World tropics.

In his book on wasps, Spradberg (1973) concluded that the family Vespidae originated in southeast Asia; it is the only area in which all the subfamilies are represented. The chrysidid wasps of the world were reviewed by Kinsey and Bohart (1990). Fossils from amber and shale indicate that the family probably arose in the Holarctic Region. It now exists in all six biogeographic regions. There are no close relationships among the Chilean, Argentinean, and Australian chrysidids. There is almost no relationship between the African and South American species, and the Australian forms are related to those of the Oriental Region. The authors concluded that the family must have evolved, or at least diversified, following the breakup of Gondwana.

The phylogeny and biogeography of the parasitoid wasp family Megalyridae were studied by Shaw (1990). His cladistic analysis showed the most primitive living genus to exist in Chile. All other living genera are southern hemisphere except one which is in the Oriental Region. Three fossil genera are from the northern cool-temperate zone. Shaw concluded that the family was possibly late Triassic in origin and that the dispersal of the genera took place along with continental rifting. One might observe that the family seems to be in the process of retreating from a once worldwide distribution towards one where it will eventually be represented only by southern hemisphere relicts.

Bees arose from specoid wasps after the appearance of probably beetle-pollinated, primitive angiosperms (Michener, 1979). The earliest fossil bees are known from the Eocene but they represent families that were already highly specialized so it is probable that they arose in the Upper Cretaceous. The richest bee fauna (315 genera and subgenera) is found in the neotropics. In comparison, the Oriental and African tropics have relatively poor bee faunas. The Nearctic Region (260 genera and subgenera) is richer than the Palearctic (243 genera and subgenera) largely because of the invasion by Neotropical elements. Michener stated that the place of origin might have been the arid interior of west Gondwana. But if the origin was in the Upper Cretaceous, the center of origin was most likely South America. There is a notable southern distribution of primitive groups. The tribe Paracolletini occurs in South America, South Africa, and Australia. The relict family Fedeliidae is found only in southern Africa and Chile.

Although many aspects of the dipterid family Drosophilidae, particularly the genus *Drosophila*, are well known, very-little work has been done on its geographical distribution. The family apparently arose in the Old World tropics (Throckmorton, 1975) and subsequently went through five major radiations. The earliest fossil has been reported from the Eocene Baltic amber but the group is probably Cretaceous in origin. According to Throckmorton, the five successive radiations of the Drosophilidae occurred before the end of the Eocene. In each case, the initial evolutionary development was considered to have taken place in the Old World tropics. Subsequently, there was an adaptation to the temperate parts of the Old World, a migration to the temperate New World (presumably via the Bering Land Bridge) and a southward movement into the New World tropics. From the Neotropical region there is evidence of considerable radiation back into the Neoarctic Region but no evidence of a reverse migration back across Beringia. The rich (about 700 species) endemic, drosophilid fauna of the Hawaiian Archipelago probably developed as the result of two introductions from the northern subtropics of East India.

The robber flies (Asilidae) are strong-flying predators, comprising one of the largest families of present day insects. Over 400 genera and subgenera have been proposed and about 4761 species are known (Hull, 1962). The earliest fossils are from the Eocene Green River shales where the two principal subfamilies are represented. This probably means that the family originated sometime in the Cretaceous. The greatest concentration of genera and species is in the Neotropical Region but the temperate parts of the world, especially the Palearctic and the Nearctic have very diverse faunas. In both the New World and Old World, it is the drier, subtropical areas rather than the mesic tropics that demonstrate the greatest species richness. The phylogenetic tree published by Hull shows that, in general, the larger, more-advanced tribes have a worldwide or almost-worldwide distribution. One relatively primitive group (Chrysopogonini) is confined to the Australian Region, another (Megapodinae) is strictly Neotropical, and a third (Phellini) is shared between Chile and Australia. Judging from such patterns, it seems probable that the family has undergone most of its evolution in the Neotropics.

In his work on the systematics and zoogeography of the Blissinae (Hemiptera, Lygaeidae) of the world, Slater (1979) decided that the group probably originated in the Upper Cretaceous and radiated together with the monocot angiosperms. The various genera and species are predominately tropical, and relatively depauperate Palearctic and Nearctic faunas appear to have been derived from the Old and New World tropics respectively. The Australian fauna is also relatively poor and three genera, restricted to that continent, are considered to be remnants of a very old element in the Australian blissine fauna. Within the rich, tropical regions, the Oriental appears to have a slightly more diverse fauna and the chief feature of that area is the large number of varied and highly advanced (apomorphic) taxa. Even so, Slater concluded that the subfamily originated in Gondwana. But if the subfamily originated in the Upper Cretaceous, Gondwana had been long before dispersed.

In a revision of another lygaeid (Hemiptera) group, the tribe Myodochini with 56 genera and more than 260 species, Harrington (1980) presented a detailed zoogeographic analysis. All members of the tribe are seed eaters and they occur in all six major biogeographic regions. Based upon some fossil material representing the subfamily to which the tribe belongs, Harrington suggested that the group arose in the Mesozoic and radiated during the Cretaceous coincident with the radiation of the angiosperms. She also suggested that the tribe originated in South America (west Gondwana) when South America and Africa were still close enough to permit some interchange. Almost all of the advanced genera occur in the Neotropics, many of them extending northward into the Nearctic, and by far the greatest diversity of species is in the Neotropics. However, the five most-primitive genera also occur in that general area.

Usinger and Matsuda (1959), in their monograph on the flat bugs (Hemiptera, Aradidae), noted that modern genera were known from the Oligocene. The three most-primitive subfamilies have a southern-temperate distribution. The Isoderminae inhabit the *Nothofagus* forests of South America, Australia, and New Zealand; the Prosympiestinae are found only in Australia and New Zealand; and the Chinamyersiinae are restricted to New Zealand. The concentration of diversity is in the tropics with the largest number of genera (44) and species (185) being found in the Neotropics. The next most diverse region is the Oriental with 35 genera and 115 species. The greatest number of subfamilies is present in the Neotropics so that this appears to be the best candidate for the center of origin.

The mosquitoes comprise a very-large group that is well known in some local areas but not on a worldwide basis (Mattingly, 1962). Belkin (1962) noted a great concentration of phylads around the Indomalayan and the adjoining Oriental and Papuan Regions. He suggested that the area (the Oriental Region out to and including New Guinea) has been the most active evolutionary center during the Tertiary. The South Pacific fauna was considered to have been formed by a series of dispersals from the west and consists largely of relict species.

In his treatise on the geographical distribution of the earwigs (Dermaptera), Popham (1963) commented that, on the whole, the more primitive groups tended to have peripheral or fragmented ranges. The earliest fossil records are from the Jurassic of Turkestan. As the result of his study, Popham reached the following general conclusion: "If the geographical distribution of a rapidly evolving group of animals is studied, it will be found that the individuals retaining the largest number of primitive characters are unable to compete with the more specialized members of the group and tend to survive mainly near the perimeter of the area in which the group is distributed." The earwigs were considered to have evolved as tropical insects probably in the Oriental Region. In a later work which considered the continental drift hypothesis, Popham and Manley (1969) thought that the center of origin may have been in eastern Gondwana because the primitive groups are there today - an interesting shift in analytical thought.

The moth family Sphingidae (Lepidoptera) is a relatively specialized group of fast fliers that can travel long distances. Their world distribution has been analyzed by Lin (1990). He found that the generic relationships were greatest within the Old World and New World respectively. There was very little evidence of contact between Africa and South America or between South America and Australia. The Palearctic and Nearctic faunas are essentially separate. He concluded that the center of origin was the Old World tropics and that dispersal took place with the continents in approximately their recent positions.

In his classic study of the Lepidoptera of New Caledonia, Holloway (1979) found that the largest pattern was formed by a widespread group that extended from southeast Asia, through northeastern Australia, and into the western Pacific as far as the Samoan Islands. About half of this group of 133 species also occurred in Africa and *all* of them in New Guinea. New Caledonia is an old island that probably first emerged in the Upper Jurassic. That island, together with the nearby Loyalty Islands has 444 lepidopteran species with 31 of them being endemic. In comparison, the New Hebrides have 364 species (29 endemic), the Fiji Islands 395 (46), and Samoa 183 (27). It appears that most of the groups with good powers of dispersal (butterflies and birds) have not been on New Caledonia for long periods of time. The progressive loss of competitiveness by populations on such small islands leaves them vulnerable to extinction when new invaders come along. Groups with less dispersal ability undergo slower turnover rates and older elements are more likely to survive.

In the Baltic amber, about 40 Ma old, two spider families have been found. The primitive Archaeidae has living members only in southern Africa, Madagascar, and Australia. The Cyatholipidae is found in both the Baltic and Dominican amber (about 20 Ma old); it now lives in southern Africa, Australia, and New Zealand (Wunderlich, 1986). The archaic Actinopodidae is found only in Australia and the Neotropical Region (Main, 1981). Main suggested that all the widespread spider families probably attained cosmopolitan distributions during Pangaean times, with some subsequently becoming more restricted.

The composition of the scorpion fauna of Australia reflects an early indirect connection to South America, followed by the long isolation of Australia until its approach to southeast Asia (Koch, 1981). The various genera of the scorpion family Bothruiridae are known only from Australia and South America. Other Australian scorpions in the families Buthridae and Scorpionidae belong to southeast Asian genera or to taxa that have clearly evolved from such genera.

Birds

Birds are a popular subject of study by amateur naturalists as well as professional biologists. Due to this interest and the fact that birds are relatively easy to observe, we have a large quantity of information about geographical distribution. Unfortunately this knowledge is not accompanied by an equal understanding of the phylogeny of the various avian groups. Compared to other large terrestrial groups such as the mammals or reptiles, the fossil record of birds is not as well documented. But avian paleontology is making rapid advances and continuing studies of morphology, behavior, biochemistry, and parasites are all contributing to the determination of evolutionary relationships.

Although birds are generally very mobile, there are few cosmopolitan species. Most land birds, despite their flying ability, seldom cross even modest stretches of seawater. It has been noted that the birds of Japan and England, though separated by 11 000 km, are more closely related than those of Africa and Madagascar, separated by only 400 km (Welty and Baptista, 1988). For some animal groups, biogeographers recognize a single Holarctic Region for the northern latitudes but bird distribution indicates the need to utilize two regions, a Nearctic for the New World and a Palearctic for the Old World. The limits of the other four regions are about the same as is usually recognized for the other terrestrial animal groups.

The modern distribution of birds, in a regional sense, has been summarized by Welty and Baptista (1988): the largest region is the Palearctic which contains 69 families (1 endemic) and 1026 species; the Nearctic has 62 families (1 endemic) and 750 species; the Neotropical has 86 families (31 endemic) and 2780 species; the Ethiopian has 73 families (6 endemic) and 1556 species; the Oriental has 66 families (1 endemic) and 961 species; and the Australian has 64 families (13 endemic) and 906 species. In considering the relationships of these regions, several facts of biogeographic importance must be kept in mind.

The first important fact is that the amount of endemism at the family level ought to reflect the long-term, geological history of the area concerned. For example, the Australian Region not only has a long history of isolation but it is still quite isolated today; about one-fifth of the families are endemic. The Ethiopian Region has a less-isolated history and only about one-twelfth of its families are endemic. As noted above, the northern continental faunas of the Oriental, Palearctic, and Nearctic Regions exhibit very little (one each) family endemism. However, the Neotropical Region does not fit in this scheme at all. It has a history of isolation throughout most of the Tertiary, but certainly not as much as Australia, yet more than one-third of its families are endemic and, furthermore, it contains an enormous number of species – about one-third of all known species of birds.

The second important fact is that many families and genera of birds are very widespread. The Palearctic encompasses the largest geographic area but, to a considerable extent, shares its fauna with the adjoining regions. Its relationship with the Nearctic, with which it was connected during most of the Tertiary, is especially close. The Palearctic shares 48 of its families, 35% of its 329 genera, and 12.5% of its species with the Nearctic (Welty, 1979). In the tropics, the Oriental and the Ethiopian faunas are closely related but the Australian and Neotropical faunas are considerably more independent. There is an interesting, and probably old, migratory relationship between the Ethiopian and Palearctic regions. In the northern winter, about one-third of all Palearctic bird species, especially the insect-eaters, migrate to Africa south of the Sahara desert.

A third important consideration is that the biogeographic regions differ in the structure of their bird faunas in a general evolutionary sense. When one looks over a list of bird families endemic to or typical of the various regions, it becomes apparent that the rich Neotropical fauna is composed, to a large extent, of primitive families. For example, such groups as the Tinamidae (tinamous), Rheidae (rheas), Cracidae (curassows), Aramidae (limpkins), Psophidae (trumpeters), Ceriamidae (seriemas), Steatornithidae (oilbirds), Nyctibiidae (potoos), Galbulidae (jacamars), Bucconidae (puffbirds), Ramphastidae (toucons), and the nine families of the more primitive passeriforms, give the Neotropics a decidedly primitive complex. The higher passeriform families (the oscines or songbirds) are found mainly in the other parts of the world. In contrast to the Neotropics, the more highly isolated Australian Region has a more modern bird fauna which, in large part, has apparently been derived from the Oriental Region (Mayr, 1944).

The interrelationships of the various regions is of interest since it emphasizes historic changes and helps us to realize that distribution patterns are changing. The Nearctic fauna has been analyzed by Mayr (1946, 1976) who identified three elements in its makeup. The first is an old indigenous element comprised of five passerine families: wrens (Troglodytidae), mockingbirds (Mimidae), vireos (Vireonidae), American wood warblers (Parulidae), and the buntings and sparrows (Fringillidae). Several other families were listed as possibly belonging to this category. The second element includes immigrants from the Old World, probably by way of Beringia: cranes (Gruidae), pigeons and doves (Columbidae), cuckoos (Culculidae), owls (Strigidae), crows (Corvidae), and thrushes (Turdidae). The third element includes two major families from South America: hummingbirds (Trochilidae), and tyrant flycatchers (Tryannidae).

In regard to the Neotropical Region, the archaic indigenous element consists of most of the 31 families that are endemic to that part of the world. In addition, there is a large element comprised of more recent immigrants from the Nearctic (Mayr, 1946): pigeons (Columbidae), jays (Corvidae), wrens (Troglodytidae), thrushes (Turdidae), vireos (Vireonidae), wood warblers (Parulidae), blackbirds (Icteridae), tanagers (Thraupinae), cardinals (Cardinalinae), and finches (Fringillidae). Most of the exchanges between South and North America probably took place after the creation of the isthmian link.

There is a surprising lack of relationship between the bird faunas of South America and Africa (Mayr, 1976). Hardly any two bird faunas could be more different with almost no sharing between the two continents. Mayr considered only three families of land birds to be circumtropical. These are the barbets (Capitonidae), trogons (Trogonidae), and the parrots (Psittacidae). These families were probably able to migrate between South or Middle America and Africa when the continents were closer to one another. Mayr suggested that the trogons probably originated in the New World while the barbets and parrots came from the Old World.

The distribution of parrots and pigeons was discussed under the Miocene account. Although it has been suggested that the early history of these groups took place on Gondwana (Cracraft, 1973), neither family appears to be that old. Their diversity, and that of related families is greatest in the southern hemisphere, particularly the Australian Region. It seems more likely that they arose in that general area after continental dispersement had taken place. It also seems apparent that the bird tribe Corvida (crows, ravens, magpies, and relatives) arose in Australia (Sibley and Ahlquist, 1986). As Australia moved northward, some corvids reached southeast Asia where they underwent a secondary radiation. Modern crows and ravens dispersed from southeast Asia to other parts of the world and also reinvaded Australia (Fig. 47).

The origin and evolution of the flightless ratite birds was discussed under the Eocene account. Although Cracraft (1973) had proposed that this group (kiwis, rheas, cassowaries, emus, ostriches, and a number of extinct families) evolved on Gondwana and achieved their present distributions as a result of continental drift, more recent research indicates that the living and extinct ratites are evidently relicts of a widespread northern hemisphere group that probably originated in the early Tertiary. The penguins are another archaic group that has a good fossil record. All indications are that they arose in the circumantarctic area where they are still numerous. The only exception to this pattern is the one penguin species that has become established on the Galápagos Islands.

A phylogenetic and distributional scheme for the order Galliformes, another relatively primitive group, has been presented by Cracraft (1973). He considered the Megapodiidae (mound birds) and the Cracidae (curassows) to be the most primitive taxa; they are confined to the Australian and Neotropical Regions respectively. Cracraft's plan involves a distribution of primitive galliforms across Gondwana in the Cretaceous. A subsequent dispersal of descendent forms to North America supposedly gave rise to the Phasianidae (grouse, quails, turkeys) which then migrated to the Old World and produced other pheasant groups (pheasants, junglefowl, peafowl) and the Numididae (Guinea fowl). This model is quite different from that given by Mayr (1946, 1976) who, noting the occurrence of two cracid species in the Tertiary (lower Miocene and lower Pliocene) of North America, felt that the family originated in the tropics of North America then moved southward. More recently, Sibley and Ahlquist (1991) determined, with DNA-DNA hybridization studies, that the megapodes and cracids were sister groups. This indicates that the two families had a common ancestor.

Both Mayr (1946) and Darlington (1957) felt that the Phasianidae originated in the Old World. In fact, Darlington pointed out that pheasants reach their greatest diversity in the Oriental Region and that the group shows an apparent pattern of radiation from that area and that the cracids in the Neotropics and the megapodids in Australia may be the survivors of one or more earlier radiations; the Guinea fowls would have easily reached Africa from the evolutionary center in the Orient. A worldwide galliform pattern of high diversity and the more advanced genera and species in the Oriental Region, with the more-primitive groups in the peripheral areas, is similar to that found for other animal groups. Cracraft's Gondwana hypothesis would require a mid-Mesozoic origin for a group that is probably not that old.

In regard to the higher groups of birds, Cracraft (1973) suggested the suboscine passeriform families had a Gondwana history. Certainly, the present distribution of the 15 families in this group is predominately southern; nine of them are typical of or confined to the Neotropics, three are confined to the Australian Region, one is found only on Madagascar, and only two (Eurylaimidae and Pittidae) extend to the northern hemisphere in large numbers. It is this group of families, together with those mentioned previously, that account for the generally primitive cast of the bird faunas of the southern hemisphere, particularly Australia and South America. Because of their much-greater development in South America, it seems reasonable to consider South America to be the most-likely center of origin for the suboscine families.

The oscine passeriforms or songbirds number about 4000 species and are considered to be the most advanced and successful of all major bird groups. They dominate the avian fauna of the Oriental, Palearctic, Nearctic, and Ethiopian Regions and are well represented in the other areas. Mayr (1946, 1976) and Darlington (1957) considered that most of the evolutionary development took place in the Old World tropics and that the New World has received most of its songbird fauna from the Old World. Also, Darlington suggested that, over the world as a whole, the suboscines are being replaced by the oscines. Many of the dominant Old World families have not yet reached America and others have done so only recently.

Mammals

The modern distribution of mammalian families has been summarized by Brown and Gibson (1983). Each of the six biogeographic regions has endemic families. But history shows that some of these families have once lived elsewhere and that many others are shared among some of the regions. The Nearctic and Palearctic are the most closely related. Of the 24 terrestrial families now present in the Nearctic, 17 occur or once occurred in the Palearctic.

In the Nearctic, most of the more important mammalian invasions took place via Beringia. The following have had notable effects on the composition of the present-day fauna of North America: Rodents arrived from Asia in the latest Paleocene. Their invasion may have been responsible for the demise of the multituberculates and the last of the therapsids. The earliest lagomorphs probably immigrated from Asia in the Eocene (Webb, 1985a). Modern lagomorphs consist of rabbits, hares, and pikas. Fossils from the Eocene also indicated the arrival from Asia of the first perissodactyls, artiodactyls, and primates. Many new Nearctic families appeared in the Oligocene and some, such as the Castoridae and Tapiridae, are traceable to Asiatic stocks.

Five different North American mammal ages are recognized for the Miocene and all five are characterized by some immigration from Asia. Included were antilocarpids, early genera of cats, bears, the first deer, modern beavers, and ochotonid hares. The highest rate of immigration took place in the Pliocene. Two major contingents from South America showed up, one occurs in deposits of about 2.5 Ma ago and the other about 1.9 Ma ago (Marshall, 1988). Most of these South American invaders eventually died out, but a porcupine, an opossum, and an armadillo still remain. From Asia came the elk, reindeer, jaguar, wolf, and advanced rodents. Pleistocene arrivals from Asia were mountain sheep, musk ox, moose, and humans. Many other large mammals, including the saber-tooth cat, mammoths, and mastodons, also invaded but did not survive the Pleistocene.

The Palearctic, with its much larger area and greater faunal diversity, exported more genera than it received. At first, there was successful (colonizing) traffic across Beringia in both directions. For example, camels (Camelidae) arose in North America in the Upper Eocene and reached Eurasia in the Pliocene. The horse family Equidae radiated in North America beginning in the Eocene and, from then until the Pleistocene, six different genera invaded Eurasia (MacFadden, 1992). By the late Pleistocene, the successful mammalian traffic became entirely one-way, with 21 genera going from Asia to North America but none in the opposite direction.

Aside from its primary relationship to the Nearctic, the Palearctic faunal makeup was influenced from both the African and Oriental Regions. The African influence arose following its connection with Eurasia in the early Miocene. This allowed many African mammals, including the proboscideans and several bovid genera, to become widespread in Eurasia. Oriental mammals have been historically able to move northward into the Palearctic, especially during times of favorable climate.

In the early Eocene, Europe was still separated from Asia so that a continuous Palearctic fauna did not exist. Tectonic rifting between North America and Europe in the Paleocene had apparently made intercontinental migration difficult. However, by the early Eocene, a migratory corridor via Greenland existed and the mammalian faunas became remarkably similar. At least half of the 61 land mammal genera of Europe had congeners in North America. This relationship was closer than any other time in history. Perissodactyls, artiodactyls, and rodents all arrived in Europe at this time. With the retreat of the Turgai Sea in the Oligocene, many species migrated from Asia to Europe. This changeover (Grand Coupure) marked the establishment of the present widespread Palearctic fauna.

The Oriental Region shares most of its families with the Ethiopian. Most of that intermigration has taken place since the early Miocene when the African connection occurred. In the late Tertiary, northern Africa had a high rainfall and a tropical vegetation. The Ethiopian Region may be divided into a subregion for mainland Africa and another for Madagascar. The latter has a peculiar history which has been alluded to in the historical chapters. Eisenberg (1981) called attention to the importance of Madagascar as a refugium. Its mammalian fauna derives from a series of invasions that probably began in the Paleocene and ended in the Pliocene. The insectivores were among the earliest invaders. They were followed by the lemuroids, probably in the Eocene. The dramatic radiation of both groups arose from primitive forms that were conservative in morphology.

As noted in the Paleocene account, there is some evidence of occasional connection between Africa and Europe by means of which primitive stocks of insectivores, primates, and ungulates may have been introduced into Africa. These dispersals may have been via an Apulian plate between the two continents. Subsequently, the first two groups were able to reach Madagascar but not the third. The tenrecoid insectivores of Madagascar, which are related to the African Potomogalidae, evolved to eventually occupy most of the small-mammal habitats that have been described for continental areas. At least a dozen genera and subgenera may be recognized (Eisenberg, 1981). The lemuroid primates on Madagascar radiated into five families with more than 20 species.

The isolation of the African mainland from probably the early Eocene to the early Miocene resulted in the evolution of 15 families that are still endemic to Africa. Also, families of proboscideans and artiodactyls first arose in Africa then migrated to Asia, Europe, and the New World. Among the primates, the first catarrhine fossils were found in the early Oligocene of North Africa. These Old World monkeys belong to the family Cercopithecidae which includes a large variety of genera and species. Some of them dispersed to South Asia where they have continued their radiation. There are now 11 living genera and well over 100 species.

As we noted in Chapter 7, members of the superfamily Hominoidea are known from the Miocene of east Africa, some 22 Ma ago. At the end of the early Miocene, hominoids migrated into Europe and Asia. The fossil record of the family Hominidae began in Africa about 4 Ma ago. The great apes and humans comprise a related group of higher primates. While it seems clear that the Hominidae originated in Africa, this is possibly true for the great apes (Pongidae) as well. The more advanced apes, *Pan* (chimpanzees) and *Gorilla* (gorillas), are confined to Africa while the more primitive *Hylobates* (gibbons)

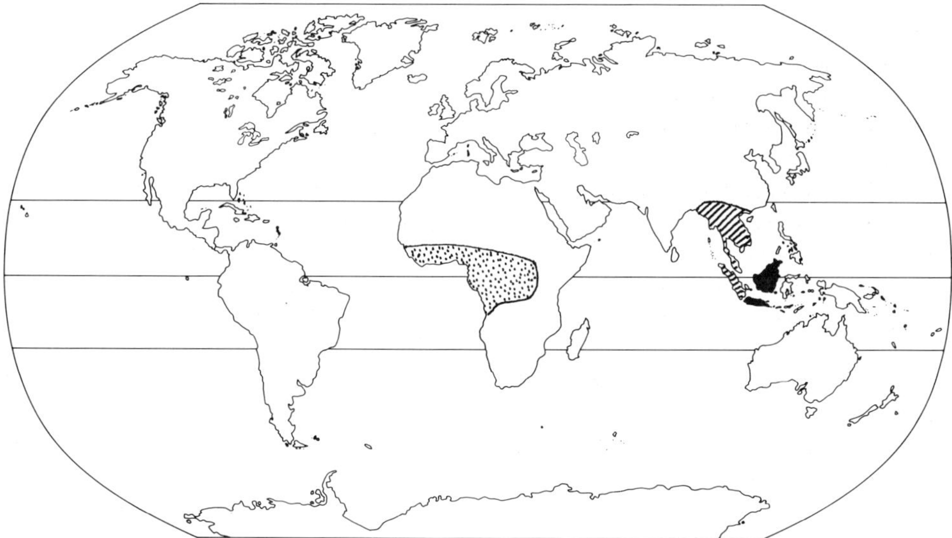

Fig. 106. Distribution of gibbons, *Hylobates* (lined), the orangutan, *Pongo* (black) and the chimpanzee and gorilla, *Pan* (stipple). After Briggs (1984).

and *Pongo* (orangutan) are restricted to southeast Asia (Fig. 106). This common geographic pattern, where the more advanced taxa occupy the putative center of origin, suggests that Africa was the center of evolutionary radiation for all the higher primates.

By the early Cretaceous, South America was separated from Africa and, by the late Cretaceous-Paleocene, the peninsular connection to North America was broken. This left South America in virtual isolation for more than 60 Ma. Just prior to the loss of the Central American connection, South America probably received immigrant marsupial, edentate, and ungulate mammals. The New World (platyrrhine) monkeys and the caviomorph rodents may have immigrated, via waif dispersal along the island chain, from the north in the early Tertiary. All of these groups underwent significant radiations.

Prior to the Great American Interchange of the Pliocene, South America contained a great variety of unique mammalian taxa. The early ungulates had diversified into six orders with many families and genera; the edentates produced armadillos, glyptodonts, anteaters, and sloths; the caviomorphs diverged into guinea pigs, chinchillas, agoutis, and capybaras; and the primates produced three families with about 25 species. The completion of the isthmian connection about 3 Ma ago resulted in a great interchange between the Americas. The impact on the composition of the South American fauna has been summarized by Marshall (1988). Today, the interchange invaders comprise about 44% of the families and 54% of the genera. So nearly half the families and genera now on the South American continent belong to groups that have emigrated from North America during the past 3 Ma.

It seems clear that Australia became separated from the rest of Gondwana in the Jurassic (Molnar, 1992). The isolation of Australia ended, to some degree, in the Paleocene-Eocene when a filter route to South America became available. The southern tip of Australia was in close proximity to Antarctica and the latter was probably connected to South America by an island chain. This permitted a limited invasion which involved the marsupials and a condylarth but none of the other mammalian groups. The discovery of a polydolopid marsupial from the late Eocene of the Antarctic Peninsula (Woodburne and Zinsmeister, 1984) helped to substantiate the theory of an Eocene marsupial dispersal via Antarctica.

In Australia, the marsupials diversified until they were able to occupy almost all of the habitats that are filled by placental mammals on the northern continents. Before the arrival of the marsupials, Australia was occupied by the egg-laying monotremes. This ancient group, represented by the modern platypus and the spiny anteaters, may have become isolated in the Australian Region by the early Jurassic. An Australian monotreme has been described from the early Cretaceous (Archer et al., 1985).

Placental rodents of the family Muridae (about 13 genera) are abundant in Australia. It seems clear that they reached there by island hopping down the East Indian chain. While Simpson (1965) suggested that the first arrivals might have entered Australia in the Miocene, more recent research on murid relationships (Hand, 1984) seems to indicate a later date, the early Pliocene. The oldest rodent fossils are 4–5 Ma in age. Invasions by the genus *Rattus* evidently took place in the Pleistocene. Throughout, the predominant movement has been from the smaller East Indian islands to the great island of New Guinea and then to Australia. The only other native placental land mammal is the dingo (*Canis dingo*) which was almost certainly brought in by the aborigines.

The mammalian history of India, prior to its contact with Asia, is very brief. A new genus of Cretaceous mammal from India was described by Prasad and Sahni (1988). Its relationships were determined to lie with genera from Morocco and North America, so it probably reached India from North Africa. Fossils from the middle to late Eocene represent a variety of orders and families with European, North American or African affinities. The mammalian migrations to India came through two gateways, one in the Assam region to the east and one to the northwest. The migratory flow through the Assam gateway appeared to have been the greater (Kurup, 1974).

Amphibians and reptiles

The salamanders (Caudata) may have evolved in the late Permian or early Triassic (Carroll, 1988) but the earliest fossils are from the middle Jurassic. Their history is completely Laurasian and their modern distribution may be correlated with continental changes that have taken place from the Cretaceous onward. The most primitive of the living families, according to Milner's (1983) cladistic analysis, is the Sirenidae. It is restricted to southern and eastern North America. The Cryptobranchidae probably arose in Asia, reached North America via Beringia, and entered Europe following the desiccation of the Turgai Sea in the Oligocene. It's sister family, the Hynobiidae, originated in Asia and remained there.

The salamander family Proteidae was probably Euramerican in origin as indicated by living genera in Europe and eastern North America. The Salamandridae probably originated in Europe but reached North America prior to the continental separation that took place in the early Eocene. The Plethodontidae evolved primarily in North America, crossed to Asia, and eventually to Europe. Plethodontids also spread southward to Central America and apparently entered South America by the late Miocene (Duellman and Trueb, 1986). The rest of the salamander families (Amphiumidae, Dicampodontidae, Ambystomatidae) apparently originated in, and have remained endemic to North America.

There are five living families of caecilians (*Gymnophiona*). Three occur in South America with two being endemic. The family distribution is primarily southern hemisphere except that three families occur in India. They were supposedly transported from Africa to Asia by the Indian plate (Duellman and Trueb, 1986). One group remained on the Seychelles when those islands became detached. An African-South American origin, prior to continental division, seems probable.

The frogs (Anura) are also an old amphibian group. The earliest-known true frog is *Vieraella* from the Lower Jurassic of Argentina. It is a member of the family Leiopelmatidae which has two living genera, one in North America and one in New Zealand. Another ancient group, the pipoids, is represented by one living family in Mexico and Central America and another in Africa-South America. Two other archaic families, the Pelobatidae and the Discoglossidae, have living species in the northern hemisphere.

The evolutionary radiation of the modern frog families has taken place in the southern hemisphere. Before their separation, Africa and South America apparently shared four families while three others were endemic to Africa. From Africa, Madagascar-India apparently received four families and these were transported to Asia by the Indian plate (Duellman and Trueb, 1986). One stock was left behind on the Seychelles when that island group broke off from India. In the late Cretaceous, four families evidently dispersed from South to Central America. Later northward migrations took place in the Tertiary.

It has been suggested that the widespread frog family Myobatrachidae was derived from the South American Leptodactylidae. The former subsequently spread to Africa, India, and Australia. The Australian Hylidae may have come from South America via Antarctica. It seems evident that the initial radiation of the modern frog families took place on Africa-South America rather than Gondwana as a whole. From those two continents, they were eventually able to reach North America, Madagascar, Eurasia, and Australia. Considering that most of the archaic frog families have a relict distribution in Laurasia or New Zealand, and the fact that frogs are considered to be a monophyletic order, their ultimate origin is likely to have been Africa-South America. The stem amphibian group, the Lissamphibia, may have given rise to the salamanders in Euramerica and the frogs and caecilians in western Gondwana (Africa-South America).

The primitive, lizard-like sphenodontids are first known from the Upper Triassic and were common in the Upper Jurassic of Europe (Carroll, 1988). A few are known from the Cretaceous but there are no Cenozoic fossils. Their general decline may be related to the success of the lizards which became progressively more diverse. Living sphenodonts are limited to a single species of *Sphenodon* that lives on small islands off the coast of New Zealand.

As noted in Chapter 5, the true lizards, or lacertilians, apparently underwent a dichotomy in the Jurassic when Laurasia and Gondwana began to separate. This vicariance produced a southern iguanian group and a northern group ancestral to all other lizards (Estes, 1983). All of the modern families had appeared by the Cretaceous. The most primitive true lizards are placed in the family Iguanidae. This family originally occupied Africa-South America and eventually extended to Madagascar. They apparently reached North America from South America in the late Cretaceous.

The present distribution of the Iguanidae includes the New World, Madagascar, and the Fiji and Tonga Islands. They are not found in Africa where they became extinct probably due to competition from the more specialized agamid and chamaeleontid lizards. The family Agamidae may be considered the Old World counterpart of the Iguanidae. The greatest agamid diversity is in the Oriental Region but they are well represented in Australia, Africa, and southern Eurasia. A gekkonid group apparently originated in Asia. From there, one group became established in Australia and another spread westward to India, Africa, Madagascar, and eventually South America.

The scincomorph lizards diversified in Laurasia. This group includes the modern families Lacertidae, Xantusiidae, Scincidae, Cordylidae, and Teiidae. The teiids are considered to be the most primitive. In the Cretaceous, they were diverse in Laurasia, but are now most common in Central and South America. Most remaining lizards belong to the Anguimorpha, which includes the modern families Xenosauridae, Anguidae, and Helodermatidae. The poisonous *Heloderma* lives in western North America. It has been recorded from the early Tertiary of France, so probably represents an invasion from Eurasia via Beringia. The varanoids have achieved a large size and a predaceous way of life. Included is the Komodo dragon and about 23 other living species in the Old World tropics. The living families of lizards have a complex history as Estes (1983) has indicated.

Snakes probably arose from an ancestral stock within the lizard groups. Nearly all Cretaceous specimens have come from the southern hemisphere, so snakes are usually considered to have originated somewhere in Gondwana (Rage, 1987). A dichotomy apparently produced two primitive groups, the boids and the aniliids. In the latest Cretaceous, both groups probably entered North America via the Central American connection.

The family Boidae includes the boas, anacondas, and pythons. The boas are found in the tropics of the New World, Madagascar, and three small species are isolated in Polynesia. The pythons have a complementary distribution occurring from Africa to the Oriental and Australian regions. It appears that the pythons have the superior competitive ability and may have eliminated the boas from Africa, southeast Asia, and Australia. The distribution of the boas (subfamily Boinae) is remarkably similar to the primitive lizard family Iguanidae. Other Boidae subfamilies have a scattered distribution, mainly in the southern hemisphere.

The great majority of the remaining snake species belong to the superfamily Colubroidea. The stem family is the Colubridae with about 2000 species. The geographical distribution of the most-primitive colubrids is predominately in the Neotropical, Nearctic, and Palearctic regions. Rabb and Marx (1973) noted that one may possibly conclude that successions of colubrids originated in the Old World tropics and, as they moved out from those areas, eliminated or forced out to the geographic periphery, their more primitive relatives.

The family Elapidae seems to represent a monophyletic line that developed a more efficient method of injecting venom than any of the Colubridae (Parker, 1977). This family includes the cobras and the coral snakes. It is most diverse in Australia and the Oriental Region. Elapids also extend in lesser numbers to Africa. To the west, three isolated genera are found on Fiji and the Solomon Islands. It extends north to the Rui Kiu Islands. In the New World, it is represented by three genera of coral snakes. An active evolutionary center in the Oriental Region may account for this pattern (Briggs, 1984).

The snakes with the most effective method for injecting venom are the vipers (Viperidae). They have long, tubular fangs which can be rotated through an angle of about 90 degrees. The group includes the true vipers, pit vipers, and rattlesnakes. The most primitive viperid is restricted to the mountains of southern China, Tibet, and Burma. The true vipers are confined to the Old World with most of the genera occurring in Africa. The genus *Vipera* extends all the way to the Arctic Circle in Eurasia. The pit vipers and rattlesnakes belong to the subfamily Crotalinae. The former are found in both the Old and New Worlds but the rattlesnakes are strictly New World. The origin of the subfamily, and possibly the family itself, was probably the Oriental Region.

Most modern turtles (Testudinata) are semi-aquatic marsh dwellers. The most primitive living group constitutes the side-necked turtles (Pleurodira). The family Pelomedusidae contains three recent genera, two in Africa and Madagascar, and one in South America and Madagascar (but not in Africa). The related family Chelidae has 10 genera, six in South America and four in Australia. Most modern turtles belong to the suborder Cryptodira. The most flourishing family is the Testudinidae. Its subfamily, the Emydinae, contains the most living genera. Of the total of 25 emydine genera, two occur in

South America, two in North America, two in Europe, and about 19 in east and southeast Asia. The Oriental Region appears to be the primary center of origin for this group.

PLANTS

Among phytogeographers, it is customary to recognize six major regions (called Kingdoms) but these are not exactly the same as the six zoogeographic regions. The latest, most authoritative work on the subject is that of Takhtajan (1986). The Holarctic Kingdom includes all of the temperate northern hemisphere. So it is equivalent to the Holarctic Region that is often recognized by zoologists. The tropics are divided into the Paleotropical Kingdom (= the Oriental and most of the Ethiopian Region), Neotropical Kingdom (= Neotropical Region), and the Cape Kingdom. The latter, with no zoological equivalent, is named for the highly distinct flora of the Cape region of southern Africa. In the temperate southern hemisphere, there is an Australian Kingdom (= Australian Region), and a Holantarctic Kingdom. The last includes southern South America, New Zealand, and the subantarctic islands. It does not have a zoological counterpart, yet many amphinotic plant and animal patterns are very similar.

The plant kingdoms are subdivided into subkingdoms, regions (Fig. 107), and provinces. Kingdoms and subkingdoms are characterized by a high percentage of endemic families and subfamilies, regions by many endemic genera, and provinces by endemism at the species level. Sometimes different districts are designated for subspecific endemism. Use of the various categories is somewhat subjective since there are no rules about the extent of endemism for the nomenclature of a given area.

Rather than attempting a description of each area and its characteristic vegetation, a task that has been accomplished by Takhtajan (1986), the distributions of some major plant groups at different evolutionary levels are discussed briefly. In plants, as in animals, biogeographic and phylogenetic patterns are interrelated so the knowledge of one can often help to unravel the other.

Bryophytes and Pteridophytes

The liverworts (Division Hepatophyta) are among the most simple land plants and may have been the first group to have left the aquatic environment. The Holantarctic has been shown by Schuster (1983 and earlier works) to contain an unusually high preponderance of primitive genera, many of which are endemic. His data indicate a high concentration in the Antipodes area (New Zealand, Tasmania, southeastern Australia, New Caledonia) and a significant, but lesser, number in southern South America. For example, all four genera of the Lepidolaenaceae plus the monotypic Goebeliellaceae show the typical pattern. Schuster pointed out that 45–50 genera could be mapped, all showing essentially the same pattern. In contrast, families belonging to more advanced orders tend to be distributed in continental areas at lower latitudes.

Other hepatophytes show disjunct distributions in the north temperate zone (Schuster, 1983). The genus *Mastigophora* (Mastigophoraceae) is of unusual interest since it is rel-

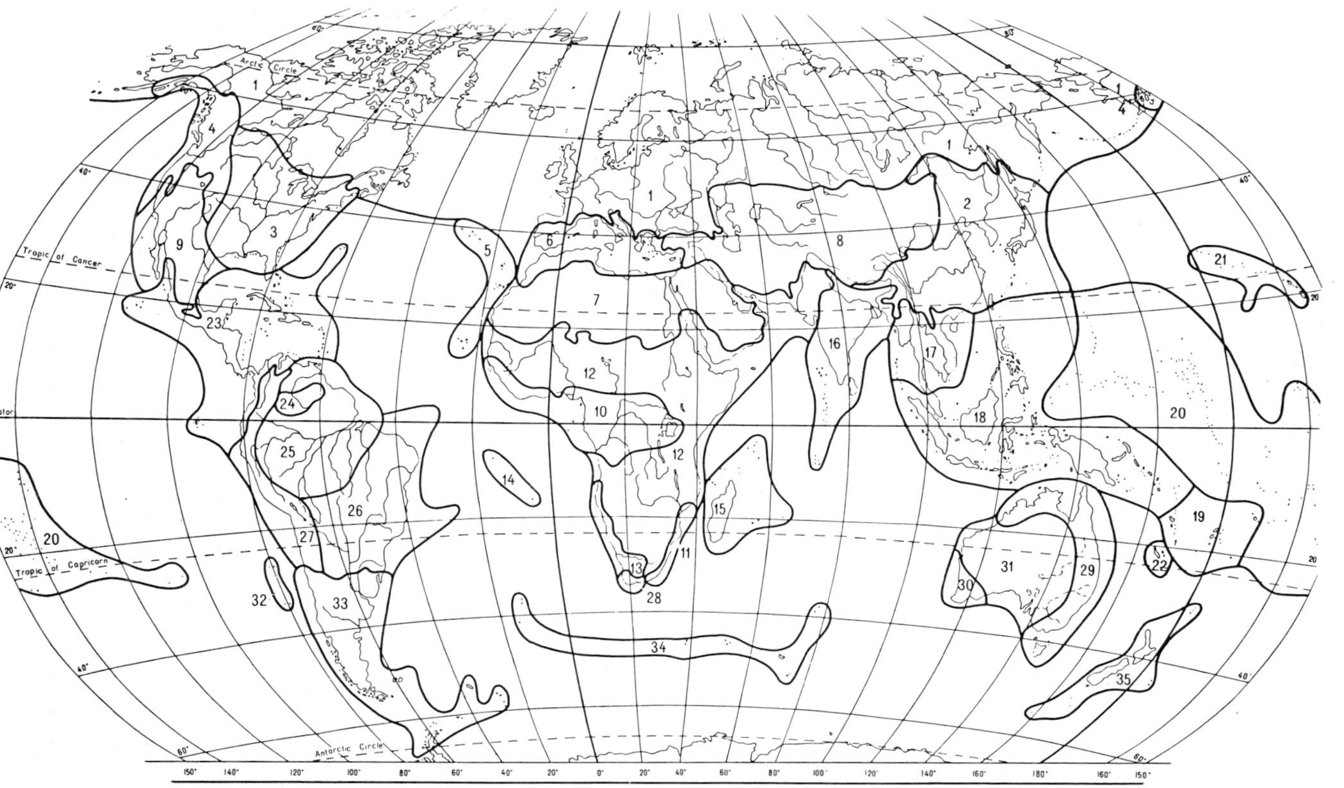

Fig. 107. Floristic regions of the world. 1, Circumboreal Region; 2, Eastern Asiatic Region; 3, North American Atlantic Region; 4, Rocky Mountain Region; 5, Macaronesian Region; 6, Mediterranean Region; 7, Saharo-Arabian Region; 8, Irano-Turanian Region; 9, Madrean Region; 10, Guineo-Congolian Region; 11, Uzambara-Zululand Region; 12, Sudano-Zambezian Region; 13, Karoo-Namib Region; 14, St. Helena and Ascension Region; 15, Madagascan Region; 16, Indian Region; 17, Indochinese Region; 18, Malesian Region; 19, Fijian Region; 20, Polynesian Region; 21, Hawaiian Region; 22, Neocaledonian Region; 23, Caribbean Region; 24, Region of the Guayana Highlands; 25, Amazonian Region; 26, Brazilian Region; 27, Andean Region; 28, Cape Region; 29, Northeast Australian Region; 30, Southwest Australian Region; 31, Central Australian or Eremaean Region; 32, Fernándezian Region; 33, Chile-Patagonian Region; 34, Region of the South Subantarctic Islands; 35, Neozeylandic Region. After Takhtajan (1986).

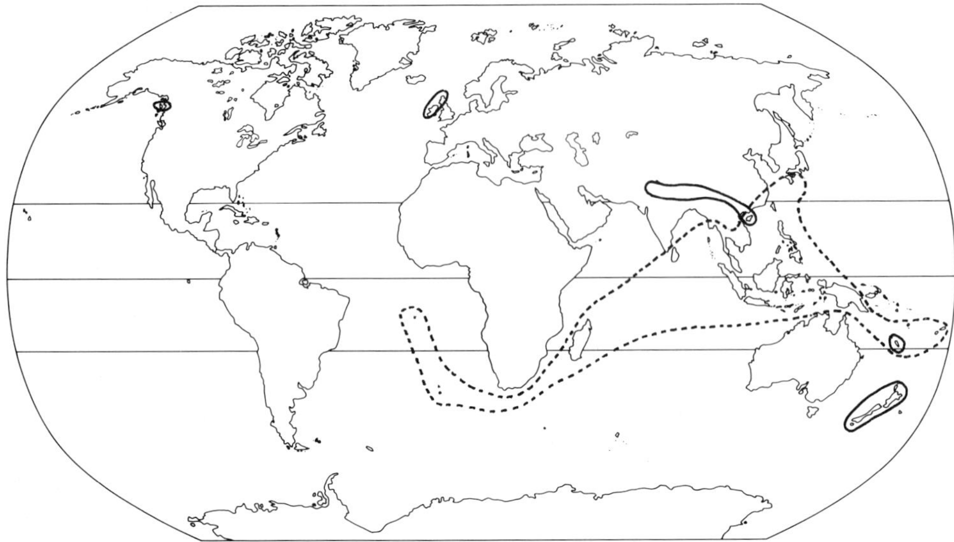

Fig. 108. Distribution of the genus *Mastigophora* showing one northern disjunct species, southern endemics on New Caledonia and New Zealand and the widespread *M. diclados*. After Schuster (1983).

ict in the north and in the south, yet one species (*M. diclados*) has become widespread from the south Pacific to some volcanic islands in the south Atlantic (Fig. 108). Schuster postulated a Gondwana origin with the ancestor of the northern species being carried there by the Indian plate. He noted that the ages of some species may be on the order of 5–50 Ma. Many of the old species have apparently become unisexual and lack asexual reproductive devices. For these reasons, they seem to provide good material for tracing old continental relationships. This brings up the question of the widespread *M. diclados*. Perhaps it suddenly developed sporophytes (which are unknown in the genus).

In his monograph of the hepatic genus *Clasmotocolea*, Engel (1980) found that all of the species of the two most-primitive subgenera occurred in southern South America. He felt that the genus had originated in South America and had been distributed to other parts of the subantarctic area by continental drift. However, recent work on Macquarie Island by Selkirk et al. (1990) has revealed the presence of five species of *Clasmotocolea*. The subaerial age of the island was determined to be under about 300 000 years. It seems that the genus must, after all, be an effective colonizer. Van Zanten (1983) felt that similarities of bryophyte taxa at the generic level between southern South America and Australasia could be explained by plate tectonics, while similarities at the species level could be explained by tectonics for slowly evolving species and by dispersal for rapidly evolving species.

Although pteridophytes in general tend to be widespread due to spore dispersal by wind, there are some exceptions. The relationships of species in the genus *Isoctes* (Isotaceae), which are found in Africa, India, and Australia, have been attributed to the Gondwana origin of the genus (Jermy, 1990). The fern family Loxomataceae includes two genera. One ranges from Bolivia to Costa Rica and the other is confined to the North Island of New Zealand (Kramer, 1990). The curious lack of diversity in Africa, compared to the Neotropics and southeast Asia, is found in ferns as well as other vascular plants. The large family Thelypteridaceae has 440 species in Malesia, 300 in the Neotropics, and only 55 in Africa.

Gymnosperms

The living gymnosperms may be divided into two subdivisions, the coniferophytes and cycadophytes (Page, 1990). The family Ginkgoaceae was very widespread in the Jurassic and Cretaceous but is now represented by a single species, *Ginkgo biloba*, native to eastern China. Another very old family is the Araucariaceae. The genus *Araucaria* has a disjunct distribution in the southern hemisphere (South America, Australia-New Guinea, and the islands of Norfolk and New Caledonia). But the fossil material indicates a much wider distribution in the Mesozoic and early Tertiary. In the southern hemisphere it once occurred in Antarctica, southern Africa, India, New Zealand, Tasmania, and Kerguelen Island. To the north, it ranged from North America to Greenland, Europe, and Russia. The one other living genus is *Agathis*, found primarily from southeast Asia to Australia and out to New Caledonia, Solomon Islands, New Hebrides, Santa Cruz Islands, and Fiji; it also reaches north to the Philippines and south to New Zealand. The small island of New Caledonia has five endemic species of *Agathis* and 13 of *Araucaria*.

The family Cupressaceae contains about 20 genera and 125 species widely distributed in both hemispheres (Page, 1990). The most wide-ranging genera are almost exclusively northern. To the south, four genera are circum-Pacific, four are African-Australian, two are in China-southeast Asia, and two are in southern South America. Single genera are isolated in Japan, Tasmania, and New Caledonia. Three small conifer families have restricted distributions in the Far East. These are the Cephalotaxaceae from the Himalayas to Japan, Taiwan, and Thailand; the Phyllocladaceae from the Philippines to New Zealand and Tasmania; and the Sciadopityaceae which is endemic to Japan.

The Pinaceae is a family of about 12 genera and 200 species almost entirely confined to the northern hemisphere (Page, 1990). Its origin is probably late Mesozoic, with the modern genera having evolved by the Oligocene. Most of the species of the large genera (*Pinus*, *Picea*, *Abies*, *Larix*) are widely distributed in both the New and Old Worlds. Two genera are found only in east Asia and North America, three are restricted to east Asia, and one is endemic to western North America. There are significant concentrations of local endemic species in eastern Asia and western North America.

The Podocarpaceae, with at least 17 genera and about 125 species, is the most important southern hemisphere conifer family (Page, 1990). The various genera are widely distributed and indicate a relationship among all the large, southern land masses. Fossils have been found on Antarctica, India, and Kerguelen Island (Florin, 1963). The genus *Podocarpus* extends across the southern hemisphere and, in the New World, northward to Mexico and the West Indies; in the Old World, it reaches to the Rui Kiu Islands and Japan. The genus *Afrocarpus* is endemic to southern Africa, but almost all the other genera are located in the area from southeast Asia to Australia, New Zealand, New Caledonia, and Fiji. Two genera demonstrate a South America-New Zealand-New Caledonia link and one is endemic to southern South America.

The Taxaceae is a small family of four genera and about 20 species (Page, 1990). *Taxus* is the only widespread genus. It occurs through North America and Eurasia and southward to Java. *Torreya* is found in China and Japan, but also has one species in California and another in Florida. The monotypic *Pseudotaxus* is confined to east central China and *Austrotaxus*, also monotypic, is endemic to New Caledonia.

The Taxodiaceae, a family of nine genera and about 16 species, is widespread in the northern hemisphere, but one genus exists in Tasmania. The family once played a conspicuous role in the forest vegetation of North America and Eurasia but has undergone extinction to such a degree that its present range is made up of discontinuous relicts (Florin, 1963; Page, 1990). *Taxodium* is found in the northeastern United States to Florida, with a discontinuous species in Mexico. *Cunninghamia* is found from northern China to Taiwan, *Taiwania* from northern China and Taiwan to Burma, and *Cryptomeria* in China and Japan. The remaining genera are highly localized: *Sequoia* and *Sequoiadendron* in western North America, *Glyptostrobus* and *Metasequoia* in south and central China, and *Arthrotaxus* in Tasmania.

The cycadophytes contain two classes, the Cycadate and the Gnetatae. There are four living cycad families (Johnson and Wilson, 1990). The Bowenaceae is endemic to northeastern Australia; the Cycadaceae is found from east Asia to Australia and Polynesia; Stangeriaceae is endemic to southeastern Africa; and, Zamiaceae ranges in North, Central, and South America as well as Africa and Australia. The Gnetatae is a bizarre group comprising three families, each with a single genus. They exhibit a mixture of gymnosperm and angiosperm characters. The Ephedraceae is widespread in the arid regions of Eurasia and America. The Gnetaceae is a pantropical, lowland family with a center of diversity in southeast Asia. The Welwitschiaceae exists as a single species in the Namib Desert of southeastern Africa (Kubitzki, 1990).

Angiosperms

The distribution of many of the most primitive dicot families has been analyzed by Schuster (1972, 1976). Of a total of 19 archaic families, he found that nine occur in the southeast Asian area. Of the nine, four are endemic to that area with the remaining five being also found in North America. Eight of the families, including seven endemics, are concentrated in the Australian-New Zealand area. Three families, two endemic, are found in South America. The one widespread family is the Winteraceae.

In his review of the orders and families of primitive living angiosperms, Smith (1972) considered the Winterales, with the single family Winteraceae, to be the most-primitive angiosperm order. The details of the distribution of the Winteraceae were published by Takhtajan (1969). The largest and most-primitive genus is *Drimys* which occurs from the East Indies to Australia, Juan Fernandez Island, and from Mexico to Cape Horn. The other six genera are much more restricted: *Bubbia* (New Guinea, Queensland, Lord Howe Island, New Caledonia, Madagascar), *Belliolum* (New Caledonia, Solomon Islands), *Pseudowinteria* (New Zealand), *Exospernum* and *Zygogynum* (New Caledonia), and *Tetrathalamus* (New Guinea).

Another very primitive dicot family is the Magnoliaceae. This family is represented by 11 genera in the general southeast Asia area (two of them extending to New Guinea); three of these genera are also known from the New World but there are no endemic genera there compared to eight for southeast Asia (Schuster, 1976). Four small, relict families of primitive angiosperms are virtually confined to Australia (Endress, 1983). The Idiospermoideae and the Austrobaileyaceae are endemic to a small section of northeastern Australia. The Himantandraceae extends from Australia-New Guinea to New Britain and the Molluccas, and the Eupomatiaceae is confined to Australia-New Guinea. New Caledonia has seven endemic angiosperm families and the Fiji Islands has one (Takhtajan, 1986).

Aside from the old families that are isolated on the periphery of southeast Asia, the montane forests of that region contain many other primitive taxa. The list includes 19 of 32 families and subfamilies of Annonales, all 11 taxa of Hammelidales, and major old lineages of Beberidales, Theales, and Cornales (Brown and Gibson, 1983). In considering the worldwide distribution of the flowering plants, it is remarkable that so many of the primitive orders and families are concentrated along the western side of the Pacific Basin. This has led to considerable speculation about the significance of this pattern in relation to southeast Asia as the putative center of origin for the angiosperms.

Thorne (1963) felt that southeastern Asia could be the primary site of angiosperm evolution because there, and in the adjacent Malayan-Papuan region, are concentrated the great majority of living angiosperm relicts as well as the richest and most varied flora in the world. It has been pointed out (Takhtajan, 1969), that during the Mesozoic and Cenozoic, climatic and physiographic conditions were no more stable in that area than in tropical America or even tropical Africa. Despite this, the extremely rich flora of tropical America is significantly poorer in primitive angiosperms than the flora of the western part of the Pacific Basin and the flora of tropical Africa is almost devoid of them. Both Takhtajan and Van Steenis (1972) felt that the geography of the primitive angiosperms indicated a center of origin in the southeast Asian-Australian region. The former said "Assam to Fiji" and the latter "Yunnan to Queensland". Smith (1972, 1973), in his work on angiosperm evolution and the relationship of the floras of Africa and America, emphasized the concept that these plants arrived on the scene too late to have been seriously affected by continental movements. He presented a generalized model of flowering plant distribution indicating probable routes of migration from a center of origin in the southeast Asian area (Fig. 109). Similarly, Thorne (1973), in his analysis, indicated that continental drift was not an important factor. In fact, Thorne found that there were, at the family level, closer ties between South America and southeast Asia than between South America and Africa. Brenan (1978) felt that long-distance dispersal was quite adequate to explain the floristic relationships between tropical Africa and tropical America.

Raven and Axelrod (1974) published a comprehensive review of angiosperm biogeography and included comparisons with various groups of vertebrate animals. They suggested that west Gondwana (proto-South America) was the area of origin for the older biota of Australasia including the austral gymnosperms, primitive orders and families of flowering plants, and probably marsupials, dinosaurs, monotremes, ratite birds, cerato-

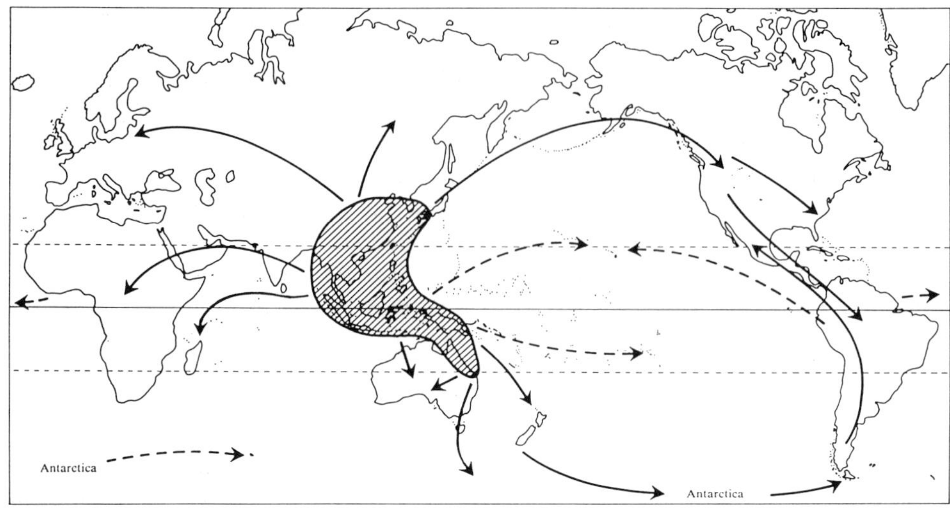

Fig. 109. Smith's (1972) model of angiosperm dispersal from their center of origin. This may be compared to Fig. 38 showing proposed dispersal routes on a late Cretaceous globe.

dont lungfish, and osteoglossomorph fish. This older biota supposedly migrated, prior to middle-Cretaceous time, across Africa-Madagascar-India-Antarctica to reach its present home in the east. Essentially, the same scheme was reiterated in later publications (Axelrod and Raven, 1978; Axelrod, 1979).

Schuster (1976) has emphasized that the high incidence of generalized taxa of Angiospermae in the region from southeastern Asia to New Caledonia, Tasmania, and New Zealand reflects the occurrence of centers of relict climatic conditions rather than a center of origin. He concluded that, for both angiosperms and gymnosperms, southeast Asia was a center of survival and not necessarily a center of origin. On the other hand, Hughes (1976), in his work on the paleobiology of angiosperms, decided that southeast Asia was a center of origin rather than a refugium since tropical Africa and tropical America were devoid of the more primitive groups.

Audley-Charles (1987) reviewed the geological evidence for the dispersal of the eastern parts of Gondwana. He determined that Burma, Thailand, Malaya, and Sumatra were comprised of continental fragments that departed from Australia-New Guinea in the late Jurassic and contacted southeast Asia in the late Eocene. He suggested the possibility that those fragments could have acted as Noah's arks to transport early angiosperms from Gondwana to Asia. Takhtajan (1987) agreed that this might have been the case. However, more recent work (Metcalfe, 1991) suggests that the rifting and movement of the continental fragments took place much earlier, from the late Paleozoic to the Triassic, prior to the angiosperm radiation.

The first angiosperms to appear in northeastern Australia did so some 10 Ma later than they did in contemporary tropical and north-temperate areas. Truswell et al. (1987) provided an extensive list of angiosperm families that appeared in southeast Asia and elsewhere before their arrival in Australia. In regard to southeast Asia, Takhtajan (1969) emphasized that the primitive taxa were found in the mountains while in the tropical lowlands, where the greatest angiosperm diversity was found, the proportion of primitive forms was notably smaller. Takhtajan concluded that the geography of the most archaic flowering plants brings us to the conclusion that their primary center of dispersal was in or near to southeast Asia.

A few of the archaic angiosperm families extended all the way to southern South America. An interesting example is the southern beech family Nothofagaceae, which is closely related to the northern Betulaceae (Nixon, 1989). By the late Cretaceous, the family had spread over an area including Australia, New Zealand, Antarctica, and southern South America (Humphries, 1981). It apparently never reached Africa or India. Other families with essentially the same pattern are the Stylidiaceae and the Winteraceae. The presence of islands between Australia and southeast Asia during the late Mesozoic and early Tertiary very likely provided stepping stones for plant dispersal from the latter to the former.

Considering that (1) a great modern diversity of angiosperm plants occurs in southeast Asia, that (2) the most primitive families are found either on the fringes of that area or in the adjacent mountains, and that (3) many phylogenetic lines leading from primitive to advanced stages are found in that area (indicating a long history of evolutionary development), it is concluded that the center of origin and evolutionary radiation for the flowering plants lies in that part of the world. The presence of *Nothofagus* and other primitive angiosperms in southern South America can be accounted for by a southern migration route greatly facilitated by the presence of an Antarctic continent with a rich, cool-temperate fauna. Even now, 60% of the seed-plant genera and all of the fern genera of the highlands of northern Chiloe (Chile) are found in New Zealand (Godley, 1975).

Finally, it may be worthwhile to speculate about the relationship among the general geographical patterns demonstrated by angiosperms, gymnosperms, and the hepatophytes. It was noted that the earliest conifers are known from the late Carboniferous and early Permian. Thus, if they were not widespread in the Paleozoic, they certainly became so during the coalition of continents that took place by the early Triassic. It seems that the early Mesozoic conifers were tropical lowland plants that probably did not separate into northern and southern groups until the early Cretaceous when the angiosperms took over the habitat to provide a substantial biological barrier (Hughes, 1976).

In his analysis of relict conifer and taxad genera, Florin (1963) called attention to the notable concentration of such genera in east and southeast Asia and to the south in the Australian-New Zealand area plus associated islands; a few were found in southern South America. This general gymnosperm pattern is extremely similar to that shown by the primitive angiosperms (Schuster, 1976). Furthermore, as Schuster (1981) pointed out, the even more ancient (back to the lower Devonian) hepatophytes also show a remarkable concentration of primitive forms in the southern Australian-New Zealand area with some extending to southern South America.

What is the meaning of these coincident patterns that extend through the major groups of higher plants? First, because of their peripheral location, they appear to place considerable emphasis on the southeast Asian area as an important theatre of evolutionary events. Therefore, one may suggest that this part of the world functioned as an evolutionary center not only for the angiosperms but for the earlier gymnosperms and perhaps even for the still earlier hepatophytes. Second, the general track that does connect the amphinotic regions (southern South America, Australia, New Zealand) seldom extends to Africa, Madagascar, and India. Thus it does not argue for a general Gondwana amalgamation in the middle and late Mesozoic but one in which the latter three areas were excluded.

As the early angiosperms evolved in the tropics of southeast Asia, they expanded northward as well as to the south. They gradually became acclimated to cooler temperatures and were also able to spread into the mountains of east Asia. With the global warming that culminated in the early-Eocene thermal maximum (Wolfe, 1987), warm-temperate conditions reached high northern latitudes. This allowed many angiosperm families to disperse across Beringia and spread across North America. Before the European connection was broken in the early Eocene, some of them reached Europe (Tiffney, 1985). The thermal maximum was succeeded by a cold-temperate climate which permitted the invasion of several conifer genera from Asia to North America (Axelrod, 1990).

The evidence for the influence of east Asia on the angiosperm vegetation of North America lies in the remarkable relationships that have been discovered. There are large numbers of disjunct families and genera, particularly between east Asia and eastern North America. The comparative poverty of the western North American flora is evidently due to orogenic activity that began near the end of the Miocene. The Rocky Mountains and adjacent areas were uplifted along with erosion, faulting, and volcanism. The Pacific coast was affected by marine deposition and uplifts (Dott and Batten, 1988). This geologic activity and associated climatic disturbances fragmented the flora of the west and resulted in high rates of species extinction.

In the east, the Appalachian chain remained relatively quiescent and therefore retained much of its original floral diversity. The close relationships between East Asia and eastern North America are exemplified by a number of recent comparative studies. These were described in detail in the Eocene account. Such relationships are the reason why a single Holarctic Kingdom (Takhtajan, 1986) is recognized rather than separate Nearctic and Palearctic regions as is customary in zoogeography.

The Neotropical Kingdom occupies the southernmost part of the Florida peninsula, parts of the lowlands of Mexico, the islands of the Antilles, all of Central America, and the greater part of South America, with the exception of the southern part. The neotropical flora appears to have its roots in the paleotropics. There are many pantropical families and probably no less than 450 genera (Takhtajan, 1986). Despite such relationships to the Old World, the neotropical flora is highly distinctive. No less than 47 families are endemic. If one includes the Bromeliaceae with about 1500 species and the Cactaceae with more than 2000 species, both virtually confined to the New World, the distinction becomes clearer. In comparison, tropical Africa has only 17 endemic families and southeast Asia 16.

Although many genera that are confined to the tropics exist in both the neo- and paleotropics, there are also many that have more restricted distributions. Good (1974) listed 97 that are shared between the neotropics and Africa, but the total number of genera unknown outside tropical America appears to approach 3000. At both the family and generic levels, it is evident that tropical America has a highly peculiar flora, one that must have evolved after its separation from Africa. A series of narrow marine and evaporite basins between Africa and South America during the early Cretaceous probably constituted a significant barrier even though the two continents were still close to one another (Briggs, 1987).

Tropical Africa is considered to be a subkingdom within the Paleotropical Kingdom (Takhtajan, 1986). Although the angiosperm flora is quite varied, it is not nearly as distinctive as that of the neotropics. Some 60 families have pantropical distributions, but only about 12 are shared exclusively between Africa and South America (Good, 1974). Raven and Axelrod (1974) and Axelrod and Raven (1978) envisioned a rather-late (mid-Cretaceous) final separation between Africa and South America. They suggested that the two continents shared a common angiosperm flora into Eocene time. However, as noted above, the family relationships show that South America has a highly distinctive flora, one that must have evolved following its separation from Africa. Smith (1973), Thorne (1973), Niklas (1981) and others have emphasized the lack of relationship between the two floras.

The Cape region of South Africa is placed in its own plant kingdom. The flora of this small region includes 8550 species, of which 73% are endemic. There are eight endemic or subendemic families and about 200 endemic genera (Takhtajan, 1986). Two of the prominent non-endemic families are the Proteaceae (Fig. 110) and the Restionaceae. Although Johnson and Briggs (1975) suggested that both families may have reached South Africa from Australia in Gondwana times, an early Mesozoic origin does not seem possible in view of the general consensus that angiosperms did not diversify until the early Cretaceous.

Madagascar constitutes a subkingdom within the paleotropics. Besides the main island, the region includes the Comores, Aldabra, Seychelles, the Mascarenes and other nearby islands. There are 12 endemic families and well over 400 endemic genera. In

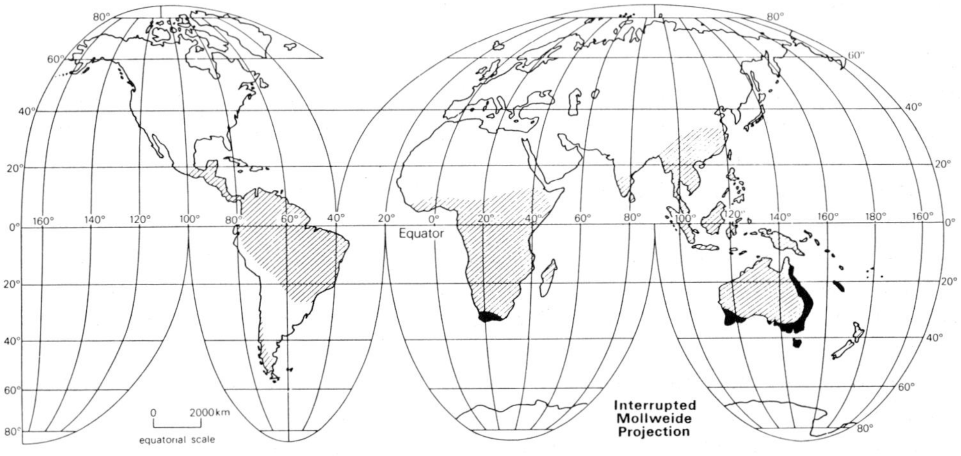

Fig. 110. Distribution of the family Proteaceae. Areas of greatest generic diversity are shown in black. After Stott (1981), based on Johnson and Briggs (1975).

addition to its endemism, the Madagascar flora is peculiar in its remote relationships (Takhtajan, 1986). There are several genera with species that also occur in South America but not in Africa. Several other genera have species in Australia and some show relationships to New Zealand, New Caledonia, and Malesia. On the other hand, the relationship to Africa is not very close. The general proportion of species endemic to Madagascar is about 90%.

Some of the major plant groups that are still very diverse on Madagascar have become relatively scarce in Africa. The palms (Order Principes) are a large, circumtropical monocot group consisting of more than 200 genera and over 3000 species. They are very well established in southeast Asia and the New World, but are scarce in Africa which has only three genera and a few species (Moore, 1972). In contrast, Madagascar has 18 genera and more than 100 species. A similar distribution may be noted for the bamboos (Bambusaceae), a large tropical group of about 50 genera and 500 species (Good, 1974). About 90% or more of the species are Asiatic or American. There are only about six genera and 14 species in Africa. Madagascar has nine genera and about 30 species.

The Indomalesian Subkingdom includes the tropics from India to Indochina and out to the Fiji Islands. Although the entire area has 16 endemic families, none of them are endemic to the Indian Region. This may be due to the fact that India did not receive most of its angiosperm flora until it achieved a broad connection to Asia in the Eocene (Krassilov, 1972). The Malesian Region includes the southern part of the Malay Peninsula, north to the Philippines and Taiwan, and east to New Guinea and the Solomon Islands plus other small island groups (Takhtajan, 1986). Its extremely rich flora comprises more than 25 000 species and it is characterized by exceptionally large numbers of archaic angiosperms. New Guinea itself contains more than 9000 species, with about 8500 of them considered to be endemic. Its floral relationships are strongly Indomalesian rather than Australian.

Earlier (p. 328), we noted that Australia probably received its primitive angiosperms from southeast Asia via terranes that existed as islands between the two continents. There is also some evidence that later, there may have been some migratory movements from Australia-New Guinea towards Indomalesia. Among the palms, Dransfield (1987) found evidence of two distinct elements in Malesia, one northern and western Laurasian-Tethyan and the other southern and eastern Gondwanic. The genus *Grevillea* in the Proteaceae seems to provide another example. There are about 190 species in Australia, New Caledonia, and Vanuatu; but four species extend into the Malesian Region.

The Australian Kingdom contains 18 endemic families and 550 endemic genera (Takhtajan, 1986). A filter route to South America via Antarctica, which was probably used during the Paleocene-Eocene by a variety of animal groups, was probably important for plants as well. This could explain the appearance of many Australasian families in southern South America. Examples are the gymnosperms Araucariaceae and Podocarpaceae and the angiosperms Winteraceae, Berberidopsidaceae, Proteaceae, Restionaceae, and others. The same relationship may be found among a number of genera.

In discussing the general relationship of the southern, cool-temperate floras, Moore (1972) noted that between 50 and 60 genera, or distinctive parts of more-widespread genera, occur principally in Australasia and in southern South America, about half of them in New Zealand alone. Most of these genera have their major diversity in the Old

World and appear to have migrated from east to west. The major differentiation of such families as the Myrtaceae, Proteaceae, and Restionaceae took place in Australia, but they may have originally come from the tropics. The earliest fossil occurrence for the Myrtaceae is from the late Cretaceous of southeast Asia (Truswell et al., 1987). Moore also observed that considerable recent or current intercontinental migration of plant species is taking place. Some 32 species, excluding cosmopolitan forms, occur in both the South American and Australian temperate regions.

According to Takhtajan (1986), the Holantarctic Kingdom includes all of the temperate southern hemisphere except southern Australia and Tasmania. This circumglobal area contains the southern part of South America and many isolated islands. There is a historical relationship among these areas which is best expressed at the family and generic levels. These reflect a time when many of these areas were in closer proximity.

The general relationships of the New Zealand flora have been summarized by Godley (1975). The fern flora has a low degree of endemism with about 50% of the species being found in southeastern Australia; of the New Zealand genera, about 85% are represented in Australia. In the seed plants, many of the species and 75% of the genera are shared with Australia. The close species relationships are attributed to trans-Tasman dispersal. On the other hand, many of the important Australian genera of Myrtaceae and Rutaceae and the Australian families Memosacaceae and Casuarinaceae do not occur in New Zealand. Relatively close relationships with Chile are shown by the sharing of 40% of the fern genera and 43% of seed-plant genera.

Temperate South America is included in the Chile-Patagonian Region along with nearby islands and the northern part of the Antarctic Peninsula (Takhtajan, 1986). There are eight endemic or subendemic families. The mixed flora is derived from ancient amphinotic elements along with some taxa of Holarctic origin. The archaic elements include the conifers *Araucaria*, *Podocarpus*, and three genera of cedars. Ancient angiosperms are represented by *Nothofagus*, *Misodendrum* (which is parasitic on *Nothofagus*), *Peumus* (Monimiaceae), *Tepualia* (Myrtaceae), and others.

A review of island biology is included in the next chapter. Only two of the smaller oceanic islands are mentioned here, and this is due to their peculiar plantlife and their geographical locations with respect to the angiosperm center of origin. The island of New Caledonia is recognized as a Subkingdom because of its highly distinctive flora (Takhtajan, 1986). There are seven endemic families and more than 130 endemic genera. The archaic angiosperm genera include six out of the 12 known genera that lack vessels in their woody stems. A valuable survey of New Caledonian biogeography and ecology has been published by Holloway (1979). The great age of the island, possibly Upper Jurassic, has enabled it to accumulate its biota over some 150 Ma. Its external relationships are primarily with the Indomalesian region, but a number of genera are shared with the New Hebrides and Fiji.

The second island of importance in regard to the general angiosperm pattern is Fiji. It has one ancient family, the Degeneriaceae, and 11 endemic genera. About 101 genera of flowering plants apparently reach their eastern terminus on Fiji. Both Fiji and New Caledonia are old islands that were in a fortuitous position to capture early stages of angiosperm evolution as they dispersed from southeast Asia. In this manner, their function was similar to that of Australia and New Zealand.

SUMMARY

1. All major groups of the freshwater and terrestrial biota had their origin in a relatively restricted part of the earth's surface and, over time, became more widely dispersed. There is no evidence that any species, or larger taxonomic group, came into being simultaneously over a worldwide or large continental area.
2. The worldwide patterns of such disparate groups as the birds, mammals, insects, and freshwater fish all conform remarkably well to the regional zoogeographic system devised by P.L. Sclater in 1858 and adopted by Wallace (1876).
3. Of all the animal groups that might be used to trace the earth's tectonic history since the mid-Mesozoic, the freshwater fish are probably the most reliable.
4. Aside from two catfish families that have become secondarily adapted to saltwater, ostariophysan fish have a purely freshwater history and have not been able to reach such places is Madagascar, Australia, New Zealand, and the West Indies. This permits us to predict that these land masses have not been connected to any larger continent since the Upper Jurassic.
5. Some fish groups, now confined to freshwater, had marine ancestors so that their present distributions may have been achieved by oceanic migration. Examples are the lungfish (Ceratodidae), the bonytongues (Osteoglossidae), the Melanotaeniidae, the Percichthyidae, and probably the Nandidae.
6. The presence of six archaic fish families that are endemic in Africa, plus the presence of four old families that are also found in South America, gives Africa a remarkably rich fauna of ancient freshwater fish. In comparison, South America is relatively depauperate. This suggests that South America may have originally received its primary freshwater fish from Africa.
7. The cypriniform fish (minnows) probably arose in southeast Asia in the early Paleogene. But this group, along with the characoids and catfish, forms a monophyletic triad. This means that all three had a common ancestor somewhere. Catfish have a long evolutionary history in southeast Asia, but the more-primitive characoids now exist only in Africa and South America. Characoid fossils are known from the Eocene and Oligocene of France. Are these fossils remnants of an earlier Eurasian distribution or did the characoids make their way from Africa across the Tethys Sea?
8. Another theory says that the three ostariophysan fish groups arose in Gondwana and were transported to Asia when portions of that supercontinent moved northward. But there is nothing in the present or fossil faunas of India and Australia to indicate that this may have happened.
9. The North Atlantic connection of the early Eocene permitted the intercontinental migration of the ancient gars and bowfins. The teleost family Percidae probably originated in Europe and invaded North America before the connection was broken.
10. Primary freshwater fish began to make their way across Beringia from Asia to North America as early as the late Cretaceous. Paleogene migrations included several archaic fish as well as the more advanced families Hypsidoridae, Ictaluri-

dae, Cyprinidae, and Catostomidae. Some reciprocal crossings have been identified.

11. The history of the secondary freshwater fish families Cyprindontidae (killifish) and Cichlidae (cichlids) is of biogeographic interest. Both families are found in Madagascar and the West Indies as well as in Africa and South America. It was suggested that the cichlids probably arose in Africa while the cyprinodontids originally evolved in Central to South America.
12. Seven major groups of insects with aquatic larval stages are old, at least early Mesozoic, and their present patterns are often assumed to have been caused by continental dispersement. These are the stoneflies (Plecoptera), caddisflies (Trichoptera), mayflies (Ephemeroptera), blackflies (Simuliidae), midges (Chironomidae), net-winged midges (Blephariceridae), and dragonflies (Petaluridae). All of these groups have antitropical tendencies and relict families and/or genera with amphinotic distributions. The adult stages are generally capable of flight.
13. In the southern hemisphere, many groups exhibit amphinotic patterns. Southern South America is invariably included and usually Australia and New Zealand. Less often, the pattern extends to southern Africa, Madagascar, or New Caledonia. The amphinotic groups usually represent the most primitive (plesiomorphic) genera or families within their respective larger groups. Their more advanced (apomorphic) relatives are located to the north, either in the tropics or more commonly at boreal latitudes.
14. Many of the amphinotic freshwater groups have been used as examples to illustrate vicariance caused by the rifting of Gondwana. Even though most of the groups considered seem to be sufficiently old, was continental movement the primary reason for the present patterns? Prior to the mid-Miocene, before Antarctica was completely glaciated, the southern land masses were closer together and migration across modest oceanic gaps must have been relatively easy.
15. Among the terrestrial insects, the beetles (Coleoptera) are a very old group with a fossil record extending back to the Lower Permian. The most archaic survivor is the Australian genus *Omma*. Fossil relatives have been described from the Lower Jurassic of England.
16. The beetle family Carabidae has been said to illustrate a world-wide pattern of the evolution of dominant groups in Africa and tropical Asia and dispersal into smaller areas, with replacement of older by more recently dominant groups. In such old families, both dispersal and vicariance are important.
17. The concept of repetitive "taxon pulses" was developed from work on carabid beetles. Taxa arise in equatorial latitudes and are driven along pathways in three directions: towards the poles, up mountains, and into the forest canopy. Various groups of carabids have repeatedly followed these paths.
18. Work with ants indicates that, since mid-Tertiary times, evolution of world-dominating new taxa has proceeded mainly from the combined tropics of Africa and southeast Asia. Older taxa are forced to retreat before the new forms and can shift only from larger to smaller land masses.

19. The dipterid family Drosophilidae probably also arose in the Old World tropics, became acclimated to temperate conditions, and crossed to the New World via Beringia. Further dispersal reached South America and, from there, a reverse migration reached North America.
20. Other major insect groups also have their greatest diversity in the Old World tropics and may have originated there. Examples are the wasp family Vespidae (southeast Asia), the mosquitos (southeast Asia), the earwigs (southeast Asia), and the moth family Sphingidae (Old World). Others are most diverse in the Neotropical Region and may have arisen there: the bees, robber flies (Asilidae), the hemipterid tribe Myodochini, and flat bugs (Aradidae).
21. Many of the large insect groups, that probably originated in the tropics, are now represented by primitive taxa at higher latitudes. The oldest elements tend to occur as isolated relicts in the southern hemisphere. Fossil relatives are sometimes found in the northern temperate zone. The same worldwide pattern is reflected by the spiders and possibly the scorpions.
22. In regard to bird distribution, it is important to note that the biogeographic regions differ in the diversity of their avian faunas. The richest by far is the Neotropical Region with 86 families (31 endemic) and 2780 species. The Ethiopian has 73 families (6 endemic) and 1556 species, the Palearctic has 69 families (1 endemic) and 1026 species, the Nearctic has 62 families (1 endemic) and 750 species, the Oriental has 66 families (1 endemic) and 961 species, and the Australian has 64 families (13 endemic) and 906 species.
23. The Neotropical bird fauna is comprised, to a large extent, of primitive families. Its diversity has been increased, probably since the creation of the isthmian link, by immigration from North America. The southward invasion may have involved about 10 families. Two Neotropical families evidently dispersed to North America.
24. The Palearctic Region encompasses the largest geographic area. The relationship of its bird fauna with that of the Nearctic is especially close. The Palearctic shares 48 of its families, 35% of its genera, and 12.5% of its species with the Nearctic.
25. The Oriental and Ethiopian bird faunas are closely related but the Australian is considerably more independent. There is a surprising lack of relationship between the Neotropical and Ethiopian regions.
26. The bird tribe Corvida (crows, ravens, magpies, etc.) probably originated in Australia. The same is probably true of the parrots and pigeons; their diversity and that of related families is greatest in the southern hemisphere, particularly the Australian Region. A Gondwana history for the latter two has been suggested, but neither family appears to be that old.
27. A Gondwana origin has been assumed for the flightless ratite birds, but they evidently originated in the northern hemisphere in the early Tertiary. They are southern hemisphere relicts of a once widely distributed northern group. The penguins, on the other hand, probably originated in the circum-Antarctic area where they are found today.
28. In the Order Galliformes, the Megapodiidae of Australia and the Cracidae of South America are considered to be the most primitive taxa. The worldwide galli-

form pattern of high diversity, with the more advanced genera and species in the Oriental Region, suggests an origin in that area with a dispersal to the New World via Beringia. The alternative of a Gondwana origin has been proposed, but this bird group is probably not pre-Tertiary.

29. The most advanced and successful of all the major bird groups are the oscine passeriforms (songbirds). They dominate the avian fauna of the Palearctic, Nearctic, Oriental, and Ethiopian regions. Most of their evolutionary development probably took place in the Old World tropics, and the New World has received most of its songbird fauna from the Old World. Over the world as a whole, the suboscines are probably being replaced by the oscines.
30. Among the mammals, the present faunal complex of the Nearctic Regions reflects a strong Palearctic influence which began in the late Cretaceous and continued through most of the Tertiary. Modern North American rodents, sheep, goats, beavers, rabbits, cats, bears, deer, elk, reindeer, and wolves have all evolved from ancestral species that came across the Bering Land Bridge. The sabre-tooth cat, mammoths, and mastodons also invaded from Asia but became extinct in the Pleistocene.
31. Although the Palearctic, with its larger area and greater mammalian diversity, exported more genera than it received, there were reciprocal migrations. From the Nearctic, it received camels, tapirs, and five horse genera. Following the connection to Africa in the early Miocene, Asia was invaded by a variety of proboscidean, bovid, and other genera. Other forms migrated northward from the Oriental Region.
32. Prior to the Oligocene, Europe was still separated from Asia so that a continuous Palearctic fauna did not exist. In the early Eocene, Europe was invaded by a large number of North American mammals by way of the North Atlantic connection. With the retreat of the Turgai Sea in the Oligocene, Europe was invaded from Asia, so that its modern mammalian fauna has been derived from two main sources.
33. The Oriental Region shares most of its mammalian families with the Ethiopian. This mixture began with the early Miocene connection of Africa and Asia through the Arabian Peninsula. The peculiar mammalian fauna of Madagascar was derived from Africa by means of waif dispersal across an oceanic barrier.
34. Primitive mammalian stocks may have been introduced to Africa by an occasional connection in the Paleocene. From that time until the early Miocene, the Ethiopian fauna developed in isolation. Among the many endemic families were the higher primates and humans.
35. By the early Cretaceous, South America was separated from Africa and, by the late Cretaceous-Paleocene, the peninsular connection to North America was broken. Prior to its 60 Ma of isolation, South America had probably received from North America primitive marsupial, edentate, and ungulate mammals. Playrrhine monkeys and caviomorph rodents may have reached South America by migrating along the Central American island chain in the early Tertiary.
36. The completion of the isthmian connection about 3 Ma ago resulted in a significant mammalian interchange between the Americas. Nearly half the families and

genera now on the South American continent belong to groups that have emigrated from North America during the past 3 Ma.

37. Australia apparently became separated from the rest of Gondwana in the Jurassic. The isolation of Australia ended, to some degree, in the Paleocene-Eocene when a filter route from South America via Antarctica became available. This permitted a limited dispersal to Australia which involved the marsupials but none of the other mammalian groups except a primitive condylarth. Before their arrival, Australia was occupied by only one other land-mammal group, the egg-laying monotremes. Placental rodents apparently arrived from Asia in the early Pliocene.
38. Evolution of the modern amphibians had evidently gotten underway by the time Pangaea had become essentially separated into its northern (Laurasian) and southern (Gondwanian) parts. The history of the salamanders has been entirely Laurasian and that of the caecilians almost entirely Gondwanian. The Asian families of caecilians may have been transported there by the Indian plate.
39. The early evolutionary history of the frogs may have been Gondwanian, although no primitive frogs or fossils have been found in Australia. The archaic living frog genera in North America and New Zealand are likely to have come from South America-Africa in the Lower Jurassic or even before that.
40. Considerable frog evolution took place in South America-Africa before those two continents split apart. Subsequently, frog genera dispersed to North America in the late Cretaceous to early Paleocene and across to Asia via Beringia. Some were transported to Asia by India and others migrated to Asia via the Arabian Peninsula. They probably first reached Australia from South America by means of the Eocene filter route. Despite their aversion to saltwater, frogs have been able to cross many minor oceanic barriers.
41. The Upper Triassic to Cretaceous sphenodonts were once widely distributed but now survive only as a single species on small islands off New Zealand. Their relatives, the true lizards, apparently underwent a dichotomy when Laurasia and Gondwana began to separate. The separation produced a southern iguanian group and a northern group ancestral to all other lizards.
42. The most primitive true lizards belong to the family Iguanidae. The iguanids are presently found in the New World, Madagascar, and the Fiji and Tonga Islands. Their extinction in Africa may have been caused by competition from more advanced lizards. Many advanced lizard families evolved in Laurasia and some spread into parts of the southern hemisphere.
43. Snakes probably arose in Gondwana, most likely in the South American-African region. Two primitive groups reached North America from South America in the late Cretaceous. The archaic family Boidae includes the boas and pythons. The boas are found in the New World, Madagascar, and three isolated localities in Polynesia. The pythons have a complimentary distribution occurring in the Ethiopian, Oriental, and Australian regions. The pythons may have outcompeted the boas so that the latter, like the iguanid lizards, now have a disjunct distribution.
44. The more advanced snake families are typical of the Old World tropics with one family (Elapidae) being numerous in Australia and along the west Pacific rim.

The poisonous elapids and viperids (Viperidae) have entered the New World via Beringia.

45. The most primitive living turtles belong to the group called the side-necked turtles (Pleurodira). There are two families, one in South America-Africa-Madagascar and the other in South America and Australia. The more advanced turtles (Cryptodira) are found mainly on the northern continents. Most of the living genera belong to the family Testudinidae which is most diverse in the Oriental Region.
46. Terrestrial plants are usually divided into a series of kingdoms, subkingdoms, regions, and provinces. In general, the kingdoms and subkingdoms are somewhat similar to the regions recognized by zoogeographers.
47. The liverworts (hepatophytes) are among the most simple of the land plants. Although they are widely distributed in mesic habitats, they are represented by many primitive, endemic genera in the Antipodes area (New Zealand, Tasmania, southeastern Australia, New Caledonia). Other genera have disjunct distributions in the north temperate zone.
48. The pteridophytes tend to be very widespread, but some genera have distributions that have been attributed to a Gondwana origin. There is a curious lack of fern diversity in Africa compared to the Neotropics and southeast Asia.
49. Among the gymnosperms, the archaic family Ginkgoaceae was widespread in the Mesozoic but now there is only a single species which is native to eastern China. Another old family, the Araucariaceae, was once broadly distributed in both hemispheres, but now has a relict distribution, primarily in the southern hemisphere. The Cupressaceae are currently widespread in both hemispheres. Three small conifer families have restricted distributions in the far east.
50. The family Pinaceae is almost entirely northern hemisphere. Most genera are widespread, but three are restricted to east Asia and one is endemic to North America. The Podocarpaceae is the most diverse southern hemisphere conifer family. Most of the genera are located in the area from southeast Asia to Australia, New Zealand, New Caledonia, and Fiji.
51. The Taxaceae has one widespread northern genus, another with isolated species in North America and east Asia, a third that is confined to east central China, and a fourth that is endemic to New Caledonia.
52. The Taxodiaceae, once widespread in the northern hemisphere, is now reduced to a series of discontinuous relicts. Three genera exist in North America, five in east Asia, and one in Tasmania.
53. The cycads have four living families. Three of them have limited ranges in the southern hemisphere while one (Zamiaceae) is widespread in the Americas as well as Africa and Australia.
54. The most primitive orders and families of the flowering plants are isolated on the periphery of southeast Asia. They extend from Australia and New Guinea out to New Caledonia, Fiji, and a few westward to Madagascar and South America. Many are also found in the mountains of east Asia.
55. There are two popular theories about the center of origin of the angiosperms. One of them argues for western Gondwana (South America-Africa) and the other for

the general southeast Asian area. The peripheral pattern of most of the primitive orders and families appears to favor the latter.

56. As the early angiosperms evolved in the tropics of southeast Asia, they expanded in all directions. During the global warming of the Eocene, many angiosperm families dispersed across Beringia and spread across North America. Some of them reached Europe before the North Atlantic connection was broken in the early Eocene. The evidence for this major migration lies in the close relationships between the modern floras of east Asia and eastern North America.
57. The Neotropical Kingdom is a distinctive floral region with 47 endemic families and almost 3000 endemic genera. A barrier to plant dispersal between South America and Africa has probably existed since the early Cretaceous.
58. The Paleotropical Kingdom includes all of the Old World tropics and is divided into five subkingdoms. While some 60 families have pantropical distributions, only 12 are shared exclusively between Africa and South America. The Cape region of South Africa is placed in a kingdom of its own.
59. The Malesian Region, which includes the southern part of the Malay Peninsula, north to the Philippines and Taiwan, and east to New Guinea and the Solomon Islands, possesses an extremely rich flora of more than 25 000 species. It is also characterized by large numbers of archaic angiosperms.
60. The Australian Kingdom contains 18 endemic families and 550 endemic genera. Some Australian families of angiosperms and conifers reached South America via Antarctica, probably in the Eocene. Between 50 and 60 genera, or distinctive parts of more-widespread genera, occur principally in Australasia and South America, about half of them in New Zealand alone.
61. The Holantarctic Kingdom includes all of the temperate southern hemisphere except southern Australia and Tasmania. The New Zealand flora is most closely related to that of Australia, but about 40% of its fern genera and about 43% of its seed-plant genera are shared with Chile.
62. The small island of New Caledonia has seven endemic families and more than 130 endemic genera. The archaic angiosperm genera include six of the twelve known taxa that lack vessels in their woody stems.

CHAPTER 12

Significant patterns

So, by the early nineteenth century there were two major biogeographical issues facing naturalists. One was the delineation and enumeration of the regions of creation, and the other was the question of disjunct distributions.

Michael Paul Kind, *Geographical Distribution and the Origin of Life*, 1980.

The continuing discoveries of apparently identical species in widely separated parts of the world presented a perplexing problem to the early naturalists. Either the same species had been created more than once or the disjuncts had some mysterious power of dispersal that could transport them over enormous distances. Prominent figures like Prichard (1826), Lyell (1830–1833), Swainson (1835), and Forbes (1846) believed that a given species could be created only once and that its present distribution must be the result of dispersal from its place of origin. But others, such as Agassiz (1857) and Sclater (1858), were proponents of multiple creations. The botanist de Candolle (1855) advocated natural dispersal when possible but resorted to special creation for some widely separated species.

The pioneering work of Prichard (1826) was important for he used the same kind of evidence that Darwin did in his later work. Prichard emphasized that disjunct distribution in plants could be explained by natural causes. He noted that seeds could be transported by winds, rivers, ocean currents, and animals. Seeds of edible fruits could be carried for great distances in the digestive tracts of birds. Finally, it was Darwin (1859) who presented a body of convincing evidence on the various means of dispersal. In so doing, he observed that ... "the simplicity of the view that each species was first produced within a single region captivates the mind. He who rejects it, rejects the *vera causa* of ordinary generation with subsequent migration, and calls in the agency of a miracle."

ANTITROPICAL DISTRIBUTIONS

The kinds of disjunct distribution that attracted the greatest interest were those cases where organisms, living in the polar or temperate parts of the north, also occurred as the same or a closely related species in the southern hemisphere. In most instances, the northern and southern populations were separated by great distances across tropical environments completely unsuited to the maintenance of species that lived in colder climates. How did such species manage to migrate across thousands of miles of inhospitable territory?

Examples of north-south, interhemispheric relationships were, at first, referred to as "bipolar distributions." According to Ekman (1935, 1953), the term probably originated from statements made by the famous polar explorer Sir James Ross. Ross, who took H.M.S. *Erebus* and *Terror* to the Antarctic in 1839 to 1843, observed that some of the marine animals taken on that voyage appeared to be the same as those he had seen in the Arctic. He assumed that such animals had been able to migrate between the two polar seas by means of the cold, abyssal regions. Hooker (1853) who reported on the botany of the same voyage, suggested that terrestrial plants with bipolar distributions had been able to transgress the tropics by migrating along the mountain tops of the west coast of the Americas.

In his *Origin of Species*, Darwin (1859) devoted considerable attention to the interhemispheric similarity of temperate species now separated by thousands of miles of tropical habitat. He emphasized that plants growing on the tops of high mountains in the tropics were temperate or alpine species and that they often ranged from the north into the southern hemisphere. Darwin also remarked on the interhemispheric similarity of temperate marine groups such as certain fish, crustaceans, and algae.

Darwin (1859) provided an explanation for the north-south disjunctions by supposing that migrations across the tropics had taken place during the Glacial Period when, "tropical productions were in a suffering state and could not have presented a firm front against intruders." He concluded that a considerable number of plants, some terrestrial animals, and some marine organisms had, during the Glacial Period, migrated south with some crossing the equator. When the glaciers receded, such temperate forms moved back to higher latitudes (or altitudes), but a few would travel farther south.

MARINE ENVIRONMENT

Antitropicality (amphitropicality) in the marine environment, with an emphasis on the shelf habitat of the Indo-West Pacific, has been recently discussed (Briggs, 1987a). The several different theories that might account for the existence of such patterns were examined. It was noted that the "relict theory" proposed by Théel (1886) seemed to best fit the evidence. He envisioned that former widespread, warm-water species became extinct in the tropical regions leaving as relicts the northern and southern parts of their populations. Although Théel's hypothesis was supported by others in the late 19th Century, it fell into disfavor and was not revived until many years later (Briggs, 1974a).

When we examine the geographic patterns shown by the better-known Indo-West Pacific antitropical species, it becomes evident that they may be divided into two categories: those that are confined to the western Pacific and those that, in addition, have relict populations in the western Indian Ocean. Examples of these two patterns are provided by the fish species *Bodianus vulpinus* (Fig. 111) and *Myripristis chryseres* (Fig. 112) (Briggs, 1987a). Some interesting variations may also be found. In the gastropod family Mitridae, *Mitra chinensis* is found only along the southeast Asian coast and in southeast Africa (Fig. 113). A close relative, *M. triplicata*, occupies the intermediate area (Cernohorsky, 1976). In the gastropod genus, *Strombus*, the species *S. vomer* is comprised of three subspecies with one of them, *S. v. vomer* being antitropical. The other two

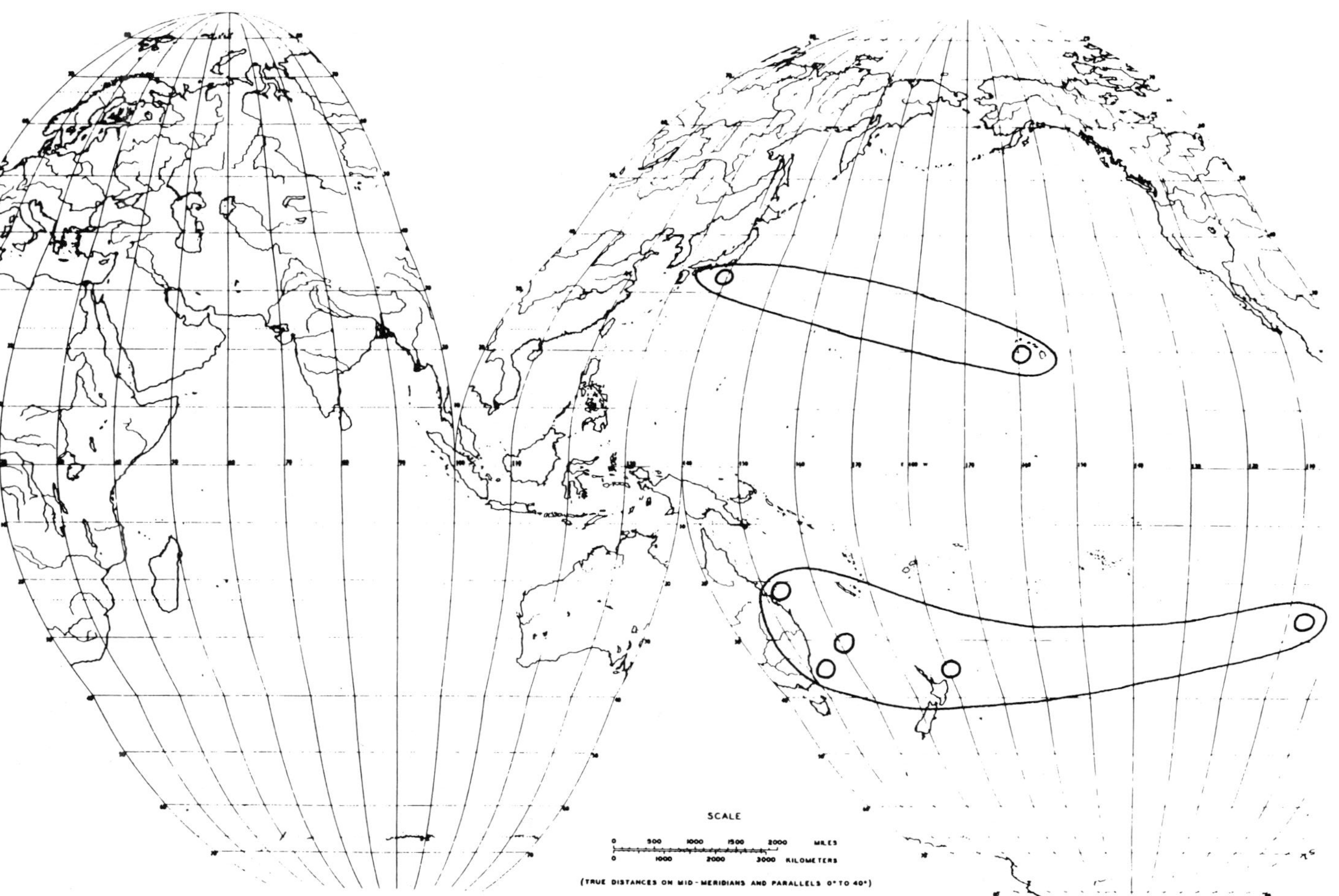

Fig. 111. Distribution of *Bodianus vulpinus*. After Randall (1982).

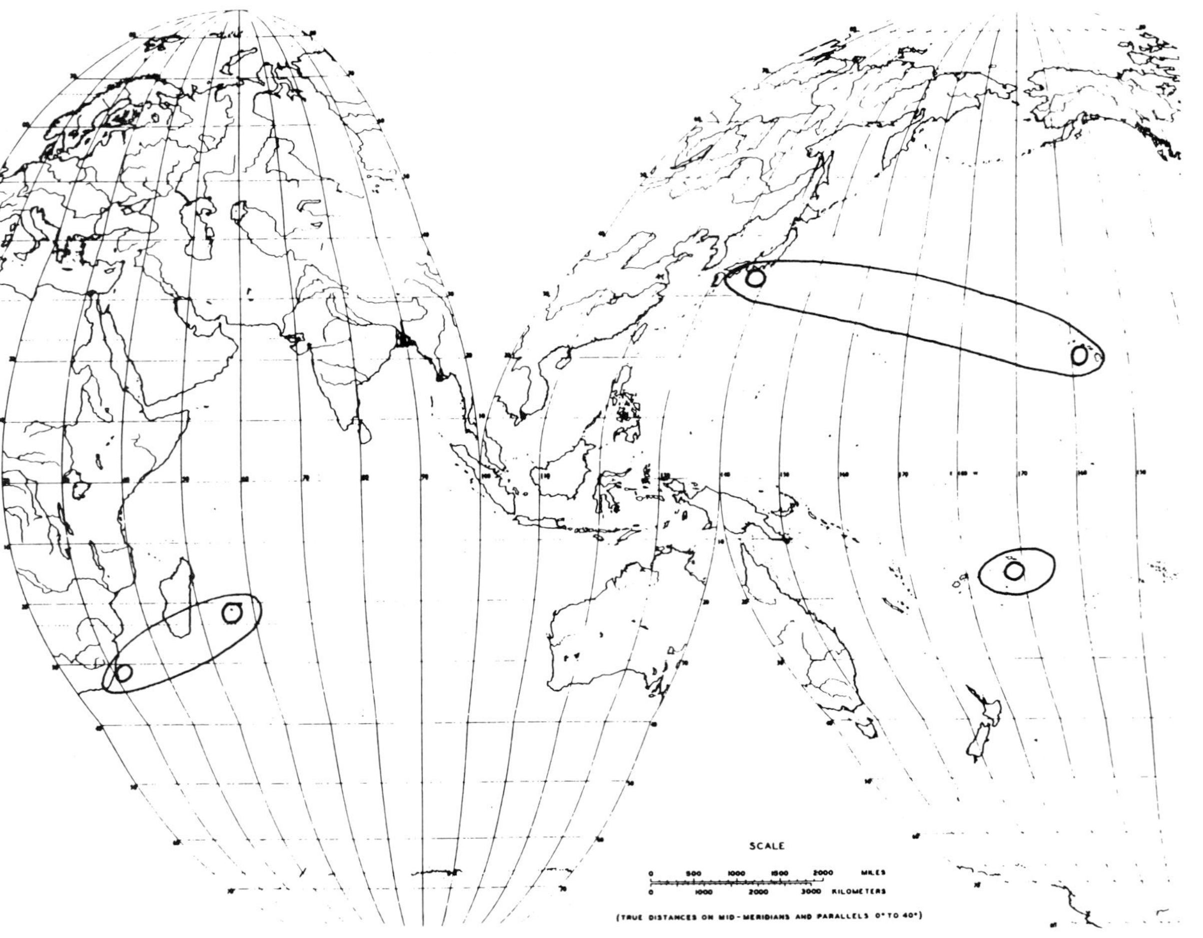

Fig. 112. Distribution of *Myripristis chryseres*. After Randall (1982).

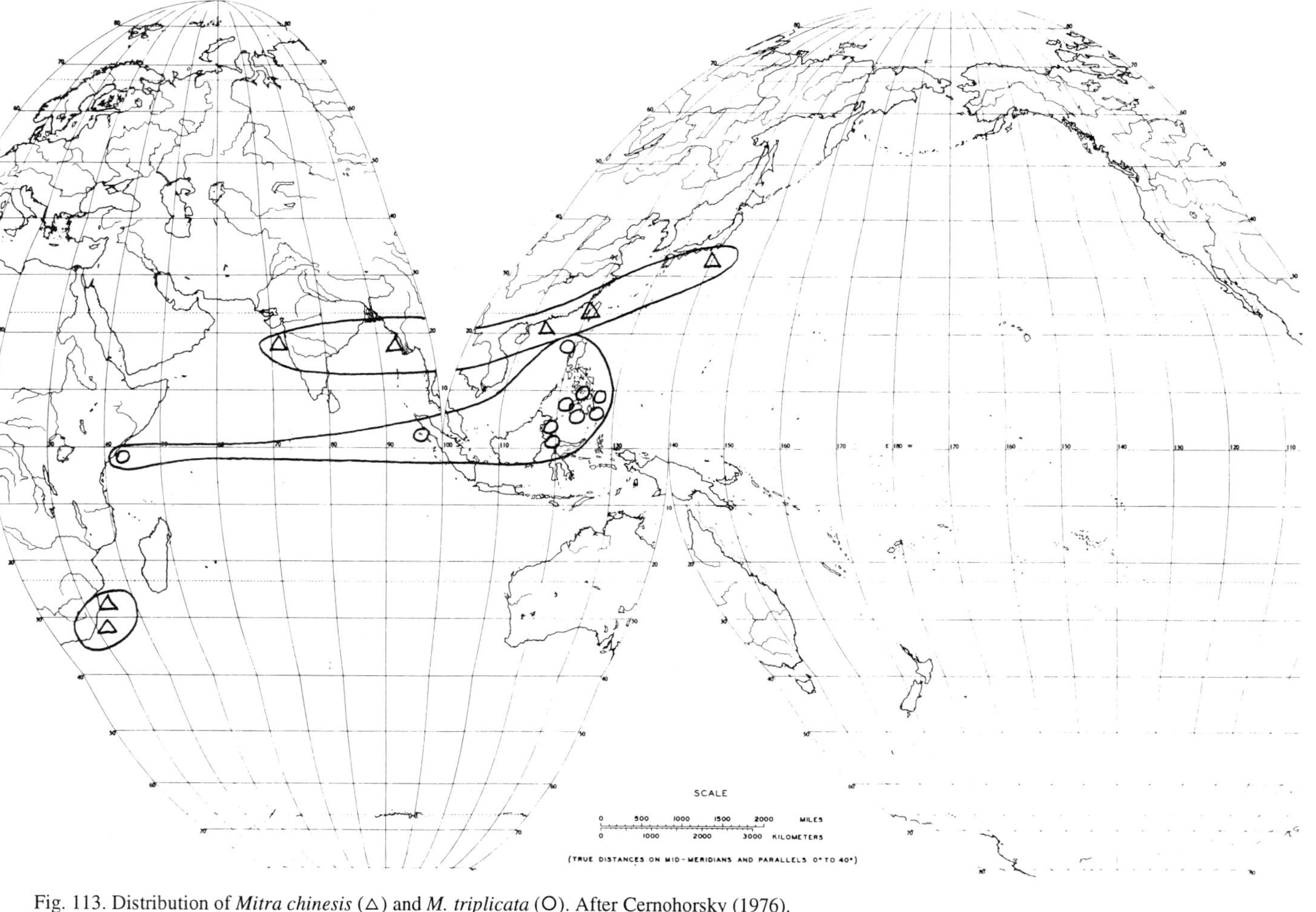

Fig. 113. Distribution of *Mitra chinesis* (△) and *M. triplicata* (O). After Cernohorsky (1976).

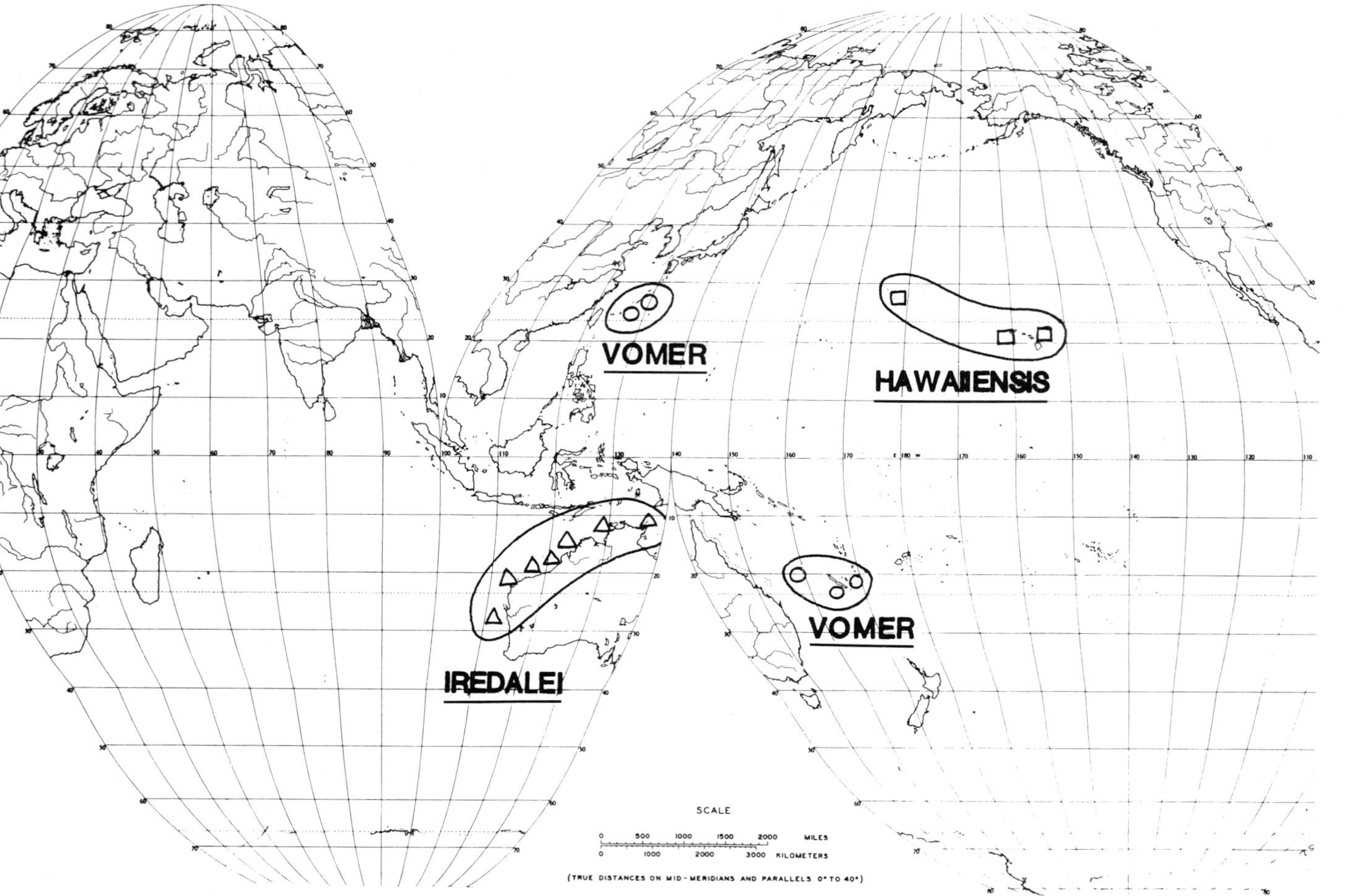

Fig. 114. Distribution of the subspecies of *Strombus vomer*. After Abbott (1960).

subspecies have a similar relationship (Fig. 114). It seems clear that this relict pattern was formed as the result of tropical extinction since Pliocene fossils from the intervening tropics show that *S. vomer* was once a widespread Indo-West Pacific species (Abbott, 1960). Other disjunct species occur at isolated island groups in the easternmost part of the Western Pacific (Newman, 1986).

Upon examination of many disjunct patterns, representing various shallow-water species, it may be seen that in almost all cases a distributional gap occurs in the central East Indies. Sometimes, the vacated areas may be very large with the affected taxon occupying only the outermost fringes of the Indo-West Pacific. In other instances, the gap may be small but will include the East Indies. Taken together, the general pattern of all the peripheral disjuncts suggests that the extinction process begins in the central East Indies. The latter may be defined as the triangular area including the Philippines, Malay Peninsula, and New Guinea (Briggs, 1987a). This is the location of greatest species richness and presumably the highest level of species competition. There is also a high diversity of predators, parasites, and disease organisms. Any of these factors could play a role in the extinction process.

Once the extinction process begins, it might be expected to spread outward from the East Indies. As this happens, the range of a species that originally extended broadly in the Western Pacific could be converted to an antitropical distribution in that ocean. Similarly, a species that originally ranged throughout the Indo-West Pacific might now be represented by antitropical populations in the western Pacific and a remnant in the western Indian Ocean.

While it seems clear that tropical extinction can result in antitropical patterns, the extinction mechanism is not easy to identify. An important clue has been provided by Stanley and Newman (1980) who gave several examples of relict distributions in the barnacle superfamily Chthamaloidea. This 70 million year old group is evidently being outcompeted and displaced by the younger Balanoidea. Consequently, once continuous distributions in the former have become disjunct with the chthamaloids surviving in areas where the balanoids are not yet dominant. Some chthamaloid genera and species demonstrate antitropical patterns while others survive only around isolated oceanic islands. Both distributions are demonstrated by the species *Chionelasmus darwini* which is found in moderately deep water only around the Hawaiian Islands (North Pacific), Kermadec Island (South Pacific), and Rodriques Island plus a seamount off Madagascar (western Indian Ocean).

It has been suggested that the antitropical split in the marine environment may have been caused by rising Neogene temperatures which followed a cool-water regime at low latitudes (Valentine, 1984; White, 1986). But the evidence for such a temperature change in the tropics appears to be weak (Briggs, 1987a). Fossil material from invertebrates and coral reefs supports the view of a stable, warm tropical ocean since the mid-Cretaceous (Horrell, 1990).

Antitropical patterns along the shores of the Eastern Pacific have been reviewed by Lindberg (1991). He concentrated on the Mollusca and, where possible, utilized fossil evidence to help determine the time when populations became separated. In contrast to the western side of the Pacific, the tropics of the eastern side occupy a relatively narrow, latitudinal stretch extending from Magdalena Bay in Baja, California to the Gulf of

Guayaquil on the Ecuador-Peru border (Briggs, 1974a). Antitropicality among the Chilean molluscs is striking. Marincovitch (1973), who studied the intertidal forms at Iquique, found that 49 genera out of a possible 68 were common to Chile and California. Crame (1993a) has published a list of Antarctic and subantarctic marine invertebrate taxa that display bipolar relationships. This includes at least 17 gastropod and five bivalve families.

Among the Crustacea, there are a number of antitropical genera. Garth (1957), in his account of the Chilean brachyurans, placed in this category the genera *Cancer*, *Taliepus*, *Hemigrapsus*, and *Cyclograpsus*. Nations (1979), who did a detailed study of the evolution and distribution of *Cancer*, concluded that this group originated in the northeastern Pacific and dispersed into the southeastern Pacific during the Pliocene.

In the fish, antitropical patterns have been noted in the genera *Girella* and *Hypsoblennius* (Stepien, 1990). Also, in the bonito *Sarda chilensis*, two antitropical subspecies are recognized. These are *S. c. chilensis* for the southeastern Pacific and *S. c. lineolata* for the northeastern Pacific (Collette and Chao, 1975). The tropical waters between the two subspecies are occupied by the wide-ranging, Indo-Pacific species *S. orientalis*. The latter species appears to be the more specialized since it possesses fewer gill rakers and a lesser number of large teeth. It appears likely that it may have migrated from the western Pacific to the eastern Pacific and, in so doing, displaced *S. chilensis* from the tropics. Therefore, this may be regarded as a possible example of tropical displacement resulting in the formation of antitropical subspecies. The sardine genus *Sardinops*, formerly believed to consist of a complex of several antitropical species, has been shown to be a monotypic genus (Parrish et al., 1989). In the eastern Pacific, *S. sagax* is present in the form of two temperate, antitropical populations. Their close relationship suggests either a recent separation or the occasional occurrence of gene flow across the tropics.

The large brown algae or kelps of the Order Laminariales belong to four families, all of them found in the North Pacific. Three of the families are large and each of these has representatives in the southern oceans. Estes and Steinberg (1988) concluded that the center of origin for the Order was in the North Pacific and that the southern species must have reached their present ranges via dispersal. Four of the southern species belong to the genus *Laminaria*. This genus also exists in the North Atlantic and has been taken in deep water off Brazil. So its route south may have been through the western Atlantic. However, two other southern genera (*Macrocystis* and *Ecklonia*) do not exist in the North Atlantic so they probably reached the southern hemisphere along the American west coast. Estes and Steinberg suggested that the dispersal may have occurred during periods of Pleistocene cooling but, as the CLIMAP (1981) project showed, the tropical waters of the eastern Pacific did not cool down appreciably during the last glaciation.

The foregoing examples are by no means a definitive list of shallow-water, antitropical organisms. Almost every marine order, well represented in temperate seas, has its share of taxa with such patterns. On the basis of antitropical and other relict patterns in the Indo-West Pacific, it was proposed that the tropical extinction theory of Théel (1886) was the most applicable (Briggs, 1987a). However, Lindberg (1991), after studying eastern Pacific distributions, rejected the extinction hypothesis. As evidence, he cited the fact that several antitropical molluscan genera first appeared in the fossil record at high instead of low latitudes. This led him to adopt a dispersal hypothesis to account for appearances in the opposite hemisphere. Most of the cited genera are typically deep water and probably dispersed by means of tropical submergence.

The problem that needs to be solved is the identification of the mechanism by which certain taxa, restricted to shallow water (either intertidal or epipelagic), managed to achieve antitropicality. Did their tropical populations become extinct, or did they undergo interhemispheric migrations? How did such organisms as the kelp genera *Laminaria* and *Macrocystis*, the molluscan genera *Fissurella* and *Tegula*, and the fish genera *Girella* and *Hypsoblennius*, all highly adapted to the intertidal environment, manage to bridge the tropics? In these few cases, an alternative to the tropical extinction hypothesis might be coastal migration from one cold, upwelling region to another. The existence of such local upwellings along the coast of Baja California has been known for many years. These support pockets of temperate organisms considerably south of their ordinary ranges. During the glacial stages of the Pleistocene, the colder sea-surface temperatures at high latitudes increased the oceanic thermohaline circulation and gave added strength to, and possibly multiplied, the coastal upwelling regions. This may have permitted some organisms to transgress the tropics by means of short migrations from one upwelling region to another. As Lindberg (1991) noted, associated sea-level changes might have also affected the positions and intensity of the upwelling areas. The idea of an upwelling shuttle is not applicable to epipelagic organisms that live well off-shore. Although many of the latter are quite eurybathic, some are not. Tropical extinction has been suggested for the fish genus *Sarda* (p. 348) and is probably the most suitable explanation for others that appear to be confined to the upper 200 m.

Although antitropical patterns commonly involve identical species or closely related species and subspecies, there are instances of antitropical genera where the separated species are quite different, of families where the separated subfamilies are different, and of orders where the separated suborders are different (Table 5). These separations cannot all have occurred at the same time. So the cause must be one that has been operating over a considerable period, possibly since the Paleozoic. For example, the jawless vertebrates (Agnatha) arose in the Ordovician and colonized freshwater habitats in the late Silurian and early Devonian. The freshwater (and anadromous) forms soon became extinct in the tropics (Halstead, 1988). They survive today in the form of antitropical subfamilies (Nelson, 1984). Crame (1993b) found that bipolarity in the molluscs could be traced back to at least the early Jurassic. Explanations that involve temperature changes within the late Cenozoic (Pleistocene decreases or Neogene increases) are clearly inadequate.

Although distributional patterns in the Pacific Ocean have been emphasized, there are many examples applicable to the shelf biota of the Atlantic. About 20% of the opisthobranch mollusc species found in the warm-temperate waters of the Cape Peninsula in southern Africa, also occur in European waters (Gosliner, 1987). And, several of the genera to which Cape endemic species belong are also antitropical. The same sets of relationships may be found among some shallow-water fish (Smith and Heemstra, 1986). Howes (1991a) found Atlantic antitropical patterns in four families of gadoid fish. He concluded that they appeared to be the result of equatorial extinctions of widespread taxa. Among the sea grasses, two of the most primitive genera in the family Potamogetonaceae are antitropical. Hartog (1970) suggested that they had been forced out of the tropical waters by more stenothermic genera.

TABLE 5

EXAMPLES OF DIFFERENT LEVELS IN THE TAXONOMIC HIERARCHY THAT DEMONSTRATE ANTITROPICAL DISTRIBUTIONS

Group	Taxon	Northern population(s)	Southern population(s)	Reference
Fish	Order Salmoniformes	Suborder Esocoidei	Suborder Lepidogalaxoidei	Nelson (1984)
Fish	Family Petromyzontidae	Subfamily Petromyzoninae	Geotriinae, Mordaciinae	Nelson (1984)
Fish	Subfamily Atherinopsinae	Tribe Atherinopsini	Tribe Basilichthyini	White (1985)
Seals	Tribe Arctocephalini	Genus *Callorhinus*	Genus *Arctocephalus*	Gaskin (1973)
Fish	Family Cottidae	70 genera	Genus *Antipodocottus*	Nelson (1984)
Barnacles	Family Catophragmidae	*Catophragmus*	*Catomerus*	Stanley and Newman (1980)
Crabs	Genus *Cancer*	22 species	*C. novaezealandi*	Nations (1979)
Plants	Genus *Laminaria*	30+ species	4 species	Estes and Steinberg (1988)
Plants	Genus *Posidonia*	*P. oceanica*	8 species	Larkum and Hartog (1989)
Cetaceans	Genus *Lissodelphis*	*L. borealis*	*L. peronii*	Gaskin (1982)
Cetaceans	Species *Globiocephala melaena*	Subspecies *melaena*	Subspecies *edwardi*	Gaskin (1982)
Fish	Species *Squalus acanthius*	*S. acanthias*	*S. acanthias*	Compagno (1984)
Fish	Species *Cetorhinus maximus*	*C. maximus*	*C. maximus*	Compagno (1984)

Information on the timing of the antitropical distributions in the Atlantic was provided by Vermeij (1992). In studying the molluscs that invaded the North Atlantic from the North Pacific at the beginning of the middle Pliocene, he found that none of them had been able to reach southern Africa. This suggested to him that trans-equatorial interchange was infrequent after the opening of the Bering Strait about 3.5 Ma ago. This information does not support the common view that the interchanges took place during glacial stages of the Pleistocene. It may also suggest that interchanges, as opposed to tropical extinctions, were rare.

Aside from a few migrations across the tropics in the eastern Pacific and eastern Atlantic, the appearance of most antitropical patterns in shelf and epipelagic organisms seem to represent a loss of territory that is probably the first step in the eventual extinction of the taxa concerned. The second step appears to be an extinction of the northern population. For relicts that have a widespread southern distribution, a third step may be a restriction to one part of its southern range. The fourth and last step would be a final extinction. One should not expect to find all of these steps included within the history of a single species. As populations become separated, speciation is apt to occur so that the sequence described would take place within a given phyletic line rather than a single species.

The suggested steps do not comprise a completely theoretical account. There are fossil data which appear to provide indications that the sequence actually takes place. Newman and Foster (1987) examined fossil and recent evidence for the geographical distribution of barnacle genera. Some were shown to be antitropical and others to be southern hemisphere endemics. Two of the southern endemics are represented by fossils in the northern hemisphere so they were considered to be former antitropicals that had become extinct in the north. The older three of the five recent antitropical genera are represented in the north only by populations in the Atlantic. The two younger genera have populations in both the North Atlantic and the North Pacific. This led Newman and Foster to suggest that the sequence might lead from a broad antitropical distribution to extinction in the North Pacific, followed by extinction in the North Atlantic. This would leave a southern hemisphere population that would, in turn, become increasingly restricted. This process evidently takes a long time. Some of the barnacle genera extend back to the Mesozoic.

Isothermic submersion

Although it appears that many, perhaps most, antitropical patterns are the result of tropical extinctions, some such distributions have a demonstrably different origin. For example, the chaetognath *Eukrohnia hamata* is found in both Arctic and Antarctic waters (Alvariño, 1965). At latitudes greater than 60° it is found in the epipelagic zone but at lower latitudes it is found in deeper water, reaching 1000 m or more in equatorial regions. This has been cited as a classic example of bipolar distribution with tropical submergence (Briggs, 1974a) but, because its population is continuous, it is not strictly comparable to antitropical distributions that involve disjunct populations.

Most instances of interhemispheric, isothermic submersion involve populations that have become separated long enough to become distinct at the species or generic level.

The fish genus *Sebastes* (family Scorpaenidae) is extraordinarily diverse in the North Pacific, being represented there by almost 100 species. The genus is also represented along the Chilean coast by two or three species and by one of these (or a related species) at Tierra del Fuego, the Falkland Islands, Tristan da Cunha, and the tip of South Africa. Eschmeyer and Hureau (1971) suggested that representatives of the genus made their way down the west coast of the Americas, via isothermic submergence, and then emerged to shallower depths in the cooler waters of the southern hemisphere. At least one species was subsequently able to extend its range around Cape Horn and from there, via the West Wind Drift, to its present distribution.

The fish family Zoarcidae is almost certainly of North Pacific origin (Briggs, 1984). It probably made its way south in the same manner as *Sebastes*. There are now 9 genera and 21 species in Antarctic and subantarctic waters (Anderson, 1988). The molluscan genera *Fusitriton*, *Argobuccinum*, and *Aforia* appear to have originated in the North Pacific and to have reached south- temperate waters by the same route (Smith, 1970). A similar dispersal has been suggested for the crab genus *Cancer* (Nations, 1975). This genus has a center of origin in the northeastern Pacific and is represented in the temperate part of the southern hemisphere by one species (*Cancer novaezealandae*).

A different tropical submersion route has been suggested for the fish family Cottidae. The southern hemisphere species of this family belong to a distinct genus (*Antipodocottus*), while the family itself is undoubtedly of North Pacific origin. It has been determined that *Antipodocottus* is most closely related to *Atopocottus tribranchius*, a Japanese species (Nelson, 1985). Nelson indicated agreement with the suggestion of Bolin (1952) that the invasion route was probably from Japan southward by way of the Philippines, New Guinea, and the New Hebrides.

The fish family Liparididae has an interesting distributional history that has been worked out by Andriashev (1986). This group is also of North Pacific origin. The shallow-water liparidids of the North Atlantic evidently reached that area via the Arctic Ocean when the Bering Land Bridge was inundated in the middle Pliocene. However, members of the genus *Paraliparis*, a deep-water group, made their way to the Antarctic by the American west coast route. From the Antarctic, the genus dispersed northward along the mid-Atlantic ridge and thence to the Arctic Basin. As a result, the liparidid fauna of the Arctic and North Atlantic owes its origin to two migratory groups, the shallow-water genera came directly northward to and through the Arctic Ocean while the deep-water paraliparids migrated all the way to the Antarctic via the eastern Pacific, then reached the Arctic Basin via the Atlantic.

Many other taxa evidently originated in the North Pacific and subsequently invaded other high-latitude areas. When the Bering Land Bridge was inundated in the middle Pliocene, huge numbers of North Pacific organisms invaded the North Atlantic via the Arctic Basin (Vermeij, 1991). In contrast, relatively few species were able to achieve reciprocal invasions. Similarly, those temperate taxa that have achieved, by means of isothermic submersion or movement along upwelling pockets, antitropical distributions in the Pacific, are almost all representatives of well-developed North Pacific groups. Examples are the fish families Cottidae, Liparididae, and Zoarcidae; the kelps of the order Laminariales; and many molluscan, crustacean, and echinoderm genera. As Lindberg (1991) noted, the arrival of both phocid and otariid seals in the southeastern Pacific dates from about 5 Ma ago, but these groups are known to have been present in the northeastern Pacific for at least 15 Ma. On the other hand, Lindberg considered three intertidal limpet taxa to have migrated from south to north; this, because the earliest fossil material was found in Chile.

In general, it may be observed that the predominate interhemispheric dispersals in the Pacific Ocean have taken place from north to south. This indicates that the rich fauna and flora of the North Pacific has produced many dominant species that have been able to transgress biogeographic boundaries to invade the cool waters of the southern hemisphere. At the same time, the North Pacific biota has resisted penetration of species from other areas. This means that the North Pacific has been functioning as an important center of origin. It has had a profound influence on the biotic composition of the Arctic, North Atlantic, and the temperate parts of the Southern Ocean.

TERRESTRIAL ENVIRONMENT

Flora

As early as the first part of the 19th Century, naturalists became concerned about the puzzle of disjunct distributions involving plant species that were said to occur towards the poles but not in the intervening tropics. The older literature on bipolar plant distributions was reviewed by Du Rietz (1940). The general subject of major disjunctions in the geographic ranges of seed plants was discussed by Thorne (1972). A detailed treatment of angiosperm biogeography, including the problem of north-south disjuncts, was published by Raven and Axelrod (1974).

In an early work on antitropical relationships in the New World, Raven (1963) distinguished three patterns: bipolar disjuncts, temperate disjuncts, and desert disjuncts. The bipolar pattern, involving the most extreme distances, is now recognized for 30 species or species pairs of vascular plants (Brown and Gibson, 1983). A number of these have broad, circumarctic or circumboreal distributions to the north but only small, scattered footholds to the south. An example is the crowberry *Empetrum*. *E. nigrum* is found in the northern hemisphere and *E. rubrum* to the south. These two species are closely related and may actually be varieties of the same species.

The category of temperate disjunctions in the New World is much larger. There are about 130 species, species pairs, and species complexes that show this pattern; and desert disjuncts are also numerous (Brown and Gibson, 1983). A desert disjunct is the creosote bush *Larrea* (Mabry et al., 1977). *L. tridentata* is widespread in the deserts of North America but, in South America, there are four related species. Most botanists believe that the northern species is the result of a northward migration. Other examples indicate migrations in the opposite direction. Brown and Gibson (1983) published a list of 75 "noteworthy genera" of angiosperms that demonstrate north-south disjunctions in the desert regions.

In their analysis of relationships among the desert disjuncts, Raven and Axelrod (1974) concluded that long-distance dispersal was the predominate mechanism. Their reasoning was based upon observations that such taxa constitute a very-small proportion of their respective floras, that the animals associated with them are different in each area, and that the plants are mainly self-compatible so that a new population could be founded by a single individual. They also noted that semi-arid to arid climates were widespread in low, tropical latitudes during the ice ages. Thus desert areas in regions that are now forested could have aided migration by permitting a series of relatively short-distance dispersals. In regard to all three categories of New World disjuncts, Brown and Gibson (1983) noted that many of the species were attractive to migratory birds or lived in habitats frequented by them. It seems that in some cases, the distances across the tropics may have been negotiated in a series of steps but, in other cases, a single long jump probably took place.

Although antitropical distribution in New World plants is an obvious and frequently discussed phenomenon, similar patterns in the Old World are rare. For instance, there is in the temperate Cape Region of southern Africa a rich and diverse flora (Goldblatt, 1978), but it is considered to have evolved almost entirely from an ancient southern African flora and a more recent tropical African flora. One of the few examples of a Cape-Mediterranean (antitropical) distribution is provided by the genus *Cytinus* of the Rafflesiaceae (Thorne, 1972). Similarly, there appears to be very-little relationship between the temperate floras of eastern China and those of Australia and New Zealand. The lack of an antitropical relationship for the latter areas may be attributable in part to their historical isolation. Africa, on the other hand, has remained close to Europe and Eurasia throughout the Tertiary.

Although the current Indo-Australian Archipelago has been essentially in place since the mid-Miocene, this route has not had much effect in terms of modern temperate plants being able to reach southern Australia and New Zealand. There is apparently very little movement by birds between those areas and Eastern China so that numerous long-distance migrations, so typical elsewhere, have not taken place. The high-mountain alpine floras of Australia comprise a special case for some of those species have immigrated from extra-Australian sources by means of dispersal events over long distances (Smith, 1986).

Although the higher-plant species do not provide many examples of antitropical distributions that have evidently been caused by tropical replacement, there are relationships at the generic and family levels which, together with fossil evidence, indicate historical patterns of this type. The old gymnosperm family Taxodiaceae was widespread in both hemispheres during the Cretaceous but now exists as antitropical relicts in eastern Asia and California and in Tasmania (Sauer, 1988). The family Cupressaceae contains 22 genera, half of which are found in the northern hemisphere and half in the southern. The Araucariaceae was once widespread in the north but now has a relict distribution in the southern hemisphere (Florin, 1963). The tree which produced much of the Dominican amber (*Hymenaea protera*) is extinct there but is related to an extant east African species (Poinar, 1993).

The Winteraceae, usually considered to be the most primitive angiosperm family, has a relict distribution in the southern hemisphere but some pollen data (Truswell et al., 1987) indicates that it may have been widespread in the late Cretaceous and early Tertiary. The distribution of the family Fagaceae (beeches and oaks), and particularly the southern beech (*Nothofagus*), has been the object of much speculation. It seemed, at first, that the family had an antitropical distribution with *Nothofagus* being the only southern hemisphere representative (Briggs, 1987b). However, it now appears that *Nothofagus* should be placed in a family of its own (Nothofagaceae) which is most closely related to the Betulaceae (alders and birches) (Nixon, 1989). The latter is restricted to the northern hemisphere except that *Alnus* extends south along the Andes to about 20°S (Crane, 1989). So, instead of *Nothofagus* being the southern representative of a northern family, it now must be considered a southern family with its closest relative being the northern Betulaceae. This results in an antitropical pattern at the family instead of the generic level.

Although most of the more-primitive families of higher plants may be said to be non-tropical, the geographic location of some archaic angiosperm families may appear to be contradictory. As Schuster (1976) noted, 10 primitive families are concentrated in and about the southeast Asian area, yet it is in the tropics of this area that the greatest angiosperm diversity and the most-advanced families are found. In reality, the locations of the primitive and advanced forms are usually quite different. Takhtajan (1969) emphasized that the primitive taxa were found in the mountains while in the tropical lowlands, where the greatest angiosperm diversity was found, the proportion of primitive forms was notably smaller.

Fauna

Antitropical patterns among terrestrial and freshwater animals occur in almost all the major groups. An important contribution towards a theory of the development and significance of such patterns was made by Darlington (1965). He examined the distribution of three tribes of beetles belonging to the family Carabidae. The first was the tribe Migadopini. It is confined to temperate areas of Australia-Tasmania, New Zealand, and southern South America (Fig. 115). Initially, the tribe was thought to be wholly flightless and to have originated on an ancient Antarctic continent from which it reached its present locations (Jeannel, 1942). However, it was subsequently discovered that two genera, one in Chile and one in eastern Australia, had well developed wings. It was also found that the most closely related living tribe is the Elephrini which is confined to temperate Eurasia and North America (Darlington, 1965; Thiele, 1977).

Another tribe of beetles that occurs in the southern parts of Australia, New Zealand, and South America is called the Broscini. But, since they are also widespread above the tropics in the northern hemisphere, they are considered to be antitropical. The southern forms are flightless while the northern ones have wings. It has been suggested that the ancestral form must have had wings (Darlington, 1965). The Trechini comprise a third tribe of Carabidae. They are best represented in the north and south temperate zones but a few genera occur in or across the tropics. Most of the tropical forms are winged while most of the temperate ones have atrophied wings. One of the subtribes, which is well represented in southern South America and in Australia-Tasmania, is represented in the northern hemisphere by a single, blind monotypic genus found in a cave in Spain.

As Darlington (1965) noted, the geographic history of the southern continent Carabidae must be deduced from various indirect clues, for pertinent fossils are lacking.

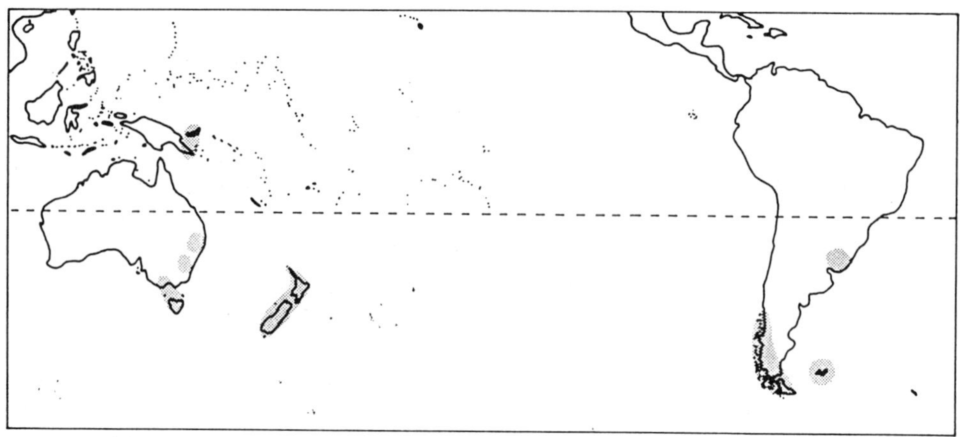

Fig. 115. Distribution of the carabid beetle tribe Migadopini. After Brown and Gibson (1983), based on Darlington (1965).

He proposed a dispersal cycle for the family as follows: rise on the large land masses in the northern hemisphere, or possibly in the tropics; dispersal southward into southern South America and southern Australia by separate routes, and to New Zealand probably from Australia; disappearance of the tropical forms, leaving an antitropical pattern; and finally extinction in the northern hemisphere leaving survivors on the three main pieces of land in the southern cold-temperate zone. However, in view of the evidence now available about other groups of antitropical organisms, it seems probable that the evolution of these beetle groups began in the tropics rather than in the northern hemisphere. Accordingly, the sequence would start with the occurrence of tropical extinctions as shown by the tribe Trechini. This would be followed by the completion of the antitropical pattern as illustrated by the Broscini. Finally, the northern group would disappear leaving a southern relict distribution as demonstrated by the Migadopini.

In addition to the Migadopini, there are other primitive groups that have a scattered distribution in the cooler parts of the southern hemisphere and have been considered as candidates for evolution on Gondwana and dispersal by continental movement. Among beetles in general, the Australian genus *Omma* is considered to be the most-archaic surviving type (Crowson, 1981). But, it is almost identical with fossils described from the Lower Jurassic of England. Crowson (1980) reviewed the evidence for antitropical patterns in seven different families of Coleoptera. Such distributions apparently exist at the subfamily, tribal, subtribal, and generic levels. He felt it to be plausible that crossings of the tropical belt took place in "exceptional circumstances" in the more or less remote past.

The family Sciadoceridae (Diptera) has one living species in New Zealand and Australia and another at the southern tip of South America. But a fossil sciadocerid has been discovered in the Oligocene Baltic amber and two others in Cretaceous amber from Canada (Watt, 1975). In the termites (Isoptera), the Mastotermitidae is the most-primitive family. An Australian species is the only living member but fossil species, belonging to the same genus, have been described from the Tertiary of Europe and North America (Poinar, 1993). A related fossil genus is known from Brazil (Krishna, 1970). In general, the older genera of the more archaic families tend to have relict distributions in the temperate parts of the northern and southern hemispheres.

As noted, the dragonfly family Petaluridae (Odonata), an ancient group that was dominant during the Jurassic, now contains only nine living species in five genera. They are found in Australia, New Zealand, Chile, eastern North America, western North America, and Japan (Watson, 1981). The archaic mayfly (Ephemeroptera) family Leptophlebiidae has a predominately southern, cool-temperate distribution but a small group of genera is found in the northern hemisphere (Peters, 1980). Three genera of caddisflies (Trichoptera) have been found in the Upper Cretaceous amber of Siberia (Botosaneanu and Wichard, 1984). They are related to modern groups with an exclusive, or almost exclusive, southern hemisphere distribution pattern. The ant genus *Leptomyrmex*, confined to Australia, has been found in early Tertiary Dominican amber. The same pattern has been found for the woodgnat genus *Valeseguya*, also from the Dominican amber (Poinar, 1993). Several genera of butterflies (Lepidoptera) have apparently been able to achieve antitropical distributions by making their way southward from the northern temperate zone by way of the higher mountains of the tropics (Holloway, 1986).

In the earthworm family Megascolecidae, all four subfamilies have a predominate southern hemisphere distribution but all also occur to some extent in the northern hemisphere. It has been stated that the origin of the megascolecids is almost certainly Gondwana (Brown and Gibson, 1983), but the present pattern may reflect a stage in which an antitropical distribution is declining in the north, and will eventually leave relict populations only in the southern hemisphere. The primitive spider family Archaeidae is confined to the southern hemisphere but fossils have been taken from the Palearctic region (Main, 1981).

The freshwater mussel family Hyriidae contains the subfamily Hyriinae which was antitropical in the New World. Living species are widespread in South America and the fossil record there extends back to the Jurassic. Fossils are known from the Triassic and Cretaceous of North America, but the group apparently died out there in the Cretaceous (Bănărescu, 1990).

Among the vertebrate animals, the side-necked turtles (Pleurodira) with a presence in South America, Africa, and Madagascar, are supposed by some to have an ancient Gondwana history. But fossils have been found in the Cretaceous of Europe and North America. The oldest fossil was taken from the Lower Cretaceous of Africa (Baez and de Gasparini, 1979). The primitive frog family Leiopelmatidae is represented by a living genus, *Ascaphus*, in northwestern North America. It is also known from an early Jurassic fossil, *Vieraella*, from Argentina. A late Jurassic fossil, *Notobatrachus*, from South America seems to be related to both *Ascaphus* and *Leiopelma* (Estes and Reig, 1973). The latter is a living genus in New Zealand. So, two very widely separated living genera are related to Jurassic fossil genera from South America.

Sphenodontids, unquestionably related to the modern *Sphenodon*, are first known from the Upper Triassic of Europe (Carroll, 1988). They became diverse and their remains are common in the European Jurassic. Some fossils are known from the Cretaceous, but the group has no fossil record in the Cenozoic. Sphenodontids are now limited to a single species of *Sphenodon* that lives on small islands off the coast of New Zealand.

It is often thought that the living lungfish, which exist only in South America, Africa, and Australia, provide a good example of a group with a single Gondwana ancestor whose population was rifted apart by continental movement. The supposed sequence is used as a logo on the cover of the journal *Cladistics*. However, as Brown and Gibson (1983) have noted, the Australian lungfish is not closely related to the other two. It belongs to the ancient family Ceratodontidae which was widespread in the marine and freshwaters of the Mesozoic. So the Australian lungfish may be considered a southern hemisphere relict of a family that was once common in both hemispheres.

In the avian fauna, the cassowaries, emus, and extinct dromorthinids (Rich, 1980) of Australia constitute an ancient group called the ratite birds. There is a distinct but somewhat distant (ordinal) relationship to similar large, flightless, living and extinct ratites of New Zealand, Africa, Madagascar, and South America. Cracraft (1973, 1974), on the basis of morphological and biochemical studies, decided that all the ratites had a common ancestry and that the South American tinamous were closest to the ancestral type. Cracraft felt that the ratites developed on Gondwana and attained their present distribution as the result of continental drift. Sibley and Ahlquist (1986) confirmed, on the basis of their DNA-DNA hybridization work, that the ratites were a monophyletic group. They also adopted Cracraft's (1974) view that the ancestral ratite was apparently distributed throughout Gondwana prior to that continent's fractionation. Sibley and Ahlquist devised a DNA "molecular clock" in which genetic distances were largely calibrated against the geologic dates for the Mesozoic breakup of Gondwana.

As noted in the historical account, paleontological research (Houde, 1988) has revealed a series of grades in ratite evolution that occurred in North America and Europe from the late Paleocene to the middle Eocene. These grades, represented by eight species in three genera, appear to be ancestral to the modern ratites. They contradict Cracraft's assumption of a single, flightless ancestor in Gondwana and they cast doubt on the validity of the dating of the avian molecular clock. The living and recently extinct ratites are southern hemisphere relicts of a widespread group that probably originated in the early Tertiary.

The Australian monotremes (the duckbill and the spiny anteaters) are famous relicts that possibly evolved from morgancuodontid ancestors (Carroll, 1988). The latter had a broad, possibly worldwide, distribution in the late Triassic and early Jurassic. Carroll had suggested that the ancestral monotremes became isolated in the Australian region by the early Jurassic. Among the more recent mammals, the tapirs, rhinoceroses, camels, and elephants are all represented by fossil forms indicating that they were once widespread in North America and northern Eurasia. The llamas (Camelidae) evolved in North America and spread into South America, probably in the late Pliocene. These modern families are apparently on their way to becoming southern hemisphere relicts.

DISCUSSION

Beginning in 1981, several vicariant theories were introduced to account for antitropical distributions. One hypothesis was that of a lost Pacific continent (Pacifica) which predated the origin of the modern Pacific Ocean (Melville, 1981; Nur and Ben-Avraham, 1981). Supposedly, the original Pacifica existed about 225 Ma ago. It then broke up into several parts which were rafted in various directions and, in the process, opened up the present Pacific Basin. In this way, a variety of trans-Pacific disjunctions could be accounted for without invoking the mechanism of dispersal by individual organisms. Another theory to accomplish the same objective was proposed by Rotondo et al. (1981) and Springer (1982). This involved an ancient tectonic movement of islands and island chains called "island integration".

The expanding earth hypothesis (Owen, 1983) has been adopted by some biologists to help explain disjunct distributions. This theory argues that the earth expanded its diameter by about 20% since the late Triassic-early Jurassic. Such expansion would create oceanic gaps in addition to those caused by plate tectonics. It has even been proposed that earth expansion was primarily responsible for the origin of the Pacific Ocean (Shields, 1983). But the expanding earth idea has received little support from geologists and virtually none from geophysicists. This theory and the Pacifica idea were reviewed by Cox (1990). He observed, "The expanding earth seems to be in the same category as the flat earth, and Pacifica in the same category as Atlantis."

In their work on cladistic biogeography, Humphries and Parenti (1986) observed that certain aspects of both the expanding earth and Pacifica hypotheses would help to explain the puzzle of amphitropical (antitropical) relationships. They proposed a new view of earth history where a pre-Pangaea or Pacifica continent existed in which the northern and southern temperate zones were adjacent to one another (Fig. 116). Then the two temperate zones split apart and the tropical zone migrated in between. Supposedly, at the completion of these movements, Pangaea was then formed.

Aside from the lack of geological evidence, there is an important reason why the Pacifica, island integration, expanding earth, and the pre-Pangean-shuffle hypotheses are not applicable. That is, the chronology apparent in the antitropical patterns argues against them. The proposed causes necessarily had their effects in ancient times, Mesozoic or earlier. Although antitropical distributions at the higher taxonomic levels might have been caused by earth movements, the majority of such patterns demonstrate relationships at the species or generic level. Oftentimes the separation involves identical species. This

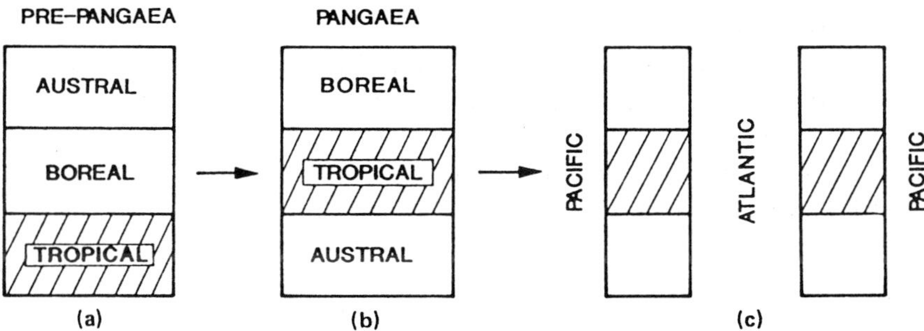

Fig. 116. Origin of antitropical relationships according to Humphries and Parenti (1986).

suggests that the causes of antitropicality have operated continuously through time. As the data in this chapter indicate, these causes are primarily tropical extinction and long distance dispersal.

When one compares the biota of the high-latitude versus the low-latitude parts of the world, it becomes at once apparent that the former usually contains the older and more primitive organisms (Briggs, 1987b). This pattern is most apparent in the ectothermic animals and is expressed most often at the higher taxonomic levels from genus up to families and orders. It is less obvious in the endothermic mammals and birds. The positive association between age and latitude extends through many levels of organization and is found in all three major divisions of the biosphere: marine, terrestrial, and freshwater. Why should this be so?

It is the tropical regions of the globe – the Indo-West Pacific and the western Atlantic in the marine environment and the Old and New World tropics for the terrestrial and freshwater areas – in which the younger, more recently evolved, families and genera are found. Newer taxa are concentrated in these regions probably because they evolved there and it takes time for them to become widespread, particularly in a latitudinal sense. As taxa gradually extend outward from their tropical centers of origin, the evolutionary process continues. Thus the older, wide-ranging forms tend to eventually become superseded by younger, newly evolved forms that may be closely related.

To judge from existing biogeographic patterns, which apparently represent different stages in the extinction and replacement process, the elimination of an older species generally begins in its center of origin. These are the places that contain high levels of biotic diversity where such factors as competition, predation, and parasitism are apt to be the most intense. As a given species becomes eliminated in the central part of its range, its geographic pattern becomes converted from a continuous to a relict distribution in which it exists only on the outer fringes of its former territory.

It is easiest for young species to spread from their tropical centers of origin to other parts of the tropics. This is evident from the usually close relationships of the tropical biotas compared to those that inhabit the temperate parts of the globe. There are many circumtropical species and genera but few of them extend to higher latitudes. When a widespread species begins to undergo extinction, it will generally first lose the central tropics. As this process goes forward, the resulting fringe distribution is likely to assume an antitropical shape in which relict populations exist north and south of the tropics but not within them. If the species undergoing extinction was originally widespread or circumtropical, the resulting antitropical pattern should also be widespread. On the other hand, if the declining species occupied only a limited longitudinal range, the eventual antitropical pattern would be similarly limited. There are numerous examples of both kinds of distribution.

Once antitropical populations are established, they may develop differences from one another depending on how long they have been separated. So, one may find young antitropical populations that belong to the same species, older ones that have been recognized as different subspecies or species, and still older ones that have been placed in higher taxonomic categories. However, the gradual extinction process does not stop at the antitropical stage, for there are many relicts in the southern hemisphere that no longer have northern counterparts. For example, there are the Araucariaceae among the plants, the beetle genus *Omma*, the dipterid family Sciadoceridae, the termite family Mastotermitidae, the side-necked turtles, *Sphenodon*, the ratite birds, the Australian lungfish, and certain barnacle genera in the marine environment. In each case, the living members have a relict distribution in the south but are represented by fossil material in the north.

There is one difficulty in the application of this scenario to the higher plants. In general, plants are better colonizers than animals, even insects. In the majority of antitropical patterns at the species level, botanists have identified long-distance dispersal as the cause rather than extinction in the tropics. Yet there are some historic trends involving higher taxa which may indicate that tropical extinction has taken place. Antitropical patterns in the Taxodiaceae, Cupressaceae, and Araucariaceae are probable examples. Hughes (1976) noted that the early Mesozoic coniferophytes were tropical lowland plants and that the first signs of separation into northern and southern groups appeared in the early Cretaceous. He suggested that when the angiosperms began to fill the equatorial lowlands they replaced the conifers and, from then on, provided a substantial biological barrier between the northern and southern groups.

The oceanic habitat poses some additional problems in evaluating the significance of antitropicality because we are dealing with a three dimensional environment. Some organisms are eurybathic to the extent that they can occupy shallow water at high latitudes and deep water in the tropics. They therefore have been able to establish antitropical populations by means of isothermic submersion. In the eastern Pacific and eastern Atlantic, where the tropics occupy relatively narrow, latitudinal bands, a few shallow-water organisms have possibly transgressed the tropics by following pockets of cool water upwelling. Also, a number of older marine taxa, that were at one time numerous on the tropical shelf, simply retreated to deeper water, so have relict populations there instead of at higher latitudes. Sometimes they did both.

Hypothesis

Antitropical distributions are a global phenomenon. They are found in all major habitats and in many major groups of higher organisms. For the most part, with the exceptions of some terrestrial plant species whose antitropical distributions may be the result of long distance dispersal, and marine taxa that have undergone isothermic submergence, they are caused by the tropical extinction of once continuous, widespread populations. They represent a stage in the extinction of older taxa and, at the same time, may signal the appearance of a new tropical replacement. The extinction begins in the tropical centers of evolutionary radiation; from there, it extends to the rest of the tropics eventually resulting in an antitropical pattern. Then the extinction spreads to the north leaving relict populations in the southern hemisphere. Finally, the southern relicts die out population by population until the extinction process is complete.

The foregoing hypothesis is similar to the dispersal cycle that Darlington (1965) proposed for the beetle family Carabidae except that he postulated an origin on the large land masses of the northern hemisphere, with origin in the tropics as a possible alternative. He then envisioned a dispersal southward followed by a disappearance of the tropical forms. Since that time, evidence for the existence of tropical centers of origin for many diverse groups has been published (Briggs, 1984). It now appears that most major groups represented by extended phyletic lines probably had their origin in the tropics or in places that had a tropical climate at the time of origin.

The hypothesis also bears a similarity to that proposed for the history of barnacle distribution by Newman and Foster (1987). They recognized the possibility of extinction in the central tropics leaving antitropical populations as relicts. They also included a number of alternatives mainly applicable to marine animals with extensive bathymetric tolerances. For widespread antitropicals that occupy more than one ocean, they predicted a history which involved (1) an extinction in the North Pacific, (2) extinction in the North Atlantic, and (3) gradual extinction of the remaining southern hemisphere populations.

Once the antitropical stage is reached, why is it that the northern populations almost always seem to disappear first? It has been suggested that in the northern hemisphere, at least in the terrestrial environment, the earlier extinction of primitive forms may be related to the fact that the geographical areas are larger enabling the species diversity to be greater and the interspecific competition more intense (Briggs, 1987b). As far as marine organisms are concerned, the better known taxa are found on the continental shelves. There is a positive relationship between shelf area and species diversity (Briggs, 1985). The amount of shelf area is greatest in the North Pacific, closely followed by that of the North Atlantic. Only a relatively small amount of shelf area is found around the ends of the southern continents. This would account for the greater diversities in the north and could explain the sequence predicted by Newman and Foster (1987).

As indicated earlier, the first person to propose the idea of tropical extinctions to account for antitropical distributions was Théel in 1886. His theory lay dormant for a long time. Vicariant theories proposed to account for antitropical patterns involve the creation of barriers based on unrealistic time scales or an unlikely history of ocean surface temperatures. It is now time to recognize that Théel's theory was essentially correct. One may conclude that antitropicality is a device whereby certain widespread species, and their descendent lines, may extend themselves through time.

ISLAND LIFE

The peculiar fauna and flora of oceanic islands has fascinated biologists for more than 150 years. Darwin (1859) devoted 18 pages of his book *On the Origin of Species* to life on such islands. His observations on the relationships of island species to those of the nearest mainland, provided powerful evidence to support his theory of evolution by natural selection. In particular, he emphasized the contrast between the Galápagos Islands in the eastern Pacific and the Cape Verde Islands in the eastern Atlantic.

Darwin (1859) noted that the Galápagos and the Cape Verdes were extremely similar in size, climate, and composition of the soil, but their biotas were markedly different. The inhabitants of the Galápagos proved to be related to those of South America, and the inhabitants of the Cape Verdes related to those of Africa. Darwin believed that this "grand fact" could not be explained by the ordinary view of independent creation, and that each island group received colonists from the nearest mainland. Such colonists would then be liable to modification, with their inheritance betraying their original birthplace.

After the publication of his great work on the geographical distribution of animals in 1876, Wallace (1881) wrote a smaller book which he entitled *Island Life*. In that work, he discussed most of the world's islands where the terrestrial biota had been described. More recently, books by Carlquist (1965, 1974), Bramwell (1979), and Williamson (1981) have been devoted to island populations. A useful work on the island biogeography of mammals was edited by Heaney and Patterson (1986). The vast Pacific Basin, with its widely scattered islands and archipelagoes, is of particular interest to island biogeographers. Important works on that area have been published by Gressitt (1963), Radovsky et al. (1984), and Stoddart (1992).

Almost all oceanic islands, that are separated from the nearest mainland by extensive stretches of deep water, are volcanic in origin. Furthermore, as Wilson (1963) first showed, there is an interesting age relationship between many islands and the plate margins as identified by the mid-ocean ridges. In the Atlantic Ocean for example, Ascension Island lies very close to the ridge and is only about 1 million years old. The Azores, St. Helena, and Gough Islands are somewhat removed from the ridge and are about 20 million years old. Finally, such islands as Bermuda, Canaries, and Cape Verde are still farther away and are much older. This distance-age pattern caused Wilson to suggest that such islands originate along the ridges where volcanic activity is the greatest. Once such an island was formed, it would be carried laterally as new sea floor was created at the ridge. However, as the sea floor moved away from the ridge, it became cooler and sunk to greater depths. This often caused the islands to sink beneath the surface to form seamounts or guyots. The larger islands would tend to persist longer but eventually all would be consumed in the trenches.

Wilson's (1963) observations were compatible with and helped to support the (then) new theory of sea-floor spreading. Wilson (1973) also proposed a mechanism for the formation of certain island chains located far from plate margins. These appear to have arisen as the result of static "hot spot" activity. As a plate gradually moves over such a hot spot or magma source, a string of volcanic islands may be produced. The chain of islands comprising the Hawaiian archipelago is a good example. The hot spot currently lies under the big island of Hawaii and the islands to the northwest become progressively older with distance from Hawaii. Other examples of island chains formed as the result of hot spot activity are the Society, Marquesas, and Galápagos in the Pacific, Tristan da Cunha and Bouvet in the Atlantic, and the Prince Edward-Marion and the St. Paul's-Amsterdam groups in the Indian Ocean (Kennett, 1982).

Island chains, once formed, may be transported from their original positions along with the movement of the plate itself. In the Pliocene, there occurred a rather sudden invasion of Indo-West Pacific hermatypic corals into the Eastern Pacific (Dana, 1975). Dana suggested that this invasion was the result of the Pacific plate having moved the Line Islands into the path of the Equatorial Countercurrent, that current then being able to transport the coral larvae eastward across East Pacific deep-water barrier.

When plate subduction occurs in the oceanic environment, it often results in the formation of three related structures: island arcs, trenches, and back-arc basins. The latter two have been discussed but island arcs form very important structures, particularly in the Pacific Ocean. Some of the island arcs occur in isolated locations far from continents or large islands. Others are continuous with or close to terrestrial areas or their continental shelves. Examples of isolated arcs are Kermadec-Tonga, New Hebrides, and the Marianas. Arcs closer to the mainland are the Kuriles, the Aleutians, and the Lesser Antilles. The relative geographic isolation of such arcs have an important bearing on their biogeographic significance.

Another rare, category of oceanic islands are those that are apparently formed from remnants of continental crust. The Seychelles comprise the best-known example. The platform from which they project was probably left behind as India separated from Madagascar and began its northward journey (Braithwaite, 1984). In the terrestrial flora there is one monotypic family, the Medusagynaceae, and nine endemic genera (Procter, 1984). Among the terrestrial animals, the amphibia are the most distinctive with a monotypic family of frogs, the Sooglossidae, and three endemic genera of caecilians (Nussbaum, 1984). Another possible example is New Caledonia with its strange endemic families and genera of plants (Holloway, 1979).

Oceanic islands and island chains that are well isolated (300 miles or more from the nearest mainland or island group) and are of sufficient age (Pliocene or older) tend to demonstrate in their terrestrial and shallow-water marine biota remarkable evolutionary changes. Once an isolated island or island group is established and, by means of fortuitous accumulation picks up its founder species, the various populations can be expected to embark immediately on their own lines of evolutionary change. Since relatively small populations are usually involved and because many aspects of the ecology are apt to be different, it may be expected that such change would occur rapidly in comparison to a mainland situation. As a result, oceanic islands that are relatively old possess faunas and floras that show a high degree of evolutionary divergence. The best indication of such divergence is the rate of endemism (Briggs, 1966).

Isolated oceanic islands with known ages are of great interest to the evolutionary biologist for they represent natural laboratories where the results of experiments in natural selection, competition, behavior, and ecology can be examined. Hubbell (1968) pointed out that isolated archipelagos such as Hawaii are the best of all because such islands, though themselves isolated, are often sufficiently close to insure occasional interchange of species. When a species of one island colonizes another, the two populations become increasingly different as time and evolution proceed. Additional colonizations will often follow and may result in the development of a species flock. Such flocks of related species as the Hawaiian honey creepers (birds) of the family Drepanididae or the giant land turtles of the Galápagos, provide interesting examples of the ways in which natural selection and random genetic changes can operate in small populations.

The huge Pacific plate, which underlies the greater part of the Pacific Ocean, is the largest tectonic plate in the world (Fig. 117). Its primary movement is from east to west so that along its western borders from the Aleutians to south of New Zealand, there occur a whole series of island arcs as the result of local subduction processes. The Pacific plate also contains many sites of intraplate volcanism, places where island chains are created above hot spots or mantle plumes. The result of these structural activities is an enormous expanse of scattered islands and archipelagos that extend more than one-third of the way around the world.

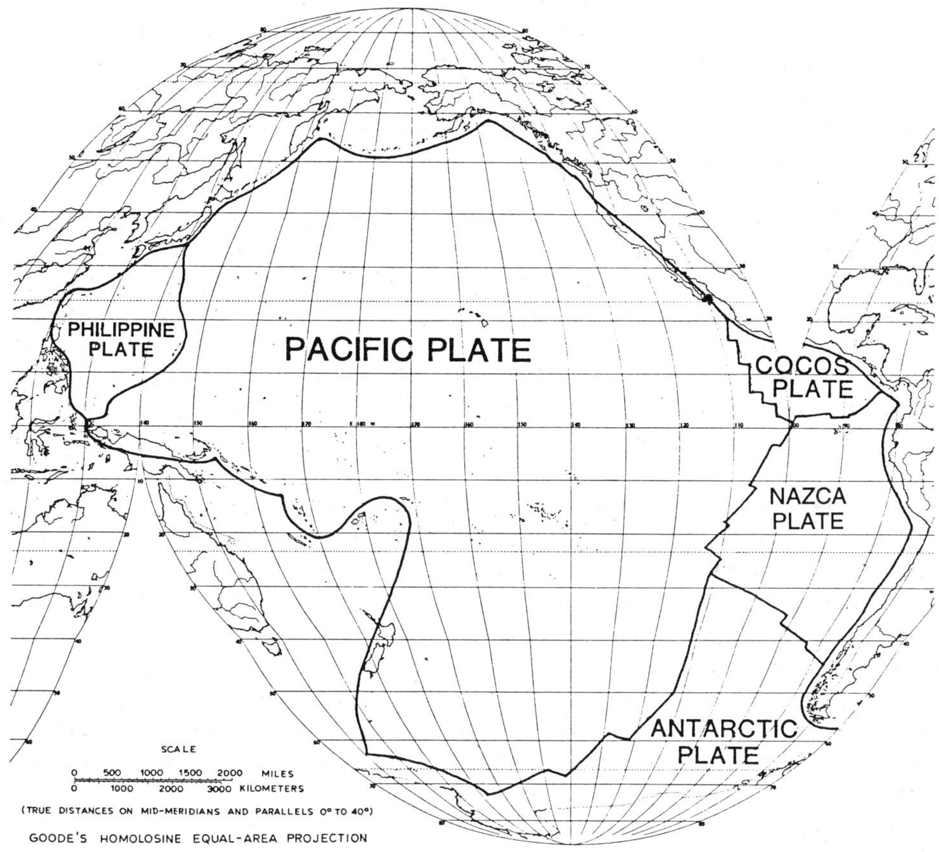

Fig. 117. The Pacific Plate, largest tectonic plate in the world.

In general, the islands of the Pacific plate demonstrate a biogeographic pattern that has three notable characteristics (Kay, 1979): (1) the flora and fauna all exhibit definite western Pacific relationships; (2) there is a west to east diversity gradient; and (3) there is sometimes an abrupt elimination of certain groups from west to east. It is evident that the third phenomenon has not occurred in a random manner for it is those groups that have the least dispersal ability over saltwater that drop out the quickest. The true (primary) freshwater fish do not extend appreciably beyond Wallace's Line which marks the edge of the mainland–large island continental shelf. Terrestrial mammals, amphibians, and reptiles extend somewhat farther east. None of these vertebrates, with the exception of bats, extends appreciably into Polynesia (Cranbrook, 1981). In contrast, land birds are more widespread as are spiders, insects, and small species of land snails. Ferns and angiosperm plants in general are good colonizers and have achieved broad distribution in Polynesia.

The three notable characteristics of the Pacific plate biogeographic pattern apply also to the shallow-water marine biota. Among the most-isolated island groups, the percentage of endemics is markedly lower than it is among the terrestrial organisms. In Hawaii for instance, 80–90% of the terrestrial snails are endemic but the endemism rate among the marine molluscs is only about 20% (Kay, 1979). In most tropical marine invertebrate groups, the great majority of species possess planktonic larvae that can be carried by ocean currents. The same is true for many shallow-water fish species. Thus, a place like the Hawaiian Archipelago, despite its isolation, is likely to receive some genetic reinforcement for its marine inhabitants while the terrestrial populations remain more deprived of such contact.

Over the years, many schemes for the biogeographic subdivision of the Pacific Ocean have been proposed. As Stoddart (1992) has shown in his review, there has been little agreement among the various authors. Because the area is so large and so many islands and archipelagos are included, a multitude of subdivisions are usually recognized. Yet, seldom have such subdivisions been defined in terms of their species endemism. As was noted for the marine environment, the area of greatest species diversity is the East Indies Triangle which includes the Philippines, the Malay Peninsula, and somewhat past New Guinea (Fig. 56). From there, eastward into the outer Pacific, there is a notable decrease in diversity which is correlated with distance. As Stoddart has shown, this same general pattern is evident among the terrestrial biota of the islands (taking into consideration the diversity contrast between the high islands and the low islands).

As terrestrial and marine groups extend away from the East Indies, there appears in most a precipitous drop in diversity just east of the Fiji and Tonga Islands. For example, several genera of seagrasses and mangroves do not progress beyond this point. Although the islands immediately beyond are relatively depauperate, they do not support many endemic species. A few endemics belonging to the less vagile groups may appear, but most of the biota is comprised of wideranging species that are also present in the East Indies. Not until one considers the biota of the extremely isolated islands, such as the Hawaiian, Marquesas, and Easter Island, do endemics appear in large numbers.

Biogeographic regions and provinces must be defined on the basis of endemism. While a decrease in species diversity may otherwise be of interest, it has no significance for the determination of biogeographic subdivisions or the placement of their boundaries. Such subdivisions can be recognized only when the area in question is found to contain a significant number of endemic species (usually 10% or more). In the marine environment, for example, only one extensive Indo-Polynesian Province can be recognized. This subdivision extends all the way from India across the Pacific to the Tuamotu Archipelago.

The distribution of terrestrial plants in the Pacific has been studied by several botanists. The most recent general scheme is that of Takhtajan (1986). His Malesian Region is virtually identical to the East Indies Triangle (Briggs, 1992). Other regions are identified for the Fiji Island area, the Hawaiian Archipelago, and Polynesia (Fig. 107). New Caledonia is placed in its own Subkingdom. It may be noted that the Polynesian region is very large, reaching all the way from Fiji to Easter Island. So, terrestrial plants and marine animals show some similarity in their distributional patterns.

The distribution of land snails on the Pacific islands has been analyzed by Peake (1981). The dispersal avenues indicated by him are probably applicable to other small organisms. The classic work on Pacific insects was that of Gressitt (1956). Schuh and

Stonedahl (1986) undertook a cladistic study of 10 monophyletic groups within the family Miridae (Heteroptera). Their results indicated area relationships beginning with Africa and extending eastward through India, southeast Asia, the Philippines, Bornea, and New Guinea. These conclusions are compatible with Brown's (1973) analysis of the historical distribution of ant genera. He concluded that, from at least mid-Tertiary times, evolution of world-dominating new taxa has proceeded mainly from the combined tropics of Africa and southeast Asia.

Many of the early works on the biology of the Pacific islands sought to explain disjunct distributions by postulating former terrestrial connections. This exercise became so popular that Zimmerman (1942) was moved to observe, "So many continents and land bridges have been built in and across the Pacific by biologists that, were they all plotted on a map, there would be little space left for water. Many of the land bridges suggested to account for the distribution of certain plants and animals in the Pacific create more problems than they solve. If the central and eastern Pacific ever included large land areas and land bridges, there should be some indication of the consequent peculiar development of the faunas and floras, but there is no such evidence." Most subsequent works, including two symposia on the biogeography of the Pacific (Gressitt, 1963; Radovsky et al., 1984) emphasized the importance of long-distance dispersal to account for island biogeographic patterns.

SUMMARY

1. How could identical species appear in widely separated parts of the world? Such disjunct distributions provoked considerable argument among the early naturalists. Some believed that a given species could be created only once, but others were proponents of multiple creations.
2. The kinds of disjunct distribution that have attracted the greatest interest are cases in which organisms living in the north temperate zone also occur as the same or a closely related species in the temperate zone of the southern hemisphere.
3. The interhemispheric disjuncts that were separated by the tropics were first referred to as "bipolar distributions." Later, the terms "antitropical" or "amphitropical" became more commonly used.
4. Within the marine environment, the surface waters of the Indo-West Pacific provide many examples of antitropical distribution. Two patterns are frequently found, one is confined to the western Pacific and one that includes, in addition, a relict population in the western Indian Ocean. Sometimes, the disjunct populations occur only along the east Asian coast and in southeast Africa. A few are found only in outer Polynesia.
5. Examination of many disjunct patterns showed that, in almost all cases, a distributional gap occurred in the central East Indies. This information, plus the fossil distributions of *Strombus vomer* and the chthamaloid barnacles, suggest that the disjuncts were caused by an extinction process that originated in the East Indies.
6. One hypothesis stated that rising Neogene temperatures, following a cool-water regime at low latitudes, caused the antitropical split. However, fossil evidence

from invertebrates and coral reefs supports the view of a stable, warm tropical ocean since the mid-Cretaceous.

7. There are many antitropical species and genera in the eastern tropical Pacific. In this area, which is latitudinally less extensive than the western Pacific, a possible alternative to tropical extinction is migration from one cool, upwelling region to another. During the glacial stages, when the general thermohaline circulation was intensified, such upwellings may have been stronger and more numerous.
8. Many antitropical patterns may be found among the shelf biota of the Atlantic Ocean. Recent work on four families of gadoid fish suggested that their antitropical distributions were the result of equatorial extinctions of formerly widespread taxa.
9. It is concluded that the appearance of antitropical patterns in most shelf and epipelagic organisms represents a loss of territory that is probably the first step in the eventual extinction of those taxa.
10. For marine animals capable of existing in deep waters beyond the continental shelf, antitropicality may be the result of isothermic submersion. The most common route, which has been followed by representatives of many different families, extends from the North Pacific along the west coast of the Americas to the south temperate and Antarctic waters.
11. It is observed that the predominant interhemispheric dispersals in the Pacific Ocean have taken place from north to south. This indicates that the North Pacific has produced many dominant species that have been able to transgress biogeographic boundaries to invade the cool waters of the southern hemisphere. At the same time, the North Pacific biota has resisted penetration of species from other areas.
12. In the terrestrial environment, botanists have distinguished three kinds of antitropical distributions: bipolar disjuncts, temperate disjuncts, and desert disjuncts. While such patterns seem to be common in the New World, they are less frequent in the Old World. Many of the species in all three categories are attractive to migratory birds or live in habitats frequented by them.
13. Fossil materials indicate that some plant families (Araucariaceae, Winteraceae) were once widespread in the North but now have only a relict distribution in the southern hemisphere. Others, such as the Nothofagaceae and the Betulaceae, demonstrate antitropical relationships at the family level.
14. Antitropical patterns in terrestrial and freshwater animals occur in almost all major groups. In many insect groups, in which the phylogeny has been investigated, the most primitive living species have relict distributions in the southern hemisphere. Such relicts often have fossil relatives in the north indicating past antitropical distributions.
15. Among the vertebrates there are many antitropical patterns, most of them involving the older taxa. Like the insects, some archaic taxa exist only as southern hemisphere relicts but were, at one time, widespread in the north. Examples are the tuatara (*Sphenodon*) and the ratite birds. Some modern mammal families are in the process of retreating from their northern ranges and may eventually become southern hemisphere relicts.

16. Since 1981, several vicariant theories to explain antitropical distributions have been introduced. These have been termed the Pacifica, island integration, expanding earth, and the pre-Pangaean shuffle. All are applicable only to mid-Mesozoic or earlier times, although the causes of antitropicality have evidently operated continuously through time. These causes appear to be primarily tropical extinction and long-distance dispersal.
17. One may conclude that in general the older, more primitive, organisms tend to be concentrated at the higher latitudes. Many survive only in the southern hemisphere. This pattern is most apparent in ectothermic animals and is expressed most often at the higher taxonomic levels.
18. The younger, more recently evolved families and genera are concentrated in the tropics, probably because they evolved there and it takes time for them to become widespread, particularly in a latitudinal sense. As taxa gradually spread from their tropical centers of origin, the evolutionary process continues. The older, wide-ranging forms eventually become superseded by younger, newly evolved forms.
19. When a widespread species begins to undergo extinction, it will first lose the central tropics. As this process goes forward, the resulting pattern is likely to assume an antitropical shape in which the relict populations exist north and south of the tropics but not within them.
20. The foregoing explanation for antitropical patterns is not applicable to marine organisms capable of living in the deep sea nor to many higher plants whose propagules can be transported by birds.
21. Antitropical distributions are a global phenomenon. They often represent a stage in the extinction of older taxa and, at the same time, may signal the appearance of a new tropical replacement. The next step is often extinction in the north leaving relict populations in the southern hemisphere. With the demise of the southern populations, the extinction process becomes complete.
22. Animals and plants that manage to reach oceanic islands and to populate them constitute a special category of disjunct distributions. There have been attempts to explain the presence of the island biotas by means other than long-distance dispersal, but these have generally been discredited.
23. The islands of the Pacific plate demonstrate a biogeographic pattern that has three notable characteristics: (1) the flora and fauna exhibit definite western Pacific relationships; (2) there is a west to east diversity gradient; and (3) there is sometimes an abrupt elimination of certain groups from west to east. It is evident that the third characteristic does not occur in a random manner for it is those groups that have the least dispersal ability over salt water that drop out the quickest.

CHAPTER 13

Species diversity: land and sea

It is this union of passionate interest in the detailed facts with equal devotion to abstract generalization which forms the novelty in our present society. This balance of mind has now become part of the tradition which infects cultivated thought. It is the salt that keeps life sweet. The main business of universities is to transmit this tradition as a widespread inheritance from generation to generation.

Alfred North Whitehead, *Science and the Modern World*, 1925

GLOBAL DIVERSITY

The world is undergoing a catastrophic loss of species diversity, but so far, we do not have a reasonable estimate of the numbers that exist in the two principal environments, land and sea. There are two generally accepted approaches to the estimation of diversity: (1) a statistical method in which a diversity figure is reached by extrapolation from small samples, and (2) an expert opinion method which sums the opinions of experts on the systematics of the component groups. The two methods have resulted in estimates that are often vastly different.

Application of the Energy-Stability-Area (ESA) theory of biodiversity, plus the additional factor of evolutionary time, predicts that the diversity of multicellular species in the sea will be far greater than that found on land. However, estimates using the expert opinion approach indicate that just the opposite is true. Species diversity in the sea may be less than 2% of that found on land. Why so few species in the sea? When one marine area is compared to another, the ESA theory seems to work. Here it is argued that the individual size of the primary producers must be taken into consideration. This is probably the critical factor that accounts for the huge discrepancy in species diversity between land and sea.

How many species of multicellular animals and plants exist in the world? This question, or parts of it, such as how many species occur in certain areas or habitats, has interested biologists for many years. Recently, because of the ongoing mass extinction of species due to human activities, the question has grown more important. We know that great numbers of species are being lost, particularly in the tropics (Wilson, 1991), but what do these losses mean in terms of local or global diversity? If we do not have a reasonable estimate of how many species are present, we have no base against which to measure the losses.

Since most terrestrial species and a sizable fraction of marine species remain undescribed, we cannot obtain a useful diversity figure by compiling a list of known species. Furthermore, even in groups that have been well studied, systematic works may not give a reliable indication of the number of species involved. The common swan mussel, *Anodonta cygnea*, of Europe has been described and named 552 times; and the North American, *Elliptio complanata*, has about 100 synonyms (Boss, 1971). On the other hand, many organisms that have been traditionally recognized as a single species have, upon close examination, turned out to be a complex of cryptic species.

In the final analysis, our knowledge of species diversity will need to be based on good systematics. There are two current methods that are being used to provide estimates. One is a statistical approach whereby small samples are taken. The samples are assumed to be typical of the entire area and the species numbers from the samples are accordingly multiplied. The other is to consider separately each higher taxon (phylum, class, order) and attempt to obtain estimates from experts who are familiar with that group. So far, at least on a global basis, the second method appears to be more consistent than the first.

The disadvantage of the extrapolated method is that it is dependent upon the samples being representative. The greater the extrapolation, the less likely that this will be true. For example, Erwin (1982) sampled beetles from the canopies of 19 specimens of one species of Panamanian tree. From these samples, he deduced that there may be 30 million species of arthropods (mainly insects) in the tropical regions of the world. This and other estimates of terrestrial species numbers have been published without taking advantage of the specialized knowledge of systematists who are familiar with the various taxa (Gaston, 1991).

An example from the marine environment is provided by Grassle (1991) and Grassle and Maciolek (1992). They obtained box-core bottom samples of sediment from a total area of 21 m^2 of the continental slope of the northwestern Atlantic. Since they found a rich fauna consisting of 798 species of tiny metazoans, and since the proportion of rare species did not diminish as the sampling proceeded, they concluded that they could extrapolate to determine the total species diversity of the deep sea below 1000 m. This involved a prodigious leap from one habitat covering 21 m^2 to multiple habitats covering 300 million km^2, and resulted in an estimate of some 10 million species. The 1991 paper was described as an act of statistical legerdemain (Briggs, 1991a).

Terrestrial

Following the terrestrial estimate of Erwin (1982), several other calculations have appeared. Stork (1988) reanalyzed Erwin's data and predicted a total of 10–80 million arthropods. Thomas (1990) did a similar exercise and emerged with a total of 6–9 million tropical arthropods. May (1990) also reexamined Erwin's (1982 and later) data and discussed methods of scaling up from insect species per tree to a global total. He also discussed direct estimates based on species-area relations, species-size relations, and food web structure. These methods are all similar in that they begin with small samples and attempt to extrapolate upward. May noted that conservative reappraisals of Erwin's estimate suggest totals of at least 7 million and possibly more. Raven (1990) referred to a minimum world total of 10 million species, and Wilson (1991) said that many tropical biologists considered the rain forest alone to support 10 million or more species. On the other hand, Hodkinson and Casson (1991) sampled the Hemiptera fauna in Indonesia and extrapolated their counts to give a global estimate of 1.84–2.57 million species of insects. While these estimates were made by knowledgeable people and need to be given proper consideration, they do not reflect opinions, published and unpublished, by systematists who have worked on the component groups.

The expert opinion method was utilized by Gaston (1991) to determine the magnitude of the global species diversity of insects. For each of the four major insect orders (Coleoptera, Diptera, Lepidoptera, and Hymenoptera), he took into consideration the number of species already described, the rate at which new species are being described, and the rate at which previously described species are being placed in synonymy. This procedure resulted in a world total of "less than 10 million". If 10 million is used as a rough estimate of living species, this may be compared to the estimate of known insect species determined by a 1976 census of the collections in the (then) British Museum of Natural History, which indicated about 827,000 (Gaston, 1991). This probably means that somewhat less than one tenth of the world's insects have been described.

Assuming that 10 million is a reasonable estimate for the insects, this leaves four other large multicellular groups to be accounted for. These are the mites (Acari), spiders (Araneae), nematodes, and vascular plants. The mites are even more poorly known than the insects. Johnston (1982) referred to various estimates which suggested 500 000–1 000 000 million species. A recent evaluation of spider diversity indicated about 170 000 living species, of which about 20% have been described (Coddington and Levi, 1991). The nematodes are another poorly known group. Brusca and Brusca (1990) referred to about 12 000 described species, but the actual number of living species is much larger. May (1988) suggested "1 million?". Finally, there appear to be about 300 000 species of flowering plants (Burger, 1981).

All other metazoan groups are comparatively small, the largest of the remainder being the gastropod molluscs with about 20 000 living species (Brusca and Brusca, 1990). All of the remaining groups would not reach another 100 000. This includes the total number of terrestrial and freshwater vertebrates which is about 30 000. This information can be summed to give a rough idea of the number of living terrestrial species (Table 6).

TABLE 6

TERRESTRIAL SPECIES DIVERSITY

Group	Species estimate	Reference
Insects	10000000	Gaston (1991)
Mites	750000[a]	Johnston (1982)
Spiders	170000	Coddington and Levi (1991)
Nematodes	1000000?	May (1988)
Plants	300000	Burger (1981)
Molluscs	20000	Brusca and Brusca (1990)
Remainder	100000	Various sources
Total	12340000	

[a]This represents the midpoint of the range given by Johnston (1982).

The foregoing number appears to be more exact than is justifiable. Perhaps 12 ± 1 million would be more appropriate. This coincides with a preliminary estimate of 12 million made 2 years ago with data that were not as timely (Briggs, 1991a). If the current rate of extinction due to rain forest destruction is about 50 000 species per year (Wilson, 1991), we stand to lose 1 million species over the next 20 years. At that rate, more than three-fourths of all terrestrial species will be gone within 200 years.

Some biologists, in warning about the magnitude of the current extinction, have compared it to the great extinction episode that took place near the Cretaceous-Tertiary boundary around 65 million years ago (Raven, 1990). But that, and the other historic extinctions, took place gradually over periods of one million to several million years, depending on the biotic group concerned (Briggs, 1990). The catastrophic rate of species extinction occurring today, has no historic precedent.

Marine

Contrary to the terrestrial habitat, in which more than 90% of all the species are members of one phylum (Arthropoda), the bulk of the diversity in the marine environment is spread out over many phyla (Table 7).

As in the case of the terrestrial number, the figure of 178 000 should be treated as a very rough approximation. It is greater than a preliminary estimate of 160 000 made several years ago (Briggs, 1991). In general terms, it would be safe to say that the marine species diversity is less than 200 000. The groups listed in the table are those phyla containing more than 8000 species. Other speciose phyla are the Platyhelminthes with about 5000 marine species and the Echinodermata with about 6000. The eight phyla listed in the table contain more than 90% of all marine species.

The range suggested for the nematodes was 20 000–50 000 (W. Duane Hope, pers. commun.) and for the bryozoans 10 000–20 000 (Alan H. Cheetham, pers. commun.). The phylum Arthropoda is a large complex comprised of six classes and many orders. The three largest orders are the decapods with about 10 000 species, the amphipods with

TABLE 7

MARINE SPECIES DIVERSITY

Phylum	Species estimate	Reference
Porifera	9000	Brusca and Brusca (1990)
Cnidaria	9000	Brusca and Brusca (1990)
Nematoda	35000[a]	W. Duane Hope (pers. commun.)
Annelida	15000	Kristian Fauchald (pers. commun.)
Arthropoda	37000	Brian F. Kensley, et al. (pers. commun.)
Mollusca	29000	Brusca and Brusca (1990)
Bryozoa	15000[a]	Alan H. Cheetham (pers. commun.)
Chordata	15000	Several sources
Remainder	14000	Several sources
Total	178000	

[a]Midpoint of range given.

8000–10 000, and the isopods with about 5000 (Brian F. Kensley, pers. commun.). Within the Chordata, the two largest groups are the Urochordata with about 3000 species (Brusca and Brusca, 1990) and the Vertebrata with about 13 000. Some vascular plants live in the sea but there are less than 100 species. Even if one included the nonvascular marine algae, those species probably do not number more than 7000 (Sterrer, 1986).

Although this analysis is focused on the species level, it may be appropriate to note that, at the level of phylum and class, the marine biota is far more diverse than the terrestrial biota. There are 34 phyla living in the marine realm, 17 in freshwater, and only 15 on land. If we carry this analysis to the class level, we see that 73 occur in the sea, 35 in freshwater, and only 33 on land (Nicol, 1971). In view of the fact that multicellular life probably originated in the sea and existed there for hundreds of millions of years before invading the land, these numbers are not unexpected. They reflect the inability of many marine groups to adapt to conditions in freshwater or on land.

Considering the greater diversity of higher taxonomic levels in the sea, why is this not accompanied by a greater species diversity? The ratio of less than 200 000 marine species to around 12 million terrestrial species means that the ocean contains less than 2% of the species on earth. So, the important question is, why so few species in the sea? Ray (1985) observed that the sea contains about 20% of animal species. It is now apparent that this figure was much too high.

Considerable research has been devoted to the question of species diversity. Why should it vary so much from one place to another? As Wilson (1992) has observed, the answer appears to be explainable by the ESA theory of biodiversity. The research leading to this theory has been accomplished in the terrestrial habitat. It has been shown that the greater the input of solar energy, the greater the species diversity – accounting for the establishment of latitudinal gradients. Climatic stability is important as illustrated by the historical record of climatic changes and extinction events. The close relationship of area and diversity is revealed by the rule of thumb, which states that a tenfold increase in area results in a doubling of the number of species. The final ingredient is evolutionary time. With plenty of time and under optimum ESA conditions, species should be able to evolve, accommodate to one another, and accumulate in large numbers.

One can apply the ESA theory to the ocean as follows: (1) the ocean absorbs a much greater portion of the sun's energy than does the land; in fact, it functions as a great heat reservoir; (2) the oceanic physical environment is highly stable compared to the land where there are great fluctuations in diurnal and seasonal temperature and humidity; and (3) the ocean covers 71% of the earth's surface and extends downward to almost 11 000 m. Thorson (1971) estimated that the total oceanic living space was roughly 300 times larger than that available for life on land. In regard to evolutionary time, the first metazoan animals appeared in the sea about 800 million years ago (Stanley, 1989). Such animals probably did not invade the land until the late Silurian (Seldon and Edwards, 1990), somewhat more than 400 million years ago.

Why does the ESA theory not hold in the sea? Well, to some extent it does. Most shallow water organisms demonstrate the familiar latitudinal gradients; waters that undergo the least seasonal changes in temperature tend to support more species; and, within a given habitat, the larger areas will have more species; and marine organisms have had some 400 million additional years in which to evolve. Has evolution in the ocean slowed down? Apparently not; there are plenty of species that have evolved within the past 3 million years. One has only to examine the multitude of geminate species that exist on each side of the Panamanian isthmus (Briggs, 1974a).

So what is the missing ingredient in the marine diversity formula? It is generally known that individual body size is correlated with species diversity in terrestrial animals. As one goes from animals whose characteristic linear dimension is a few meters down to those of around 1 cm, there is an approximate empirical rule which says that for each tenfold reduction in length there are 100 times the number of species (May, 1990). The next step is to apply the size factor to the primary producers.

Although the sizes of terrestrial plants vary over a wide range, they are generally very much larger than the insects which prey upon them. In the tropics, the various parts of the large trees and their epiphytes offer an incredible number of habitats that can be occupied by small animals. The tropical forests occupy only 6% of the land surface but contain more than half the world's species (Wilson, 1991). Hutchinson (1959) suggested that "the extraordinary diversity of the terrestrial fauna, which is much greater than that of the marine fauna, is clearly due to the diversity provided by terrestrial plants."

Hutchinson's statement is correct as far as it goes, but the key lies in the size differential of the individual primary producers. In the sea, almost all photosynthesis is accomplished by single-celled organisms. The large marine algae and angiosperm sea grasses exist only on the fringes and contribute but a small fraction of the total primary production. The small phytoplankters do not provide physical support for metazoans and other plants. They serve only as a food source for a variety of pelagic and benthic animals.

Although energy, stability, area, and time are important parts of the species diversity formula, the individual size of the primary producers must not be neglected. This is probably the critical factor that accounts for the huge discrepancy in species diversity between land and sea.

Conclusions

The sampling approach to the determination of species diversity, where numbers for very large areas are calculated from very small samples, is apt to be highly inaccurate or the variance is likely to be so high as to make the figure meaningless. A deduction of the diversity of the 300 million km^2 ocean floor from a 21 m^2 sample (Grassle and Maciolek, 1992) is the equivalent of my determining the terrestrial species diversity of North America by counting the organisms in samples of soil from my backyard. It takes many samples to cover all habitats and to account for patchy distributions within habitats. Even so, the greater the extrapolation, the less dependable the result.

People who spend years doing systematic work on various groups become familiar with the rates at which new species are being discovered and previously described species are being placed in synonymy. They are acquainted with the habitats in which their organisms are found, and are in the best position to provide educated guesses about diversity. At the present state of our knowledge, a summary of expert opinions is probably the better approach to the determination of global diversity.

Published and unpublished opinions from various sources indicate that terrestrial species diversity is about 12 million, plus or minus 1 million. In contrast, marine species diversity appears to be less than 200 000. This huge discrepancy exists contrary to the prediction of the ESA theory. The sea also has had the advantage of about 400 million years more time for diversity to accumulate.

It is suggested that the missing factor in the marine diversity formula is the size of the individual primary producer. In the sea, almost all photosynthetic production is accomplished by single-celled organisms. The large marine algae and sea grasses contribute but a small fraction. The size of the individual phytoplankter is so small that it cannot provide physical support for metazoans and other plants.

As May (1990) has stated, ... "we need to understand the diversity of living things for the same reasons that compel us to reach out toward understanding the origins and eventual fate of the universe, or the structure of the elementary particles that it is built from, or the sequence of molecules within the human genome.... Unlike these other quests, understanding and conserving biological diversity is a task with a time limit. The clock ticks faster and faster as human numbers continue to grow." At the current rate of species extinction, more than three-fourths of all terrestrial species will be gone within 200 years.

LATITUDINAL GRADIENTS

Although there is an enormous contrast in global species diversity between marine and terrestrial environments, some of the diversity patterns within each habitat are similar and may be due to common causes. One of these is the latitudinal diversity gradient. In most widespread groups of animals and plants there is a negative relationship between latitude and species diversity. Some terrestrial examples have been illustrated by Stevens (1992) (Fig. 118) and marine examples by Angel (1992) (Fig. 119).

While the existence of latitudinal gradients is no longer disputed, there are some notable exceptions to that pattern. For example, in the sea, there are only about 1000 tropical amphipod species compared to about 5000 in the temperate zones and the deep sea (Barnard, 1991). Platnick (1992) has pointed out that in the spiders, and perhaps other terrestrial invertebrate groups, the species diversity pattern is pear-shaped. That is, from northern high latitudes there is a gradual increase into the tropics, but the diversity increase continues into the southern temperate latitudes. The spider diversity of Australia is apparently about double that of the United States.

The first comprehensive review of latitudinal gradient causes was published by Pianka (1966). He managed to fit all previous hypotheses into six general categories. These have continued to be recognized, in a slightly modified form, by later workers (Brown and Gibson, 1983) (Table 8). A new factor was introduced by Stevens (1989, 1992) when he pointed out the effect of Rapaport's rule on the generation of latitudinal gradients. Rapaport (1982) had noted that the latitudinal ranges of individual species became less as latitudes became lower. Thus more species could be accommodated at lower latitudes because each required less space.

By combining the Rapaport rule with explanations for the continued existence of competitively inferior species, and the mass effect of Shmida and Wilson (1985), Stevens

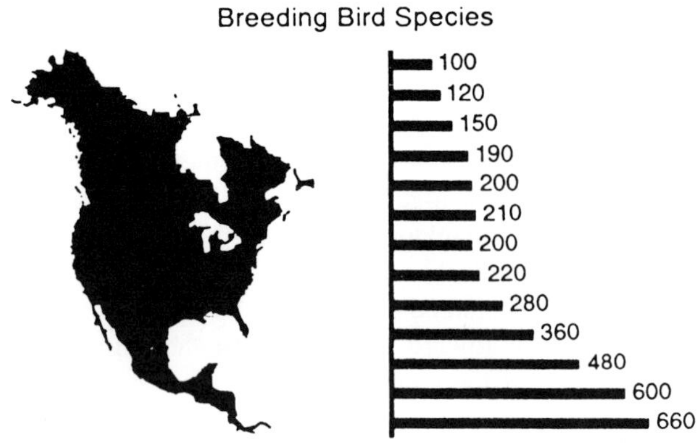

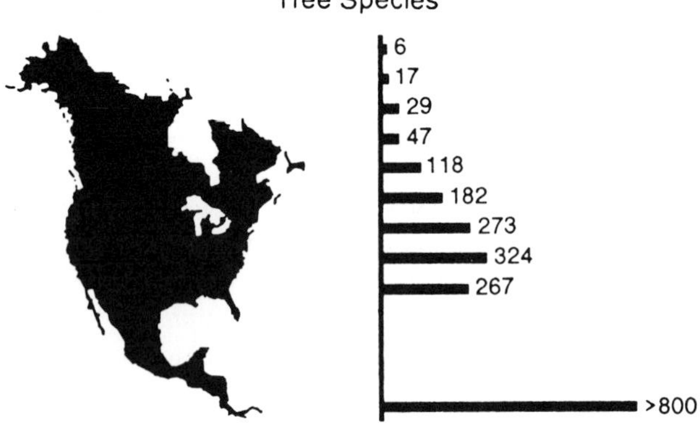

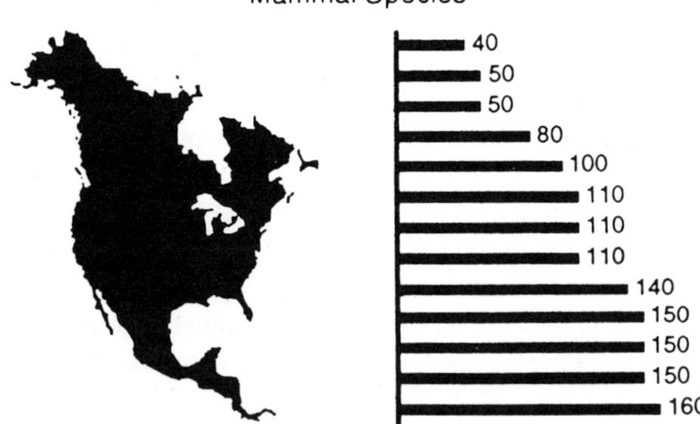

Fig. 118. Numbers of breeding bird species, tree species and mammal species for different latitudes in North and Central America. After Stevens (1992).

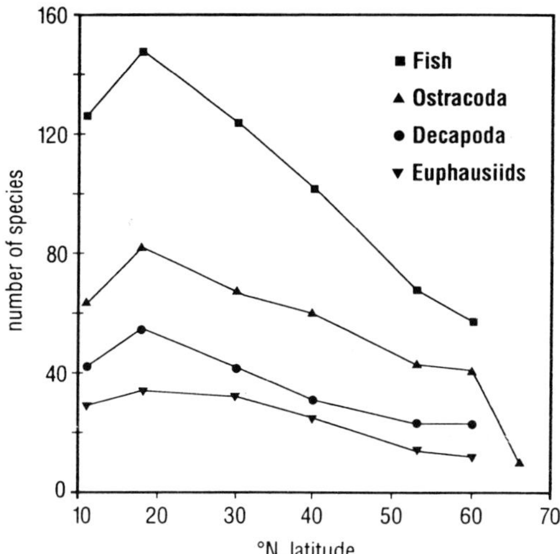

Fig. 119. Latitudinal variation in numbers of species collected in the water column to depths of 2000 m. After Angel (1992).

(1989, 1992) envisioned a considerable increase in tropical diversity. On the other hand, Rohde et al. (1993) found that Rapaport's rule did not apply to marine fish. They also observed that the rule has not been shown to apply to any group of organisms below a latitude of approximately 25° to 35°N.

TABLE 8

PROCESSES THAT HAVE BEEN HYPOTHESIZED TO ACCOUNT FOR GEOGRAPHIC PATTERNS OF SPECIES DIVERSITY (AFTER BROWN AND GIBSON 1983)

Hypothesis or theory	Mechanism of action
Historical perturbation	Habitats that have experience historical changes are undersaturated because of inadequate time for species to colonize and adapt
Productivity	The greater the availability of usable energy, the larger the number of species that can be supported and the greater the specialization of coexisting species
Harshness	Small isolated, ephemeral, or physically extreme habitats have lower colonization rates or higher extinction rates than large, continuous, permanent, and physically equable habitats
Climatic stability	A fluctuating environment may preclude specialization or increase the extinction rate, whereas in a constant environment species can specialize on predictable resources and persist when rare
Habitat heterogeneity	Diverse physical habitat structure permits finer subdivision of limiting resources and hence greater specialization
Competition, predation, or mutualism	One or more of these classes of interspecific interactions promote coexistence and specialization

At this stage in our knowledge, we may observe that Rapaport's rule and the six general theories may all have some effect on latitudinal distribution. Depending on the biotic group concerned, some factors will be more important than others.

VERTICAL GRADIENTS

Species diversity also responds to changes in depth or altitude. In the sea, in the temperate and cold waters of the higher latitudes, the diversity of benthic animals will increase from the inner shelf down to the middle part of the continental slope. This gradient was first emphasized by Sanders (1968) and reiterated by many others. The most speciose group in the soft bottom habitat is the polychaete worms. For example, along the northeast coast of the United States some 226 species are known from the shelf, but this increases to more than 700 species on the slope (Grassle, 1991). From about 2000 m downward, there is a negative relationship with depth and the species diversity drops off rapidly (Vinogradova, 1962).

It is sometimes assumed that species diversity is always greater in the deep sea (slope) than on the continental shelves. However, on a global basis, this is not true. While the diversity of the infauna of the soft substrate of the northeastern Atlantic slope is impressively high (Grassle and Maciolek, 1992), it does not match that of the shallow tropics (Sanders, 1968). The greatest species numbers of all are found on the tropical shelf where there are well developed coral reefs in conjunction with lagoons and estuaries. Within this habitat, the world's greatest concentration of marine species is found in the East Indies Triangle (Briggs, 1992). In the tropics, there is apparently a decrease in diversity from the inner shelf downward. The decrease appears to become accelerated at the recognized depth boundaries (200 m, 2000 m, 6000 m).

In the pelagic realm, most phytoplankton species and many small herbivores are found in the upper 200 m, but the greatest zooplankton diversity is often found at greater depths. But diversity concentrations in the zooplankton and neckton can undergo significant diurnal fluctuations. Within a given group, some species may undergo extensive vertical migrations while others are relatively sedentary. This may be illustrated by the diurnal activity of the sergestid prawns (Fig. 120) which tend to feed on euphausiids (Marshall, 1980). The most common mesopelagic fish are lanternfish (Myctophidae) which are inveterate migrators. As a result, the diversity of animal life in the epipelagic zone will undergo a considerable diurnal fluctuation. So the depth of the diversity peak may shift depending on the time of day.

In regard to altitudinal change in the terrestrial environment, this was first publicized by Alexander von Humboldt (1805) in his famous work on plant geography. He described the series of floral belts that he saw on the slopes of Mt. Chimborazo in the Andes. He found that, as he climbed higher, the tropical forests gave way consecutively to temperate woodlands, boreal forests, and finally arctic communities. Each association showed a decrease in species diversity. Later, C. Hart Merriam (1894) made similar observations in the United States and proposed a theory of altitudinal life zones that would apply to animals as well as plants. This idea eventually developed into the modern ecological concept of biomes. Although altitudinal

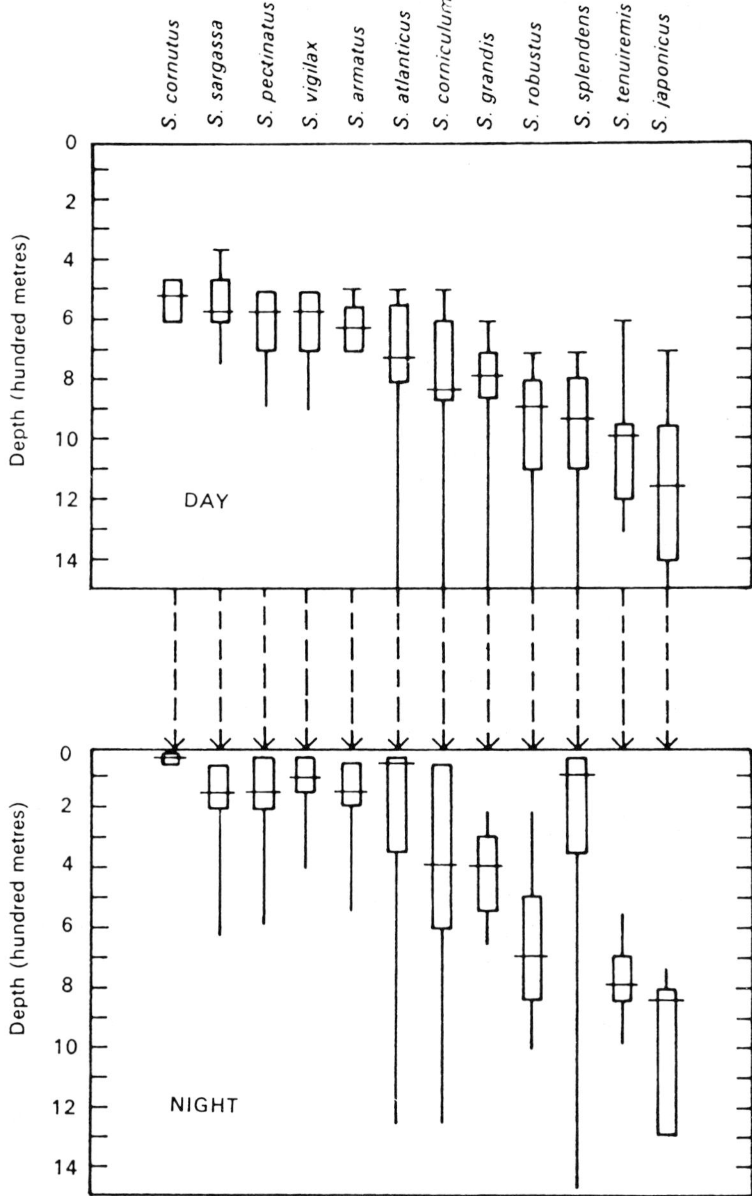

Fig. 120. Daily vertical migrations of the prawn genus *Sergestes*. The distribution by night is shown immediately below the daytime depth range. Each rectangle represents at least 75% of the catch. After Marshall (1980).

changes are important, they are often obscured by changes in precipitation and soil structure.

The diversity changes that occur along with increases in depth or altitude cannot be said to have common causes. In the sea, almost all light absorption and photosynthetic production takes place in the upper 200 m. As the depth increases, the food supply decreases and so does the diversity. This general observation holds despite the effects of climatic stress on the shelf at high latitudes. The altitudinal gradient is primarily a response to climatic conditions, but local conditions can cause major alterations.

LONGITUDINAL GRADIENTS

Although gradients in longitudinal diversity are seldom discussed, they become obvious when species diversity in general, or in individual groups of organisms, is considered. In the marine environment, an attempt has been made to compare the diversity of the shelf fauna of the four great tropical regions (Briggs, 1985). Knowledge of all the animal groups was not sufficient to attempt a direct count but, in several groups that had been subjected to recent systematic work, a reasonable measure of the species diversity was possible. The groups utilized were the echinoderms, molluscs, some crustaceans, reef corals, and fish. The data were then compiled in order to illustrate the differences among the four regions. The faunal richness of the Indo-West Pacific proved to be 2.5 times that of the western Atlantic, 3.5 times that of the eastern Pacific, and 7.3 times that of the eastern Atlantic.

When the diversity data were compared to the area of continental shelf in each region, an interesting relationship became apparent (Fig. 121). The species/area curve for the Indo-West Pacific, western Atlantic, and eastern Pacific indicated a fairly consistent relationship but the eastern Atlantic fell far below the line. The relatively depauperate state of the latter required an explanation. It was concluded that its impoverishment was due to glacial stage temperature declines which were considerably more severe than in the other tropical regions (Briggs, 1985). On the other hand, Abele (1982) plotted the species/area data for the crustaceans alone and found a virtually straight-line relationship.

On a broad scale, the foregoing data demonstrate significant longitudinal changes in marine diversity and at the same time show a good relationship between diversity and geographic area. On a smaller scale, species diversity will vary considerably with such factors as substrate heterogeneity, proximity to land, and physical disturbance. Another way of demonstrating diversity changes over longitude is to examine the distribution patterns of individual higher taxa. In the shallow tropics, most families, orders, and classes of animals will exhibit a broad-scale pattern similar to that shown by the estimates of total diversity.

The Indo-West Pacific is a special case for its longitudinal axis extends across almost two-thirds of the globe. Within this enormous expanse, there is a great variation in species diversity. Almost all biologists, who have worked on groups widespread in that region, have found a peak of diversity in a relatively small area extending from the Philippines to the Malay Peninsula to somewhat beyond New Guinea (Briggs, 1992). From this East Indies triangle, diversity tends to drop off in all directions. This has been illustrated

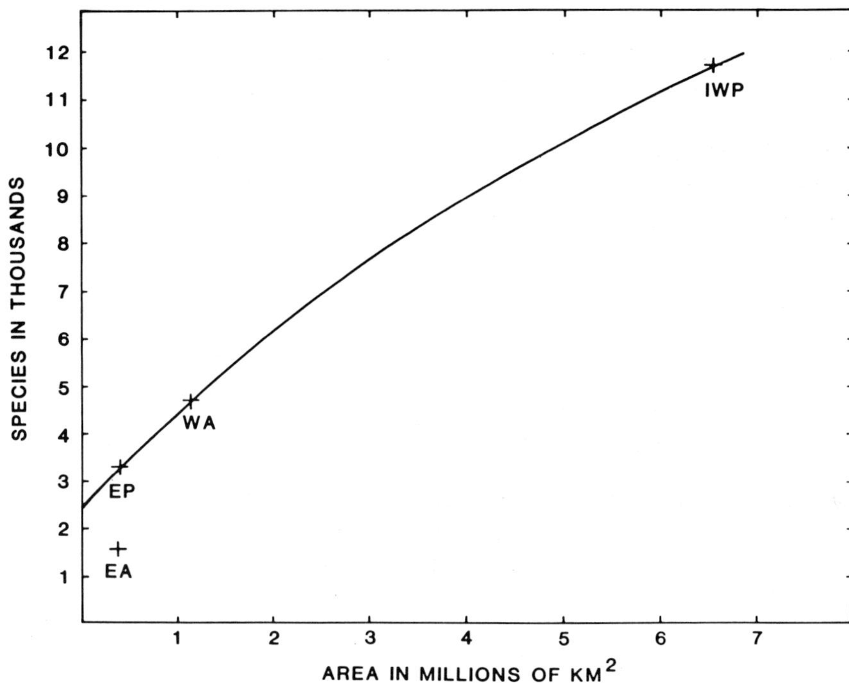

Fig. 121. Species/area curve illustrating the relationships among the four marine tropical regions: Indo-West Pacific (IWP), Western Atlantic (WA), Eastern Pacific (EP) and Eastern Atlantic (EA). After Briggs (1985).

for the molluscan genus *Strombus* (Fig. 58) by Abbott (1960) and the fish family Pomacentridae (Fig. 57) by Allen (1975a). Veron (1986) has illustrated the pattern of generic diversity in hermatypic corals, which is probably a reflection of the species richness pattern. The reasons for the existence of this diversity peak have been discussed elsewhere (pp. 213–223).

A longitudinal contrast in the northern hemisphere has been recognized. In general, the North Pacific Ocean is much richer in species than the North Atlantic. The former has about twice as many molluscan species (Vermeij, 1991) and possibly three times the number of fish species (Briggs, 1974a). The differences can probably be accounted for by the contrast in the climatic history of the two oceans. The North Atlantic, being a smaller ocean with less heat storage capacity, was more severely affected by the atmospheric temperature changes of the Pliocene and Pleistocene. Also, during the ice ages, the North Atlantic was penetrated by ice and cold currents from the Arctic Ocean while the North Pacific was protected by the Bering Land Bridge. The resulting climatic stress has probably prevented the North Atlantic from achieving a higher diversity. Although the North Atlantic is the smaller of the two oceans, its area of continental shelf is only slightly smaller than that of the North Pacific.

On land, many of the longitudinal patterns in species diversity appear to be related to area, but some of them do not. For the largest geographic areas, the species/area ratio appears more or less constant. The Palearctic bird fauna consists of 1026 species while the Nearctic possesses 750 species. Considering the size of the two regions, this diversity contrast seems appropriate. However, if one compares the number of species in the Neotropics (2780) with the Ethiopian (1556), areas that are about equivalent in size, there is a notable contrast (Welty and Baptista, 1988). The high diversity in the Neotropics is probably due to the survival of primitive families during the long Tertiary isolation of South America, plus the addition of modern birds since the elevation of the Panamanian isthmus.

Similarly, if one compares species diversity of the mammals in northern South America (an area of 2 575 400 km^2) with that of Zaire (2 336 889 km^2) the number of species is about the same (424 and 427) (Eisenberg, 1981). But, if southern India (1 813 000 km^2) is compared to east Africa (1 768 450 km^2) the former has only 102 species while the latter has 351. India probably had very few mammals prior to the Tertiary. After the connection to Asia, mammalian immigration through the east and west gateways was still restricted (Kurup, 1974). Cuba, an island of 114 522 km^2, has only 38 mammalian species while much smaller Panama (74 009 km^2) possesses 201. Obviously, Cuba's history as an oceanic island prevented it from achieving a higher diversity.

Although it is not possible to assemble useful longitudinal diversity figures for the insects as a group, the family of tiger beetles (Cicindelidae) has been suggested as an appropriate taxon for determining regional patterns of biodiversity (Pearson and Cassola, 1992). The taxonomy of the family is stabilized, individuals are readily observed and

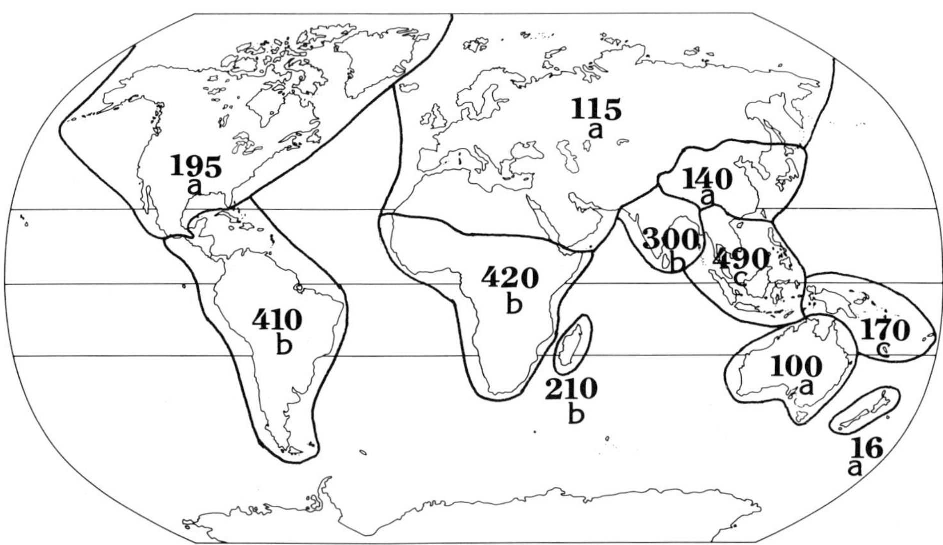

Fig. 122. Estimated total number of tiger beetle species for each indicated region. Lower case letters indicate author's assessment of precision of estimate: excellent (a), good (b), poor (c). After Pearson and Cassola (1992).

captured, and patterns of species richness appear to be highly correlated with those of other vertebrate and invertebrate taxa. Inspection of the diversity map (Fig. 122) shows the greatest richness to be in the East Indies followed closely by the Ethiopian and Neotropical regions. To the north, the Palearctic is less diverse than the Nearctic, which might be unexpected considering the difference in area. The cause is probably due to China being considered a separate region rather than part of the Palearctic.

Although the distribution pattern shown by the Cicindelidae is useful for comparative purposes, it should not be considered to be a proxy for all insect families. A number of them have centers of diversity in the neotropics and probably originated there. Examples are the bees (Michener, 1979), robber flies (Hull, 1962), the hemipteran tribe Myodochini (Harrington, 1980), and the flat bug family Aradidae (Usinger and Matsuda, 1959). For other groups, such as the Carabidae (Darlington, 1971) and the ants (Brown, 1973), the Old World tropics in general were identified as the diversity and radiation center.

DIVERSITY AND CONSERVATION

The loss of species diversity is proceeding at a rapid pace, possibly at the rate of 50 000 species per year (Wilson, 1991). Why should we care? The other species that share the earth with us are our only known companions in the universe. Many people think that we have a moral responsibility to protect them. Every species, no matter how small or insignificant in appearance, represents a unique evolutionary accomplishment. Each one (except for a few produced by polyploidy) is generally the result of more than a million years of natural selection. As Mayr (1970) observed, "By showing us that each species is uniquely different from every other species and thus irreplaceable, the student of evolution has taught us reverence for every single product of evolution, one of the most important components of conservation thinking."

The human population is expected to double in the next half century to more than 10 billion people. A five- to tenfold increase in global economic activity is projected to meet the demands of the exploding population (Ehrlich and Wilson, 1991). With that much human alteration of the environment, it is hopeless to expect to save more than one quarter of the worlds' species. Most current conservation activity consists of setting aside relatively small patches of acreage in order to rescue certain endangered species or to preserve a complex of native species. Most parks and wildlife refuges contain only very small fractions of the native habitat that once existed. Their effectiveness in helping to maintain species diversity is severely limited by the species/area rule (a habitat that is reduced to one-tenth its former size will lose half its species).

One cannot blame people for their interest in saving certain admirable species. It has been argued that the largest marine and terrestrial mammals should be saved, that special attention should be given to the primates, and that many other species (redwood trees, bald eagles, penguins) have special amenity value (Morowitz, 1991). But, is this piecemeal attention to individual species the desired direction for a global conservation effort? Would it not be better to concentrate on the preservation of habitats where the greatest number of species are found?

In both terrestrial and marine environments, the highest species diversities are found in the tropics. And, within the worldwide tropics there are important concentrations of diversity. These concentrations, which are found in the rain forests and in the marine East Indies, are unfortunately located in regions of rapid human population growth. Before human impact, rain forests covered a total area of 6 million square miles. They have now been reduced to an area of 3 million square miles. The coral reefs of the East Indies are being severely impacted by erosion from deforestation, coral mining, pollution, and destructive fishing methods.

Tropical centers of diversity also have an important evolutionary function. They apparently act as centers of radiation from which newly evolved species can spread into other parts of the tropics and, eventually, to higher latitudes. Species and genera tend to demonstrate a positive relationship between age and distance from the evolutionary centers. Some people have attributed the high diversity of the centers to an accumulation of species from other areas, but this idea is refuted by the evidence of age gradients, barrier effects, and the phylogenetic patterns in certain groups (Briggs, 1992).

The production of new species in the tropical centers, resulting in the formation of new phylogenetic lines, has evidently been going on for a long time. A paleontological analysis of post-Paleozoic marine orders showed significantly more first appearances in tropical waters than expected from sampling alone (Jablonski, 1993). This provides direct evidence that tropical regions have been a major source of evolutionary novelty, and not simply a refuge that accumulated diversity owing to low extinction rates. Similar histories are suggested by work on insects where taxa were found to have arisen in equatorial latitudes and are driven in taxon pulses toward the poles, up mountains, and into the forest canopy (see Chapter 11).

At present, there is no accepted scientific agreement on conservation strategy. The emphasis is on saving unusual or rare species. Most often, these are relicts that are well on the road to extinction. Many are endemic to islands or other small habitats. Interesting as some of them are, they have no phyletic future. By focusing our attention on the centers of origin, we can save the newer, more advanced species that have the most evolutionary potential. They are the products of highly competitive associations and they have the best chances for widespread success.

Although the phylogeny of many families and orders still needs to be worked out, making it difficult to identify the more advanced forms, species richness is a readily observable index for the location of the centers of origin. This index should be used in setting policy and making decisions about biotic conservation and management (Erwin, 1991). The more area that can be conserved within the tropical centers of diversity, the less the interference with the long-term process of global evolutionary change, and the greater the benefit to future human generations.

In the higher latitudes, important steps can also be taken. In the more developed countries, where the human population pressure has begun to abate, it would be relatively easy to stop the more destructive forms of development. When new facilities are needed, they should replace the deteriorating old ones. Forestry should be practiced on a sustained yield basis, with attention to the conservation of some virgin stands. Biodiversity studies, including national biological inventories and cataloging of the genetic library, need scientific and public support (Ehrlich and Wilson, 1991).

SUMMARY

1. The world is undergoing a catastrophic loss of species diversity, but so far, we do not have a reasonable estimate of the numbers that exist in the two principal environments, land and sea.
2. There are two generally accepted approaches to the estimation of diversity: (1) a statistical method by which a diversity figure is reached by extrapolation from small samples, and (2) an expert opinion method which sums the opinions of experts on the systematics of the component groups.
3. Application of the Energy-Stability-Area (ESA) theory of biodiversity, plus the additional factor of evolutionary time, predicts that the diversity of multicellular species in the sea will be far greater than that found on land. However, estimates using the expert opinion approach indicate that just the opposite is true. Why so few species in the sea?
4. In the final analysis, estimates of species diversity will need to be based on good systematics. In the meantime, estimates from experts, who are familiar with the individual biotic groups, appear to be the best approach.
5. In the terrestrial habitat, the insects comprise by far the largest group. Considering the many estimates that have been made, it is suggested that about 10 million species is a reasonable figure.
6. When estimates are made for the mites, spiders, nematodes, vascular plants, and a multitude of smaller taxa, a total of more than 12 million terrestrial species is reached. This is a very rough estimate and it is suggested that the figure of 12 million, plus or minus one million, would be appropriate.
7. Contrary to the terrestrial habitat, in which more than 90% of all the species are members of one phylum (Arthropoda), the bulk of the diversity in the marine environment is spread over many phyla.
8. Of the most speciose marine phyla, the Arthropoda is probably the most diverse followed by the Nematoda and Mollusca. Eight phyla contain more than 90% of the marine species.
9. The total number of marine species is approximately 178 000. This figure should also be used with caution. It may only be safe to say that the total number of species in the sea is less than 200 000.
10. It should be noted that, at the level of phylum and class, the marine biota is far more diverse than the terrestrial biota. There are 34 phyla living in the marine realm, 17 in freshwater, and only 15 on land. At the class level, 77 occur in the sea, 35 in freshwater, and 33 on land.
11. The ratio of less than 200 000 marine species to around 12 million terrestrial species means that the ocean contains less than 2% of the species on earth.
12. Within the terrestrial habitat, the ESA theory appears to work very well. Within limited marine areas it also appears to be applicable, but it cannot be applied to the entire ocean.
13. It is suggested that the missing factor in the marine diversity formula is the size of the individual primary producer. In the sea, almost all photosynthetic production is accomplished by single-celled organisms. The size of the individual phyto-

plankter is so small that it cannot provide physical support for metazoans and other plants.

14. A latitudinal gradient in species diversity is commonly found in both marine and terrestrial environments. The causes of such gradients appear to fit into six general categories: (1) historical perturbation, (2) productivity, (3) harshness, (4) climatic stability, (5) habitat heterogeneity, and (6) competition, predation, or mutualism.
15. A seventh category may be the effect of Rapaport's rule which says that the latitudinal ranges of individual species become less as latitudes become lower.
16. In the sea, in the temperate or cold waters of the higher latitudes, the diversity of benthic animals will increase from the inner shelf down to about the middle of the slope.
17. In tropical seas, especially where there are well developed reefs, the greatest species diversity is found in the shallow waters of the inner shelf. From there, there is a decrease in diversity with depth. From about 2000 m down, benthic diversity at all latitudes apparently decreases with depth.
18. In the pelagic region, most phytoplankton species are found in the upper 200 m, but the maximum zooplankton diversity is often found at greater depths. However, many zooplankton and nektonic organisms are vertical migrators, so that their diversity near the surface will be much greater at night.
19. On land, there is a notable negative correlation of species diversity and altitude. This was first described by the early botanists. Later, animals were included in a theory of altitudinal life zones. The idea of life zones then became incorporated into the ecological concept of biomes.
20. Gradients or changes in longitudinal diversity are conspicuous on a global basis but are seldom discussed. In the sea, the diversity among the four great tropical regions was compared. The species richness of the Indo-West Pacific is approximately about 2.5 times that of the western Atlantic, 3.5 times that of the eastern Pacific, and 7.5 times that of the eastern Atlantic. The species/area curve indicated a fairly consistent relationship among the first three, but the eastern Atlantic proved to be relatively depauperate.
21. The impoverishment of the eastern Atlantic is likely due to its history of severe temperature declines during the glacial stages.
22. The Indo-West Pacific is a special case for its longitudinal axis extends across almost two thirds of the globe. There is a conspicuous peak of diversity within a relatively small area called the East Indies Triangle.
23. The North Pacific Ocean is much richer in species diversity than the North Atlantic. Although the two areas have close to the same amount of continental shelf, the latter has been subjected to more severe ice age temperature declines.
24. On land, many of the longitudinal patterns appear to be closely related to area, but some do not. The very high diversity of neotropical bird species is probably due to the survival of many primitive families that evolved during the isolation of the South American continent.
25. Northern South America and Zaire have about the same geographic area and the

same number of mammal species. But oceanic islands such as Cuba have far fewer than would be expected on the basis of their areas.

26. The tiger beetle family Cicindelidae has been proposed as an appropriate taxon to illustrate terrestrial patterns of species diversity. The greatest diversity is in the East Indies followed by the Ethiopian and Neotropical Regions. While this may be true for most insect families, there are several that demonstrate their highest diversity in the neotropics.
27. The problem of conserving as much as one fourth of the world's present species diversity will be difficult. Species may be disappearing at the rate of 50 000 per year. In 50 years, the human population will double to reach 10 billion.
28. The greatest species diversities are found in the tropics. Before human impact, the rain forests covered six million square miles. They have now been reduced by half. The coral reefs of the East Indies are being severely impacted by erosion from deforestation, coral mining, pollution, and destructive fishing methods.
29. The tropical centers of diversity have an important evolutionary function. They apparently act as centers of radiation from which newly evolved species can spread to other parts of the tropics and, eventually, to higher latitudes.
30. Present conservation strategy is focused on saving unusual or rare species. Most often, these are relicts that are well on the road to extinction. Many are endemic to islands or other small habitats. They have no phyletic future.
31. If a significant number of species are to be saved, conservation efforts must concentrate on the tropical centers of origin. This would help preserve the more advanced species that have the best evolutionary potential.
32. In the industrial countries at higher latitudes, the more destructive forms of development should be halted. When new facilities are needed, they should replace the deteriorating old ones.

CHAPTER 14

Epilogue

... that grand subject, that almost keystone of the laws of creation, geographical distribution.
Charles Darwin, *in a letter to J. D. Hooker*, 1845.

When a new species evolves, if it does not immediately become extinct, it will proceed to enlarge its territory. As it does so, it produces a geographic pattern. This pattern will change and persist through time as long as that species, and the phyletic line formed by it, lasts. If the pattern in its entirety could be recovered, it would reveal the origin, the expansion, the contraction, and end of that evolutionary line. This biogeographic ideal cannot be realized because the fossil record is never that complete. For phyletic lines represented by living species, it is possible to gain insight into the historical pattern by working out the phylogeny and comparing it to the present distribution. If the living group has a fossil record, the chances of working out its geographic history may be enhanced.

Recent progress in paleontology, geophysics, climatology, and in the systematics of various groups of animals and plants, has provided us with a new perspective of biogeography. The emerging view is one of continuous change, a history of dynamic shifts in the biota in response to alterations in the physical environment. The changes in patterns of dispersal and evolution remain hazy and uncertain for much of the Paleozoic, but become sharper and more informative in the Mesozoic and Cenozoic. Two related series of physical events have had profound effects and are, in large part, responsible for the biogeography of today's world. These are the making and breaking of land and sea barriers and global sea-level fluctuations.

The first momentous event was the fragmentation of the (theoretical) Proterozoic supercontinent called Rodinia. This tectonic activity may have caused a sea-level rise, flooding of the continental shelves, and a climatic amelioration. These changes possibly set the stage for the most important evolutionary radiation in the history of earth's metazoan biota. As many as 100 different phyla may have evolved during the Cambrian Period.

The Paleozoic continents remained separated until the latest Silurian, about 410 Ma ago. This marked a consolidation between Baltica and Laurentia which formed the new continent of Laurussia. At the same time, some continental terranes became detached from Gondwana and moved northward. The next significant collision was that between Gondwana and Laurussia which took place in late Devonian about 365 Ma ago. During the Carboniferous and early Permian, the principal continents of Gondwana, Laurussia,

Siberia, and Kazakhstania moved closer together. By the late Permian, the formation of Pangaea was virtually complete.

The breakup of Pangaea began, in about the mid-Jurassic, with the separation of Madagascar from Africa and the creation of the Turgai Sea which separated Euramerica from Asia. By the late Jurassic-early Cretaceous, Africa was separating from Euramerica, South America, and Australia-Antarctica. The latter pair may have already moved apart. By the mid-Cretaceous, rising sea level had formed the great Mid-continental Sea which divided Euramerica. In the late Cretaceous, Westamerica became connected to Asia and an isthmus attached South America to Central and North America.

By the end of the Cretaceous, the Mid-continental Sea had dried up. The early Central American isthmus may have persisted into the Paleocene. In the early Eocene, the North Atlantic Ocean expanded, breaking the link between Europe and North America. In the Paleocene-Eocene, a filter bridge (probably an archipelago) became available between South America and Australia-New Zealand via Antarctica. India became firmly fused with Asia by the Eocene. With the desiccation of the Turgai Sea in the Oligocene, Europe was attached to Asia.

By the early Miocene, Africa became connected to Asia via the Arabian Peninsula. This marked the end of the Tethys Sea and the beginning of the Mediterranean. With the flooding of Beringia about 3.5 Ma ago, a marine Transarctic Biotic Interchange took place. Some 0.5 Ma later, the formation of the Panamanian isthmus permitted the terrestrial Great American Biotic Interchange. All of these continental breakages and linkages created dispersal corridors in one environment and, at the same time, dispersal barriers in the other.

The series of historic extinction events had important biogeographic and evolutionary consequences. We now live in an era of neocatastrophism in which it is widely believed that the great extinctions were sudden, catastrophic events caused by extraterrestrial impacts or volcanic eruptions or both. On the other hand, the fossils of many organisms tell us that the extinction episodes developed gradually over periods of one to several Ma. The recovery periods tended to take even longer. Throughout its history, the earth has been struck by many missiles from outer space and numerous volcanic eruptions have taken place. But these sudden, high-energy events are not consistent with the tempo of the global extinctions. The one event that is consistent is the regression in sea level.

As Hallam (1992) has pointed out, the most important cause of sea-level change involves alteration in the volume of the oceanic ridges. These structures may vary in both spreading rate and overall length. The changes may take place at a rate of about 1 cm per 10^3 years. Continental compression can cause the sea level to fall at a comparably slow rate. Glacioeustatic changes can take place much more rapidly but their importance throughout the Phanerozoic is controversial because of limited evidence for the existence of major ice sheets. Ocean-trench subsidence and the rise of mantle plumes appear to be of negligible importance. Neither oceanic sedimentation nor the desiccation of isolated basins is likely to have played a major role. The volume of ocean water has probably remained constant through the Phanerozoic, the addition of juvenile water being more or less counteracted by the subtraction of pore water in sediments by subduction.

The implication of sea-level regressions as the primary cause of the historic extinctions is only part of the story. The regressions themselves did not have much direct effect, except on the marine biota of the continental shelves. As the sea level dropped, the continents became larger and higher. These changes caused the continental climates to become colder and dryer. The cessation of tectonic-plate movement produced less volcanism and a drop in the production of CO_2 from that source. Increased weathering of the exposed land absorbed greater amounts of CO_2. The drop in atmospheric CO_2 meant more heat lost to space by radiation. The result was a decrease in global temperature related to the extent of the regression.

The larger, higher continents caused severe weather patterns which affected the entire globe. So, sea-level regressions, depending on their magnitude, caused climatic changes which had a detrimental effect on diversity on land and in the sea. Conversely, sea-level rises had the opposite effect. Long-term sea level cycles, called supercycles, appear to be the proximal causes of the historic alteration between greenhouse and icehouse conditions. The relative positions of the continents, considering their effects on the major oceanic currents, are also important factors in the global climatic cycles.

There has been considerable speculation about the evolutionary effects of extinction episodes. These provide a consistent message which gives the impression that a "wiping out of the old forms to make way for the new" has long-term evolutionary benefits. But, the extinction episodes have had a much greater effect on the tropical organisms than on those at higher latitudes. When this happens, the advanced tropical species tend to be replaced by more primitive forms from higher latitudes or other refuges. It has been argued that the historical extinction episodes were disastrous interruptions to evolutionary progress. They set back the clock of evolutionary time and destroyed communities that took millions of years to reassemble.

In the preface, it was noted that a complete biogeography should offer a prognosis for the future. This is a painful task, for we are in the midst of the most-rapid decrease in species diversity ever recorded. It means that most species now living on earth will be lost within the next 200 years. This will nullify all the diversity increase of the past 65 Ma, leaving a depauperate world to be studied by future biogeographers. As the species diversity declines, it will become increasingly difficult to reconstruct historic trends in biogeography and evolution. Although many causes of this catastrophic decline have been identified, the primary one is seldom mentioned. This is the effect of the exploding human population which increases by 95 million each year, or about 260 000 per day. Religious opposition has made human population control a taboo subject. Most conservation groups have avoided the problem. It is the missing agenda (Meffe et al., 1993).

This work is primarily an examination of historic changes in the marine and terrestrial biotas. The scope of the subject is so large and the literature so voluminous, that one cannot possibly find every publication of value and give it the recognition it deserves. Also, every biologist interested in biogeography has his or her own concept of how the subject should be addressed. There are very few generalists, so that each person will look at biogeography from a different perspective, one that has been shaped by specialty and experience. While this work is a book of fact, as opposed to fiction, my interpretation of the facts may be quite different than the conclusions of others. As John Steinbeck observed, "The design of a book is the pattern of a reality controlled and shaped by the mind of the writer. This is completely understood about poetry or fiction, but it is too seldom realized about books of fact."

References

Abbott, R.T. 1960. The genus Strombus in the Indo-Pacific. Indo-Pacific Mollusca 1:33–146.

Abele, L.G. 1982. Biogeography. *in*: *The biology of Crustacea*, Vol. I (L.G. Abele, ed.). Academic Press, New York. pp. 241–304.

Agassiz, L. 1857. *Contribution to the natural history of the United States of America*. Little, Brown and Co., Boston. Vols. 1–5.

Ager, D.V. 1973. Mesozoic Brachiopoda. *in*: *Atlas of paleobiogeography* (A. Hallam, ed.). Elsevier, Amsterdam. pp. 431–436.

Ager, D.V. 1988. Extinctions and survivals in the brachiopods and the dangers of data bases. *in*: *Extinction and survival in the fossil record* (G.P. Larwood, ed.). Clarendon Press, Oxford. pp. 89–97.

Aiello, L.C. 1993. The fossil evidence for modern human origins in Africa: a revised view. *Am. Anthropol. 95*:73–96.

Allaby, M. and J. Lovelock. 1983. *The great extinction*. Sacker and Warburg, London.

Allen, G.R. 1975a. *Damselfishes of the South Seas*. TFH Publications, Neptune City, NJ.

Allen, G.R. 1975b. *Anemonefishes*, 2nd edn. TFH Publications, Neptune City, NJ.

Allen, G.R. and F.H. Talbot. 1985. Review of the snappers of the genus *Lutjanus* (Pisces: Lutjanidae) from the Indo–Pacific, with the description of a new species. *Indo–Pacific Fishes 11*:1–87.

Alvarez, L.W. 1983. Experimental evidence that an asteroid impact led to the extinction of many species 65 Myr ago. *Proc. Natl. Acad. Sci. USA 80*:627–642.

Alvarez, L.W., W. Alvarez, F. Asaro and H.V. Michel. 1980. Extraterrestrial cause for the Cretaceous–Tertiary extinction. *Science 208*:1095–1108.

Alvarez, W. and F. Asaro. 1990. An extraterrestrial impact. *Sci. Am. 263*:78–84.

Alvariño, A. 1965. Chaetognaths. *Oceanogr. Mar. Biol. Annu. Rev. 3*:115–194.

Anderson, M.E. 1988. Studies on the Zoarcidae (Teleostei: Perciformes) of the Southern Hemisphere. II. Two new genera and a new species from temperate South America. *Proc. Calif. Acad. Sci. 45*:267–276.

Anderson, M.E. 1990. The origin and evolution of the Antarctic ichthyofauna. *in*: *Fishes of the Southern Ocean* (O. Gon and P.C. Heemstra, eds.). J.L.B. Smith Inst. Ichthyology, Grahamstown, South Africa. pp. 28–33.

Andriashev, A.P. 1986. *Review of the snailfish genus Paraliparis (Scorpaeniformes: Liparididae) of the Southern Ocean*. Koeltz, Koenigstein.

Angel, M.V. 1992. Managing biodiversity in the Oceans. *in*: *Diversity of Oceanic Life* (N.A. Peterson, ed.). Cent. Strategic and Internat. Studies. Washington, DC. pp. 23–59.

Anstey, R.L. 1987. Taxonomic survivorship and morphologic complexity in Paleozoic bryozoan genera. *Paleobiology 4*:407–418.

Archer, M., T.F. Flannery, A. Ritchie and R.E. Molnar. 1985. First Mesozoic mammal from Australia – an early Cretaceous monotreme. *Nature 318*:363–366.

Archibald, J.D. 1991. Survivorship patterns on non–marine vertebrate species across the Cretaceous–Tertiary (K/T) boundary in the western U.S. *Abs. Geol. Soc. Am.*, San Diego Meeting, p. A359.

Archibald, J.D. 1992. Dinosaur extinction: how much and how fast? *J. Vert. Paleo. 12*:263–264.

Archibald, J.D. and L.J. Bryant. 1990. Differential Cretaceous/Tertiary extinctions of nonmarine vertebrates; evidence from northeastern Montana. *in*: *Global catastrophes and earth history* (V.L. Sharpton and P.D. Ward, eds.). Geol. Soc. Am., Spec. Pap. 247:549–562.

Archibald, J.D. and W.A. Clemens. 1984. Mammal evolution near the Cretaceous– Tertiary boundary. *in*: *Catastrophes and earth history* (W.A. Berggren and J.A. Van Couvering, eds.). Princeton Univ. Press, Princeton,. NJ. pp. 339–371.

Ash, S. 1986. Fossil plants and the Triassic–Jurassic boundary. *in*: *The beginning of the age of dinosaurs* (K. Padian, ed.). Cambridge Univ. Press, Cambridge. pp. 21–30.

Aubry, M.–P. 1983. Late Eocene and early Oligocene calcareous nannoplankton biostrategraphy and biogeography. *Am. Assoc. Petrol. Geol. Bull. 67*:415.

Aubry, M.–P. 1992. Late Paleogene calcareous nannoplankton evolution: a tale of climatic deterioration. *in*: *Eocene–Oligocene climatic and biotic evolution* (D.R. Prothero and W.A. Berggren, eds.). Princeton Univ. Press, Princeton, NJ. pp. 272–309.

Audley–Charles, M.G. 1987. Dispersal of Gondwanaland: relevance to evolution of the angiosperms. *in*: *Biogeographical evolution of the Malay Archipelago* (T.C. Whitmore, ed.). Clarendon Press, Oxford. pp. 5–25.

Axelrod, D.I. 1979. The roles of plate tectonics in angiosperm history. *in*: *Historical biogeography, plate tectonics and the changing environment* (J. Gray and A.J. Boucot, eds.). Oregon State Univ. Press, Corvallis, OR. pp. 435–447.

Axelrod, D.I. 1990. Age and origin of subalpine forest zone. *Paleobiology 16*:360–369.

Axelrod, D.I. and P.H. Raven. 1978. Late Cretaceous and Tertiary vegetation history of Africa. *in*: *Biogeography and ecology of southern Africa* (M.J.A. Werger, ed.). W. Junk, the Hague. pp. 77–130.

Babinot, J.F. and J.P. Colin. 1988. Paleobiogeography of Tethyan Cretaceous marine ostracods. *in*: *Evolutionary biology of Ostracods* (T. Hanai, et al., eds.). Elsevier, Amsterdam. pp. 823–839.

Babinot, J.F. and J.–P. Colin. 1992. Marine ostracode provincialism in the late Cretaceous of the Tethyan Realm and the Austral Province. *Palaeogeogr., Palaeoclimatol., Palaeoecol. 92*:283–293.

Baez, A.M. and Z.B. de Gasparini. 1979. The South American herpetofauna: an evaluation of the fossil record. *in*: *The South American herpetofauna: its origin, evolution and dispersal* (W.E. Duellman, ed.). Mus. Nat. Hist. Univ. Kansas, Monogr. 7:29–54.

Bakker, R.T. 1978. Dinosaur feeding behavior and the origin of flowering plants. *Nature, 274*:661–663.

Baldauf, J.G. 1992. Middle Eocene through early Miocene diatom floral turnover. *in*: *Eocene–Oligocene climatic and biotic evolution* (D.R. Prothero and W.A. Berggren, eds.). Princeton Univ. Press, Princeton, NJ. pp. 310–326.

Ball, G.E. 1985. Reconstructed phylogeny and geographical history of genera of the tribe Galeritini (Coleoptera: Carabidae). *in*: *Taxonomy, phylogeny and zoogeography of beetles and ants* (G.E. Ball, ed.). W. Junk, Dordrecht. pp. 276–321.

Ball, I.R. 1974. A contribution to the phylogeny and biogeography of the freshwater triclads (Platyhelminthes: Turbellaria). *in*: *Biology of the Turbellaria* (N.W. Riser and M.P. Morse, eds.). McGraw–Hill, New York. pp. 339–401.

Bambach, R.K. 1986. Phanerozoic marine communities. *in*: *Patterns and processes in the history of life* (D.M. Raup and D. Jablonski, eds.). Springer–Verlag, Berlin. pp. 407–428.

Bambach, R.K. 1990. Late Paleozoic provinciality in the marine realm. *in*: *Palaeozoic palaeography and biogeography* (W.S. McKerrow and C.R. Scotese, eds.). Geol. Soc. London, Mem. 12:307–323.

Bambach, R.K., C.R. Scotese and A.M. Ziegler. 1980. Before Pangea: the geographics of the Paleozoic world. *Am. Sci. 68*:26–38.

Bănărescu, P. 1990. *Zoogeography of fresh waters, Vol. I.* AULA–Verlag, Wiesbaden.

Bănărescu, P. 1992. *Zoogeography of fresh waters, Vol. II.* AULA–Verlag, Wiesbaden.

Barnard, J.L. 1991. Amphipodological agreement with Platnick. *J. Nat. Hist. 25*:1675–1676.

Barron, E.J., C.G.A. Harrison, J.L. Sloan, II and W.W. Hay. 1981. Paleogeography, 180 million years ago to the present. *Eclogae Geol. Helv. 74*:443–470.

Bartek, L.R., L.C. Sloan, J.B. Anderson and M.I. Ross. 1992. Evidence from the Antarctic continental margin of late Paleogene ice sheets: a manifestation of plate reorganization and synchronous changes in atmospheric circulation over the emerging Southern Ocean? *in*: *Eocene–Oligocene climatic and biotic evolution* (D.R. Prothero and W.A. Berggren, eds.). Princeton Univ. Press, Princeton, NJ. pp. 131–159.

Beauvias, L. 1979. Paleobiogeography of the Middle Jurassic corals. *in*: *Historical biogeography, plate tectonics and the changing environment* (J. Gray and A.J. Boucot, eds.). Oregon State Univ. Press, Corvallis, OR. pp. 289–303.

Beauvias, L. 1992. Paleobiogeography of the early Cretaceous corals. *Palaeogeogr., Palaeoclimatol., Palaeoecol. 92*:233–247.

Belkin, J.N. 1962. *The mosquitoes of the South Pacific*, 1 and 2. Univ. of California Press, Berkeley, CA.

Belyaev, G.M. 1966. *Hadal bottom fauna of the World Ocean*. English translation by Israel Program for Scientific Translations, Jerusalem, 1972.

Belyaev, G.M. 1989. *Deep–sea ocean trenches and their fauna*. Inst. of Oceanology, Moscow (in Russian).

Benson, R.H. 1984. The Phanerozoic "crisis" as viewed from the Miocene. *in*: *Catastrophes and earth history* (W.A. Berggren and J.A. Van Couvering, eds.). Princeton Univ. Press, Princeton, NJ. pp. 437–446.

Benson, R.H., R.E. Chapman and L.T. Deck. 1985. Evidence from the Ostracoda of major events in the South Atlantic and world–wide over the past 80 million years. *in*: *South Atlantic paleoceanography* (K.J. Hsü and H.J. Weissert, eds.). Cambridge Univ. Press, Cambridge. pp. 325–350.

Benton, M.J. 1988. Mass extinction in the fossil record of reptiles: paraphyly, patchiness and periodicity (?). *in*: *Extinction and survival in the fossil record* (G.P. Larwood, ed.). Clarendon Press, Oxford. pp. 269–294.

Benton, M.J. 1990a. End–Triassic. *in*: *Palaeobiology: a synthesis* (D.E.G. Briggs and P.R. Crowther, eds.). Blackwell, Oxford. pp. 194–198.

Benton, M.J. 1990b. Mass extinctions in the fossil record of late Paleozoic and Mesozoic tetrapods. *in*: *Extinction events in earth history* (E.G. Kauffman and O.H. Walliser, eds.). Springer-Verlag, Berlin. pp. 239–251.

Benton, M.J. 1993. Late Triassic extinctions and the origin of the dinosaurs. *Science 260*:769–770.

Berdan, J.M. 1990. The Silurian and early Devonian biogeography of ostracods in North America. *in*: *Palaeozoic palaeogeography and biogeography* (W.S. McKerrow and C.R. Scotese, eds.). Geol. Soc. London, Mem. 12:223–231.

Berenbaum, M. and D. Seigler. 1992. Biochemicals engineering problems for natural selection. *in*: *Insect chemical ecology: an evolutionary approach* (B.D. Roitberg and M.B. Isman, eds.). Routledge, Chapman and Hall, New York. pp. 89–121.

Bergström, S.M. 1990. Relations between conodont provincialism and the changing paleogeography during the early Paleozoic. *in*: *Palaeozoic palaeogeography and biogeography* (W.S. McKerrow and C.R. Scotese, eds.). Geol. Soc. London, Mem. 12:105–121.

Berra, T.M. 1981. *An atlas of distribution of the freshwater fish families of the world*. Univ. of Nebraska Press, Lincoln, NE.

Berry, W.B.N. and P. Wilde. 1990. Graptolite biogeography: implications for palaeogeography and palaeoceanography. *in*: *Palaeozoic palaeogeography and biogeography* (W.S. McKerrow and C.R. Scotese, eds.). Geol. Soc. London, Mem. 12:129–137.

Bertelsen, E. 1951. *The ceratioid fishes: ontogeny, taxonomy, distribution and biology*. Dana Report 39, 281 pp.

Beurlen, K. 1956. Der Faunenschnitt an der Perm–Triasgrenze. *Z. Dtsch. Geol. Ges. 108*:88.

Bianco, P.G. 1990. Potential role of the palaeohistory of the Mediterranean and Paratethys basins on the early dispersal of Euro–Mediterranean freshwater fishes. *Ichthyol. Explor. Freshwaters 1*:167–184.

Bishop, G.A. 1986. Occurrence, preservation and biogeography of the Cretaceous crabs of North America. *in*: *Crustacean biogeography* (R.H. Gore and K.L. Heck, eds.). A.A. Balkema, Rotterdam. pp. 111–142.

Blodgett, R.B., D.M. Rohr and A.J. Boucot. 1990. Early and Middle Devonian gastropod biogeography. *in*: *Palaeozoic palaeogeography and biogeography* (W.S. McKerrow and C.R. Scotese, eds.). Geol. Soc. London, Mem. 12:277–284.

Blot, J. and J.C. Tyler. 1990. New genera and species of fossil surgeon fishes and their relatives (Acanthuroidei, Teleostei) from the Eocene of Monte Bolca, Italy, with application of the Blot formula to both fossil and recent forms. *Miscellanea Paleontologica*, 1990, Verona. pp. 13–92.

Blum, S.D. 1989. Biogeography of the Chaetodontidae: an analysis of allopatry among closely related species. *Environ. Biol. Fishes* 25:9–31.

Bolin, R.L. 1952. Description of a new genus and species of cottid fish from the Tasman Sea, with a discussion of its derivation. *Vidensk. Medd. Dansk. Naturh. Foren. Kbh. 114*:431–441.

Bonaparte, J.F. 1990. New late Cretaceous mammals from the Los Alamitos Formation, northern Patagonia. *Natl. Geogr. Res. 6*:63–93.

Boschi, E.E. 1979. Geographic distribution of Argentinean marine decapod crustaceans. *Bull. Biol. Soc. Wash. 3*:134–143.

Bose, M.N., E.L. Taylor and T.N. Taylor. 1990. Gondwana floras of India and Antarctica – a survey and appraisal. *in*: *Antarctic paleobiology* (T.N. Taylor and E.L. Taylor, eds.). Springer–Verlag, New York. pp. 118–148.

Boss, K.J. 1971. Critical estimate of the number of recent Mollusca. *Occas. Paps. on Mollusks, Harvard Mus. 3*:81–135.

Botosaneanu, L. and W. Wichard. 1984. Upper–Cretaceous amber Trichoptera. *in*: *Proc. 4th Internat. Symp. on Trichoptera* (J.C. Morse, ed.). W. Junk, The Hague. pp. 43–48.

Boucot, A.J. 1988. Devonian biogeography: an update. *in*: *Devonian of the World, Vol. III* (N.J. McMillan, A.F. Embry, and D.J. Glass, eds.). Canadian Soc. Petrol. Geol., Calgary, Canada. pp. 211–227.

Boucot, A.J. 1990a. Silurian biogeography. *in*: *Palaeozoic palaeogeography and biogeography* (W.S. McKerrow and C.R. Scotese, eds.). Geol. Soc. London, Mem. 12:191–196.

Boucot, A.J. 1990b. Phanerozoic extinctions: how similar are they to each other? *in*: *Extinction events in earth history* (E.G. Kaufman and O.H. Walliser, eds.). Springer–Verlag, Berlin. pp. 5–30.

Boufford, D.E. and S.A. Spongberg. 1983. Eastern Asian–eastern North American phytogeographical relationships – a history from the time of Linnaeus to the twentieth century. *Ann. Mo. Bot. Gard. 70*:423–439.

Boulter, M.C., R.A. Spicer and B.A. Thomas. 1988. Patterns of plant extinction from some palaeobotanical evidence. *in*: *Extinction and survival in the fossil record* (G.P. Larwood, ed.). Clarendon Press, Oxford. pp. 1–36.

Braithwaite, C.J.R. 1984. Geology of the Seychelles. *in*: *Biogeography and ecology of the Seychelles Islands* (D.R. Stoddart, ed.). W. Junk, The Hague. pp. 17–38.

Braithwaite, C.J.R. 1987. Geology and palaeogeography of the Red Sea region. *in*: *Red Sea* (A.J. Edwards and S.M. Head, eds.). Pergamon Press, Oxford. pp. 22–44.

Bramlette, M.N. 1965. Massive extinctions in biota at the end of Mesozoic times. *Science 148*:1696–1699.

Bramwell, D. 1979. *Plants and islands*. Academic Press, London.

Brasier, M.D. 1975. An outline history of seagrass communities. *Paleontology 18*:681–702.

Brasier, M.D. 1988. Foraminiferid extinction and ecological collapse during global biological events. *in*: *Extinction and survival in the fossil record* (G.P. Larwood, ed.). Clarendon Press, Oxford. pp. 37–64.

Brenan, J.P.M. 1978. Some aspects of the phytogeography of tropical Africa. *Ann. Mo. Bot. Gard. 65*:437–478.

Brenchley, P.J. 1989. The late Ordovician extinction. *in*: *Mass extinctions: processes and evidence* (S.K. Donovan, ed.). Columbia Univ. Press, New York. pp. 104–132.

Brenchley, P.J. 1990. End Ordovician. *in*: *Palaeobiology: a synthesis* (D.E.G. Briggs and P.R. Crowther, eds.). Blackwell, Oxford. pp. 181–184.

Briggs, D.E.G., R.A. Fortey and E.N.K. Clarkson. 1988. Extinction and the fossil record of the arthropods. *in*: *Extinction and survival in the fossil record* (G.P. Larwood, ed.). Clarendon Press, Oxford. pp. 171–209.

Briggs, J.C. 1955. A monograph of the clingfishes (order Xenopterygii). *Stanford Ichthyol. Bull. 6*:1–224.

Briggs, J.C. 1966. Oceanic islands, endemism and marine paleotemperatures. *Syst. Zool. 15*:153–163.

Briggs, J.C. 1967. Relationship of the tropical shelf regions. *Stud. Trop. Oceanog. Miami 5*:569–578.

Briggs, J.C. 1970. Tropical shelf zoogeography. *Proc. Calif. Acad. Sci. 38*:131–137.

Briggs, J.C. 1974a. *Marine Zoogeography*. McGraw–Hill, New York.

Briggs, J.C. 1974b. Operation of zoogeographic barriers. *Syst. Zool. 23*:248– 256.

Briggs, J.C. 1979. Ostariophysan zoogeography: an alternative hypothesis. *Copeia 1979:*111–118.

Briggs, J.C. 1984. Centres of origin in biogeography. *Biogeogr. Monogr. No. 1*. Univ. of Leeds, Leeds.

Briggs, J.C. 1985. Species richness among the tropical shelf regions. *Biologiya Morya 6*:3–11 (in Russian; English translation, Plenum, New York, 1986).

Briggs, J.C. 1987a. Antitropical distribution and evolution in the Indo–West Pacific Ocean. *Syst. Zool. 36*:237–247.

Briggs, J.C. 1987b. *Biogeography and plate tectonics*. Elsevier, Amsterdam.

Briggs, J.C. 1989. The historic biogeography of India: isolation or contact? *Syst. Zool. 38*:322–332.

Briggs, J.C. 1990. Global extinctions, recoveries and evolutionary consequences. *Evolutionary Monograph No. 13*, Univ. of Chicago, Chicago, IL.

Briggs, J.C. 1991a. Global species diversity. *J. Nat. Hist. 25*:1403–1406.

Briggs, J.C. 1991b. A Cretaceous–Tertiary mass extinction? *BioScience 41*:619– 624.

Briggs, J.C. 1992. The marine East Indies: centre of origin? *Global Ecol. Biogeogr. Lett. 2*:149–156.

Briggs, J.C. 1994. Species diversity: land and sea compared. *Syst. Biol. 43*:130–135.

Brown, J.H. and A.C. Gibson. 1983. *Biogeography*. Mosby, St. Louis, MO.

Brown, W.L., Jr. 1973. A comparison of the Hylean and Congo–West African rain forest and faunas. *in*:

Tropical forest ecosystems in Africa and South America: a comparative review (B.L. Meggers, E.S. Ayensu and W.D. Duckworth, eds.). Smithsonian Press, Washington, DC. pp. 161–185.

Browne, J. 1983. *The secular ark: studies in the history of biogeography*. Yale Univ. Press, New Haven, CT.

Brundin, L. 1966. Transantarctic relationships and their significance, as evidenced by chironomid midges. *Kungl. Svenska Vetenskapsakademiens Handlingar 2*:1–472.

Brundin, L. 1967. Insects and the problem of austral disjunctive distribution. *Annu. Rev. Entomol. 12*:149–168.

Brundin, L. 1975. Circum–antarctic distribution patterns and continental drift. *Mem. Mus. Nat. d'Hist. Nat. Paris, 88*:19–27.

Brundin, L. 1988. Phylogenetic biogeography. *in*: *Analytical biogeography* (A.A. Myers and P.S. Giller, eds.). Chapman and Hall, London. pp. 343–369.

Bruneau de Miré, P. 1979. Trans–Atlantic dispersal: several examples of colonization of the Gulf of Biafra by Middle American stocks of Carabidae. *in*: *Carabid beetles, their evolution, natural history and classification* (T.L. Erwin, et al., eds.). W. Junk, The Hague. pp. 327–330.

Brusca, R.C. and G.J. Brusca. 1990. *Invertebrates*. Sinauer Assoc., Sunderland, MA.

Brusca, R.C. and B.R. Wallerstein. 1979. Zoogeographic patterns of idoteid isopods in the northeast Pacific, with a review of shallow water zoogeography of the area. *Bull. Biol. Soc. Wash. 3*:67–105.

Bruun, A.F. 1956. The abyssal fauna: its ecology, distribution and origin. *Nature 177*:1105–1108.

Bull, P.C. and A.H. Whitaker. 1975. The amphibians, reptiles, birds and mammals. *in*: *Biogeography and Ecology in New Zealand* (G. Duschel, ed.). W. Junk, The Hague. pp. 231–276.

Burger, W.C. 1981. Why are there so many kinds of flowering plants? *BioScience 31*:572, 577–581.

Burgess, W.E. 1978. *Butterflyfishes of the world*. TFH Publications, Neptune City, NJ.

Burrett, C., J. Long and B. Stait. 1990. Early–middle Paleozoic biogeography of Asian terranes derived from Gondwana. *in*: *Palaeozoic palaeogeography and biogeography* (W.S. McKerrow and C.R. Scotese, eds.). Geol. Soc. London, Mem. 12:163–174.

Bussing, W.A. 1985. Patterns of distribution of the Central American ichthyofauna. *in*: *The great American interchange* (F.G. Stehli and S.D. Webb, eds.). Plenum Press, New York. pp. 453–473.

Calabrese, D.M. 1980. Zoogeography and cladistic analysis of the Gerridae (Hemiptera; Heteroptera). *Miscel. Pub. Ent. Soc. Am. 11*:1–119.

Campbell, A.C. 1987. Echinoderms of the Red Sea. *in*: *Red Sea* (A.J. Edwards and S.M. Head, eds.). Pergamon Press, Oxford. pp. 215–232.

Cande, S.C. and J.C. Mutter. 1982. A revised identification of the oldest sea– floor spreading anomalies between Australia and Antarctica. *Earth Planet. Sci. Lett. 58*:151–160.

Canudo, J.I., G. Keller and E. Molina. 1991. Cretaceous/Tertiary boundary extinction pattern and faunal turnover at Agost and Caravaca, S.E. Spain. *Mar. Micropaleo. 17*:319–341.

Carey, S.W. 1956. *The expanding earth*. Elsevier, Amsterdam.

Carey, S.W. 1976. The expanding earth – an essay review. *Earth Sci. Rev. 11*:105–143.

Carlquist, S. 1965. *Island life*. Natural History Press, Garden City, New York.

Carlquist, S. 1974. *Island biology*. Columbia Univ. Press, New York.

Carroll, R.L. 1988. *Vertebrate paleontology and evolution*. W.H. Freeman, New York.

Case, T.J. 1990. Invasion resistance arises in strongly interacting species– rich model competition communities. *Proc. Natl. Acad. Sci. USA 87*:9610–9614.

Cavender, T.M. 1986. Review of the fossil history of North American freshwater fishes. *in*: *The Zoogeography of North American freshwater fishes* (C.H. Hocutt and E.O. Wiley, eds.). John Wiley, New York. pp. 699–724.

Cerling, T.E., J. Quade, Y. Wang and J.R. Bowman. 1989. Carbon isotopes in paleosol carbonates as paleoecology indicators. *Nature 341*:138–139.

Cernohorsky, W.O. 1976. The Mitridae of the world. Part I. *Indo–Pac. Mollusca 3*:273–528.

Chapman, V.J. 1976. *Mangrove vegetation*. Cramer, Vaduz.

Charig, A.J. 1984. Competition between therapsids and archosaurs during the Triassic Period: a review and synthesis of current theories. *in*: *The structure, development and evolution of reptiles* (M.W.J. Ferguson, ed.). Academic Press, London. pp. 597–628.

Charig, A.J. 1989. The Cretaceous–Tertiary boundary and the last of the dinosaurs. *Phil. Trans. R. Soc. Lond. B325*:387–400.

Charpentier, R.R. 1984. Conodonts through time and space. *Geol. Soc. Am., Spec. Paper 196*:11–35.

Chatterjee, S. 1992. A kinematic model for the evolution of the Indian plate since the late Jurassic. *in*: *New concepts in global tectonics* (S. Chatterjee and N. Hotton, III, eds.). Texas Tech. Univ. Press, Lubbock, TX.

Chatterjee, S. and N. Hotton, III. 1986. The paleoposition of India. *J. Southeast Asian Earth Sci. 1*:145–189.

Chatterton, B.D.E. and S.E. Speyer. 1989. Larval ecology, life history strategies and patterns of extinction and survivorship among Ordovician trilobites. *Paleobiology 15*:118–132.

Chavez, F.P. and R.C. Brusca. 1991. The Galápagos Islands and their relation to oceanographic processes in the tropical Pacific. *in*: *Galápagos marine invertebrates* (M.J. James, ed.). Plenum Press, New York. pp. 9–33.

Cheng, Z. 1983. A comparative study of the vegetation in Hubei Province, China and in the Carolinas of the United States. *Ann. Mo. Bot. Gard. 70*:571– 575.

Cifelli, R.L. 1983. The origin and affinities of the South American Condylarthra and early Tertiary Litopterna (Mammalia). *Am. Mus. Novit. 2772*:1– 49.

Clarke, A.H., Jr. 1962. Annotated list and bibliography of the abyssal marine molluscs of the world. *Bull. Nat. Mus. Can. 181*:1–114.

Clemens, W.A. 1986. Evolution of the terrestrial vertebrate fauna during the Cretaceous–Tertiary transition. *in*: *Dynamics of extinction* (D.K. Elliott, ed.). John Wiley, New York. pp. 63–85.

Clemens, W.A., J.D. Archibald and L.J. Hickey. 1981. Out with a whimper not a bang. *Paleobiology 7*:293–298.

CLIMAP Project Members. 1981. Seasonal reconstruction of the earth's surface at the last glacial maximum. *Geol. Soc. Am., Map and Chart Ser., Map. MC.* 36 pp.

CLIMAP Project Members. 1984. The last interglacial ocean. *Quat. Res. (N.Y.) 21*:123–224.

Cloud, P. 1983. The biosphere. *Sci. Am. 249*:176–187.

Coates, A.C. 1973. Cretaceous Tethyan coral–rudist biogeography related to the evolution of the Atlantic Ocean. *Paleontology Spec. Papers. 12*:169–174.

Coburn, M.M. and T.M. Cavender. 1992. Interrelationships of North American cyprinid fishes. *in*: *Systematics, historical ecology and North American freshwater fishes* (R.L. Mayden, ed.). Stanford Univ. Press, Stanford. pp. 328–373.

Cocks, L.R.M. and R.A. Fortey. 1990. Biogeography of Ordovician and Silurian faunas. *in*: *Palaeozoic palaeogeography and biogeography* (W.S. McKerrow and C.R. Scotese, eds.). Geol. Soc. London, Mem. 12:97–104.

Coddington, J.A. and H.W. Levi. 1991. Systematics and evolution of spiders (Araneae). *Annu. Rev. Ecol. Syst. 22*:565–592.

Coetzee, J.A. and J. Muller. 1984. The phytogeographic significance of some extinct Gondwana pollen types from the Tertiary of the southwestern Cape (South Africa). *Ann. Mo. Bot. Gard. 71*:1088–1099.

Cohen, D.M., T. Inada, T. Iwamoto and N. Scialabba. 1990. Gadiform fishes of the world. *FAO Species Catalog, Vol. 10*. FAO, Rome.

Colbert, E.H. 1986. Mesozoic tetrapod extinctions: a review. *in*: *Dynamics of extinction* (D.K. Elliott, ed.). John Wiley, New York. pp. 49–62.

Colin, J.–P. and B. Andreu. 1990. Cretaceous halocyprid Ostracods. *in*: *Ostracoda and global events* (R. Whatley and C. Maybury, eds.). Chapman and Hall, London. pp. 515–523.

Collette, B.B. and P. Bănărescu. 1977. Systematics and zoogeography of the fishes of the family Percidae. *J. Fish. Res. Bd. Can. 34*:1450–1463.

Collette, B.B. and L.N. Chao. 1975. Systematics and morphology of the bonitas (*Sarda*) and their relatives (Scombridae, Sardini). *Fish. Bull. 73*:516–725.

Collette, B.B. and N.V. Parin. 1991. Shallow–water fishes of Walters Shoals, Madagascar Ridge. *Bull. Mar. Sci. 48*:1–22.

Collinson, M.E. 1992. Vegetational and floristic changes around the Eocene/Oligocene boundary in western and central Europe. *in*: *Eocene–Oligocene climatic and biotic evolution* (D.R. Prothero and W.A. Berggren, eds.). Princeton Univ. Press, Princeton, NJ. pp. 437–450.

Collinson, M.E., K. Fowler and M.C. Boulter. 1981. Floristic changes indicate a cooling climate in the Eocene of southern England. *Nature 291*:315–317.

Compagno, L.J.V. 1984. Sharks of the world. *FAO Species Catalog, Vol. 4*. FAO, Rome.

Coomans, H.E. 1962. The marine mollusk fauna of the Virginian area as the basis for defining zoogeographical provinces. *Beaufortia 9*:83–104.

Courtillot, V.E. 1990. A volcanic eruption. *Sci. Am. 263*:85–92.

Cox, C.B. 1974. Vertebrate paleodistributional patterns and continental drift. *J. Biogeogr. 1*:75–94.

Cox, C.B. 1990. New geological theories and old biogeographical problems. *J. Biogeogr. 17*:117–130.

Cracraft, J. 1973. Continental drift, paleoclimatology and the evolution and biogeography of birds. *J. Zool. London 169*:455–545.

Cracraft, J. 1974. Phylogeny and evolution of the ratite birds. *Ibis 116*:494– 521.

Cracraft, J. 1983. Cladistic analysis and vicariance biogeography. *Am. Sci. 71*:273–281.

Crame, J.A. 1987. Late Mesozoic bivalve biogeography of Antarctica. *in*: *Gondwana six: stratigraphy, sedimentology and paleontology* (G.D. McKenzie, ed.). Am. Geophys. Union, Washington, DC. pp. 93–102.

Crame, J.A. 1993a. Latitudinal range fluctuation in the marine realm through geological time. *Trends Ecol. Evol. 8*:162–166.

Crame, J.A. 1993b. Bipolar molluscs and their evolutionary implications. *J. Biogeogr. 20*:145–161.

Cranbrook, E. 1981. The vertebrate faunas. *in*: *Wallace's Line and plate tectonics* (T.C. Whitmore, ed.). Clarendon Press, Oxford. pp. 57–69.

Crane, P.R. 1989. Early fossil history and evolution of the Betulaceae. *in*: *Evolution, systematics and fossil history of the Hamamelidae. Vol. 2* (P.R. Crane, ed.). Clarendon Press, Oxford. pp. 87–116.

Crane, P.R. and S. Lidgard. 1990. Angiosperm radiation and patterns of Cretaceous palynological diversity. *in*: *Major evolutionary radiations* (P.D. Taylor and G. P. Larwood, eds.). Clarendon Press, Oxford. pp. 377–407.

Craw, R. 1988. Panbiogeography: method and synthesis in biogeography. *in*: *Analytical biogeography* (A.A. Myers and P.S. Giller, eds.). Chapman and Hall, London. pp. 405–435.

Crick, R.E. 1990. Cambro–Devonian biogeography of nautiloid cephalopods. *in*: *Palaeozoic palaeogeography and biogeography* (W.S. McKerrow and C.R. Scotese, eds.). Geol. Soc. London, Mem. 12:147–161.

Croizat, L., G. Nelson and D.E. Rosen. 1974. Centers of origin and related concepts. *Syst. Zool. 23*:265–287.

Crosskey, R.W. 1990. *The natural history of blackflies*. John Wiley, Chichester.

Crossman, E.J. and D.E. McAllister. 1986. Zoogeography of freshwater fishes of the Hudson Bay drainage, Ungava Bay and the Arctic archipelago. *in*: *The zoogeography of North American freshwater fishes* (C.H. Hocutt and E.O. Wiley, eds.). John Wiley, New York. pp. 53–104.

Crowley, T.J. and G.R. North. 1988. Abrupt climate change and extinction events in earth history. *Science 240*:996–1002.

Crowley, T.J. and G.R. North. 1991. *Paleoclimatology*. Oxford Univ. Press, New York.

Crowson, R.A. 1980. On amphipolar distribution patterns in some climate groups of Coleoptera. *Entomologia Generalis 6*:281–292.

Crowson, R.A. (ed.). 1981. *The biology of the Coleoptera*. Academic Press, London.

Dana, J.D. 1853. On an isothermal oceanic chart illustrating the geographical distribution of marine animals. *Am. J. Sci. 16*:314–327.

Dana, T.F. 1975. Development of contemporary Eastern Pacific coral reefs. *Mar. Biol. 33*:355–374.

Darlington, P.J., Jr. 1957. *Zoogeography: the Geographical Distribution of Animals*. John Wiley, New York.

Darlington, P.J., Jr. 1959. Area, climate and evolution. *Evolution 13*:488–510.

Darlington, P.J., Jr. 1965. *Biogeography of the southern end of the world*. Harvard Univ. Press, Cambridge, MA.

Darlington, P.J., Jr. 1971. The carabid beetles of New Guinea. Part. IV. General considerations; analysis and history of fauna. *Bull. Mus. Comp. Zool. 142*:129–337.

Darwin, C. 1859. *On the origin of species by means of natural selection*. John Murray, London.

Davis, G.M. 1979. The origin and evolution of the gastropod family Pomatiopsidae, with emphasis on the Mekong River Triculinae. *Monogr. 20*:1–120. Acad. Nat. Sci., Philadelphia, PA.

Dawson, C.E. 1985. *Indo–Pacific pipefishes (Red Sea to the Americas)*. Gulf Coast Research Laboratory, Ocean Springs, MS.

de Candolle, A. 1855. *Géographie botanique raisonée*. Paris. 2 Vols. Masson, Paris.

De Rietz, G.E. 1940. Problems of bipolar plant distribution. *Acta Phytogeogr. Suec. 13*:215–282.

Dettmann, M.E. 1981. The Cretaceous flora. *in*: *Ecological biogeography of Australia* (A. Keast, ed.). W. Junk, The Hague. pp. 357–375.

De–Yuan, H. 1983. The Distribution of Scrophulariaceae in the Holarctic with special reference to the floristic relationships between Eastern Asia and Eastern North America. *Ann. Mo. Bot. Gard. 70*:701–712.

Dhondt, A.V. 1992. Cretaceous inoceramid biogeography: a review. *Palaeogeogr., Palaeoclimatol., Palaeoecol. 92*:217–232.

Diamond, J.M. 1992. Twilight of the pygmy hippos. *Nature 359*:15.

Diehl, S.R. and G.L. Bush. 1989. The role of habitat preference in adaptation and speciation. *in*: *Speciation and its consequences* (D. Otte and J.A. Endler, eds.). Sinauer Associates, Sunderland, MA. pp. 345–365.

Dietz, R.S. and J.C. Holden. 1970. Reconstruction of Pangea: breakup and dispersal of continents. *Sci. Am. 223*:30–41.

Dilley, F.C. 1973. Larger Foraminfera and seas through time. *in*: *Organisms and continents through time* (N.F. Hughes, ed.). Palaeontological Assoc., London. Spec. Paper No. 12. pp. 155–168.

Dingle, R.V. 1988. Marine ostracode distribution during the early breakup of southern Gondwanaland. *in*: *Evolutionary biology of Ostracoda, its fundamentals and applications* (T. Hanai, T. Ikeya and K. Iswizaki, eds.). Proc. 9th Internat. Symp. on Ostracoda. Kodansha Ltd., Tokyo. pp. 841–854.

Dockery, D.T. 1984. Crisis events for Paleogene molluscan faunas in the southeastern United States. *Miss. Geol. 5*:1–7.

Dodson, P. 1991. Maastrichtian dinosaurs. *Abs. Geol. Soc. Am., San Diego Meeting*. pp. A184–A185.

Domning, D.P. 1981. Sea cows and sea grasses. *Paleobiology 7*:417–420.

Doré, A.G. 1991. The structural foundation and evolution of Mesozoic seaways between Europe and the Arctic. *Palaeogeogr., Palaeoclimatol., Palaeoecol. 87*:441– 492.

Dott, R.N. and Batten, R.L. 1988. *Evolution of the earth*. 4th edn. McGraw Hill, New York.

Doyle, P. 1987. Lower Jurassic–lower Cretaceous belemnite biogeography and the development of the Mesozoic boreal realm. *Palaeogeogr., Palaeoclimatol., Palaeoecol. 61*:237–254.

Doyle, P. 1992. A review of the biogeography of Cretaceous belemnites. *Palaeogeogr., Palaeoclimatol., Palaeoecol. 92*:207–216.

Dransfield, J. 1987. Bicentric distribution in Malesia as exemplified by palms *in*: *Biogeographical evolution of the Malay Archipelago* (T.C. Whitmore, ed.). Clarendon Press, Oxford. pp. 60–72.

Duellman, W.E. 1993. Amphibians in Africa and South America: evolutionary history and ecological comparisons. *in*: *Biological relationships between Africa and South America* (P. Goldblatt, ed.). Yale Univ. Press, New Haven, CT. pp. 200–243.

Duellman, W.E. and L. Trueb. 1986. *Biology of amphibians*. McGraw Hill, New York.

Dunbar, C.O. 1960. *Historical geology*. 2nd edn. John Wiley, New York.

Durazzi, J.T. and F.G. Stehli. 1973. Average generic age, the planetary temperature gradient and pole location. *Syst. Zool. 21*:384–389.

Eastman, J.T. 1991. Evolution and diversification of Antarctic notothenoid fishes. *Am. Zool. 31*:93–109.

Edmunds, G.F., Jr. 1972. Biogeography and evolution of Ephemeroptera. *Annu. Rev. Entomol. 17*:21–42.

Edmunds, G.F. 1982. Historical and life history factors in the biogeography of mayflies. *Am. Zool. 22*:371–374.

Edwards, A.J. 1993. New records of fishes from the Bonaparte Seamount and Saint Helena Island, South Atlantic. *J. Nat. Hist. 27*:493–503.

Edwards, A.J. and C.W. Glass. 1987. The fishes of Saint Helena Island, South Atlantic Ocean. I. the shore fishes. *J. Nat. Hist. 21*:617–686.

Edwards, A.J. and R. Lubbock. 1983. Marine zoogeography of St. Paul's Rocks. *J. Biogeogr. 10*:65–72.

Edwards, D. 1973. Devonian floras. *in*: *Atlas of palaeobiogeography* (A. Hallam, ed.). Elsevier, Amsterdam. pp. 105–115.

Edwards, D. 1990. Constraints of Silurian and early Devonian phytogeographic analyses based on megafossils. *in*: *Palaeozoic palaeogeography and biogeography* (W.W. McKerrow and C.R. Scotese, eds.). Geol. Soc. London, Mem. 12:233–242.

Ehrlich, P.R. and E.O. Wilson. 1991. Biodiversity studies: science and policy. *Science 253*:758–762.

Eisenberg, J.F. 1981. *The mammalian radiation*. Univ. of Chicago Press, Chicago, IL.

Ekman, S. 1935. *Tiergeographie des meeres*. Akad. Verlagsges, Leipzig.

Ekman, S. 1953. *Zoogeography of the sea*. Sidgwick and Jackson, London.

Elder, W.P. 1989. Molluscan extinction patterns across the Cenomanian–Turonian Stage boundary in the western interior of the United States. *Paleobiology 15*:299–320.

Eldredge, N. 1987. *Life pulse: episodes from the story of the fossil record.* Facts on File, New York.

Emerson, W.K. 1991. First records for *Cymatium mundum* (Gould) in the Eastern Pacific Ocean, with comments on the zoogeography of the tropical trans–Pacific tonnacean and non–tonnacean prosobranch gastropods with Indo–Pacific faunal affinities in west American waters. *The Nautilus 105(2)*:62–80.

Endler, J.A. 1982. Problems in distinguishing historical from ecological factors in biogeography. *Am. Zool. 22*:441–452.

Endress, P.K. 1983. Dispersal and distribution in some small archaic relic angiosperm families (Austrobaileyaceae, Euponatiaceae, Himandandraceae, Idospermiodeae–Calycanthaceae). *in*: *Dispersal and distribution* (K. Kutitzki, ed.). Verlag Paul Parey, Hamburg. pp. 201–217.

Endrödi, S. 1985. *The Dynastinae of the world.* W. Junk, Dordrecht.

Engel, J.J. 1980. A monograph of *Clasmatocolea* (Hepaticae). *Fieldiana Botany 3*:1–229.

Erwin, D.H. 1990. The end–Permian mass extinction. *Annu. Rev. Ecol. Syst. 21*:69–91.

Erwin, D.H. 1993. *The great Paleozoic crisis.* Columbia Univ. Press, New York.

Erwin, D.H., J.W. Valentine and J. J. Sepkoski, Jr. 1987. A comparative study of diversification events: the early Paleozoic versus the Mesozoic. *Evolution 41*:1177–1186.

Erwin, T.L. 1981. Taxon pulses, vicariance and dispersal: an evolutionary synthesis illustrated by carabid beetles. *in*: *Vicariance biogeography: a critique* (G. Nelson and D.E. Rosen, eds.). Columbia Univ. Press, New York. pp. 159–183.

Erwin, T.L. 1982. Tropical forests: their richness in Coleoptera and other arthropod species. *Coleopt. Bull. 36*:74–82.

Erwin, T.L. 1991. An evolutionary basis for conservation strategies. *Science 253*:750–752.

Erwin, T.L. and J. Adis. 1982. Amazonian inundation forests, their role as short–term refuges and generators of species richness and taxon pulses. *in*: *Biological diversification in the tropics* (G.T. Prance, ed.). Columbia Univ. Press, New York. pp. 358–371.

Eschmeyer, W.N. and J.C. Hureau. 1971. *Sebastes mouchezi*, a senior synonym of *Helicolenus tristanensis*, with comments on *Sebastes capensis* and zoogeographical considerations. *Copeia 1971*:576–579.

Estes, J.A. and P.D. Steinberg. 1988. Predation, herbivory and kelp evolution. *Paleobiology 14*:19–36.

Estes, R. 1983. The fossil record and early distribution of lizards. *in*: *Advances in herpetology and evolutionary biology* (A.G.J. Rhodin and K. Miyata, eds.). Mus. Comp. Zool. Harvard University, Cambridge, MA. pp. 365–398.

Estes, R. and O.A. Reig. 1973. The early fossil record of frogs. *in*: *Evolutionary biology of the anurans* (J.E. Vail, ed.). Univ. of Missouri Press, Columbia. pp. 11–63.

Estes, R. and A. Báez. 1985. Herpetofaunas of North and South America during the late Cretaceous and Cenozoic: evidence for interchange? *in*: *The great American interchange* (F.G. Stehli and S.D. Webb, eds.). Plenum Press, New York. pp. 139–197.

Fagerstrom, J.A. 1987. *The evolution of reef communities.* John Wiley, New York.

Fedonkin, M.A. 1990. Precambrian metazoans. *in*: *Palaeobiology: a synthesis* (D.E.G. Briggs and P.R. Crowther, eds.). Blackwell, Oxford. pp. 17–24.

Feldmann, R.M. 1986. Paleobiogeography of two decapod crustacean taxa in the southern hemisphere: global conclusions with sparse data. *in*: *Crustacean biogeography* (R.H. Gore and K. L. Heck, eds.). A.A. Balkema, Rotterdam. pp. 5–19.

Fell, H.B. 1962. West–wind–drift dispersal of echinoderms in the southern hemisphere. *Nature 193*:759–761.

Finney, S.C. and C. Xu 1990. The relationship of Ordovician graptolite provincialism to palaeogeography. *in*: *Palaeozoic palaeogeography and biogeography* (W.S. McKerrow and C. R. Scotese, eds.). Geol. Soc. London, Mem. 12:123–128.

Fischer, A.G. 1964. Brackish oceans as the cause of the Permo–Triassic marine faunal crisis. *in*: *Problems in palaeoclimatology* (A.E.M. Nairn, ed.). John Wiley, New York. pp. 566–579.

Fischer, A.G. 1984. The two Phanerozoic supercycles. *in*: *Catastrophes and earth history* (W.A. Berggren and J.A. Van Couvering, eds.). Princeton Univ. Press, Princeton, NJ. pp. 129–150.

Fleminger, A. and K. Hulsemann. 1973. Relationships of the Indian Ocean epiplanktonic calanoids to the world oceans. *in*: *The biology of the Indian Ocean* (B. Zeitzchel and S.A. Gerlah, eds.). Springer–Verlag, Berlin. pp. 339– 348.

Flessa, K.W. 1980. The biological effects of plate tectonics and continental drift. *BioScience 30*:518–523.

Flessa, K.W. and J.J. Sepkoski, Jr. 1978. On the relationship between Phanerozoic diversity and changes in habitable area. *Paleobiology 4*:359–366.

Florin, R. 1963. The distribution of conifer and taxad genera in time and space. *Acta Horti Bergiani 20*:121–312.

Flynn, L.J., R.H. Tedford and Q. Zhanxiang. 1991. Enrichment and stability in the Pliocene mammalian fauna of north China. *Paleobiology 17*:246–265.

Foin, T.C. 1976. Plate tectonics and the biogeography of the Cypraeidae (Mollusca: Gastropoda). *J. Biogeogr. 3*:19–34.

Forbes, E. 1846. On the connection between the distribution of the existing fauna and flora of the British Isles and the geological changes which have affected their area, especially during the epoch of the northern drift. *Mem. Geol. Surv. G.B. 1*:336–432.

Forbes, E. 1859. *The natural history of European seas*.....edited and continued by Robert Goodwin-Austen. John Van Voorst, London.

Fordyce, R.E. 1992. Cetacean evolution and Eocene/Oligocene environments. *in*: *Eocene–Oligocene climatic and biotic evolution* (D.R. Prothero and W.A. Berggren, eds.). Princeton Univ. Press, Princeton, NJ. pp. 368–381.

Forest, J.M., M. de Saint Laurent and F.A. Chace, Jr. 1976. *Neoglyphea inopinata*: a crustacean "living fossil" from the Philippines. *Science 192*:884.

Fortey, R.A. and R.M. Owens. 1990. Evolutionary radiations in the Trilobita. *in*: *Major evolutionary radiations* (P.D. Taylor and G.P. Larwood, eds.). Clarendon Press, Oxford. pp. 139–164.

Fox, R.C., G.P. Youzwyshyn and D.W. Krause. 1992. Post–Jurassic mammal–like reptile from the Paleocene. *Nature 358*:233–235.

Fricke, R. 1988. *Systematik und historische Zoogeographie der Callionymidae (Teleostei) des Indischen Ozeans*. 2 vols. Doctoral Dissertation, Albert– Ludwigs–Universität, Freiburg im Breisgau, Germany.

Gage, J.D. and P.A. Tyler. 1991. *Deep–sea biology*. Cambridge Univ. Press, New York.

Gallagher, W.B. 1993. The Cretaceous/Tertiary mass extinction event in the northern Atlantic coastal plain. *The Mosasaur 5*:75–154.

Gardiner, B.G. 1990. Placoderm fishes: diversity through time. *in*: *Major evolutionary radiations* (P.D. Taylor and G.P. Larwood, eds.). Clarendon Press, Oxford. pp. 305–319.

Garth, J.S. 1957. Reports of the Lund University Chile expedition 1948–49. The Crustacea Decapoda Brachyura of Chile. *Lunds Universitets Årsskrift, N.F. Avd. 2. Bd 53. 7*:1–130.

Gaskin, D.E. 1982. *The ecology of whales and dolphins*. Heinemann, London.

Gaston, K.J. 1991. The magnitude of global insect species richness. *Conserv. Biol. 5*:283–296.

Gayet, M. 1987. Lower vertebrates from the early–middle Eocene Kuldana Formation of Kohat (Pakistan): Holostei and Teleostei. *Contrib. Mus. Paleontol. Univ. Mich. 27*:151–168.

Gayet, M. and F.J. Meunier. 1991. First discovery of Polypteridae (Pisces, Cladistia, Polypteriformes) outside of Africa. *Geobios, Note brève no. 24*:463–466.

Gayet, M., J.–C. Rage, T. Sempere and P.–Y Gagnier. 1992. Modalités des échanges de vertébrés continentaux entre l'Amérique du Nord et l'Amérique du Sud au Crétacé supérieur et au Paléocène. *Bull. Soc. Géol. France 163*:781–791.

George, W. 1981. Wallace and his line. *in*: *Wallace's line and plate tectonics* (T.C. Whitmore, ed.). Clarendon Press, Oxford. pp. 3–8.

Gerstel, J., R.C. Thunell, J.C. Zachos and M.A. Arthur. 1986. The Cretaceous/Tertiary boundary event in the North Pacific: planktonic foraminiferal results from DSDP site 577, Shatsky Rise. *Paleoceanography 1*:97– 117.

Gill, T. 1884. The principles of zoogeography. *Proc. Biol. Soc. Wash. 2*:1–39.

Gingerich, P.D. 1980. *Tytthaena parrisi*, oldest known oxyaenid (Mammalia, Creodonta) from the late Paleocene of western North America. *J. Paleontol. 54*:570–576.

Gingerich, P.D. 1985. South American mammals in the Paleocene of North America. *in*: *The great American interchange* (F.G. Stehli and S.D. Webb, eds.). Plenum Press, New York. pp. 123–137.

Glen, W. 1990. What killed the dinosaurs? *Am. Sci. 78*:354–370.

Glynn, P.W. and G.M. Wellington. 1983. *Corals and coral reefs of the Galápagos Islands*. Univ. of California Press, Berkeley, CA.

Godley, E.J. 1975. Flora and vegetation. *in*: *Biogeography and ecology in New Zealand* (G. Kuschel, ed.). W. Junk, The Hague. pp. 177–229.

Godthelp, H., M. Archer, R. Cifelli, S.J. Hand and C.F. Gilkeson. 1992. Earliest known Australian Tertiary mammal fauna. *Nature 356*:514–515.

Goldblatt, P. 1978. An analysis of the flora of southern Africa: its characteristics, relationships and origins. *Ann. Mo. Bot. Gard. 65*:369– 436.

Golikov, A.N. and O.A. Scarlato. 1989. Evolution of the Arctic ecosystems during the Neogene period. *in*: *The Arctic seas: climatology, oceanography, geology and biology* (Y. Herman, ed.). Van Nostrand Reinhold, New York. pp. 257–279.

Gomez, E. 1988. Philippines. *in*: *Coral reefs of the world*, vol. 2 (C. Sheppard and S.M. Wells, eds.). VNEP, Nairobi and IUCN, Gland. pp. 229–260.

Good, R. 1974. *The geography of the flowering plants*. 4th edn. Longmans, London.

Goodfellow, W.D., H. Geldsetzer, D.J. McLaren, M.J. Orchard and G. Klapper. 1989. Geochemical and isotopic anomalies associated with the Frasnian– Famennian extinction. *Hist. Biol. 2*:51–72.

Goodwin, A.M. 1991. *Precambrian geology*. Academic Press, London.

Gosliner, T. 1987. *Nudibranchs of southern Africa*. Sea Challengers, Monterey, California.

Gould, S.J. 1984. The cosmic dance of Siva. *Nat. Hist. 93*:14–19.

Gould, S.J. 1990. Enigmas of the small shellies. *Nat. Hist.*, October:8–17.

Gould, S.J. and C.B. Calloway. 1980. Clams and brachiopads – ships that pass in the night. *Paleobiology 6*:383–396.

Grande, L. 1985. The use of paleontology in systematics and biogeography and a time control refinement for historical biogeography. *Paleobiology 11*:234–243.

Grande, L. 1987. Redescription of *Hypsidoris farsonensis* (Teleostei: Siluriformes), with a reassessment of its phylogenetic relationships. *J. Vert. Paleontol. 7*:24–54.

Grassle, J.F. 1986. The ecology of deep–sea hydrothermal vent communities. *Adv. Mar. Biol. 23*:301–362.

Grassle, J.F. 1991. Deep–sea benthic biodiversity. *BioScience 41*:464–469.

Grassle, J.F. and N.J. Maciolek. 1992. Deep–sea species richness: regional and local diversity estimates from quantitative bottom samples. *Am. Nat. 139*:313– 341.

Gray, A. 1859. Diagnostic characters of new species of phanerogamous plants collected in Japan by Charles Wright, botanist of the U.S. North Pacific Exploring Expedition. *Mem. Am. Acad. Arts 6*:377–452.

Grehan, J.R. 1991. Panbiogeography 1981–91: development of an earth/life synthesis. *Prog. Phys. Geogr. 15*:331–363.

Gressitt, J.L. 1956. Some distribution patterns of Pacific island faunae. *Syst. Zool. 6*:12–32.

Gressitt, J.L. (ed.). 1963. *Pacific Basin biogeography: a symposium*. Bishop Museum Press, Honolulu.

Grismer, L.L. 1988. Phylogeny, taxonomy, classification and biogeography of eublepharid geckos. *in*: *Phylogenetic relationships of the lizard families* (R. Estes and G. Pregill, eds.). Stanford Univ. Press, Stanford, IL. pp. 369–469.

Gruszczyński, M., S. Halas, A. Hoffman and K. Malkowski. 1989. A brachiopod calcite record of the oceanic carbon and oxygen isotope shifts at the Permian/Triassic transition. *Nature 337*:64–68.

Gurnis, M. 1988. Large–scale mantle convection and the aggregation and dispersal of supercontinents. *Nature 332*:695–699.

Haedrich, R.L. and D.C. Judkins. 1979. Macrozooplankton and its environment. *in*: *Zoogeography and diversity of plankton* (S. van der Spoel and A.C. Pierrt– Bults, eds.). John Wiley, New York. pp. 4–28.

Haedrich, R.L. and N.R. Merritt. 1990. Little evidence for faunal zonation in the deep sea demersal fish faunas. *Prog. Oceanogr. 24*:239–250.

Hallam, A. 1981. The end–Triassic bivalve extinction event. *Palaeogeogr., Palaeoclimatol., Palaeoecol. 35*:1–44.

Hallam, A. 1983. Early and mid–Jurassic molluscan biogeography and the establishment of the Central Atlantic Seaway. *Palaeogeogr., Palaeoclimatol., Palaeoecol. 43*:181–193.

Hallam, A. 1984. Pre–Quaternary sea–level changes. *Annu. Rev. Earth Planet. Sci. 12*:205–243.

Hallam, A. 1986. The Pliensbachian and Tithonian extinction events. *Nature 319*:765–768.

Hallam, A. 1989a. A possible connection between mantle plumes, sea–level changes and mass extinctions. *Geol. Soc. Am., Abstracts with Programs 21*:A92.

Hallam, A. 1989b. The case for sea–level change as a dominant causal factor in mass extinction of marine invertebrates. *Phil. Trans. R. Soc. London B325*:437– 455.

Hallam, A. 1992. *Phanerozoic sea level changes*. Columbia Univ. Press, New York.

Hallam, A. and A.I. Miller. 1988. Extinction and survival in the Bivalvia. *in*: *Extinction and survival in the fossil record* (G.P. Larwood, ed.). Clarendon Press, Oxford. pp. 121–138.

Hallock, P. 1987. Fluctuations in the trophic resource continuum: a factor in global diversity cycles? *Paleoceanography 2*:457–471.

Hallock, P. and W. Schlager. 1986. Nutrient excess and the demise of coral reefs and carbonate platforms. *Palaios 1*:389–398.

Halstead, L.B. 1988. Extinction and survival of the jawless vertebrates, the Agnatha. *in*: *Extinction and survival in the fossil record* (G.P. Larwood, ed.) Clarendon Press, Oxford. pp. 257–267.

Hand, S. 1984. Australia's oldest rodents: master mariners from Malaysia. *in*: *Vertebrate zoogeography and evolution in Australasia* (M. Archer and G. Clayton, eds.). Hesperian Press, Carlisle, Western Australia. pp. 905–919.

Hanger, R.A. 1990. A probabilistic global paleobiogeography for Permian brachiopods with implications for North American terrane analysis. *Geol. Soc. Am., Abstracts with Programs 22*:A–304.

Hansen, H.J. 1990. Diachronous extinctions at the K/T boundary; a scenario. *in*: *Global catastrophes and earth history* (V.L. Sharpton and P.D. Ward, eds.). Geol. Soc. Am., Spec. Paper 247:417–423.

Hansen, T.A. 1988. Early Tertiary radiation of marine molluscs and the long– term effects of the Cretaceous–Tertiary extinction. *Paleobiology 14*:37–51.

Hansen, T.A. 1992. The patterns and causes of molluscan extinction across the Eocene/Oligocene boundary. *in*: *Eocene–Oligocene climatic and biotic evolution* (D.R. Prothero and W.A. Berggren, eds.). Princeton Univ. Press, Princeton, NJ. pp. 341–348.

Harland, W.B. 1983. The Proterozoic glacial record. *Geol. Soc. Am., Mem. 161*:270–288.

Harrington, B.J. 1980. Generic level revision and cladistic analysis of the Myodochini of the world (Hemiptera, Lygaeidae, Rhyparochrominae). *Bull. Am. Mus. Nat. Hist. 167*:49–116.

Hartnady, C.J.H. 1991. About turn for supercontinents. *Nature 352*:476–478.

Hartog, C. den 1970. *The sea grasses of the world.* Proc. Nederl. Acad. Wetenschap. Tweede Reeks, Deel 59, No. 1.

Heaney, L.R. and B.D. Patterson. 1986. *Island biogeography of mammals*. Academic Press, London.

Hecht, A.D. and B. Agan. 1972. Diversity and age relationships in recent and Miocene bivalves. *Syst. Zool. 21*:308–312.

Heck, K.L. and E.D. McCoy. 1978. Long–distance dispersal and the reef–building corals of the Eastern Pacific. *Mar. Biol. 48*:349–356.

Heck, K.L. and E.D. McCoy. 1979. Biogeography of seagrasses: evidence from associated organisms. *Proc. 1st Int. Symp. Mar. Biogeogr. Evol. Southern Hemis. 1*:109–127.

Hedger, S.B., C.A. Haas and L.R. Maxon. 1992. Caribbean biogeography: molecular evidence for dispersal in West Indian territorial vertebrates. *Proc. Natl. Acad. Sci. USA 89*: 1909–1913.

Hedgpeth, J.W. 1957. Marine biogeography. *in*: *Treatise on marine ecology and paleoecology*. Vol. 1. Mem. Geol. Soc. Am. 67:359–382.

Hennig, W. 1966. *Phylogenetic systematics*. Univ. of Illinois Press, Urbana, IL.

Hessler, R.R. and G.D.F. Wilson. 1983. The origin and biogeography of malacostracan crustaceans in the deep sea. *in*: *Evolution, time and space: The emergence of the biosphere* (R.W. Sims, J.H. Price and P.E.S. Whalley, eds.). Academic Press, London. pp. 227–254.

Hickey, L.J. 1981. Land–plant evidence compatible with gradual, not catastrophic, change at the end of the Cretaceous. *Nature 292*:529–531.

Hill, R.S. 1992. *Nothofagus*: evolution from a southern perspective. *Trends Ecol. Evol. 7*:190–194.

Hobbs, H.H., Jr. 1974. Synopsis of the families and genera of crayfishes (Crustacea: Decapoda). *Smithson. Contrib. Zool. 164*:1–32.

Hocutt, C.H. 1987. Evolution of the Indian Ocean and the drift of India: a vicariant event. *Hydrobiologia 150*:203–223.

Hodell, D.A. and J.P. Kennett. 1985. Miocene paleoceanography of the South Atlantic ocean at 22, 16 and 8 Ma. *in*: *The Miocene ocean: paleoceanography and biogeography* (J.P. Kennett, ed.). Geol. Soc. Am. Mem. 163:317–337.

Hodkinson, I.D. and D. Casson. 1991. A lesser prediction for bugs: Hemiptera (Insecta) diversity in tropical rain forests. *Biol. J. Linn. Soc. 43*:101–109.

Hoffman, A. 1985. Patterns of family extinction depend on definition and geological time scale. *Nature 315*:659–662.

Hoffman, A. 1989a. *Arguments on evolution: a paleontologist's perspective*. Oxford Univ. Press, New York.

Hoffman, A. 1989b. What, if anything, are mass extinctions? *Phil. Trans. R. Soc. London B325*:253–261.

Hoffmann, R.S. 1985. An ecological and zoogeographical analysis of animal migration across the Bering Land Bridge during the Quaternary Period. *in*: *Beringia in the Cenozoic Era* (V.L. Katrimavichus, ed.). A.A. Balkema, Rotterdam.

Holland, H.D. 1984. *The chemical evolution of the atmosphere and oceans*. Princeton Univ. Press, Princeton, NJ.

Holloway, J.D. 1979. *A survey of the Lepidoptera, biogeography and ecology of New Caledonia*. W. Junk, The Hague.

Holloway, J.D. 1986. Origins of lepidopteran faunas in high mountains of the Indo–Australian tropics. *in*: *High altitude tropical biogeography* (F. Vuilleumier and M. Monasterio, eds.). Oxford Univ. Press, New York. pp. 533– 556.

Holser, W.T., M. Magaritz and J. Wright. 1986. Chemical and isotopic variations in the world ocean during Phanerozoic time. *in*: *Global bio–events* (S. Bhattacharji, G.M. Freidman, H.J. Neugebauer and A. Seilacher, eds.) Springer–Verlag, Berlin. pp. 63–74.

Holthuis, L.B. 1991. Marine lobsters of the world. *FAO Species Catalog 13*:1–292.

Hooker, J.D. 1853. *The botany of the Antarctic voyage of H.M. Discovery Ships Erebus and Terror in the Years 1839–1843*. II. Flora Novae–Zelandiae. Lovell Reeve, London.

Hooker, J.J. 1992. British mammalian paleocommunities across the Eocene– Oligocene transition and their environmental implications. *in*: *Eocene– Oligocene climatic and biotic evolution* (D.R. Prothero and W.A. Berggren, eds.). Princeton Univ. Press, Princeton, NJ. pp. 494–515.

Hora, S.L. 1937. Geographical distribution of Indian freshwater fishes and its bearing on the probable land connections between India and adjacent countries. *Curr. Sci. Bangalore 5*:351–356.

Hora, S.L. 1939. On some fossil fish scales from the Inter–trappean beds at Deothan and Kheri, Central Province. *Rec. Geol. Surv. India 73*:267–294.

Horn, M.H. and L.G. Allen. 1978. A distributional analysis of California coastal marine fishes. *J. Biogeogr. 5*:23–42.

Horrell, M.A. 1990. Energy balance constraints on $^{18}0$ based paleo–sea surface temperature estimates. *Paleoceanography 5*:339–348.

Houde, P.W. 1988. Paleognathous birds from the early Tertiary of the northern hemisphere. *Publ. Nuttall Ornithol. Club, No. 22*:1–148.

House, M.R. 1988. Extinction and survival in the Cephalopoda. *in*: *Extinction and survival in the fossil record* (G.P. Larwood, ed.). Clarendon Press, Oxford. pp. 139–154.

Howes, G.J. 1991a. Systematics and biogeography: an overview. *in*: *Cyprinid fishes* (I.J. Winfield and J.S. Nelson, eds.). Chapman and Hale, London. pp. 18–30.

Howes, G.J. 1991b. Biogeography of gadoid fishes. *J. Biogeogr. 18*:595–622.

Hsü, K.J., et al. 1982. Mass mortality and its environmental and evolutionary consequences. *Science 216*:249–256.

Hsü, K.J. 1986. *The great dying*. Harcourt, Brace, Jovanovich, San Diego.

Hubbell, T.H. 1968. The biology of islands. *Proc. Nat. Acad. Sci. USA 60*:22–32.

Hughes, N.F. 1976. *Paleobiology of angiosperm origins*. Cambridge Univ. Press, Cambridge.

Hull, F.M. 1962. Robber flies of the world, Pt. 1. *Bull. U.S. Nat. Mus. 224*:1– 430.

Humphries, C.J. 1981. Biogeographical methods and the southern beeches (Fagaceae: *Nothofagus*). *in*: *Advances in cladistics* (V.A. Funk and D.R. Brooks, eds.). New York Botanical Garden. pp. 177–207.

Humphries, C.J. and L.R. Parenti. 1986. *Cladistic biogeography*. Clarendon Press, Oxford.

Hutchins, J.B. 1991. Dispersal of tropical fishes to temperate seas in the southern hemisphere. *J. Roy. Soc. West. Austr. 74*:79–84.

Hutchinson, G.E. 1959. Homage to Santa Rosalia or why are there so many kinds of animals? *Am. Nat. 93*:145–159.

Hutchison, J.H. 1992. Western North American reptile and amphibian record across the Eocene/Oligocene boundary and its climatic implications. *in*: *Eocene–Oligocene climatic and biotic evolution* (D.R. Prothero and W.A. Berggren eds.). Princeton Univ. Press, Princeton, NJ. pp. 451–463.

Imbrie, J. and K.P. Imbrie. 1979. *Ice Ages*. Enslow, NJ.

Ivany, L.C., R.W. Portell and D.S. Jones. 1990. Animal–plant relationships and paleobiogeography of an Eocene seagrass community from Florida. *Palaios 5*:244–258.

Iwami, T. 1985. Osteology and relationships of the family Channichthyidae. *Mem. Nat. Inst. Polar Res., Tokyo 36*:1–69.

Jablonski, D. 1986. Evolutionary consequences of mass extinctions. *in*: *Patterns and processes in the history of life* (D.M. Raup and D. Jablonski, eds.). Springer–Verlag, Berlin. pp. 313–329.

Jablonski, D. 1993. The tropics as a source of evolutionary novelty through geological time. *Nature 364*:142–144.

Jablonski, D., J.J. Sepkoski, Jr., D.J. Bottjer and P.M. Sheehan. 1983. Onshore–offshore patterns in the evolution of Phanerozoic shelf communities. *Science 222*:1123–1125.

Jablonski, D., K.W. Flessa and J.W. Valentine. 1985. Biogeography and paleobiology. *Paleobiology 11*:75–90.

Jacobs, L.L., et al. 1988. Mammal teeth from the Cretaceous of Africa. *Nature 336*:158–160.

James, M.I. (ed.). 1991. *Galápagos marine invertebrates*. Plenum Press, New York.

Jamieson, B.G.M. 1971. Geographical distribution of the Glossoscolecidae. *in*: *Aquatic Oligochaeta of the world* (R.D. Brinkhurst and B.G.M. Jamieson, eds.). Univ. of Toronto Press, Toronto. pp. 147–164.

Janis, C.M. 1993. Tertiary mammal evolution in the context of changing climates, vegetation and tectonic events. *Annu. Rev. Ecol. Syst. 24*:467–500.

Jeannel, R. 1942. *La genèse des faunes terrestres, éléments de biogéographie*. Presses Universitaires de France.

Jermy, A.C. 1990. Isoetaceae. *in*: *The families and genera of vascular plants* (K. Kubitzki, ed.). Springer–Verlag, Berlin. pp. 26–31.

Johnson, L.A. and M.J. Sims. 1989. The timing and cause of the late Triassic marine invertebrate extinctions: evidence from scallops and crinoids. *in*: *Mass extinctions: processes and evidence* (S.K. Donovan, ed.). Columbia Univ. Press, New York. pp. 174–194.

Johnson, L.A.S. and B.G. Briggs. 1975. On the Proteaceae: the evolution and classification of a southern family. *Bot. J. Linn. Soc. 70*:83–102.

Johnson, L.A.S. and K.L. Wilson. 1990. Cycadatae. *in*: *The families and genera of vascular plants* (K. Kubitzki, ed.). Springer–Verlag, Berlin. pp. 362–377.

Johnston, D.E. 1982. Acara. *in*: *Synopsis and classification of living organisms* (S.P. Parker, ed.). 2 Vols. McGraw Hill, New York. Vol. 2:111–169.

Jokiel, P. and F.J. Martinelli. 1992. The vortex model of coral reef biogeography. *J. Biogeogr. 19*:449–458.

Jones, D.S. and P.F. Hasson. 1985. History and development of the marine invertebrate faunas separated by the Central American Isthmus. *in*: *The great American interchange* (F.G. Stehli and S.D. Webb, eds.). Plenum Press, New York. pp. 325–355.

Kafanov, A.I. 1987. The Wills "age and area" rule and latitudinal heterochrony of marine biota. *J. Gen. Biol. 48*:105–114 (in Russian).

Kalandadze, N.N. and A.S. Rautian. 1991. Late Triassic zoogeography and a reconstruction of the terrestrial tetrapod fauna of North Africa. *Paleontol. J. 25*:1–12 (transl. from Russian).

Kauffman, E.G. 1973. Cretaceous Bivalvia. *in*: *Atlas of palaeobiogeography* (A. Hallam, ed.). Elsevier, Amsterdam. pp. 353–383.

Kauffman, E.G. 1979a. Cretaceous. *in*: *Treatise on invertebrate paleontology* (R.A. Robison and C. Teichert, eds.). Part A. Introduction. Geol. Soc. Am. Univ. Kansas. pp. 418–487.

Kauffman, E.G. 1979b. The ecology and biogeography of the Cretaceous–Tertiary extinction event. *in*: *Cretaceous/Tertiary boundary events* (W. Kegel Christensen and T. Birkelund, eds.). Univ. of Copenhagen Press, Copenhagen. pp. 29–37.

Kauffman, E.G. 1984. The fabric of Cretaceous marine extinctions. *in*: *Catastrophes and earth history* (W.A. Berggren and J.A. Van Couvering, eds.). Princeton Univ. Press, Princeton, NJ. pp. 151–246.

Kauffman, E.G. 1986. High–resolution event stratigraphy: regional and global Cretaceous bio–events. *in*: *Global bio–events* (S. Battacharji, G.M. Friedman, H.J. Neugebauer and A. Seilacher, eds.). Springer–Verlag, Berlin. pp. 279–335.

Kauffman, E.G. and O.H. Walliser, (eds.) 1990. *Extinction events in earth history*. Springer–Verlag, Berlin.

Kay, E.A. 1979. *Little worlds of the Pacific: an essay on Pacific Basin biogeography*. Harold L. Lyon Arboretum, Univ. of Hawaii, HI.

Kay, E.A. 1990. The Cypraeidae of the Indo–Pacific: Cenozoic fossil history and biogeography. *Bull. Mar. Sci. 47*:23–34.

Kearey, P. and F.J. Vine. 1990. *Global tectonics*. Blackwell, Oxford.

Keen, M. 1990. Ostracoda, sea–level changes and the Eocene–Oligocene boundary. *in*: *Ostracoda and global events* (R. Whatley and C. Maybury, eds.). Chapman and Hall, London. pp. 153–160.

Keller, G. 1985. Depth stratification of planktonic foraminifers in the Miocene Ocean. *in*: *The Miocene ocean: paleoceanography and biogeography* (J.P. Kennett, ed.). Geol. Soc. Am., Mem. 163:177–193.

Keller, G. 1988. Extinction, survivorship and evolution of planktic Foraminifera across the Cretaceous/Tertiary boundary at El Kef, Tunisia. *Mar. Micropaleo. 13*:239–263.

Keller, G. 1989. Extended Cretaceous/Tertiary boundary extinctions and delayed population change in planktonic Foraminifera from Brazos River, Texas. *Paleoceanography 4*:287–332.

Keller, G. and E. Barrera. 1990. The Cretaceous/Tertiary boundary impact hypothesis and the paleontological record. *in*: *Global catastrophes and earth history* (V.L. Sharpton and P.D. Ward, eds.). Geol. Soc. Am., Spec. Paper 247:563–575.

Keller, G., N. MacLeod and E. Barrers. 1992. Eocene–Oligocene faunal turnover in planktic Foraminifera and Antarctic glaciation. *in*: *Eocene–Oligocene climatic and biotic evolution* (P.R. Prothero and W.A. Berggren, eds.). Princeton Univ. Press, Princeton, NJ. pp. 218–291.

Kelly, D.A. 1990. A phylogenetic and biogeographic analysis of Carnosauria (Theropoda: Saurischia). *Abstract, J. Vert. Paleo. 10* (Suppl. 3):30A.

Kennett, J.P. 1982. *Marine geology*. Prentice–Hall, Englewood Cliffs, NJ.

Kennett, J.P. and L.D. Stott. 1991. Abrupt deep–sea warming, palaeoceanographic changes and benthic extinctions at the end of the Palaeocene. *Nature 353*:225–229.

Kennett, J.P., G. Keller and M.S. Srinivasan. 1985. Miocene planktonic foraminiferal and palaeoceanographic development of the Indo–Pacific region. *in*: *The Miocene ocean: paleoceanography and biogeography* (J.P. Kennett, ed.). Geol. Soc. Am. Mem. 163:197–236.

Kent, D.V. and R. Van der Voo. 1990. Palaeozoic palaeogeography from palaeomagnetism of the Atlantic–bordering continents. *in*: *Palaeozoic palaeogeography and biogeography* (W.S. McKerrow and C.R. Scotese, eds.). Geol. Soc. London, Mem. 12:49–56.

Kinch, M.P. 1980. Geographical distribution and the origin of life: the development of early nineteenth century British explanations. *Hist. Biol. 13*:91–119.

King, G. 1990. *The dicynodonts: a study in palaeobiology*. Chapman and Hall, London.

Kinsey, L.S. and R.M. Bohart. 1990. *The chrysidid wasps of the world*. Oxford Univ. Press, Oxford.

Kirsch, J.A.W. 1977. The comparative serology of Marsupialia and a classification of marsupials. *Aust. J. Zool. Supp. Ser. 52*:1–152.

Kitchell, J.A. and T.R. Carr. 1985. Nonequilibrium model of diversification: faunal turnover dynamics. *in*: *Phanerozoic diversity patterns* (J.W. Valentine, ed.). Princeton Univ. Press, Princeton, NJ. pp. 277–309.

Knoll, A.H. 1984. Patterns of extinction in the fossil record of vascular plants. *in*: *Extinctions* (M.H. Nitecki, ed.). Univ. of Chicago Press, Chicago, IL. pp. 21–68.

Koch, L.E. 1981. The scorpions of Australia: aspects of their ecology and zoogeography. *in*: *Ecological biogeography of Australia* (A. Keast, ed.). W. Junk, The Hague. pp. 875–884.

Kozhov, M.M. 1963. *Lake Baikal and its life*. W. Junk, The Hague.

Kramer, K.V. 1990. Loxomataceae. *in*: *The families and genera of vascular plants* (K. Kubitzki, ed.). Springer–Verlag, Berlin. pp. 172–174.

Krassilov, V.A. 1972. Phytogeographical classification of Mesozoic floras and their bearing on continental drift. *Nature 237*:49–50.

Krassilov, V.A. 1981. Changes of Mesozoic vegetation and the extinction of dinosaurs. *Palaeogeogr., Palaeoclimatol., Palaeoecol. 34*:207–224.

Krause, D.W. 1986. Competitive exclusion and taxonomic displacement in the fossil record: the case of rodents and multituberculates in North America. *in*: *Vertebrates, phylogeny and philosophy* (K.M. Flanagan and J.A. Lillegraven, eds.). Contrib. to Geol., Univ. of Wyoming, WY. pp. 95–117.

Krishna, K. 1970. Taxonomy, phylogeny and distribution of termites. *in*: *Biology of termites*. (K. Krishna and F.M. Wessner, eds.). Vol. II. Academic Press, New York. pp. 127–152.

Kubitzki, K. 1990. Gnetatae. *in*: *The families and genera of vascular plants* (K. Kubitzki, ed.). Springer–Verlag, Berlin. pp. 378–391.

Kummel, B. 1973. Lower Triassic (Scythian) molluscs. *in*: *Atlas of palaeobiogeography* (A. Hallam, ed.). Elsevier, Amsterdam. pp. 225–233.

Kurup, G.V. 1974. Mammals of Assam and the mammal–geography of India. *in*: *Ecology and biogeography in India* (M.S. Mani, ed.). W. Junk, The Hague. pp. 585–613.

Kussakin, O.G. 1973. Peculiarities of the geographical and vertical distribution of marine isopods and the problem of deep–sea fauna origin. *Mar. Biol. 23*:19–34.

Kussakin, O.G. 1990. Biogeography of isopod crustaceans in the boreal Pacific. *Bull. Mar. Sci. 46*:620–639.

Labandeira, C.C. and J.J. Sepkoski, Jr. 1993. Insect diversity in the fossil record. *Science 261*:310–315.

Ladd, H.S. 1960. Origin of the Pacific island molluscan fauna. *Am. J. Sci. 258A*:137–150.

Landini, W. and L. Sorbini. 1989. Ichthyofauna of the evaporitic Messinian in the Romagna and Marche regions. *Boll. Soc. Paleo. Ital. 28*:287–293.

Latham, R.E. and R.E. Ricklefs. 1993. Continental comparisons of temperate– zone tree species diversity. *in*: *Species diversity in ecological communities* (R.E. Ricklefs and D. Schluter, eds.). Univ. of Chicago Press, Chicago, IL. pp. 294–314.

Laufeld, S. 1979. Biogeography of Ordovician, Silurian and Devonian chitinozoans. *in*: *Historical biogeography, plate tectonics and the changing environment* (J. Gray and A.J. Boucot, eds.). Oregon State Univ. Press, Corvallis, OR. pp. 75–90.

Lawver, L.A., L.M. Gahagan and M.F. Coffin. 1992. The development of paleoseaways around Antarctica. *Antarct. Res. Ser. 56*:7–30.

Legendre, S. and J.–L Hartenberger. 1992. Evolution of mammalian faunas in Europe during the Eocene and Oligocene. *in*: *Eocene–Oligocene climatic and biotic evolution* (D.R. Prothero and W.A. Berggren, eds.). Princeton Univ. Press, Princeton, NJ. pp. 516–528.

Leis, J.M. 1984. Larval fish dispersal and the East Pacific Barrier. *Oceanogr. Trop. 19*:181–192.

Leopold, E.B., G. Liu and S. Clay–Poole. 1992. Low–biomass vegetation in the Oligocene? *in*: *Eocene–Oligocene climatic and biotic evolution* (D.R. Prothero and W.A. Berggren, eds.). Princeton Univ. Press, Princeton, NJ. pp. 399–420.

Lillegraven, J.A., M.J. Kraus and T.M. Brown. 1979. Paleogeography of the world of the Mesozoic. *in*: *Mesozoic mammals* (J.A. Lillegraven, Z. Kielan– Naworowska, W.A. Clemens, eds.). Univ. Calif. Press, Berkeley, CA. pp. 277–308.

Lin, C. 1990. Distribution of sphingid moths. *J. Taiwan Mus. 43*:41– 94.

Lindberg, D.R. 1991. Marine biotic interchange between northern and southern hemispheres. *Paleobiology 17*:308–324.

Lindemann, B. 1927. *Kettengebirge, Kontinentale Zersphaltung und Erdexpansion*. Gustav Fischer, Jena.

Lindsey, C.C. and J.D. McPhail. 1986. Zoogeography of the fishes of the Yukon and Mackenzie basins. *in*: *The zoogeography of North American freshwater fishes* (C.H. Hocutt and E.O. Wiley, eds.). John Wiley, New York. pp. 639–674.

Lipps, J.H. 1986. Extinction dynamics in pelagic ecosystems. *in*: *Dynamics of extinction* (D.K. Elliott, ed.). John Wiley, New York. pp. 87–104.

Lipps, J.H. and C.S. Hickman. 1982. Origin, age and evolution of Antarctic and deep–sea faunas. *in*: *The environment of the deep sea* (W.G. Ernst and J.G. Morin, eds.). Rubey Vol. II, Prentice–Hall, Englewood Cliffs, NJ. pp. 325–356.

Little, E.L. 1983. North American trees with relationships in eastern Asia. *Ann. Mo. Bot. Gard. 70*:605–615.

Lovén, S. 1846. Malacologiska notiser. Nagra anmärkningar öfver de Skandinaviska Hafs Molluskernas geografiska utbredning. *Övers K. Svenska. Vet.–Acad. Förhandl. Stockholm*. pp. 252–274.

Lubbock, R. 1980. The shore fishes of Ascension Island. *J. Fish Biol. 17*:283–303.

Lundberg, J.G. 1986. Miocene characid fishes from Columbia: evolutionary stasis and extirpation. *Science 234*:208–209.

Lundberg, J.G. 1988. *Phractocephalus hemileopterus* (Pimelodidae, Siluriformes) from the Upper Miocene Urumaco Formation, Venezuela: a further case of evolutionary stasis and local extinction among South American fishes. *J. Vert. Paleo. 8*:131–138.

Lundberg, J.G. 1993. African–South American freshwater fish clades and continental drift: problems with a paradigm. *in*: *Biological relationships between Africa and South America* (P. Goldblatt, ed.). Yale Univ. Press, New Haven, CT. pp. 156–198.

Lyell, C. 1830–33. *Principles of geology.* 3 Vols. Murray, London.

Lynch, J.D. 1989. The gauge of speciation: on the frequencies of modes of speciation. *in*: *Speciation and its consequences* (D. Otte and J.A. Endler, eds.). Sinauer Associates, Sunderland, MA. pp. 527–553.

Mabry, T.J., J.H. Hunziker and D.R.D. Feo, Jr. (eds.). 1977. *Creosote bush: biology and chemistry of Larrea in New World deserts.* Dowden, Hutchinson and Ross, Stroudsburg, Pennsylvania, PA. 284 p.

MacFadden, B.J. 1992. *Fossil horses.* Cambridge Univ. Press, Cambridge.

Main, B.Y. 1981. Australian spiders: diversity, distribution and ecology. *in*: *Ecological biogeography of Australia* (A. Keast, ed.). W. Junk, The Hague. pp. 809–852.

Maisey, J.G. 1993. Tectonics, the Santana Lagerstätten and the implications for late Gondwanan biogeography. *in*: *Biological relationships between Africa and South America* (P. Goldblatt, ed.). Yale Univ. Press, New Haven, CT. pp. 435– 454.

Manning, R.B., R.K. Kropp and J. Dominquez. 1990. Biogeography of deep–sea stomapod Crustacea, family Bathysquillidae. *Proc. Oceanogr. 24*:311–316.

Marincovitch, L., Jr. 1973. Intertidal marine mollusks of Iquique, Chile. *Sci. Bull., Los Angeles Mus. Nat. Hist. 16*:1–49.

Marshall, H.G., J.C.G. Walker and W.R. Kuhn. 1988. Long–term climate change and the geochemical cycle of carbon. *J. Geophys. Res. 93*:791–801.

Marshall, L.G. 1988. Land mammals and the great American interchange. *Am. Sci. 96*:380–389.

Marshall, L.G. and C. de Muizon. 1988. The dawn of the age of mammals in South America. *Natl. Geogr. Res. 4*:23–55.

Marshall, N.B. 1954. *Aspects of deep sea biology.* Hutchinson, London.

Marshall, N.B. 1980. *Deep–sea biology: developments and perspectives.* Garland STPM Press, New York.

Martin, P.S. 1984. Prehistoric overkill. *in*: *Pleistocene extinctions* (P.S. Martin and H.E. Wright, Jr., eds.). Yale Univ. Press, New Haven, CT. pp. 75–120.

Martin, R.D. 1990. *Primate origins and evolution.* Princeton Univ. Press, Princeton, NJ.

Masse, J.P. 1992. The Lower Cretaceous Mesogean benthic ecosystems: Palaeoecologic aspects and palaeobiogeographic implications. *Palaeogeogr., Palaeoclimatol., Palaeoecol. 91*:331–345.

Matthew, W.D. 1915. Climate and evolution. *Ann. New York Acad. Sci. 24*:171– 318.

Matthew, W.D. 1918. Affinities and origin of the Antillean mammals. *Bull. Geol. Soc. Am. 29*:657–666.

Mattingly, P.F. 1962. Towards a zoogeography of the mosquitoes. *in*: *Taxonomy and Geography* (D. Nichols, ed.). The Systematics Assoc., London, Pub. No. 4:17–36.

Maxwell, W.D. 1989. The end–Permian mass extinction. *in*: *Mass extinctions: processes and evidence* (S.K. Donovan, ed.). Columbia Univ. Press, New York. pp. 152–173.

May, R.M. 1988. How many species are there on earth? *Science 241*:1441–1449.

May, R.M. 1990. How many species? *Phil. Trans. R. Soc. London, Ser. B 330*:171–182.

Mayr, E. 1944. Timor and the colonization of Australia by birds. *Emu 44*:113– 130.

Mayr, E. 1946. History of North American bird fauna. *Wilson Bull. 58*:3–41.

Mayr, E. 1970. *Populations, species and evolution.* Harvard Univ. Press, Cambridge, MA.

Mayr, E. 1976. History of the North American bird fauna. *in*: *Evolution and the diversity of life* (E. Mayr, ed.). Harvard Press, Cambridge, MA. pp. 565–588.

McCoy, E.D. and K.L. Heck. 1976. Biogeography of corals, seagrasses and mangroves: an alternative to the center of origin concept. *Syst. Zool. 25*:201– 210.

McCoy, E.D. and K.L. Heck. 1983. Centers of origin revisited. *Paleobiology 9*:17–19.

McCune, A.R. and B. Schaeffer. 1986. Triassic and Jurassic fishes: patterns of diversity. *in*: *The beginning of the age of dinosaurs* (K. Padian, ed.). Cambridge Univ. Press, Cambridge. pp. 171–181.

McDowall, R.M. 1988. *Diadromy in fishes: migrations between freshwater and marine environments.* Croom Helm, London.

McElhinny, M.W., S.R. Taylor and D.J. Stevenson. 1978. Limits to the expansion of Earth, Moon, Mars and Mercury and to changes in the gravitational constant. *Nature 271*:316–321.

McGhee, G.R., Jr. 1988. The Late Devonian extinction event: evidence for abrupt ecosystem collapse. *Paleobiology 14*:250–257.

McGhee, G.R., Jr. 1990a. Frasnian–Famennian. *in*: *Palaeobiology: a synthesis* (D.E.G. Briggs and D.R. Crowther, eds.). Blackwell, Oxford. pp. 184–187.

McGhee, G.R., Jr. 1990b. Catastrophes in the history of life. *in*: *Evolution and the fossil record* (K. Allen and D. Briggs, eds.). Belhaven Press, London. pp. 26–50.

McIntyre, A. 1967. Coccoliths as paleoclimatic indicators of Pleistocene glaciation. *Science 158*:1314–1317.

McKenna, M.C. 1975. Toward a phylogenetic classification of the Mammalia. *in*: *Phylogeny of the primates: a multidisciplinary approach* (W.P. Luckett and F.S. Szalay, eds.). Plenum Press, New York. pp. 43–77.

McKinney, M.L. 1985. Mass extinction patterns of marine invertebrate groups and some implications for a causal phenomenon. *Paleobiology 11*:227–233.

McKinney, M.L., K.J. McNamara, B.D. Carter and S.K. Donovan. 1992. Evolution of Paleogene echinoids: a global and regional view. *in*: *Eocene–Oligocene climatic and biotic evolution* (D.R. Prothero and W.A. Berggren, eds.). Princeton Univ. Press, Princeton, NJ. pp. 349–367.

McLaren, D.J. 1983. Bolides and biostrategraphy. *Geol. Soc. Am. Bull. 94*:318–324.

McLaren, D.J. and W.D. Goodfellow. 1990. Geological and biological consequences of giant impacts. *Annu. Rev. Earth Planet. Sci. 18*:123–171.

McLellan, I.D. 1975. The freshwater insects. *in*: *Biogeography and Ecology in New Zealand* (G. Kuschel, ed.). W. Junk, The Hague. pp. 537–559.

McMenamin, M.A.S. 1982. A case for two late Proterozoic – earliest Cambrian faunal province loci. *Geology 10*:290–292.

McMenamin, M.A.S. 1989. The origins and radiation of the early Metazoa. *in*: *Evolution and the fossil record* (K. Allen and D. Briggs, eds.). Belhaven Press, London. pp. 73–98.

McMenamin, M.A.S. 1990a. Vendian. *in*: *Palaeobiology: a synthesis* (D.E.G. Briggs and P.R. Crowther, eds.). Blackwell, Oxford. pp. 179–181.

McMenamin, M.A.S. 1990b. The origins and radiation of the early Metazoa. *in*: *Evolution and the fossil record* (K. Allen and D. Briggs, eds.). Smithsonian Press, Washington, DC. pp. 73–98.

McMenamin, M.A.S. and D.L.S. McMenamin. 1990. *The emergence of animals: the Cambrian breakthrough*. Columbia Univ. Press, New York.

Meffe, G.K., A.H. Ehrlich and D. Ehrenfeld. 1993. Human population control: the missing agenda. *Conserv. Biol. 7*:1–2.

Melville, R. 1981. Vicarious plant distributions and paleogeography of the Pacific region. *in*: *Vicariance biogeography* (G. Nelson and D.E. Rosen, eds.). Columbia Univ. Press, New York. pp. 238–274.

Menard, H.W. 1986. *The ocean of truth*. Princeton Univ. Press, Princeton, NJ.

Menni, R.C., A. Ringuelet and R. Haramburu. 1984. *Peces marinos de la Argentina y Uruguay*. Editorial Hemisferio, Buenos Aires.

Menon, A.G.K. 1977. A systematic monograph of the tongue soles of the genus *Cynoglossus* Hamilton–Buchanen (Pisces: Cynoglossidae). *Smithson. Contrib. Zool. 238*:1–129.

Merrett, N.R. 1986. Biogeography and the oceanic rim: a poorly known zone of ichthyofaunal interaction. *in*: *Pelagic biogeography* (A.C. Pierrot–Bults, S. van der Spoel, B.J. Zahuranec and R.K. Johnson, eds.). UNESCO, Paris. pp. 201–209.

Merriam, C.H. 1894. Laws of temperature control of the geographic distribution of terrestrial animals and plants. *Natl. Geogr. Mag. 6*:229–238.

Metcalfe, I. 1991. Late Palaeozoic and Mesozoic palaeogeography of Southeast Asia. *Palaeogeogr., Palaeoclimatol., Palaeoecol. 87*:211–221.

Meyer, D.L. and D.B. Macurda, Jr. 1977. Adaptive radiation of the comatulid crinoids. *Paleobiology 3*:74–82.

Michanek, G. 1979. Phytogeographic provinces and seaweed distribution. *Bot. Mar. 22(6)*:375–392.

Michener, C.D. 1979. Biogeography of the bees. *Ann. Mo. Bot. Gard. 66*:277–347.

Milner, A.C. 1990. Terrestrialization: vertebrates. *in*: *Palaeobiology: a synthesis* (D.E.G. Briggs and P.R. Crowther, eds.). Blackwell, Oxford. pp. 68–72.

Milner, A.R. 1983. The biogeography of salamanders in the Mesozoic and early Cenozoic: a cladistic–vicariance model. *in*: *Evolution, time and space: the emergence of the biosphere* (R.W. Sims, J.H. Price and P.E.S. Whalley, eds.). Academic Press, New York. pp. 431–468.

Milner, A.R. 1990. The radiations of temnospondyl amphibians. *in*: *Major evolutionary radiations* (P.D. Taylor and G.P. Larwood, eds.). Clarendon Press, Oxford. pp. 321–349.

Molnar, P. 1984. Structure and tectonics of the Himalaya. *Annu. Rev. Earth Planet. Sci. 12*:489–518.

Molnar, R.E. 1981. A dinosaur from New Zealand. *in*: *Gondwana Five, Proc. 5th Int. Gondwana Symp.*, (M.M. Crasswell and P. Vella eds.). Wellington. pp. 91–96.

Molnar, R.E. 1992. Paleozoogeographic relationships of Australian Mesozoic tetrapods. *in*: *New concepts in global tectonics* (S. Chatterjee and N. Hotton III, eds.). Texas Tech. Univ. Press, Lubbock, TX. pp. 259–266.

Monod, T. 1957. Scarides et milieu corralien: notes biogeographiques. *Proc. 8th Pac. Sci. Congr. 3*:971–978.

Moore, D.M. 1972. Connections between cool temperate floras with particular reference to southern South America. *in*: *Taxonomy, phytogeography and evolution* (D.H. Valentine, ed.). Academic Press, New York. pp. 115–138.

Morowitz, H.J. 1991. Balancing species preservation and economic considerations. *Science 253*:752–754.

Myers, G.S. 1938. Fresh–water fishes and West Indian zoogeography. *Smithson. Rep. 1937*:339–364.

Myers, R.F. 1989. *Micronesian reef fishes*. Coral Graphics, Guam, GU.

Nance, R.D., T.R. Worsley and J.B. Moody. 1988. The supercontinent cycle. *Sci. Am. 259*:72–79.

Nations, J.D. 1975. The genus *Cancer* (Crustacea: Brachyura): systematics, biogeography and fossil record. *Sci. Bull. Nat. Hist. Mus. Los Angeles County 23*:1–104.

Nations, D. 1979. The genus *Cancer* and its distribution in time and space. *Bull. Biol. Soc. Wash. 3*:153–187.

Nelson, G. 1978. From Candolle to Croizat: comments on the history of biogeography. *J. Hist. Biol. 11*:269–305.

Nelson, G. and N.I. Platnick. 1980. A vicariance approach to historical biogeography. *BioScience 30*:339–343.

Nelson, G. and D.E. Rosen (eds.). 1981. *Vicariance biogeography*. Columbia Univ. Press, New York.

Nelson, J.S. 1984. *Fishes of the world*. John Wiley, New York.

Nelson, J.S. 1985. On the relationships of the New Zealand marine fish *Antipodocottus galatheae* with the Japanese *Stlengis misakia* (Scorpaeniformes: Cottidae). *Rec. N.Z. Oceanogr. Inst. 5*:1–12.

Nestor, H. 1990. Biogeography of Silurian Stromatoporoids. *in*: *Palaeozoic palaeogeography and biogeography* (W.S. McKerrow and C.R. Scotese, eds.). Geol. Soc. London, Mem. 12:215–221.

Newell, N.D. 1967. Revolution in the history of life. *Geol. Soc. Am. Spec. Paper 89*:63–91.

Newman, W.A. 1986. Origin of the Hawaiian marine fauna: dispersal and vicariance as indicated by barnacles and other organisms. *in*: *Crustacean biogeography* (R.H. Gore and K.L. Heck, eds.). A.A. Balkema, Rotterdam. pp. 21–49.

Newman, W.A. and B.A. Foster. 1987. Southern hemisphere endemism among the barnacles: explained in part by extinction of northern members of amphitropical taxa? *Bull. Mar. Sci. 41*:361–377.

Newman, W.A., P.A. Jumars and A. Ross. 1976. Diversity trends in coral– inhabiting barnacles (Cirripedia, Pyrgomatinae). *Micronesica 12*:69–82.

Nichols, D.J. 1991. Palynology supports catastrophic event at Cretaceous– Tertiary boundary: an overview of ten years of research in the United States. *Abs. Geol. Soc. Am., San Diego Meeting*, p. A357.

Nicol, D. 1971. Species, class and phylum diversity of animals. *Q. J. Florida Acad. Sci. 34*:191–194.

Niklas, K.J. 1981. Discussion. *in*: *Vicariance biogeography*: a critique (G. Nelson and D.E. Rosen, eds.). Columbia Univ. Press. pp. 428–435.

Niklas, K.J. 1986. Large–scale changes in animal and plant terrestrial communities. *in*: *Patterns and processes in the history of life* (D.M. Raup and D. Jablonski, eds.). Springer–Verlag, Berlin. pp. 383–405.

Niklas, K.J., B.H. Tiffney and A.H. Knoll. 1985. Patterns in vascular land plant diversification: an analysis at the species level. *in*: *Phanerozoic diversity patterns* (J.W. Valentine, ed.). Princeton Univ. Press, Princeton, NJ. pp. 97–128.

Nixon, K.C. 1989. Origins of Fagaceae. *in*: *Evolution, systematics and fossil history of the Hamamelidae*. (P.R. Crane, ed.). Vol. 2. Clarendon Press, Oxford. pp. 23–43.

Noonan, G.R. 1985. The influence of dispersal, vicariance and refugia on patterns of biogeographical distributions of the beetle family Carabidae. *in*: *Taxonomy, phylogeny and zoogeography of beetles and ants* (G.E. Ball, ed.). W. Junk, Dordrecht. pp. 322–349.

Nott, J.F. and J.A.K. Owen. 1992. An Oligocene palynoflora from the middle Shoalhaven catchment N.S.W. and the Tertiary evolution of flora and climate in the southeast Australian highlands. *Palaeogeogr., Palaeoclimatol., Palaeoecol. 95*:135–151.

Novacek, M.J. and L.G. Marshall. 1976. Early biogeographic history of ostariophysan fishes. *Copeia 1976*:1–12.

Nur, A. and Z. Ben–Avraham. 1981. Lost Pacifica continent: a mobilistic speculation. *in*: *Vicariance biogeography, a critique* (G. Nelson and D.E. Rosen, eds.). Columbia Univ. Press, New York. pp. 341–358.

Nussbaum, R.A. 1984. Amphibians of the Seychelles. *in*: *Biogeography and ecology of the Seychelles Islands* (D.R. Stoddart, ed.). W. Junk, The Hague. pp. 379–415.

Officer, C.B. and C.L. Drake. 1985. Terminal Cretaceous environmental events. *Science 227*:1161–1167.

Olsen, P.E., N.H. Shubin and M.H. Anders. 1987. New early Jurassic tetrapod assemblages constrain Triassic–Jurassic tetrapod extinction event. *Science 237*:1025–1029.

Olson, E.C. 1989. Problems of Permo–Triassic terrestrial vertebrate extinctions. *Hist. Biol. 2*:17–35.

Olson, S.L. 1985. The fossil record of birds. *in*: *Avian biology*.(D.S. Farner, J.R. King and K.C. Parkes, eds.). Vol. VIII. Academic Press, Orlando, FL. pp. 79–238.

Olsson, R.K. and C. Liu. 1993. Controversies on the placement of Cretaceous– Paleogene boundary and the K/P mass extinction of planktonic Foraminifera. *Palaios 8*:127–139.

Ormond, R. and A. Edwards. 1987. Red Sea fishes. *in*: *Red Sea* (A.J. Edwards and S.M. Head, eds.). Pergamon Press, Oxford. pp. 251–287.

Otte, D. and J.A. Endler (eds.). 1989. *Speciation and its consequences*. Sinauer Associates, Sunderland, MA.

Owen, H.G. 1983. *Atlas of continental displacement 200 million years to the present*. Cambridge Univ. Press, Cambridge.

Padian, K. 1986. Introduction. *in*: *The beginning of the age of dinosaurs* (K. Padian, ed.). Cambridge Univ. Press, Cambridge. pp. 1–7.

Padian, K. 1988. Triassic–Jurassic extinctions. *Science 241*:1358–1359.

Padian, K. and W.A. Clemens. 1985. Terrestrial vertebrate diversity: episodes and insights. *in*: *Phanerozoic diversity patterns* (J.W. Valentine, ed.). Princeton Univ. Press, Princeton, NJ. pp. 41–96.

Page, C.N. 1990. Coniferophyta. *in*: *The families and genera of vascular plants* (K. Kubitzki, ed.). Springer–Verlag, Berlin. pp. 279–361.

Palmer, A.R. 1973. Cambrian trilobites. *in*: *Atlas of palaeobiography* (A. Hallam, ed.). Elsevier, Amsterdam. pp. 3–11.

Palmer, A.R. 1984. The biomere problem: evolution of an idea. *J. Paleontol. 58*:599–611.

Parenti, L.R. 1981. A phylogenetic and biogeographic analysis of Cyprinodontiform fishes (Teleostei, Atherinomorpha). *Bull. Am. Mus. Nat. Hist. 168*:335–557.

Parenti, L.R. 1991. Ocean basins and the biogeography of freshwater fishes. *Aust. Syst. Bot. 4*:137–149.

Parin, N.V. 1986. Distribution of mesobenthopelagic fishes in slope waters and around submarine rises. *in*: *Pelagic biogeography* (A.C. Pierrot–Bults, S. van der Spoel, B.J. Zahuranec and R.K. Johnson, eds.). UNESCO, Paris. pp. 226–229.

Parin, N.V. 1991. Fish fauna of the Nazca and Sala y Gomez submarine ridges, the easternmost outpost of the Indo–West Pacific Zoogeographic Region. *Bull. Mar. Sci. 49*:671–683.

Parker, H.W. 1977. *Snakes: a natural history*. 2nd edition, rev. by A.G.C. Grandison. Cornell Univ. Press, Ithaca, NY.

Parrish, J.T. 1987. Global palaeogeography and palaeoclimate of the late Cretaceous and early Tertiary. *in*: *The origins of angiosperms and their biological consequences* (E.M. Friis, W.G. Chaloner and P.R. Crane, eds.). Cambridge Univ. Press, Cambridge. pp. 51–73.

Parrish, J.T. 1990. Gondwanan paleogeography and paleoclimatology. *in*: *Antarctic paleobiology* (T.N. Taylor and E.L. Taylor, eds.). Springer–Verlag, New York. pp. 15–26.

Parrish, R.H., R. Serra and W.S. Grant. 1989. The monotypic sardines, *Sardina* and *Sardinops*: their taxonomy, distribution, stock structure and zoogeography. *Can. J. Fish. Aq. Sci. 46*:2019–2036.

Pascual, R., M. Archer, E. Ortiz Jaurequizar, J.C. Prado, H. Godthelp and S.J. Hand. 1992. First discovery of monotremes in South America. *Nature 256*:704–706.

Patrait, P. and J. Achache. 1989. India–Eurasia collision chronology has implications for a crustal shortening and driving mechanism of plates. *Nature 311*:615–621.

Patterson, C. 1975. The distribution of Mesozoic freshwater fishes. *Mém. Mus. Nat. d'Hist. Nat., Paris 88*:156–173.

Patterson, C. 1981. The development of the North American fish fauna – a problem of historical biogeography. *in*: *The evolving biosphere* (P.L. Forey, ed.). Cambridge Univ. Press, Cambridge. pp. 265–281.

Patterson, C. 1993. An overview of the early fossil record of acanthomorphs. *Bull. Mar. Sci. 52*:29–59.

Patterson, C. and H.G. Owen. 1991. Indian isolation or contact? a response to Briggs. *Syst. Zool. 40*:96–100.

Patterson, C. and A.B. Smith. 1987. Is the periodicity of extinctions a taxonomic artifact? *Nature 330*:248–252.

Paul, C.R.C. 1988. Extinction and survival in the echinoderms. *in*: *Extinction and survival in the fossil record* (G.P. Larwood, ed.). Clarendon Press, Oxford. pp. 155–170.

Peake, J.F. 1981. The land snails of islands – a dispersalist's viewpoint. *in*: *The evolving biosphere* (P.L. Forey, ed.). Cambridge Univ. Press, Cambridge. pp. 247–263.

Pearson, D.L. and F. Cassola. 1992. World–wide species richness patterns of tiger beetles (Coleoptera: Cicindelidae): indicator taxon for biodiversity and conservation studies. *Conserv. Biol. 6*:376–391.

Pedder, A.E.H. 1982. The rugose coral record across the Frasnian–Famennian boundary. *Geol. Soc. Am., Spec. Paper 190*:485–489.

Peden, A.E. and D.E. Wilson. 1977. Distribution of intertidal and subtidal fishes of northern British Columbia and southeastern Alaska. *Syesis 9*:221–248.

Penman, H.L. 1970. The water cycle. *Sci. Am. 223*:98–108.

Percival, S.F. and A.G. Fischer. 1977. Changes in calcareous nannoplankton in the Cretaceous–Tertiary biotic crisis at Zumaya, Spain. *Evol. Theory 2*:1–35.

Perfit, M.R. and E.E. Williams. 1989. Geological constraints and biological retrodictions in the evolution of the Caribbean Sea and its islands. *in*: *Biogeography of the West Indies* (C.A. Woods, ed.). Sandhill Crane Press, Gainesville, FL. pp. 47–102.

Pescador, M.L. and W.L. Peters. 1980. Phylogenetic relationships and zoogeography of cool–adapted Leptophlebiidae (Ephemeroptera) in southern South America. *in*: *Advances in Ephemeroptera biology* (J.F. Flannagan and K.E. Marshall, eds.). Plenum Press, New York. pp. 43–56.

Peters, W.L. 1980. Phylogeny of the Leptophlebiidae (Ephemeroptera): an introduction. *in*: *Advances in Ephemeroptera biology* (J.J. Flannagan and K.E. Marshall, eds.). Plenum Press, New York. pp. 33–41.

Petuch, E.J. 1988. *Neogene history of tropical American mollusks*. Coastal Education and Research Foundation, Charlottesville, VA.

Phillips, R.C. and E.G. Meñez. 1988. Seagrasses. *Smithson. Contrib. Mar. Sci. 34*:1–91.

Phipps, J.B. 1983. Biogeographic, taxonomic and cladistic relationships between East Asiatic and North American *Crataegus*. *Ann. Mo. Bot. Gard. 70*:667–700.

Pianka, E.R. 1966. Latitudinal gradients in species diversity: a review of concepts. *Am. Nat. 100*:33–46.

Piccoli, G., S. Sartori and A. Franchino. 1987. Benthic molluscs of shallow Tethys and their destiny. *in*: *Shallow–Tethys 2* (K.G. McKenzie, ed.). Proc. Internat. Symp. on Shallow Tethys 2, A.A. Balkema, Rotterdam. pp. 333–373.

Pielou, E.C. 1979. *Biogeography*. John Wiley, New York.

Pierrot–Bults, A.C. and V.R. Nair. 1991. Distribution patterns in Chaetognatha. *in*: *The biology of chaetognaths* (Q. Bone, H. Kapp and A.C. Pierrot–Bults, eds.). Oxford Univ. Press, New York. pp. 86–116.

Pierrot–Bults, A.C. and S. van der Spoel. 1979. Speciation in macrozooplankton. *in*: *Zoogeography and diversity of plankton* (S. van der Spoel and A.C. Pierrot–Bults, eds.). John Wiley, New York. pp. 144–167.

Piper, J.D.A. 1987. *Palaeomagnetism and the continental crust*. John Wiley, New York.

Pitrat, C.W. 1973. Vertebrates and the Permo–Triassic extinctions. *Palaeogeol., Palaeoclimatol., Palaeoecol. 14*:249–264.

Platnick, N.I. 1992. Patterns of biodiversity. *in*: *Systematics, ecology and the biodiversity crisis* (N. Eldredge, ed.). Columbia Univ. Press, New York. pp. 15–24.

Poinar, G.O., Jr. 1993. Insects in amber. *Annu. Rev. Entomol. 46*:145–159.

Poncet, J. 1990. Biogeography of Devonian algae. *in*: *Palaeozoic palaeogeography and biogeography* (W.S. McKerrow and C.R. Scotese, eds.). Geol. Soc. London, Mem. 12:285–289.

Popham, E.J. 1963. The geographical distribution of the Dermaptera. *Entomologist 96*:131–144.

Popham, E.J. and B.F. Manly. 1969. Geographical distribution of the Dermaptera and the continental drift hypothesis. *Nature 222*:981–982.

Por, F.D. 1978. *Lessepsian migration*. Springer–Verlag, Berlin.

Potts, R. and A.K. Behrensmeyer. 1992. Late Cenozoic terrestrial ecosystems. *in*: *Terrestrial ecosystems through time* (A.K. Behrensmeyer, et al., eds.). Univ. of Chicago Press, Chicago, IL. pp. 424–541.

Prasad, G.V.R. and A. Sahni. 1988. First Cretaceous mammal from India. *Nature 332*:638–640.

Preston, F.W. 1962. The canonical distribution of commonness and rarity. *Ecology 43*:185–215.

Prichard, J.C. 1826. *Researches into the physical history of mankind*. 2nd Ed., 2 Vols. J. and A. Arch, London.

Procter, J. 1984. Floristics of the granitic islands of the Seychelles. *in*: *Biogeography and ecology of the Seychelles Islands* (D.R. Stoddart, ed.). W. Junk, The Hague. pp. 209–220.

Prothero, D.R. 1986. North American mammalian diversity and Eocene–Oligocene extinctions. *Paleobiology 11*:389–405.

Rabb, G.B. and H. Marx. 1973. Major ecological and geographic patterns in the evolution of colubroid snakes. *Evolution 27*:69–83, 4 figs.

Rabinowitz, P.D., M.L. Coffin and D. Falvey. 1983. The separation of Madagascar and Africa. *Science 220*:67–69.

Radovsky, F.J., P.H. Raven and S.H. Sohmer. 1984. Biogeography of the tropical Pacific. *Bull. Bernice P. Bishop Mus. 72*:1–221.

Raffi, S., S.M. Stanley and R. Marasti. 1986. Biogeographic patterns and Plio– Pleistocene extinction of Bivalvia in the Mediterranean and southern North Sea. *Paleobiology 11*:368–388.

Rage, J.–C. 1986. The amphibians and reptiles at the Eocene–Oligocene transition in western Europe: an outline of faunal alterations. *in*: *Terminal Eocene events* (C. Pomerol and I. Premola–Silva, eds.). Elsevier, Amsterdam. pp. 309–310.

Rage, J.–C. 1987. Fossil history. *in*: *Snakes: ecology and evolutionary biology* (R.A. Seigel, J.T. Collins and S.S. Novak, eds.). MacMillan, New York. pp. 51–76.

Rage, J.–C. 1988. Gondwana, Tethys and terrestrial vertebrates during the Mesozoic and Cenozoic. *in*: *Gondwana and Tethys* (M.G. Audley–Charles and A. Hallam, eds.). Geol. Soc. Spec. Paper No. 37:255–273.

Randall, J.E. 1982. Examples of antitropical and antiequatorial distributions of Indo–West Pacific fishes. *Pac. Sci. 35*:197–209.

Randall, J.E. 1992. Endemism of fishes in Oceania. Coastal resources and systems of the Pacific basin: investigation and steps toward protective management. *UNEP Regional Seas Reports and Studies No. 147*:55–67.

Rapaport, E.H. 1982. *Areography: geographical strategies of species*. Pergamon Press, New York.

Rasmussen, D.T., T.M. Bown and E.L. Simons. 1992. The Eocene–Oligocene transition in continental Africa. *in*: *Eocene–Oligocene climatic and biotic evolution* (D.R. Prothero and W.A. Berggren, eds.). Princeton Univ. Press, Princeton, NJ. pp. 548–566.

Rass, T.S. 1986. Vicariance ichthyogeography of the Atlantic Ocean pelagical. *in*: *Pelagic biogeography* (A.C. Pierrot–Bults, S. van der Spoel, B.J. Zahuranec and R.K. Johnson, eds.). UNESCO, Paris. pp. 237–241.

Raup, D.M. 1979. Size of the Permo–Triassic bottleneck and its evolutionary implications. *Science 206*:217–218.

Raup, D.M. 1986. *The nemesis affair*. Norton, New York.

Raup, D.M. 1988. Extinction in the geologic past. *in*: *Origins and extinctions* (D.E. Osterbrock and P.H. Raven, eds.). Yale Univ. Press, New Haven, CT. pp. 109– 119.

Raup, D.M. and D. Jablonski. 1993. Geography of end–Cretaceous marine bivalve extinctions. *Science 260*:971–973.

Raup, D.M. and J.J. Sepkoski, Jr. 1984. Periodicity of extinctions in the geologic past. *Proc. Natl. Acad. Sci. USA 81*:801–805.

Raven, P.H. 1963. Amphitropical relationships in the floras of North and South America. *Q. Rev. Biol. 38*:151–177.

Raven, P.H. 1990. The politics of preserving biodiversity. *BioScience 40*:769–774.

Raven, P.H. and D.I. Axelrod. 1974. Angiosperm biogeography and past continental movements. *Ann. Mo. Bot. Gard. 61*:539–673.

Rawson, P.F. 1981. Early Cretaceous ammonite biostratigraphy and biogeography. *in*: *The Ammonoidea* (M.R. House and J.R. Senior, eds.). Academic Press, London. pp. 499–529.

Ray, G.C. 1985. Man and the sea – the ecological challenge. *Am. Zool. 25*:451– 468.

Raymo, M.E. and W.F. Ruddiman. 1992. Tectonic forcing of late Cenozoic climate. *Nature 359*:117–122.

Raymond, A. 1987. Paleogeographic distribution of early Devonian plant traits. *Palaios* 2:113–132.

Repenning, C.A. 1987. Biochronology of the microtine rodents of the United States. *in*: *Cenozoic mammals of North America* (M.O. Woodburne, ed.). Univ. of California Press, Berkeley, CA. pp. 236–268.

Retallack, G.J. 1992. Paleosols and changes in climate and vegetation across the Eocene/Oligocene boundary.

in: *Eocene–Oligocene climatic and biotic evolution* (D.R. Prothero and W.A. Berggren, eds.). Princeton Univ. Press, Princeton, NJ. pp. 382–398.

Riccardi, A.C. 1991. Jurassic and Cretaceous marine connections between the southeast Pacific and Tethys. *Palaeogeogr., Palaeoclimatol., Palaeoecol. 87*:155– 189.

Rich, P.V. 1980. The Australian Dromorthinidae: a group of extinct large ratites. *Contrib. Sci. Nat. Hist. Mus. Los Angeles County 330*:93–103.

Rich, T.H. and P.V. Rich. 1989. Polar dinosaurs and biotas of the early Cretaceous of southeastern Australia. *Nat. Geogr. Res. 5*:15–53.

Ricklefs, R.E. and R.E. Latham. 1993. Global patterns of diversity in mangrove floras. *in*: *Species diversity in ecological communities* (R.E. Ricklefs and D. Schluter, eds.). Univ. of Chicago Press, Chicago, IL. pp. 215–229.

Rohde, K., M. Heap and D. Heap. 1993. Rapaport's rule does not apply to marine teleosts and cannot explain latitudinal gradients in species richness. *Am. Nat. 142*:1–16.

Romine, K. 1985. Radiolarian biogeography and paleoceanography of the North Pacific at 8 Ma. *in*: *The Miocene ocean: paleoceanography and biogeography* (J.P. Kennett, ed.). Geol. Soc. Am., Mem. 163:237–272.

Romine, K. and G. Lombari. 1985. Evolution of Pacific circulation in the Miocene: radiolarian evidence from DSDP Site 289. *in*: *The Miocene Ocean: paleoceanography and biogeography* (J.D. Kennett, ed.). Geol. Soc. Am., Mem. 163:273–290.

Rosa, D. 1923. Qu'est–ce que l'hologénèse? *Scientia 33*:113–124.

Rosen, B.R. 1984. Reef coral biogeography and climate through the late Cainozoic: just islands in the sun or a critical pattern of islands? *in*: *Fossils and climate* (P. Brenchley, ed.). John Wiley, Chichester. pp. 201–260.

Rosen, D.E. 1976. A vicariance model of Caribbean biogeography. *Syst. Zool. 24*:431–464.

Rosen, D.E. and R.M. Bailey. 1963. The poeciliid fishes (Cyprinodontiformes), their structure, zoogeography and systematics. *Bull. Am. Mus. Nat. Hist. 126*:1–176.

Rosenblatt, R.H. and R.S. Waples. 1986. A genetic comparison of allopatric populations of shore fish species from the eastern and central Pacific Ocean: dispersal or vicariance? *Copeia 1986*:275–284.

Rosenzweig, M.L. and R.D. McCord. 1991. Incumbent replacement: evidence for long–term evolutionary progress. *Paleobiology 17*:202–213.

Ross, H.H. 1958. Affinities and origins of the northern and montane insects of western North America. *in*: *Zoogeography* (C.L. Hubbs, ed.). Am. Assoc. Adv. Sci., Washington, DC. pp. 231–252.

Ross, J.R.P. and C.A. Ross. 1990. Late Palaeozoic bryozoan biogeography. *in*: *Palaeozoic palaeogeography and biogeography* (W.S. McKerrow and C.R. Scotese, eds.). Geol. Soc. London, Mem. 12:353–362.

Rotondo, G.M., V.G. Springer, G.A.J. Scott and S.O. Schlanger. 1981. Plate movement and island intergration – a possible mechanism in the formation of endemic biotas, with special reference to the Hawaiian Islands. *Syst. Zool. 30*:12–21.

Roux, M. 1987. Evolutionary ecology and biogeography of recent stalked crinoids as a model for the fossil record. *in*: *Echinoderm studies*.(M. Jangoux and J.M. Lawrence, eds.). Vol. 2. A.A. Balkema, Rotterdam. pp. 1–53.

Rowley, D.B., A. Raymond, J.T. Parrish, A.L. Lottes, C.R. Scotese and A.M. Ziegler. 1985. Carboniferous paleogeographic, phytogeographic and paleoclimate reconstructions. *Int. J. Coal Geol. 5*:7–42.

Sahni, A. 1984. Cretaceous–Paleocene terrestrial faunas of India: lack of endemism during drifting of the Indian plate. *Science 226*:441–443.

Sahni, A., R.S. Rana and G.V.R. Prasad. 1987. New evidence for paleobiogeographic intercontinental Gondwana relationships based on late Cretaceous–earliest Paleocene coastal faunas from peninsular India. *in*: *Gondwana six: stratigraphy, sedimentology and paleontology* (G.D. McKenzie, ed.). Am. Geophysical Union, Washington, DC. pp. 207–218.

Sanders, H.L. 1968. Marine benthic diversity: a comparative study. *Am. Nat. 103*:253–282.

Sanders, H.L. and R.R. Hessler. 1969. Ecology of the deep–sea benthos. *Science 163*:1419–1424.

Sandy, M.R. 1991. Aspects of Middle–Late–Jurassic–Cretaceous Tethyan brachiopod biogeography in relation to tectonic and paleoceanographic developments. *Palaeogeogr., Palaeoclimatol., Palaeoecol. 87*:137–154.

Sarnthein, M. and J. Fenner. 1988. Global wind–induced change of deep–sea sediment budgets, new ocean production and CO_2 reservoirs about 3.3–2.35 Ma B.P. *in*: *The past three million years: evolution of climatic variability in the North Atlantic region* (N.J. Shackleton, R.G. West and D.Q. Bowen, eds.). Philos. Trans. R. Soc. London, Ser. B, 318:487–504.

Sarnthein, M. and K. Winn. 1988. Global variations of surface ocean productivity in low and mid latitudes: influence on CO_2 reservoirs of the deep ocean and atmosphere during the last 21,000 years. *Paleoceanography 3*:361–399.

Sauer, J.D. 1988. *Plant migration: the dynamics of geographic patterning in seed plant species.* Univ. of California Press, Berkeley, CA.

Savage, J.M. 1982. The enigma of the Central American herpetofauna. *Ann. Mo. Bot. Gard. 69*:464–547.

Savage, R.J.G. 1988. Extinction and the fossil mammal record. *in*: *Extinction and survival in the fossil record* (G.P. Larwood, ed.). Clarendon Press, Oxford. pp. 319–334.

Savage, R.J.G. and D.E. Russell. 1983. *Mammalian paleofaunas of the world.* Addison–Wesley, Reading, MA.

Scarabino, V. 1977. Moluscos del Golfo San Matias. *Com. Soc. Malacológia del Uruguay, 4 (31–32)*:177–285 (11 pls).

Schäfer, P. and E. Fois–Erickson. 1986. Triassic Bryozoa and the evolutionary crisis of Paleozoic Stenolaemata. *in*: *Global bio–events* (S. Battacharji, G.M. Freidman, H.J. Neugebauer and A. Seilacher, eds.). Springer–Verlag, Berlin. pp. 251–255.

Scherer, G. 1988. The origins of the Alticinae. *in*: *Biology of the Chrysomelidae* (P. Jolivet, E. Petitpierre, T.H. Hsiao, eds.). Kluwer, Dordrecht. pp. 115–130.

Schindewolf, O. 1962. Neokatastrophismus? *Dtsch. Geol. Ges. Leitschr. Jahrg. 114*:430–445.

Schlinder, E. 1990. The late Frasnian (Upper Devonian) Kellwasser crisis. *in*: *Extinction events in earth history* (E.G. Kauffman and O.H. Walliser, eds.). Springer–Verlag, Berlin. pp. 151–159.

Schminke, H.K. 1974. Mesozoic intercontinental relationships as evidenced by bathynellid Crustacea (Syncarida: Malacostraca). *Syst. Zool. 23*:157–164.

Schopf, T.J.M. 1980. *Paleoceanography.* Harvard Press, Cambridge, MA.

Schram, F.R. 1977. Palaeozoogeography of late Palaeozoic and Triassic Malacostraca. *Syst. Zool. 26*:367–379.

Schuh, R.T. and G.M. Stonedahl. 1986. Historical biogeography in the Indo– Pacific: a cladistic approach. *Cladistics 2*:337–355.

Schuster, R.M. 1972. Continental movements, "Wallace's Line" and Indomalayan– Australasian dispersal of land plants: some eclectic concepts. *Bot. Rev. 38*:3–86.

Schuster, R.M. 1976. Plate tectonics and its bearing on the geographical origin and dispersal of angiosperms. *in*: *Origin and early evolution of angiosperms* (C.B. Beck, ed.). Columbia Univ. Press, New York. pp. 48–138.

Schuster, R.M. 1981. Paleoecology, origin, distribution through time and evolution of Hepaticae and Anthrocerotae. *in*: *Paleobotany, paleoecology and evolution* (K.J. Niklas, ed.). Vol. 2. Praeger, New York. pp. 129–191.

Schuster, R.M. 1983. Reproductive biology, dispersal mechanisms and distribution patterns in Hepaticae and Anthocerotae. *in*: *Dispersal and distribution* (K. Kubitzki, ed.). Verlag Paul Parey, Hamburg. pp. 119–162.

Sclater, J.G. and C. Tapscott. 1979. The history of the Atlantic. *Sci. Am. 240*:156–174.

Sclater, P.L. 1858. On the general geographical distribution of the members of the Class Aves. *J. Linn. Soc. (Zool.) 2*:130–145.

Scotese, C.R. 1991. Jurassic and Cretaceous plate tectonic reconstructions. *Palaeogeogr., Palaeoclimatol., Palaeoecol. 87*:493–501.

Scotese, C.R. 1992. *Paleomap project.* Progress Rept. No. 36, Univ. of Texas, Arlington, TX.

Scotese, C.R. and W.S. McKerrow. 1990. Palaeozoic base maps. *in*: *Palaeozoic palaeogeography and biogeography* (W.S. McKerrow and C.R. Scotese, eds.). Geol. Soc. London, Mem. 12.

Scrutton, C.T. 1988. Patterns of extinction and survival in Paleozoic corals. *in*: *Extinction and survival in the fossil record* (G.P. Larwood, ed.). Clarendon Press, Oxford. pp. 65–88.

Selden, P.A. 1990. Terrestrialization: invertebrates. *in*: *Palaeobiology: a synthesis* (D.E.G. Briggs and P.R. Crowther, eds.). Blackwell, Oxford. pp. 64–68.

Selden, P.A. and D. Edwards. 1989. Colonisation of the land. *in*: *Evolution and the fossil record* (K. Allen and D. Briggs, eds.). Belhaven Press, London. pp. 122–152.

Selkirk, P.M., R.D. Seppelt and D.R. Selkirk. 1990. *Subantarctic Macquarie Island: environment and biology*. Cambridge Univ. Press, Cambridge.

Sepkoski, J.J., Jr. 1986a. An overview of Phanerozoic mass extinctions. *in*: *Pattern and process in the history of life* (D. Jablonski and D. Raup, eds.). Springer–Verlag, Berlin. pp. 277–295.

Sepkoski, J.J., Jr. 1986b. Global bioevents and the question of periodicity. *in*: *Global bio–events* (O.H. Walliser, ed.). Springer–Verlag, Berlin. pp. 47–61.

Sepkoski, J.J., Jr. 1989. Periodicity in extinction and the problem of catastrophism in the history of life. *J. Geol. Soc. 146*:7–19.

Sepkoski, J.J., Jr. 1990. Evolutionary faunas. *in*: *Palaeobiology: a synthesis* (D.E.G. Briggs and P.R. Crowther, eds.). Blackwell, Oxford. pp. 37–41.

Sepkoski, J.J., Jr. 1992. Phylogenetic and ecologic patterns in the Phanerozoic history of marine biodiversity. *in*: *Systematics, ecology and the biodiversity crisis* (N. Eldredge, ed.). Columbia Univ. Press, New York.

Sepkoski, J.J., Jr. and A.I. Miller. 1985. Evolutionary faunas and the distribution of Paleozoic marine communities in time and space. *in*: *Phanerozoic diversity patterns* (J.W. Valentine, ed.). Princeton Univ. Press, Princeton, NJ. pp. 153–180.

Shaw, S.R. 1990. Phylogeny and biogeography of the parasitoid wasp family Megalyridae (Hymenoptera). *J. Biogeogr. 17*:569–581.

Sheehan, P.M. 1982. Brachiopod macroevolution at the Ordovician–Silurian boundary. *Third North Am. Paleont. Conv. Proc.* 2:477–481.

Sheehan, P.M. and P.J. Coorough. 1990. Brachiopod zoogeography across the Ordovician–Silurian extinction event. *in*: *Palaeozoic palaeogeography and biogeography* (W.S. McKerrow and C.R. Scotese, eds.). Geol. Soc. London, Mem. 12:181–187.

Sheehan, P.M. and D.E. Fastovsky. 1992. Major extinctions of land–dwelling vertebrates at the Cretaceous–Tertiary boundary. *Geology 20*:556–560.

Sheehan, P.M., D.E. Fastovsky, R.G. Hoffman, C.B. Berghaus and D.L. Gabriel. 1991. Sudden extinction of the dinosaurs: latest Cretaceous, upper Great Plains. U.S.A. *Science 254*:835–839.

Shields, O. 1983. Trans–Pacific links that suggest earth expansion. *in*: *Expanding earth symposium* (S.W. Carey, ed.). Univ. of Tasmania. pp. 199–205.

Shmida, A. and M.V. Wilson. 1985. Biological determinants of species diversity. *J. Biogeogr. 12*:1–20.

Shoemaker, E.M. and R.F. Wolfe. 1986. Mass extinctions, crater ages and comet showers. *in*: *The galaxy and the solar system* (R. Smoluchowski, J.N. Bahcall and M.S. Matthews, eds.). Univ. Arizona Press, Tucson, AZ. pp. 338–386.

Shubin, N.H. and H.–D. Sues. 1991. Biogeography of early Mesozoic continental tetrapods: patterns and implications. *Paleobiology 17*:214–230.

Sibley, C.G. and J.E. Ahlquist. 1981. The phylogeny and relationships of the ratite birds as indicated by DNA–DNA hybridization. *Proc. 2nd Int. Congr. Syst. Evol. Biol.* pp. 301–335.

Sibley, C.G. and J.E. Ahlquist. 1986. Reconstructing bird phylogeny by comparing DNA's. *Sci. Am. 254*:82–92.

Sibley, C.G. and J.E. Ahlquist. 1991. *Phylogeny and classification of birds*. Yale Univ. Press, New Haven, CT.

Signor, P.W. 1990. Patterns of diversification. *in*: *Palaeobiology: a synthesis* (D.E.G. Briggs and P.R. Crowther, eds.). Blackwell, Oxford. pp. 130–135.

Sigognau–Russell, D. 1989. *Handbuch der palaeherpetologie (Teil 17B/1 part 17B/1) Theriodontia*. Gustav Fisher Verlag, Stuttgart.

Simons, E.L. and T.M. Bown. 1985. *Afrotarsius chatrathi*, first tarsiiform primate (Tarsiidae) from Africa. *Nature 313*:475–477.

Simpson, B.B. and J.L. Neff. 1985. Plants, their pollinating bees and the great American interchange. *in*: *The great American interchange* (F.G. Stehli and S.D. Webb, eds.). Plenum Press, New York. pp. 427–452.

Simpson, G.G. 1965. *The geography of evolution*. Chilton Books, Philadelphia, PA.

Simpson, G.G. 1980. *Splendid isolation: the curious history of South American mammals*. Yale Univ. Press, New Haven, CT.

Sing–chi, C. 1983. A comparison of orchid floras of temperate North America and Eastern Asia. *Ann. Mo. Bot. Gard. 70*:713–723.

Slater, J.A. 1979. The systematics, phylogeny and zoogeography of the Blissinae of the world. *Bull. Am. Mus. Nat. Hist. 165*:1–180.

Sloan, L.C. and E.J. Barron. 1992. Paleogene climatic evolution: a climate model investigation of the influence of continental elevation and sea–surface temperature upon continental climate. *in*: *Eocene–Oligocene climatic and biotic evolution* (D.R. Prothero and W.A. Berggren, eds.). Princeton Univ. Press, Princeton, NJ. pp. 202–217.

Sloan, R.E., K.J. Rigby, Jr., L. Van Valen and D. Gabriel. 1986. Gradual dinosaur extinction and simultaneous ungulate radiation in the Hell Creek Formation. *Science 232*:629–633.

Smiley, C.J. 1979. Pre–Tertiary phytogeography and continental drift – some apparent discrepancies. *in*: *Historical biogeography, plate tectonics and the changing environment* (J. Gray and A.J. Boucot, eds.). Oregon State Univ. Press, Corvallis, OR. pp. 311–319.

Smit, J. 1990. Meteorite impact, extinctions and the Cretaceous–Tertiary boundary. *Geol. Mijnbouw 69*:187–204.

Smith, A.B. 1984. *Echinoid paleobiology*. George Allen and Unwin, Boston, MA.

Smith, A.B. and C. Patterson. 1988. The influence of taxonomic method on the perception of patterns of evolution. *in*: *Evolutionary biology*. (M.K. Hect and B. Wallace, eds.). Vol. 23. Plenum Press, New York. pp. 127–216.

Smith, A.C. 1972. An appraisal of the orders and families of primitive extant angiosperms. *J. Ind. Bot. Soc. 50A*:215–226.

Smith, A.C. 1973. Angiosperm evolution and the relationship of the floras of Africa and America. *in*: *Tropical forest ecosystems in Africa and South America: a comparative review* (B.L. Meggers, E.S. Ayensu and W.D. Duckworth, eds.). Smithsonian Press, Washington, DC. pp. 49–61.

Smith, G.R. 1981. Late Cenozoic freshwater fishes of North America. *Annu. Rev. Ecol. Syst. 12*:163–193.

Smith, J.M.B. 1986. Origins of Australasian tropicalpine and alpine floras. *in*: *Flora and fauna of alpine Australasia* (B.A. Barlow, ed.). CSIRO, Australia. pp. 109–128.

Smith, J.T. 1970. Taxonomy, distribution and phylogeny of the cymatiid gastropods *Argobuccinum*, *Fusitriton*, *Mediargo* and *Priene*. *Bull. Am. Paleo. 56*:443–573.

Smith, M.M. and P.C. Heemstra (eds.) 1986. *Smith's sea fishes*. Springer– Verlag, Berlin.

Smith, P.L. and H.W. Tipper. 1986. Plate tectonics and paleobiogeography: Early Jurassic (Pliensbachian) endemism and diversity. *Palaios 1*:399–412.

Sonnenfeld, P. and I. Finetti. 1985. Messinian evaporites in the Mediterranean: a model of continuous inflow and outflow. *in*: *Geological evolution of the Mediterranean Basin* (D.J. Stanley and F. Wezel, eds.). Springer–Verlag, New York. pp. 347–353.

Sorbini, L. 1988. Biogeography and climatology of Pliocene and Messinian fossil fish of eastern–central Italy. *Boll. Mus. Civ. St. Nat. Verona 14*:1–85.

Spaulding, W.G. 1990. Vegetation dynamics during the last deglaciation, southeastern Great Basin, U.S.A. *Quat. Res. 33*:188–203.

Specht, R.L. 1981. Biogeography of holophytic angiosperms (salt–marsh, mangrove and sea–grass). *in*: *Ecological biogeography of Australia* (A. Keast, ed.). W. Junk, The Hague. pp. 577–589.

Spinar, Z.V. and M. Hodrova. 1986. *Indobatrachus* (Anura: Leptodactylidae) and its significance for the conformation of the theory of the detachment of the Indian subcontinent. *Bull. Geolog. Survey, Prague 3(61)*:179–181.

Spradberg, J.P. 1973. *Wasps*. Sidgwick and Jackson, London.

Springer, V.G. 1982. Pacific plate biogeography, with special reference to shore fishes. *Smithson. Contrib. Zool. 367*:1–182.

Springer, V.G. 1988. The Indo–Pacific blenniid fish genus *Ecsenius*. *Smithson. Contrib. Zool. 465*:1–134.

Springer, V.G. and J.T. Williams. 1990. Widely distributed Pacific plate endemics and lowered sea–level. *Bull. Mar. Sci. 47*:631–640.

Stanley, S.M. 1984. Temperature and biotic crises in the marine realm. *Geology 12*:205–208.

Stanley, S.M. 1986. Anatomy of a regional mass extinction: Plio–Pleistocene decimation of the western Atlantic bivalve fauna. *Palaios. 1*:17–36.

Stanley, S.M. 1987. *Extinction*. Scientific American, New York.

Stanley, S.M. 1988. Climatic cooling and mass extinction of Paleozoic reef communities. *Palaios 3*:228–232.

Stanley, S.M. 1989. *Earth and life through time*. 2nd ed. W.H. Freeman, New York.

Stanley, S.M. and W.A. Newman, 1980. Competitive exclusion in evolutionary time. *Paleobiology 6*:173–183.

Stearn, C.W. and R.L. Carroll. 1989. *Paleontology: the record of life*. John Wiley, New York.

Stechell, W.A. 1920. Stenothermy and zone–invasion. *Am. Nat. 54*:385–397.

Stehli, F.G. and J. W. Wells. 1971. Diversity and age patterns in hermatypic corals. *Syst. Zool. 20*:115–126.

Stehlin, H.G. 1909. Remarques sur les faunules de Mammifers des couches eocenes et oligocenes du Bassin de Paris. *Bull. Soc. Geol. France 9*:488–520.

Stephensen, K. 1937. Crustacean groups. *Zool. Faroes 2*: various parts.

Stepien, C.A. 1990. Population structure, diets and biogeographic relationships of a rocky intertidal fish assemblage in central Chile: high levels of herbivory in a temperate system. *Bull. Mar. Sci. 47*:598–612.

Sterrer, W. 1986. *Marine fauna and flora of Bermuda*. Wiley–Interscience, New York.

Stevens, C.H. 1977. Was development of brackish oceans a factor in Permian extinctions? *Geol. Soc. Am. Bull. 88*:133–138.

Stevens, G.C. 1989. The latitudinal gradient in geographical range: how so many species coexist in the tropics. *Am. Nat. 133*:240–256.

Stevens, G.C. 1992. Spilling over the competitive limits to species coexistence. *in*: *Systematics, ecology and the biodiversity crisis* (N. Eldredge, ed.). Columbia Univ. Press, New York. pp. 40–58.

Stevens, G.R. 1973. Cretaceous belemnites. *in*: *Atlas of palaeobiogeography* (A. Hallam, ed.). Elsevier, Amsterdam. pp. 385–401.

Stiassney, M.L.J. 1992. Phylogenetic analysis and the role of systematics in the biodiversity crisis. *in*: *Systematics, ecology and the biodiversity crisis* (N. Eldredge, ed.). Columbia Univ. Press, New York. pp. 109–120.

Stigler, S.M. and M.J. Wagner. 1987. A substantial bias in nonparametric tests for periodicity in geophysical data. *Science 238*:940–945.

Stock, C.W. 1990. Biogeography of the Devonian stromatroporoids. *in*: *Palaeozoic palaeogeography and biogeography* (W.S. McKerrow and C.R. Scotese, eds.). Geol. Soc. London, Mem. 12:257–265.

Stoddart, D.R. 1992. Biogeography of the tropical Pacific. *Pac. Sci. 46:* 276–293.

Stork, N.E. 1988. Insect diversity: facts, fiction and speculation. *Biol. J. Linn. Soc. 35*:321–337.

Stott, P. 1981. *Historical Plant Geography*. George Allen and Unwin, London.

Streel, M., M. Fairon–Demaret and S. Loboziak. 1990. Givetian–Frasnian phytogeography of Euramerica and western Gondwana based on microspore distribution. *in*: *Palaeozoic palaeogeography and biogeography* (W.S. McKerrow and C.R. Scotese, eds.). Geol. Soc. London, Mem. 12:291–296.

Stromberg, P.C. and J.L. Crites. 1974. Specialization, body volume and geographical distribution of Camallanidae (Nematoda). *Syst. Zool. 23*:189–201.

Stucky, R.K. 1990. Evolution of land mammal diversity in North America during the Cenozoic. *in*: *Current mammalogy* (H.H. Genoways, ed.). Plenum, New York. pp. 375–432.

Stucky, R.K. 1992. Mammalian faunas in North America of Bridgerian to early Arikareean "ages" (Eocene and Oligocene). *in*: *Eocene–Oligocene climatic and biotic evolution* (D.R. Prothero and W.A. Berggren, eds.). Princeton Univ. Press, Princeton, NJ. pp. 464–493.

Sullivan, R.M. 1979. Revision of the Paleogene genus *Glyptosaurus* (Reptilia, Anguidae). *Bull. Am. Mus. Nat. Hist. 163*:1–72.

Swainson, W. 1835. *A treatise on the geography and classification of animals*. Longman, London.

Sweet, A.R., D.R. Braman and J.F. Lerbekmo. 1990. Palynofloral response to K/T boundary events: a transitory interruption within a dynamic system. *in*: *Global catastrophes and earth history* (V.L. Sharpton and P.D. Ward, eds.). Geol. Soc. Am., Spec. Paper 247:457–469.

Takhtajan, A. 1969. *Flowering plants: origin and dispersal*. Oliver, Edinburgh.

Takhtajan, A.L. 1986. *Floristic regions of the world*. Univ. California Press, Berkeley, CA.

Takhtajan, A. 1987. Flowering plant origin and dispersal: the cradle of the angiosperms revisited. *in*: *Biogeographical evolution of the Malay Archipelago* (T.C. Whitmore, ed.). Clarendon Press, Oxford. pp. 26–31.

Tasch, P. 1987. Fossil Conchostraca of the southern hemisphere and continental drift. *Geol. Soc. Am., Mem. 165*:1–290.

Taylor, D.W. 1988. Aspects of freshwater mollusc ecological biogeography. *Palaeogeogr., Palaeoclimatol., Palaeoecol.* 62:511–576.

Taylor, J.D., N.J. Morris and C.N. Taylor. 1980. Food specialization and the evolution of predatory prosobranch gastropods. *Palaeontology 23*:375–409.

Théel, H. 1886. Report on the Holothuroidea dredged by H.M.S. Challenger during the years 1873–76, Part II. The voyage of H.M.S. Challenger. *Zoology 14*:1–290.

Thiele, H. 1977. *Carabid beetles in their environments*. Springer–Verlag, Berlin.

Thomas, C.D. 1990. Fewer species. *Nature 347*:237.

Thomas, E. 1990. Late Cretaceous–early Eocene mass extinctions in the deep sea. *in*: *Global catastrophes and earth history* (V.L. Sharpton and P.D. Ward, eds.). Geol. Soc. Am., Spec. Paper 247:481–495.

Thomas, E. 1992. Middle Eocene–late Oligocene bathyal benthic Foraminifera (Weddell Sea): faunal changes and implications for ocean circulation. *in*: *Eocene–Oligocene climatic and biotic evolution* (D.R. Prothero and W.A. Berggren, eds.). Princeton Univ. Press, Princeton, NJ. pp. 245–271.

Thomas, J.H. 1969. Botanical explorations in Washington, Oregon, California and adjacent regions. *Huntia 3*:11.

Thorne, R.F. 1963. Biotic distribution patterns in the tropical Pacific. *in*: *Pacific basin biogeography* (J.J. Gressitt, ed.). Bishop Museum Press, Honolulu, HI. pp. 311–354.

Thorne, R.F. 1972. Major disjunctions in the geographic ranges of seed plants. *Q. Rev. Biol. 47*:365–411.

Thorne, R.F. 1973. Floristic relationships between tropical Africa and tropical America. *in*: *Tropical forest ecosystems in Africa and South America: a comparative review* (B.L. Meggers, E.S. Ayensu and W.D. Duckworth, eds.). Smithsonian Press, Washington. pp. 22–47.

Thorson, G. 1971. *Life in the sea*. World University Library, McGraw–Hill, New York.

Throckmorton, L.H. 1975. The phylogeny, ecology and geography of *Drosophila*. *in*: *Handbook of genetics* (R.C. King, ed.). Plenum Press, New York. pp. 421– 469.

Thulborn, R.A. 1986. Early Triassic tetrapod faunas of southeastern Gondwana. *Alcheringa 10*:297–313.

Tiffney, B.H. 1984. Seed size, dispersal syndromes and the rise of the angiosperms: evidence and hypothesis. *Ann. Mo. Bot. Gard. 71*:551–576.

Tiffney, B.H. 1985. The Eocene North Atlantic land–bridge: its importance in Tertiary and modern phytogeography of the Northern Hemisphere. J. *Arnold Arbor. 66*:243–274.

Timm, T. 1980. Distribution of aquatic oligochaetes. *in*: *Aquatic oligochaete biology* (R.O. Brinkhurst and D.G. Cook, eds.). Plenum Press, New York. pp. 55–77.

Titterton, R. and R.C. Whatley. 1988. The provincial distribution of shallow water Indo–Pacific marine Ostracoda: origins, antiquity, dispersal routes and mechanisms. *in*: *Evolutionary biology of Ostracoda* (T. Hanai, N. Ikeya and K. Ishizaki, eds.). Proc. 9th Int. Sym. on Ostracoda, Kodansha Ltd., Tokyo. pp. 759–786.

Tokioka, T. 1979. Neritic and oceanic plankton. *in*: *Zoogeography and diversity of plankton* (S. van der Spoel and A.C. Pierrot–Bults, eds.). John Wiley, New York. pp. 126–143.

Towns, D.R. and W.L. Peters. 1980. Phylogenetic relationships of the Leptophlebiidae of New Zealand (Ephemeroptera). *in*: *Advances in Ephemeroptera biology* (J.L. Flannagan and K.E. Marshall, eds.). Plenum Press, New York. pp. 57–69.

Truswell, E.M. 1990. Cretaceous and Tertiary vegetation of Antarctica: a palynological perspective. *in*: *Antarctic paleobiology* (T.N. Taylor and E.L. Taylor, eds.). Springer–Verlag, New York. pp. 71–88.

Truswell, E.M., A.P. Kershaw and I.R. Sluiter. 1987. The Australian–South–East Asian connection: evidence from the palaeobotanical record. *in*: *Biogeographical evolution of the Malay Archipelago* (T.C. Whitmore, ed.). Clarendon Press, Oxford. pp. 32–49.

Tschudy, R.H. 1984. Palynological evidence for change in continental floras at the Cretaceous Tertiary boundary. *in*: *Catastrophes and earth history* (W.A. Berggren and J.A. Van Couvering, eds.). Princeton Univ. Press, Princeton, NJ. pp. 315–337.

Tschudy, R.H., C.L. Pillmore, C.J. Orth, J.S. Gilmore and J.D. Knight. 1984. Disruption of the terrestrial plant ecosystem at the Cretaceous–Tertiary boundary, western interior. *Science 225*:1030–1032.

Tuckey, M.E. 1990a. Biogeography of Ordovician bryozoans. *Palaeogeogr., Palaeoclimatol., Palaeoecol. 77*:91–126.

Tuckey, M.E. 1990b. Distributions and extinctions of Silurian Bryozoa. *in*: *Palaeozoic palaeogeography and biogeography* (W.S. McKerrow and C.R. Scotese, eds.). Geol. Soc. London, Mem. 12:197–206.

Udvardy, M.D.F. 1969. *Dynamic zoogeography*. Van Nostrand and Reinhold, New York.

Udvardy, M.D.F. 1975. World biogeographical provinces. *IUCN, Occas. Paper No. 18*:map.

Upchurch, G.R., Jr. 1989. Terrestrial environmental changes and extinction patterns at the Cretaceous–Tertiary boundary, North America. *in*: *Mass extinctions: processes and evidence* (S.K. Donovan, ed.). Columbia Univ. Press, New York. pp. 195–216.

Upchurch, G.R., Jr. and J.A. Wolfe. 1987. Mid–Cretaceous to early Tertiary vegetation and climate: evidence from fossil leaves and woods. *in*: *The origins of angiosperms and their biological consequences* (E.M. Friis, W.G. Chaloner and P.R. Crane, eds.). Cambridge Univ. Press, Cambridge. pp. 75–105.

Usinger, R.L. and R. Matsuda. 1959. *Classification of the Aradidae (Hemiptera– Heteroptera)*. Br. Mus. Nat. Hist. London.

Vail, P.R., R.M. Mitchum and S. Thompson. 1978. Seismic stratigraphy and global changes in sea level. *in*: *Seismic stratigraphy: application to hydrocarbon exploration* (C.E. Payton, ed.). Mem. Am. Assoc. Pet. Geol. 26:83–97.

Vakhrameev, V.A. 1987. Cretaceous paleogeography of the U.S.S.R. *Palaeogeogr., Palaeoclimatol., Palaeoecol. 59*:57–67.

Vakhrameev, V.A. 1991. *Jurassic and Cretaceous floras and climates of the earth.* Cambridge Univ. Press, Cambridge.

Valentine, J.W. 1966. Numerical analysis of marine molluscan ranges on the extratropical northeastern Pacific shelf. *Limnol. Oceanogr. 11*:198–211.

Valentine, J.W. 1984. Neogene marine climate trends: implications for biogeography and evolution of the shallow–sea biota. *Geology 12*:647–650.

Valentine, J.W. and D. Jablonski. 1991. Biotic effects of sea level change: the Pleistocene test. *J. Geophys. Res. 96*:6873–6878.

Van Couvering, J.A.H. 1977. Early records of freshwater fishes in Africa. *Copeia 1977*:163–166.

Van Couvering, J.A.H. 1982. Fossil cichlid fish of Africa. Palaeontol. Assoc. London. *Spec. Papers Palaeo. 29*:1–103.

Van der Hammen, T. 1982. Paleoecology of tropical South America. *in*: *Biological diversification in the tropics* (G.T. Prance, ed.). Columbia Univ. Press, New York. pp. 60–66.

van der Spoel, S. 1983. Patterns in plankton distribution and their relation to speciation: the dawn of pelagic biogeography. *in*: *Evolution in time and space: the emergence of the biosphere* (R.W. Sims, J.H. Price and P.E.S. Whalley, eds.). Academic Press, London. pp. 291–334.

van der Spoel, S. and R.P. Heyman. 1983. *A comparative atlas of zooplankton*. Bunge, Utrecht.

Van der Voo, R. 1988. Paleozoic paleogeography of North America, Gondwana and intervening displaced terranes: comparisons of paleomagnetism with paleoclimatology and biogeographical patterns. *Geol. Soc. Am. Bull. 100*:311– 324.

Van Steenis, C.G.G.J. 1962. The distribution of mangrove plant genera and its significance for paleogeography. *Proc. Ned. Acad. Wetenschap. 65*:164–169.

Van Steenis, C.G.G.J. 1972. *Nothofagus*, key genus to plant geography. *in*: *Taxonomy, phytogeography and evolution* (D.H. Valentine, ed.). Academic Press, New York. pp. 275–288.

Van Valen, L. 1976. Energy and evolution. *Evol. Theory 1*:179–229.

Van Valen, L. 1978. The beginning of the age of mammals. *Evol. Theory 4*:45–80.

Van Valen, L.M. 1984. A resetting of Phanerozoic community evolution. *Nature 307*:50–52.

Van Valen, L.M. 1988. Paleocene dinosaurs or Cretaceous ungulates in South America. *Evol. Monogr. 10*:1–79.

van Zanten, B.O. 1983. Possibilities of long–range dispersal in bryophytes with special reference to the southern hemisphere. *in*: *Dispersal and distribution* (K. Kubizki, ed.). Verlag Paul Parey, Hamburg. pp. 49–64.

Vanzolini, P.E. and W.R. Heyer. 1985. The American herpetofauna and the interchange. *in*: *The great American interchange* (F.G. Stehli and S.D. Webb, eds.). Plenum Press, New York. pp. 475–487.

Vermeij, G.J. 1987. *Evolution and escalation*. Princeton Univ. Press, Princeton, NJ.

Vermeij, G.J. 1989. Geographical restriction as a guide to the causes of extinction: the case of the cold northern oceans during the Neogene. *Paleobiology 15*:335–356.

Vermeij, G.J. 1991. Anatomy of an invasion: the trans–Arctic interchange. *Paleobiology 17*:281–307.

Vermeij, G.J. 1992. Trans–equatorial connections between biotas in the temperate eastern Atlantic. *Mar. Biol. 112*:343–348.

Veron, J.E.N. 1986. *Corals of Australia and the Indo–Pacific*. Angus and Robertson, North Ryde, Australia.

Vidal, G. and A.H. Knoll. 1982. Radiations and extinctions of plankton in the late Proterozoic and early Cambrian. *Nature 297*:57–60.

Vinogradova, N.G. 1962. Vertical zonation in the distribution of the deep–sea benthic fauna of the ocean. *Deep–Sea Res. 8*:245–250.

Vinogradova, N.G. 1964. Geographical distribution of the deep–water bottom fauna of the Antarctic. *Soviet Antarctic Expedition 1*:121–122.

Vinogradova, N.G. 1979. The geographical distribution of the abyssal and hadal (ultra–abyssal) fauna in relation to the vertical zonation of the ocean. *Sarsia 64*:41–50.

von Humboldt, A. 1805. Essai sur la Géographie des Plantes. *in*: *Voyage aux Régions Equinoxiales du Nouveau Continent* (A. von Humboldt and A.J.A. Bonpland, authors). Paris.

Vrba, E.S. 1987. Ecology in relation to speciation rates: some case histories of Miocene–Recent mammal clades. *Evol. Ecol. 1*:283–300.

Vrba, E.S. 1993. Mammal evolution in the African Neogene and a new look at the Great American Interchange. *in*: *Biological relationships between Africa and South America* (P. Goldblatt, ed.). Yale Univ. Press, New Haven, CT. pp. 393–432.

Vuilleumier, F. 1985. Fossil and recent avifaunas and the interamerican exchange. *in*: *The great American biotic exchange* (F.G. Stehli and S.D. Webb, eds.). Plenum Press, New York. pp. 387–424.

Wake, D.B. and J.F. Lynch. 1976. The distribution, ecology and evolutionary history of plethodontid salamanders in tropical America. *Bull. Nat. Hist. Mus. Los Angeles County 25*:1–65.

Wallace, A.R. 1855. On the law which has regulated the introduction of new species. *Ann. Mag. Nat. Hist., Ser. 2. 16*:184–196.

Wallace, A.R. 1876. *The geographical distribution of animals*. 2 Vols. Macmillan and Co., London.

Wallace, A.R. 1881. *Island life*. Harper and Brothers, New York.

Walliser, O.H. 1990. How to define "Global bio–events". *in*: *Extinction events in earth history* (E.G. Kauffman and O.H. Walliser, eds.). Springer–Verlag, Berlin. pp. 1–3.

Walsh, J.J. 1988. *On the nature of continental shelves*. Academic Press, San Diego, CA.

Wang, B. 1992. The Chinese Oligocene: a preliminary review of mammalian localities and local faunas. *in*: *Eocene–Oligocene climatic and biotic evolution* (D.E. Prothero and W.A. Berggren, eds.). Princeton Univ. Press, Princeton, NJ. pp. 529–547.

Wang, X. 1984. The palaeoenvironment of China from the Tertiary. *in*: *The evolution of the east Asian environment* (R.O. Whyte, ed.). Vol. 2. Univ. of Hong Kong, Hong Kong. pp. 472–482.

Ward, P.D., W.J. Kennedy, K.G. MacLeod and J.F. Mount. 1991. Ammonite and inoceramid bivalve extinction patterns in the Cretaceous/Tertiary boundary sections of the Biscay region (southwestern France, northern Spain). *Geology 19*:1181–1184.

Warnke, D.A., et al. 1992. Miocene–Pliocene Antarctic glacial evolution: a synthesis of ice–rafted debris, stable isotope and planktonic foraminiferal indicators, ODP leg 114. *Antarct. Res. Ser. 56*:311–325.

Watson, J.A.L. 1981. Odonata (dragonflies and damselflies). *in*: *Ecological biogeography of Australia* (A. Keast, ed.). W. Junk, The Hague. pp. 1141–1167.

Watt, J.C. 1975. The terrestrial insects. *in*: *Biogeography and ecology in New Zealand* (G. Kuschel, ed.). W. Junk, The Hague. pp. 507–535.

Webb, S.D. 1985a. Main pathways of mammalian diversification in North America. *in*: *The great American interchange* (F.G. Stehli and S.D. Webb, eds.). Plenum Press, New York. pp. 201–247.

Webb, S.D. 1985b. Late Cenozoic mammal dispersals between the Americas. *in*: *The great American interchange* (F.G. Stehli and S.D. Webb, eds.). Plenum Press, New York. pp. 357–386.

Webb, S.D. 1991. Ecogeography and the great American interchange. *Paleobiology 17*:266–280.

Webb, S.D. and A.D. Barnovsky. 1989. Faunal dynamics of Pleistocene mammals. *Annu. Rev. Earth Planet. Sci. 17*:413–438.

Webb, T. and P.J. Bartlein. 1992. Global changes during the last 3 million years. *Annu. Rev. Ecol. Syst. 23*:141–173.

Weijermars, R. 1989. Global tectonics since the breakup of Pangea 180 million years ago: evolution maps and lithospheric budget. *Earth–Science Rev. 26*:113– 162.

Weishampel, D.B. 1990. Dinosaurian distribution. *in*: *The Dinosauria* (D.B. Weishampel, P. Dodson and H. Osmólska, eds.). Univ. Calif. Press, Berkeley, CA. pp. 63–139.

Wells, G.P. 1963. Barriers and speciation in lugworms. *Syst. Assoc. Publ. 5*:79–98.

Welty, J.C. 1979. *The life of birds*. Saunders, Philadelphia, PA.

Welty, J.C. and L. Baptista. 1988. *The life of birds*. 4th edn. Saunders, New York.

Westermann, G.E.G. 1973. The late Triassic bivalve *Monotis*. *in*: *Atlas of palaeobiogeography* (A. Hallam, ed.). Elsevier, Amsterdam. pp. 251–258.

Whalley, P. 1987. Insects and Cretaceous mass extinction. *Nature 327*:562.

Whatley, R. 1987. The southern end of Tethys: an important locus for the origin and evolution of both deep and shallow water Ostracoda. *in*: *Shallow Tethys 2* (K.G. McKenzie, ed.). A.A. Balkema, Rotterdam. pp. 461–474.

White, B.N. 1986. The isthmian link, antitropicality and American biogeography: distributional history of the Atherinopsinae (Pisces: Atherinidae). *Syst. Zool. 35*:176–194.

White, M.E. 1990. Plant life between two ice ages down under. *Am. Sci. 78*:253–262.

White, P.S. 1983. Eastern Asian – eastern North American floristic relations: the plant community level. *Ann. Mo. Bot. Gard. 70*:734–747.

Widmark, J.G.V. and B.A. Malmgren. 1988. Cretaceous/Tertiary boundary – benthonic foraminiferal changes in the deep sea. *Geol. Soc. Am., Abstracts with Programs 20*:223.

Wiedmann, J. 1973. Upper Triassic heteromorph ammonites. *in*: *Atlas of palaeobiogeography* (A. Hallam, ed.). Elsevier, Amsterdam. pp. 235–249.

Wiedmann, J. 1986. Macro–invertebrates and the Cretaceous–Tertiary boundary. *in*: *Global bio–events* (O.H. Walliser, ed.). Springer–Verlag, Berlin. pp. 397–409.

Wiley, E.O. 1976. The phylogeny and biogeography of fossil and recent gars (Actinopterygii: Lepisosteidae). *Miscel. Publ. Univ. Kansas Mus. Nat. Hist. 64*:1–111.

Wiley, E.O. 1988. Vicariance biogeography. *Annu. Rev. Ecol. Syst. 19*:513–542.

Wiley, E.O. 1992. Phylogenetic relationships of the Percidae (Teleostei: Perciformes): a preliminary hypothesis. *in*: *Systematics, historical ecology and North American freshwater fishes* (R.L. Mayden, ed.). Stanford Univ. Press, Stanford, IL. pp. 247–267.

Williams, C.A. 1986. An oceanwide view of Palaeogene plate tectonic events. *Palaeogeogr., Palaeoclimatol., Palaeoecol. 57*:3–25.

Williams, E.E. 1989. Old problems and new opportunities in West Indian biogeography. *in*: *Biogeography of the West Indies* (C.A. Woods, ed.). Sandhill Crane Press, Gainesville, FL. pp. 1–46.

Williams, W.D. 1981. Aquatic insects: an overview. *Monogr. Biol. 41*:1213–1229.

Williamson, M. 1981. *Island populations*. Oxford Univ. Press, Oxford.

Wilson B.R. and G.R. Allen. 1988. Major components and distribution of the marine fauna. *in*: *Fauna of Australia* (G.R. Dyne and D.W. Walton, eds.). General Articles, Vol. 1A, Australian Govt. Pub. Serv., Canberra. pp. 43–68.

Wilson, E.O. 1961. The nature of the taxon cycle in the Melanesian ant fauna. *Am. Nat. 95*:169–193.

Wilson, E.O. 1985. The search for faunal dominance. *in*: *Taxonomy, phylogeny and zoogeography of beetles and ants* (G.E. Ball, ed.). W. Junk, Dordrecht. pp. 489–493.

Wilson, E.O. 1991. Rain forest canopy: the high frontier. *Natl. Geogr. 180*:78–107.

Wilson, E.O. 1992. *The diversity of life*. Belknap Press, Harvard Univ., Cambridge, MA.

Wilson, J.T. 1963. Continental drift. *Sci. Am.*, pp. 1–16.

Wilson, J.T. 1973. Mantle plumes and plate motions. *Tectonophysics 19*:149–164.

Wilson, M.V.H., D.B. Brinkman and A.G. Neuman. 1992. Cretaceous Esocoidei (Teleostei): early radiation of the pikes in North American fresh waters. *J. Paleontol. 66*:839–846.

Winterbottom, R. 1985. Revision and vicariance biogeography of the subfamily Congrogadinae (Pisces: Perciformes: Pseudochromidae). *Indo–Pac. Fishes 9*:1–34.

Wolfe, J.A. 1981. Vicariance biogeography of angiosperms in relation to paleobotanical data. *in*: *Vicariance biogeography* (G. Nelson and D.E. Rosen, eds.). Columbia Univ. Press, New York. pp. 413–427.

Wolfe, J.A. 1985. Distribution of major vegetational types during the Tertiary. *in*: *The carbon cycle and atmospheric* CO_2 (E.T. Sundquist and W.S. Broecker, eds.). Am. Geophys. Union, Monogr. 32.

Wolfe, J.A. 1987. An overview of the origins of the modern vegetation and flora of the northern Rocky Mountains. *Ann. Mo. Bot. Gard. 74*:785–803.

Wolfe, J.A. 1992. Climatic, floristic and vegetational changes near the Eocene/Oligocene boundary in North America. *in*: *Eocene–Oligocene climatic and biotic evolution* (D.R. Prothero and W.A. Berggren, eds.). Princeton Univ. Press, Princeton, NJ. pp. 421–436.

Wood, A.E. 1985. Northern waif primates and rodents. *in*: *The great American interchange* (F.G. Stehli and S.D. Webb, eds.). Plenum Press, New York. pp. 267–282.

Woodburne, M.O. and W.J. Zinsmeister. 1984. The first land mammal from Antarctica and its biogeographic implications. *J. Paleontol. 58*:913–948.

Woodland, D.J. 1983. Zoogeography of the Siganidae (Pisces): an interpretation of distribution and richness patterns. *Bull. Mar. Sci. 33*:713–717.

Woodland, D.J. 1986. Wallace's Line and the distribution of marine inshore fishes. *in*: *Indo–Pacific Fish Biology* (T. Uyeno, R. Arai, T. Taniuchi and K. Matsuura, eds.). Ichthyological Society of Japan, Tokyo. pp. 453–460.

Woodruff, F. 1985. Changes in Miocene deep–sea benthic foraminiferal distribution in the Pacific Ocean: relationships to Paleoceanography. *in*: *The Miocene ocean: paleoceanography and biogeography* (J.P. Kennett, ed.). Geol. Soc. Am., Mem. 163:131–175.

Wootton, R.J. 1990. Major insect radiations. *in*: *Major evolutionary radiations* (P.D. Taylor and G.P. Larwood, eds.). Clarendon Press, Oxford. pp. 187–208.

Worsley, T.R., R.M. Nance and J.B. Moody. 1986. Tectonic cycles and the history of the earth's biogeochemical and paleoceanographic record. *Paleoceanography 1*:233–263.

Wunderlich, J. 1986. *Spinnenfauna gestern und heute*. Erich Bauer, Wiesbaden.

Ying, T. 1983. The floristic relationships of the temperate forest regions of China and the United States. *Ann. Mo. Bot. Gard. 70*:597–604.

Young, G.C. 1990. Devonian vertebrate distribution patterns and cladistic analysis of palaeogeographic hypotheses. *in*: *Palaeozoic palaeogeography and biogeography* (W.S. McKerrow and C.R. Scotese, eds.). Geol. Soc. London, Mem. 12:243–255.

Zachos, J.C., M.A. Arthur and W.E. Dean. 1989. Geochemical evidence for suppression of pelagic marine productivity at the Cretaceous/Tertiary boundary. *Nature 337*:61–64.

Zenkevitch, L.A. 1963. *Biology of the seas of the U.S.S.R.* George Allen and Unwin, London.

Zenkevitch, L.A. and Y.A. Birstein. 1956. Studies of the deep–water fauna and related problems. *Deep–Sea Res. 4*:54–64.

Ziegler, A.M. 1990. Phytogeographic patterns and continental configurations during the Permian Period. *in*: *Palaeozoic palaeogeography and biogeography* (W.S. McKerrow and C.R. Scotese, eds.). Geol. Soc. London, Mem. 12:363–379.

Ziegler, P.A. 1989. *Evolution of Laurussia: a study in late Paleozoic plate tectonics*. Kluwer, Dordrecht.

Zimmerman, E.C. 1942. Distribution and origin of some eastern oceanic insects. *Am. Nat. 76*:282–307.

Zinsmeister, W.J. and R.M. Feldmann. 1984. Cenozoic high latitude heterochroneity of southern hemisphere marine faunas. *Science 224*:281–283.

Zwick, P. 1981a. Blephariceridae. *in*: *Ecological biogeography of Australia* (A. Keast, ed.). W. Junk, The Hague. pp. 1185–1193.

Zwick, P. 1981b. Des Mittelmeergebiet als glaziales Refugium für Plecoptera. *Acta Entomol. Jugosl. 17*:107–111.

Zwick, P. 1990. Transantarctic relationships in the Plecoptera. *in*: *Mayflies and stoneflies: life histories and biology* (I.C. Campbell, ed.). Kluwer, Dordrecht. pp. 141–148.

Appendix: Biogeographer's Maps

These biogeographer's maps are intended to show the approximate configurations of land and sea for the indicated time period. The Lambert equal-area projections allow a minimum distortion of continental shapes and have the additional advantage of showing both poles on the same map. They are modified from the continental drift maps produced by the Paleomap project of the International Lithosphere Program (Scotese, 1992).

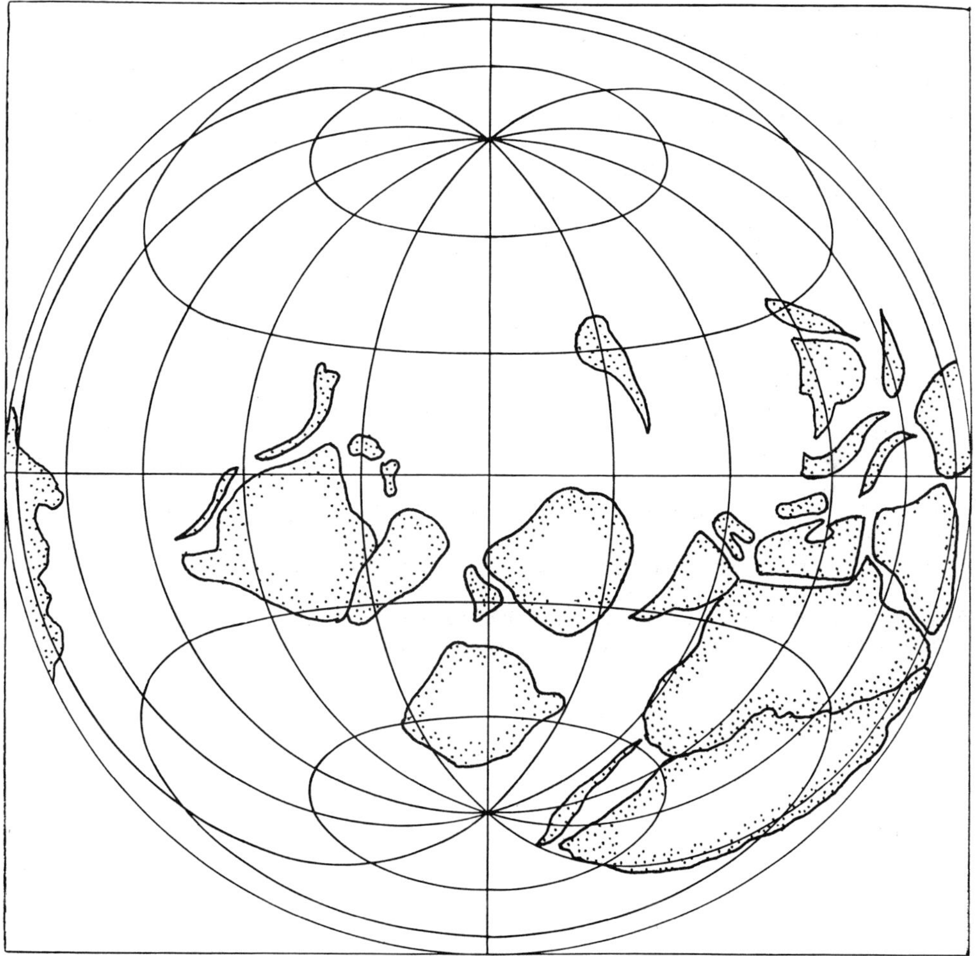

Map 1: Mid-Cambrian. By this time, the various continental blocks and terranes had apparently undergone considerable dispersal. Their possible Precambrian amalgamation is illustrated in text Figs. 1 and 2.

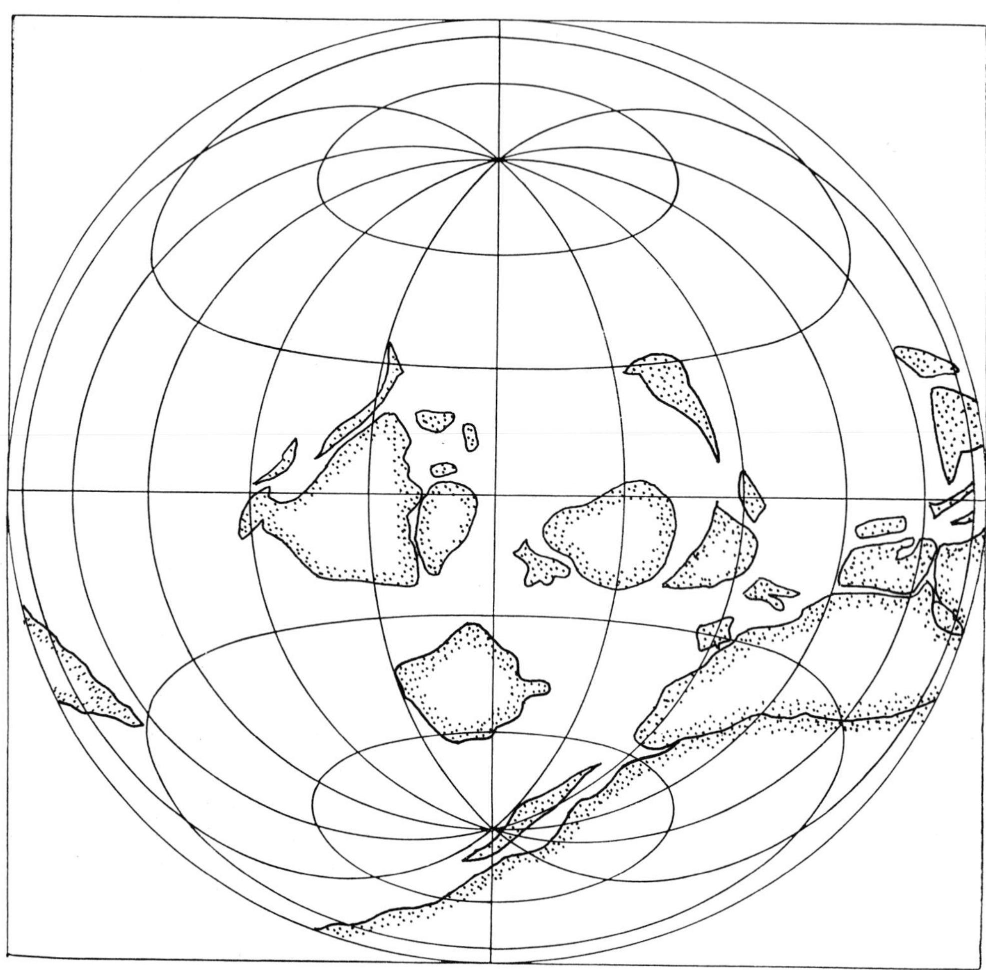

Map 2: Late Cambrian. The continental dispersal continues. The presumptive northern continents are widely separated while the Gondwana group is clustered to the right.

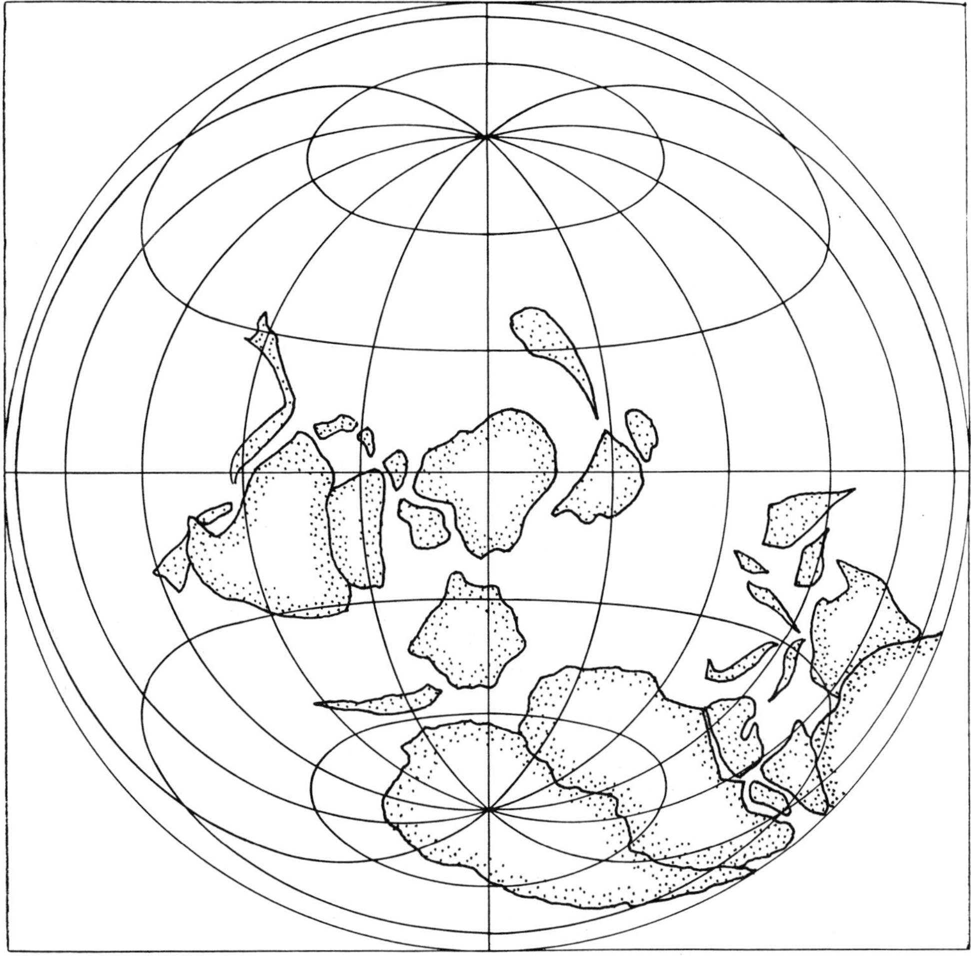

Map 3: Mid-Ordovician. The Gondwanan continents have moved southward and northern South America is now centered over the South Pole. The northern continents have moved closer to one another.

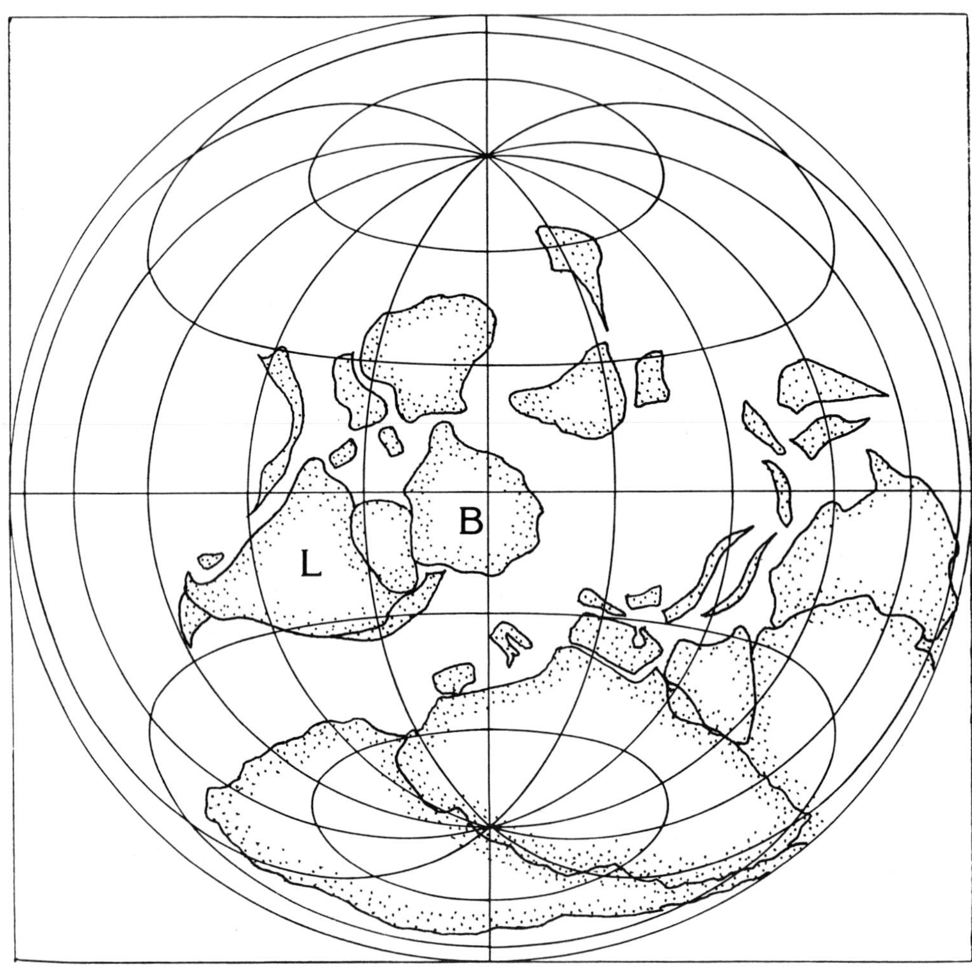

Map 4: Mid-Silurian. A collision has now taken place between Baltica (B) and Laurentia (L) to form the new continent of Laurussia.

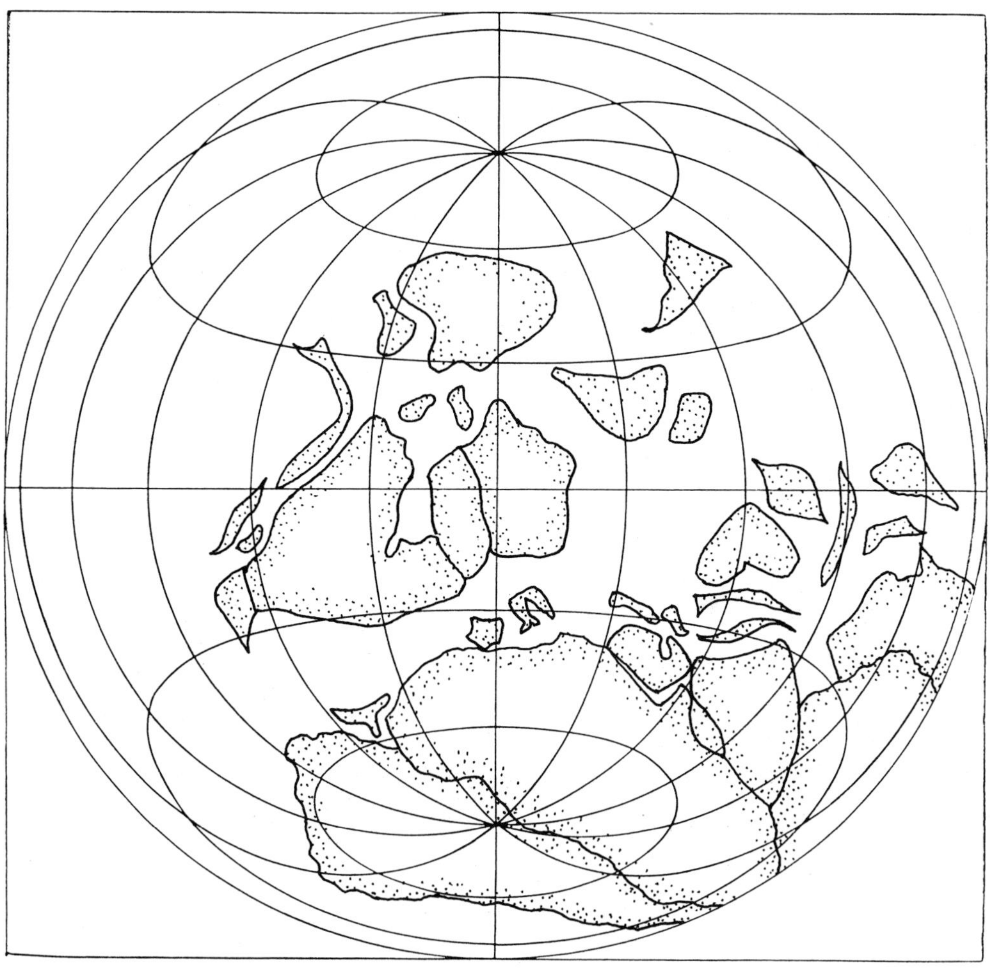

Map 5: End-Devonian. Both the Gondwanan group and northern continents have shifted to the north. A closure is taking place between Gondwana and Laurussia. The two may have been in contact by this time.

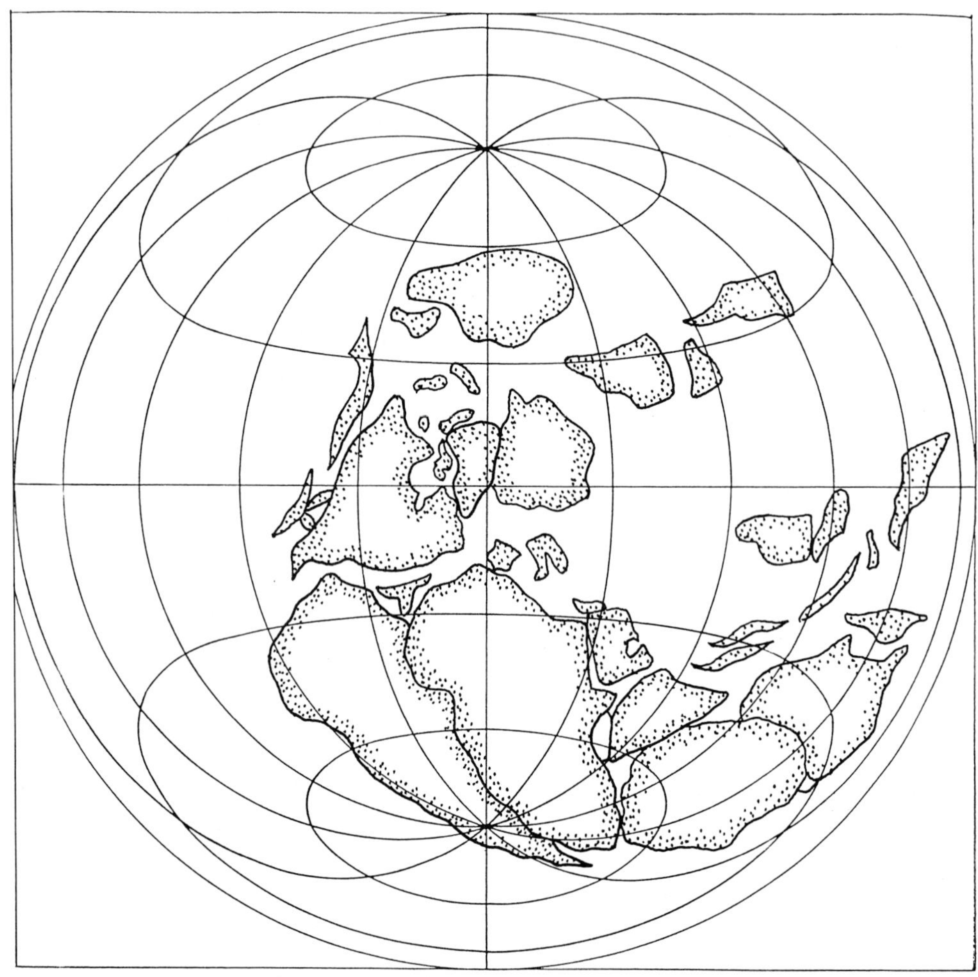

Map 6: Early Carboniferous. The general northward shift has continued, with the gap between the northern and southern continents just about closed.

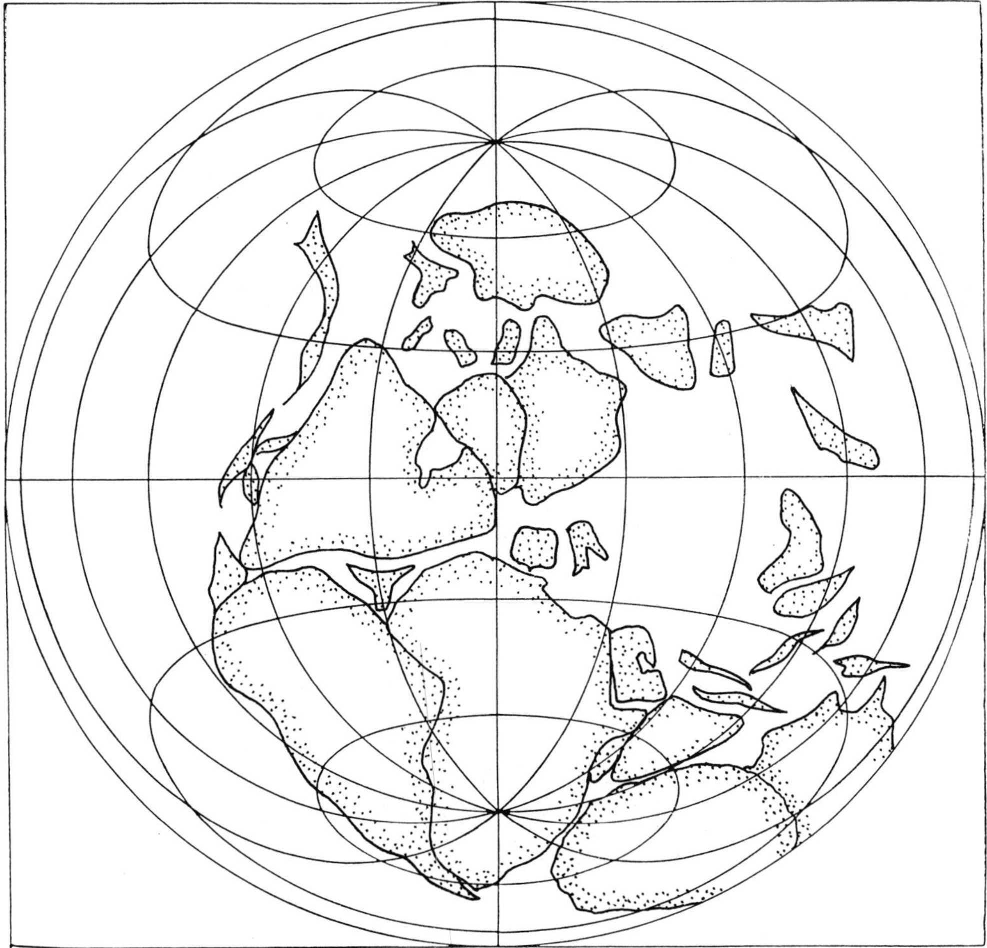

Map 7: Late Carboniferous. A number of smaller terranes, which were originally part of Gondwana, have moved to the northwest. Their identities and positions are not well constrained.

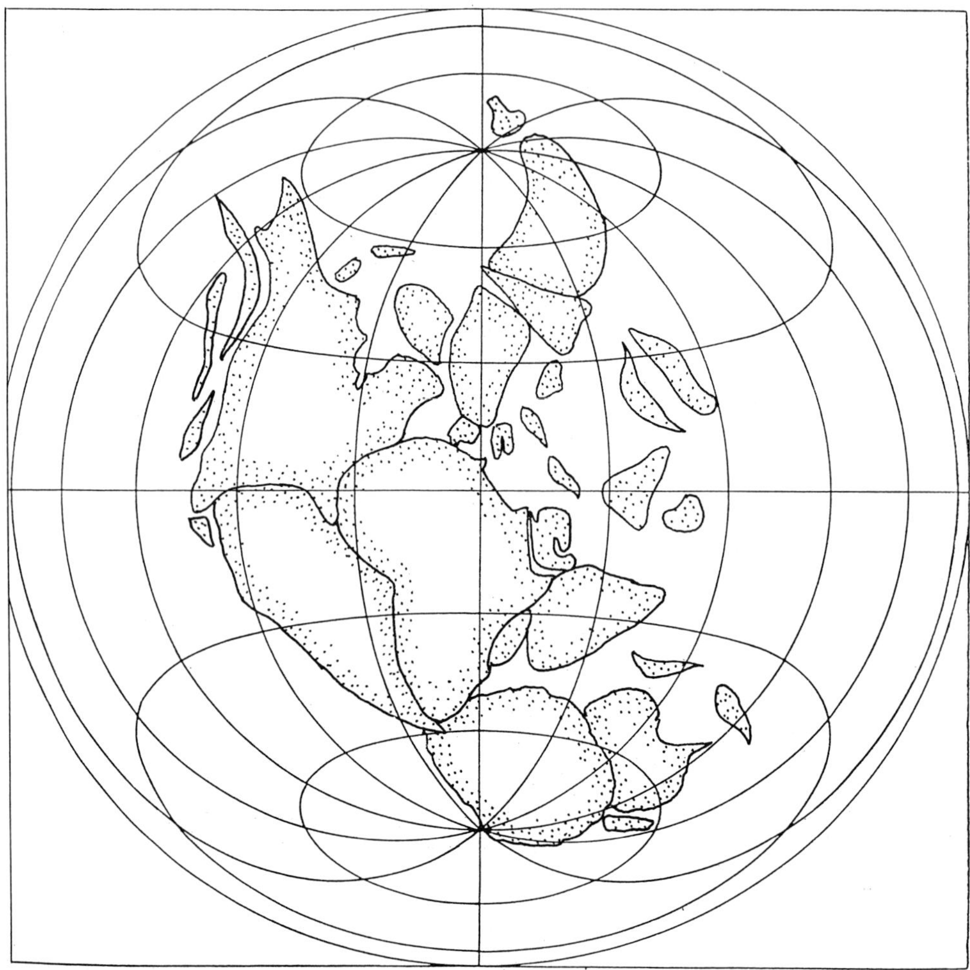

Map 8: Late Permian. Antarctica-Australia has moved to the southwest. The northern continents and smaller terranes have closed on one another.

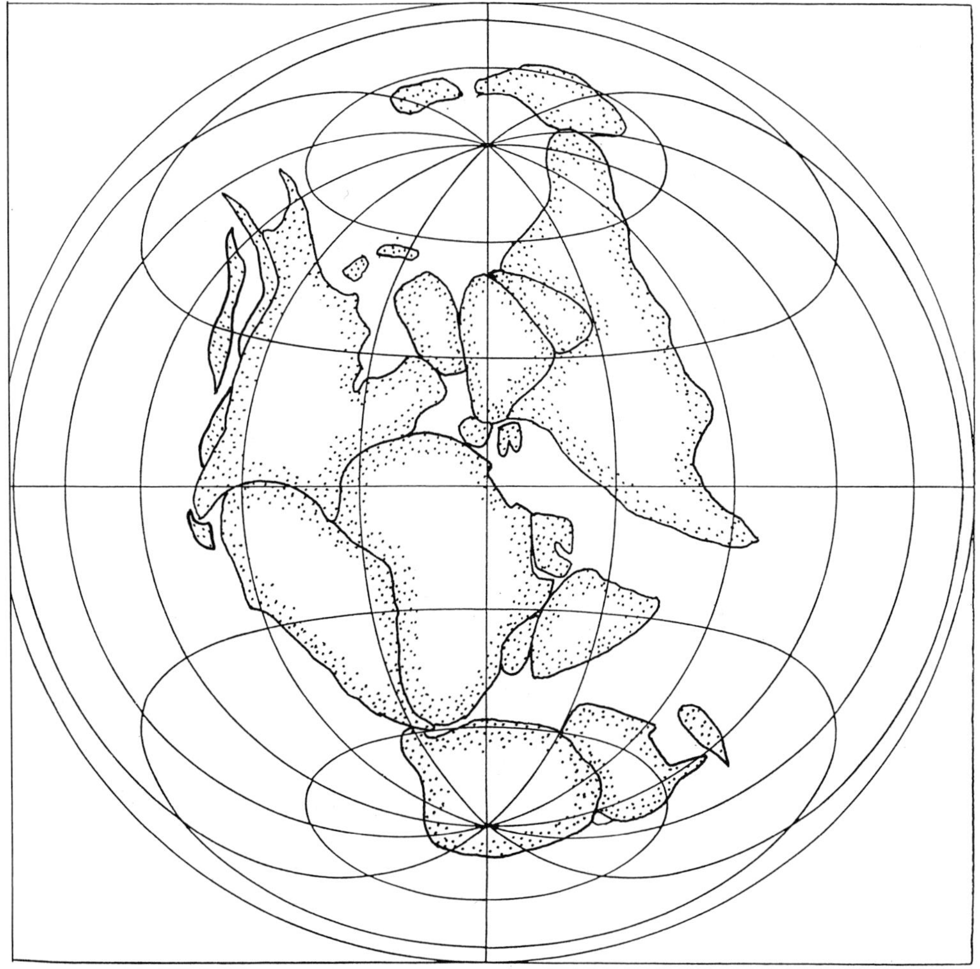

Map 9: Mid-Triassic. Pangaea is now fully assembled. Almost all tetrapod families were cosmopolitan, with no apparent latitudinal variation.

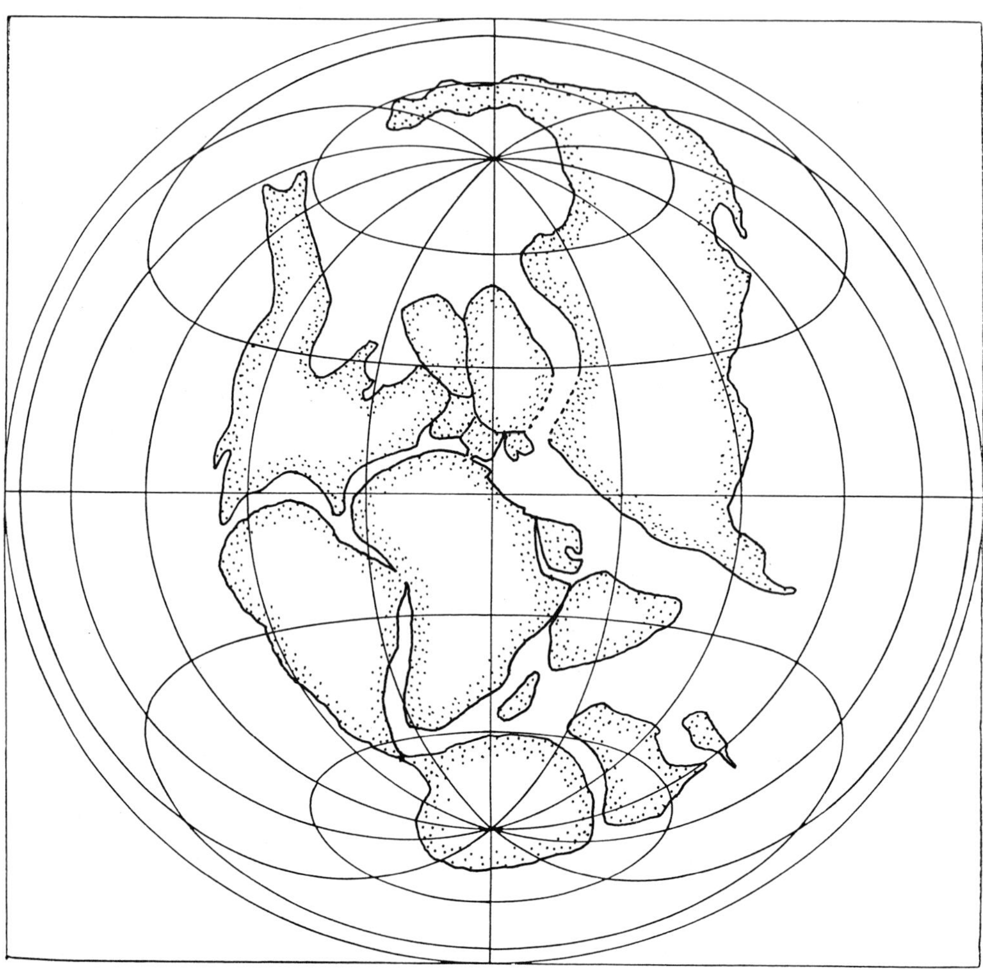

Map 10: Late Jurassic. Pangaea is beginning to break apart. The Turgai Sea has separated Asia from Euramerica. South America and Africa are separating. Madagascar has departed from Africa. Euramerica is almost separated from Africa.

Map 11: Early Cretaceous. The complete separation of the northern (Laurasian) continents from Gondwana has created the circumtropical Tethys Sea. Africa and South America may have separated but are located close together.

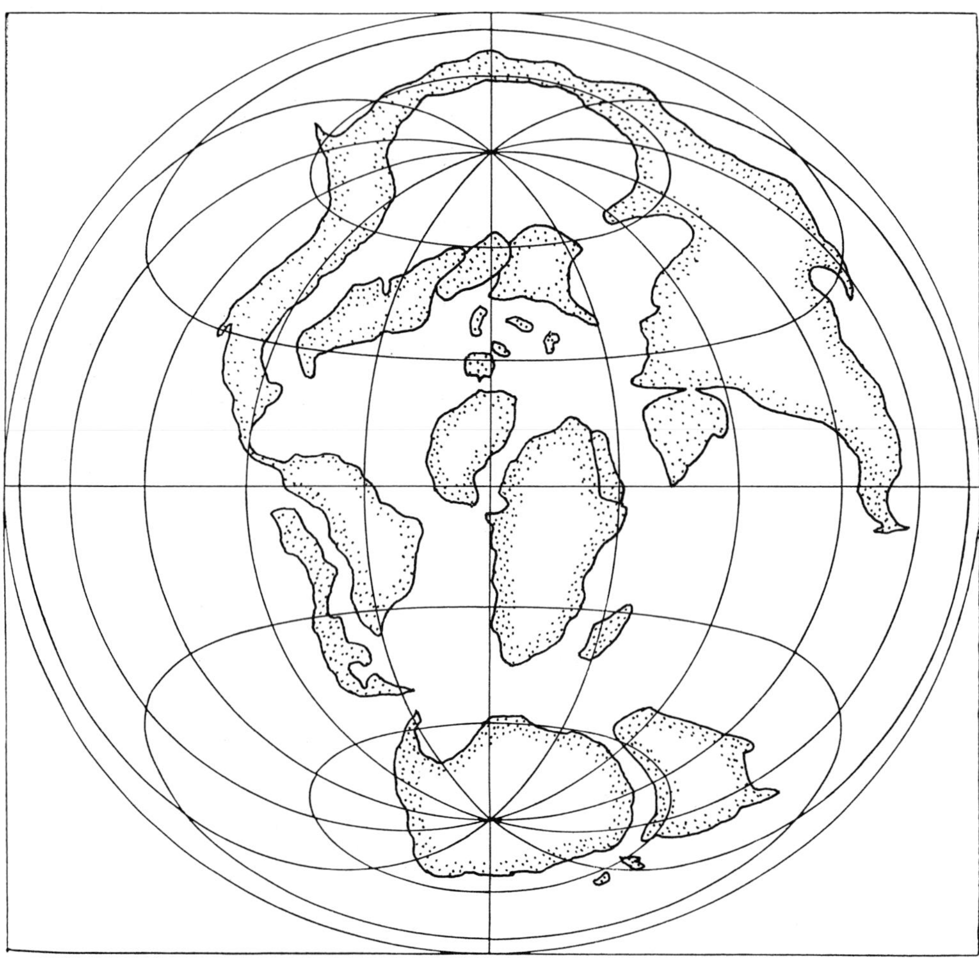

Map 12: Late Cretaceous. High sea level has resulted in the formation of extensive epicontinental seas. Asia and Westamerica have joined to form a single continent separate from Euramerica. India has probably contacted Asia. An early Central American isthmus was probably formed.

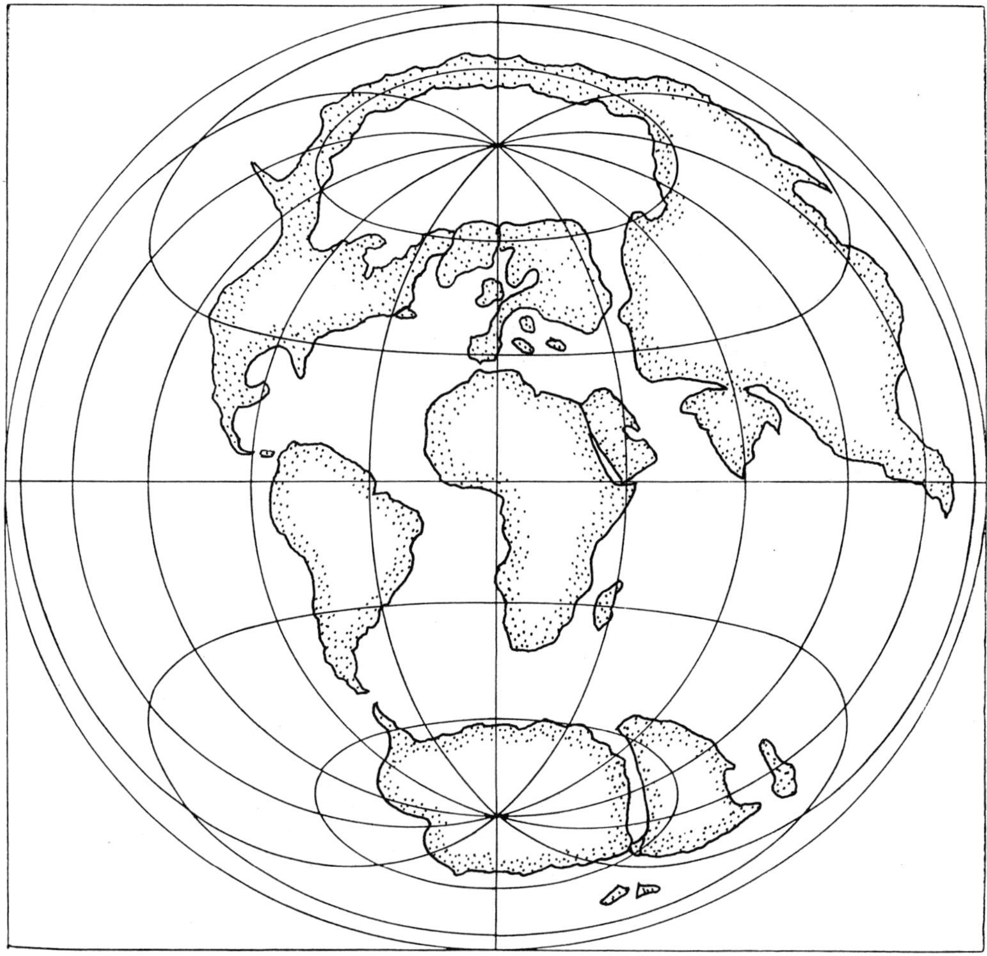

Map 13: Paleocene. Epicontinental seas have receded. There is now only one northern continent. The Central American isthmus probably persisted. India is more firmly attached to Asia. The close proximity of South America, Antarctica, and Australia provided a filter bridge for vagile organisms.

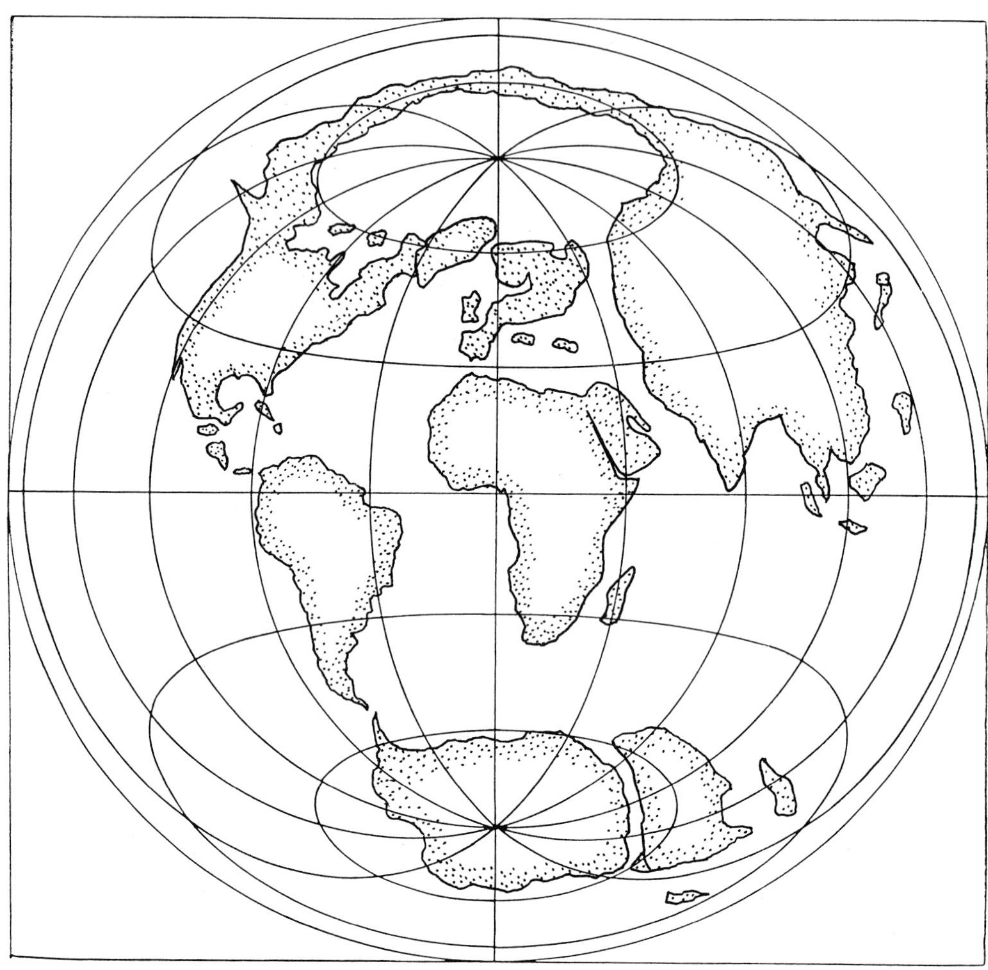

Map 14: Late Eocene. Northern Europe has separated from Greenland. Central America exists as an island archipelago. The Antilles have formed and have begun accumulating their biota from overseas. The Atlantic Ocean has continued to widen.

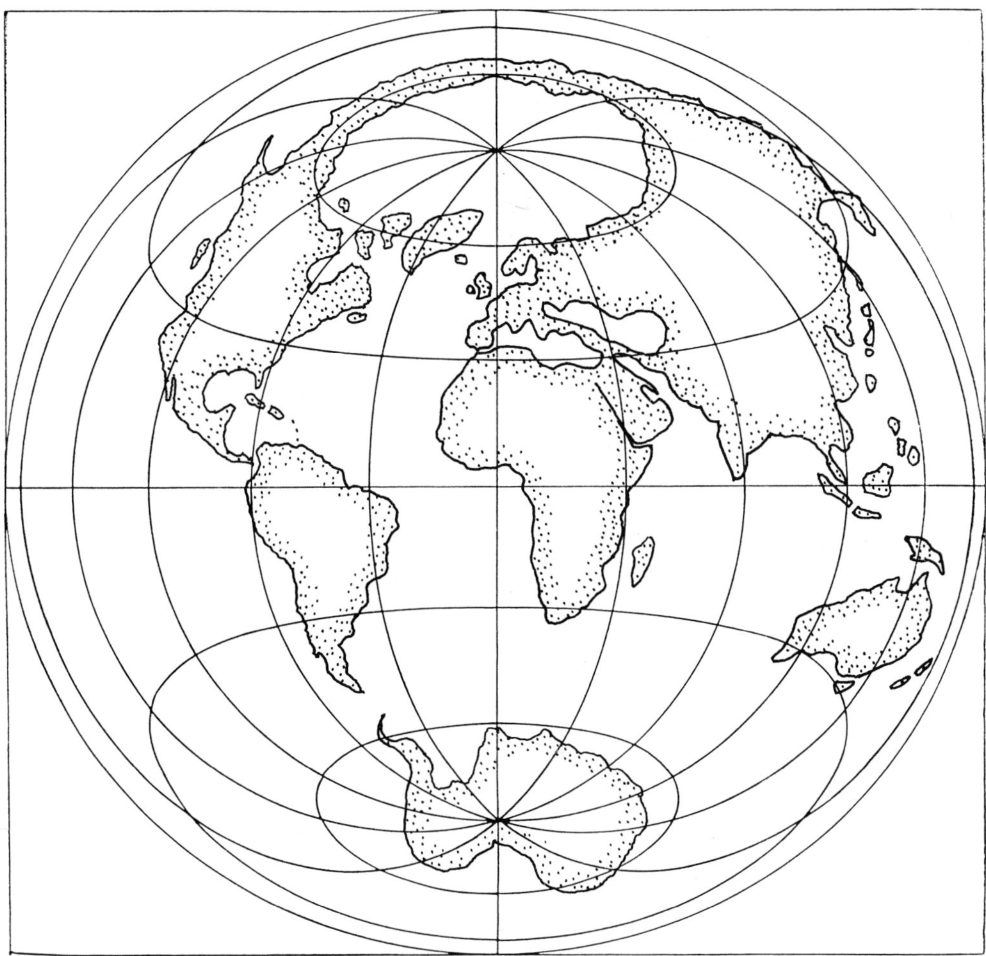

Map 15: Early Miocene. Africa has contacted Asia via the Arabian Peninsula eliminating the Tethys Sea and creating the Mediterranean. Central America has become restored except for a small gap between Panama and Columbia. Australia-New Guinea has moved northward to almost its present position.

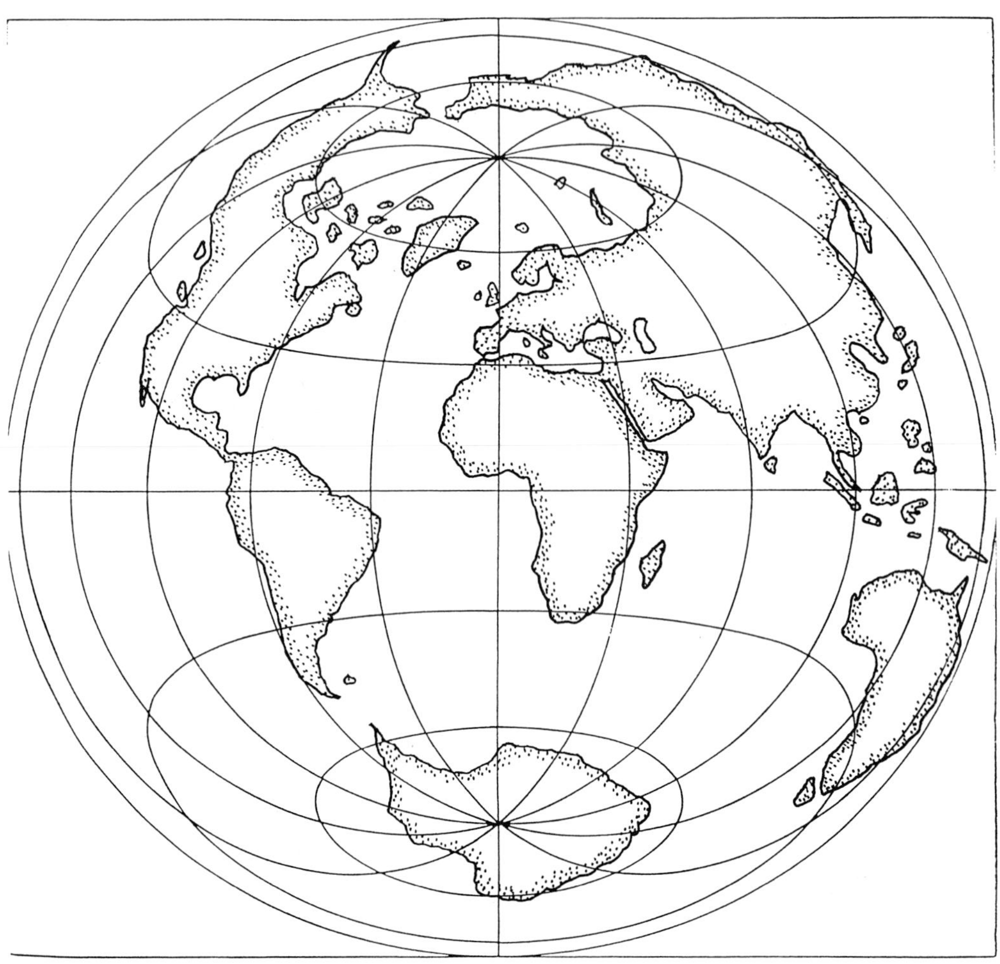

Map 16: Present Day. The Bering Land Bridge is inundated, the Central American isthmus is complete, and the Black and Caspian seas are formed.

Subject Index

GEOLOGIC TIME SCALE

ERA	PERIOD	EPOCH	STAGE	Ma
CENOZOIC	NEOGENE	PLEISTOCENE		1.64
		PLIOCENE	PIACENZIAN	
			ZANCLIAN	5.2
		MIOCENE	MESSINIAN	
			TORTONIAN	
			SERRAVALIAN	
			LANGHLIAN	
			BURDIGALIAN	
			AQUITANIAN	23.3
	PALEOGENE	OLIGOCENE	CHATTIAN	
			RUPELIAN	35.4
		EOCENE	PRIABONIAN	
			BARTONIAN	
			LUTETIAN	
			YYPRESIAN	56.5
		PALEOCENE	THANETIAN	
			DANIAN	65.0
MESOZOIC	CRETACEOUS		MAASTRICTIAN	
			CAMPANIAN	
			SANTONIAN	
			CONIACIAN	
			TURONIAN	
			CENOMANIAN	
			ALBIAN	
			APTIAN	
			BARREMIAN	
			HAUTERIVIAN	
			VALANGINIAN	
			BERRIASIAN	145
	JURASSIC		TITHONIAN	
			KIMMERIDGIAN	
			OXFORDIAN	
			CALLOVIAN	
			BATHONIAN	
			BAJOCIAN	
			AALEMIAN	
			TOARCIAN	
			PLEINSBACHIAN	
			SINEMURIAN	
			HETTANGIAN	208
	TRIASSIC		RHAETIAN	
			NORIAN	
			CARNIAN	
			LADINIAN	
			ANISIAN	
			SPATHIAN	

ERA	PERIOD	EPOCH	STAGE	Ma
			NAMMALIAN GRIESBACHIAN	245
PALEOZOIC	PERMIAN		CHANGXINGIAN LONGTANIAN CAPITANIAN WORDIAN UFIMANIAN KUNGURIAN ARTINSKIAN SAKMARIAN ASSELIAN	290
	CARBONIFEROUS	STEPHANIAN WESTPHALIAN NAMURIAN VISEAN TOURNAISIAN		362
	DEVONIAN		FAMMENIAN FRASNIAN GIVETIAN EIFELIAN EMSIAN PRAGIAN LOCHOVIAN	408
	SILURIAN	PRIDOLI LUDLOW WENLOCK LLANDOVERY		439
	ORDOVICIAN	ASHGILL CARADOC LLANDEILO LLANVIRN ARENIG TREMADOC		510
	CAMBRIAN	MERIDNETH ST. DAVID'S CAERFAI		570
PROTEROZOIC	VENDIAN	EDIACARA VARANGER		610
	STURTIAN